178346

THEORY AND APPLICATIONS
OF
DISTANCE GEOMETRY

THEORY AND APPLICATIONS
OF
DISTANCE GEOMETRY

BY

LEONARD M. BLUMENTHAL

PROFESSOR OF MATHEMATICS
UNIVERSITY OF MISSOURI

CHELSEA PUBLISHING COMPANY
BRONX, NEW YORK

SECOND EDITION

THE PRESENT SECOND EDITION IS A REPRINT, WITH VARIOUS SMALL IMPROVEMENTS AND CORRECTION OF ERRATA, OF A WORK FIRST PUBLISHED IN 1953 BY OXFORD UNIVERSITY PRESS AT OXFORD. THE PRESENT EDITION IS PUBLISHED AT NEW YORK, 1970, AND IS PRINTED ON ALKALINE PAPER

LIBRARY OF CONGRESS CATALOG CARD NUMBER 79-113117

INTERNATIONAL STANDARD BOOK NUMBER 0-8284-0242-6

PRINTED IN THE UNITED STATES OF AMERICA

PREFACE

THE mappings (homeomorphisms) with which topology is principally concerned are biuniform and bicontinuous, but not, in general, distance-preserving when applied to spaces in which 'distance' is defined. In distance geometry we deal, for the most part, with that subgroup of the group of homeomorphisms (the group of *congruences* or *isometries*) for which distance *is* invariant. The subject in which the extensive theory of this subgroup is developed is also known as *metric topology* and *abstract metrics*, and these terms will be used, together with 'distance geometry', in referring to the material treated here.

During the decade that has elapsed since the field was surveyed in an earlier work by the author, metric topology has been greatly extended in both content and application. Numerous articles in widely different mathematical disciplines offer ample evidence that abstract metric methods are playing a considerable part in developing the mathematics of our day, and one may venture the opinion that their role in the future will be no less significant. The unification achieved by the association of 'distances' with pairs of elements, with the resulting subordination of diverse parts of mathematics to a common theory, is sufficient in itself, it would seem, to justify such an opinion.

The results of the extensive development of distance geometry since Menger's *Untersuchungen* of 1928 have been published in many journals throughout the world—some of which are no longer easily obtainable—and no attempt has hitherto been made to combine the numerous contributions into a unified theory. The individual approaches used by the many writers (differences of fundamental assumptions, methods, and notations) make it difficult even for the mature mathematician to obtain a proper perspective of the subject, while the multiplicity of languages employed provides another obstacle for the graduate student to overcome. If the views of the present writer regarding the intrinsic interest of distance geometry and its usefulness as a *modus operandi* are not completely erroneous, perhaps the time has come to give a detailed connected introduction to the subject. It is the purpose of this book to furnish such an account.

In writing the book the author has had in mind its use as a text for second-year graduate students as well as a reference for workers in the field. A course in general topology is a useful preliminary to a study of distance geometry but it is not indispensable. Any previous work that gives the student some experience in dealing with abstract mathematics (for example, a course in abstract algebra) would be almost equally valuable, since nearly all of the set topology used is developed in the text. Distance geometry, as offered at the University of Missouri, forms a cycle of three courses of one semester each. There is more in this book than is covered by such a cycle, so those using it for two semesters will probably have a choice of topics, even if the preliminary material in Chapter I be omitted.

A large part of the book's contents is treated in English for the first time. To coordinate the work of numerous articles with our general plan it was found necessary to supply new proofs for many theorems. Frequently the proofs follow more readily from the present setting of the theorems than from their original environments. A notable example of this is the simple proof (with weaker hypotheses) of Wilson's metric characterization of n-dimensional euclidean space given in § 50.

The reader is well provided with exercises. While this is hardly a novelty, it is unfortunately the case that many texts intended for the advanced graduate student are deficient in that respect. The exercises are of various degrees of difficulty. In some the reader is asked to supply the details of proofs given in the text, while others invite him to develop additional theory to be used later. A few of the exercises might form the bases of masters' theses.

In the references concluding each chapter the writer acknowledges the source of the material discussed and occasionally suggests papers dealing with matters of collateral interest. These papers furnish good material for seminar reports and are useful guides to those seeking further information about a topic which they find of special interest.

Though the book contains the latest word about many of the matters with which it deals (but the last word about none), it is not intended to be encyclopaedic. For example, the recent work of the Russian school of differential geometers (in particular, that of A. D. Aleksandrov and A. V. Pogorelov) has not found a place in this book, nor have results based upon theorems of combinatorial topology been included. These and other aspects of distance geometry, neglected here, might well be treated in another volume by one more competent to do so.

The author is indebted to all those investigators whose work has been incorporated in this book. It is a pleasure to acknowledge the many useful suggestions made by the writer's students (particularly by David O. Ellis, J. W. Gaddum, and L. M. Kelly) in courses in which preliminary drafts of the manuscript were used. To all these his sincere thanks are extended.

L. M. B.

Columbia, Missouri

CONTENTS

Part I
METRIC SPACES

Chapter I. PRELIMINARY NOTIONS
1. Abstract sets. 1
2. Operations on sets. Lattices 2
3. Distributivity. Boolean algebra 4
4. Abstract spaces 5
5. Semimetric spaces 7
6. Topology of semimetric spaces 8
7. An anomalous property of distance in semimetric spaces . . 12
8. The triangle inequality. Metric spaces 13
9. Examples of metric spaces 15
10. Topology of metric spaces 26

Chapter II. METRIC SEGMENTS AND LINES
11. Historical remarks 32
12. Betweenness in metric spaces. Congruence 33
13. Characterization of metric betweenness 36
14. Convexity in metric spaces. Metric segments. . . . 40
15. Two characteristic properties of metric segments . . . 44
16. Lemmas concerning closed and compact subsets. An equilateral triple theorem 46
17. An equilateral triple characterization of metric segments . . 48
18. Criteria for unique metric segments 49
19. Convex extension of a set 51
20. Passing points and terminal points 53
21. Metric lines 55

Chapter III. CURVE THEORY
22. Arcs and arc length 59
23. Homogeneous ϵ-chains and δ-density 60
24. First sharpening of length concept 61
25. Second sharpening of length concept. Lower semi-continuity of arc length 63
26. A relaxing of the length concept 67
27. Continuous curves 68
28. Existence of geodesic arcs 70
29. The n-lattice theorem 73
30. Metric definitions of curvature 74
31. Some properties of K_M, K_A, and K_H 77
32. Ptolemaic spaces. A lemma 78
33. Segments characterized as arcs with vanishing Menger curvature . 80
34. Metrization of torsion 84
35. Wald's metrization of Gauss curvature 88

CONTENTS

PART II
EUCLIDEAN AND HILBERT SPACES

CHAPTER IV. CONGRUENT IMBEDDING IN EUCLIDEAN SPACE

36. Two fundamental problems of distance geometry	90
37. Congruence indices	91
38. Congruence order of the E_n	93
39. A set of sufficient conditions for congruence indices $(m+3, k)$	95
40. The Cayley–Menger determinant	97
41. Imbedding a semimetric $(r+1)$-tuple irreducibly in E_r	99
42. A solution of the euclidean imbedding problem	101
43. Additional criteria for imbedding finite semimetric spaces in E_r	105
44. An example of a pseudo-E_n space	109
45. The structure of pseudo-E_n $(n+3)$-tuples	110
46. Quasi-congruence order of E_n	115
47. Free $(n+2)$-tuples	118

CHAPTER V. METRIC AND VECTOR CHARACTERIZATIONS OF EUCLIDEAN AND HILBERT SPACES. IMBEDDING IN HILBERT SPACE. NORMED LINEAR SPACES

48. First metric characterization of the E_n	122
49. The weak euclidean four-point property. Some lemmas	123
50. Second metric characterization of the E_n	126
51. The pythagorean property	129
52. Metric transforms and the euclidean four-point property	130
53. Congruent imbedding in Hilbert space	132
54. Metric transforms of euclidean and Hilbert spaces	134
55. Metric characterizations of Hilbert space	136
56. Normed linear spaces, inner product, and quasi inner product space	137
57. Quasi inner product space satisfying Schwarz inequalities	139
58. Generalized euclidean space in terms of a quasi inner product	144

CHAPTER VI. CONGRUENCE INDICES OF SOME EUCLIDEAN SUBSETS

59. Introductory remarks	148
60. Hyperfinite and transfinite congruence orders of linear sets. A general theorem	149
61. One-dimensional subsets of E_2	152
62. Two-dimensional subsets of E_2. Monomorphic sets. Characterization of the circular disk	156

PART III
THE NON-EUCLIDEAN SPACES

CHAPTER VII. IMBEDDING AND CHARACTERIZATION THEOREMS FOR SPHERICAL SPACE. CHARACTERIZATION OF PSEUDOSPHERICAL SETS

63. Solution of the imbedding problem for $S_{n,r}$	162
64. The Σ_r space. Some lemmas	163

65. Additional lemmas. Subspaces of Σ_r .	167
66. Metric characterization of $S_{n,r}$	168
67. Basic properties of $S_{n,r}$	171
68. Derived properties of $S_{n,r}$	172
69. Pseudo-$S_{k,r}$ $(k+3)$-tuples, $k \leqslant n$	175
70. Some properties of pseudo-$S_{k,r}$ $(k+4)$-tuples without diametral point-pairs .	181
71. Characterization of pseudo-$S_{n,r}$ sets without diametral point-pairs	186
72. Characterization of general pseudo-$S_{n,r}$ sets	189

CHAPTER VIII. INTERSECTION THEOREMS FOR CONVEX SUBSETS OF $S_{n,r}$. CONGRUENCE INDICES OF SPHERICAL CAPS

73. Preliminary definitions and remarks .	192
74. Some lemmas	194
75. Intersection theorems	198
76. Congruence indices of hemispheres and small caps	202

CHAPTER IX. IMBEDDING AND CHARACTERIZATION THEOREMS FOR ELLIPTIC SPACE

77. The notion of δ-supplementation	206
78. Imbedding theorems for $\mathscr{E}_{n,r}$.	207
79. Some metric peculiarities of elliptic space	209
80. Equilateral subsets	211
81. The space problem for the $\mathscr{E}_{n,r}$. A lemma	214
82. Definition and elementary properties of an $\mathscr{E}_r$ space	215
83. One-dimensional subspaces (lines) of $\mathscr{E}_r$	219
84. Imbedding line-sums of $\mathscr{E}_r$ in the elliptic plane	221
85. Linear subspaces of $\mathscr{E}_r$	225
86. Characterization theorems	227

CHAPTER X. CONGRUENCE AND SUPERPOSABILITY IN ELLIPTIC SPACE

87. Preliminary remarks and definitions .	231
88. Congruent subsets, one of which is contained in $\mathscr{E}_{1,r}$.	232
89. Non-superposable congruent subsets of $\mathscr{E}_{n,r}$	233
90. Spherical sets and matrices associated with an elliptic m-tuple	235
91. First superposability theorems	237
92. Congruent m-tuples with corresponding triples superposable .	239
93. Superposability of infinite subsets of $\mathscr{E}_{n,r}$	241
94. First reduction theorem	243
95. Two lemmas	245
96. Second reduction theorem	247
97. Superposability order. Pseudo f-superposable sets	251

Chapter XI. METRIC-THEORETIC PROPERTIES OF THE ELLIPTIC PLANE. CONGRUENCE ORDER

98. Preliminary remarks 255
99. Orthocentric quadruples 255
100. Singular loci 258
101. Freely movable quintuples 259
102. A 'crowding' theorem for the elliptic plane 262
103. Congruence order of $\mathscr{E}_{2,r}$ with respect to the class of semimetric spaces 264
104. Congruence invariance of metric bases. Congruence indices of the cross 270

Chapter XII. GENERALIZED HYPERBOLIC SPACE

105. Introductory remarks 273
106. The $(n+2)$-point relation in $\mathscr{H}_{n,r}$ 273
107. Generalized hyperbolic space. Three postulates . . . 275
108. Subspaces of $\mathscr{H}_{n,r}^{\phi}$ 276
109. Some metric properties of $\mathscr{H}_{n,r}^{\phi}$ 278
110. Additional metric properties of $\mathscr{H}_{k,r}^{\phi}$ 282
111. Two existence postulates. Imbedding theorems for $\mathscr{H}_{n,r}^{\phi}$. 284

Part IV
APPLICATIONS OF DISTANCE GEOMETRY

Foreword 288

Chapter XIII. METRIC METHODS IN DETERMINANT THEORY

112. Cayley–Menger determinants 290
113. Determinants of type Δ 294
114. Determinants of types Δ_{NP}, Δ_{NN}, and Λ . . . 297
115. Quasi rank of Δ_{NN} determinants 301

Chapter XIV. METRIC METHODS IN LINEAR INEQUALITIES

116. Coefficient sets C and solution sets $\Sigma(C)$ 304
117. Some elementary properties of C and $\Sigma(C)$ . . . 305
118. The generalized Minkowski theorem 308
119. Coincidence of C and $\Sigma(C)$ 311
120. Existence theorems for solutions of a system (I) . . . 312
121. Systems of strict inequalities 313

Chapter XV. METRIC METHODS IN LATTICE THEORY

122. Introduction 315
123. Lemmas from lattice theory 315
124. Lattice characterizations of metric betweenness . . . 317
125. Betweenness in normed distributive lattices. Betweenness in arbitrary lattices 320
126. Lattice characterization of pseudo-linear quadruples . . 323
127. Congruent imbedding of normed lattices in convex normed lattices . 325
128. Congruence of $D(L)$ with a euclidean subset . . . 327

129. Properties of the associated metric space of a normed lattice . 328
130. Characterization of $D(L)$ 330
131. Autometrized Boolean algebras. Elementary properties of distance . 331
132. Betweenness in autometrized Boolean algebras . . . 333
133. The group of motions of B 334
134. Free mobility in an autometrized Boolean algebra . . . 336
135. Congruence order of B with respect to the class of B-metrized spaces . 337

BIBLIOGRAPHY 339

INDEX 344

PART I
METRIC SPACES

CHAPTER I
PRELIMINARY NOTIONS

1. Abstract sets

In the study of mathematics beyond the usual calculus courses, the word 'set' is frequently encountered. The entities forming a set are called its *elements*, and sets whose elements are real numbers (for example, the set of positive integers or the set of rational numbers) as well as sets whose elements are functions or curves are met with in algebra, analysis, and geometry. It is a matter of common experience that many properties of sets are quite independent of the nature of the constituent elements, and this suggests developing a theory of *abstract* sets—that is, of sets the nature of whose elements is unspecified or voluntarily ignored. Such a theory is obviously of very wide application.

Though the concept of a set seems simple and intuitive, closer examination reveals that in its fullest generality it is probably not a fit object for mathematical study. A completely unrestricted notion of set leads to antinomies or paradoxes which may be avoided by limiting the scope of the concept. Without entering into the logical difficulties of 'set', we shall apply the term only to those collections of objects such that (1) there is no other alternative for an entity than to belong or not to belong to the collection, and (2) there is no other alternative for each pair of objects of the collection than to be distinct or not.

Membership of an entity x in a set A is symbolized by writing $x \in A$. If A, B are two sets such that $x \in A$ implies $x \in B$, then A is called a *subset* of B, and we write $A \subset B$. This relationship is also expressed by saying that A is part of B or that B contains A.

If $A \subset B$ and $B \subset A$ then clearly the sets A, B consist of the same elements, and we write
$$A = B.$$
When $A \subset B$ but $A \neq B$ then A is said to be a *proper* subset of B.

EXERCISES

1. A number is called *transcendental* provided it is not a root of an algebraic equation with integer coefficients. Do all such numbers constitute a set even though one may be unable to determine whether or not a given number is transcendental?

2. A set may be called *ordinary* or *extraordinary* according as it is not or is a member of itself. Give examples of ordinary and extraordinary sets.
3. *The Russell paradox.* Let Σ denote the collection of all ordinary sets. Show that if Σ is an ordinary set then it is also an extraordinary set, and conversely.
4. Show that if each set consisting of a single element may be *identified* with that element, then every set is an extraordinary set. Hence such an identification is not allowable.

2. Operations on sets. Lattices

We consider now the set of all subsets of a given set S. Corresponding to each two of these sets A, B there is a set called the *sum* (union, join) of A and B, denoted by $A+B$, such that $x \in A+B$ if and only if $x \in A$ or $x \in B$. It is easily seen that the operation of set addition has the following properties:

(i) if A and B are subsets of S, then $A+B \subset S$,

(ii) $(A+B)+C = A+(B+C)$,

(iii) $A+B = B+A$,

(iv) $A+A = A$.

Thus set addition is *monotone*, *associative*, *commutative*, and *idempotent*.

Another way of combining two sets A and B to produce a third is to form the set of all elements common to A and to B. This set, called the *product* (intersection, meet) of A and B, is denoted by AB. Thus $x \in AB$ if and only if $x \in A$ *and* $x \in B$. In order that a product may be defined for each two sets, we agree to consider a collection containing no elements as a set—the null or vacuous set 0. Clearly the operation of set multiplication is also monotone, associative, commutative, and idempotent (that is, $AA = A$).

It is observed that the equality $AB = A$ is equivalent to the equality $A+B = B$, for if $AB = A$ then $p \in A$ implies $p \in B$, and consequently $A+B = B$; while if $A+B = B$, then again $p \in A$ implies $p \in B$, and so $AB = A$. Each of the relations $AB = A$, $A+B = B$ is equivalent to the relation $A \subset B$ of set inclusion.

DEFINITION 2.1. *If two associative, commutative, idempotent, binary operations, 'sum' and 'product', can be uniquely defined in an abstract set so that for each two of its elements a, b the equalities $ab = a$ and $a+b = b$ are equivalent, the set is said to form a lattice.*

We have therefore proved:

THEOREM 2.1. *The set of all subsets of an arbitrary set S forms a lattice (with respect to set-sum and set-product).*

DEFINITION 2.2. *A set is partially ordered provided a relation $\prec$ is defined such that* (i) *for each element x of the set $x \prec x$,* (ii) *if $x \prec y$ and $y \prec x$ then $x = y$, and* (iii) *if $x \prec y$ and $y \prec z$, then $x \prec z$.*†

The relation $\prec$ is called a relation of precedence or inclusion (in the wide sense). As stated in the above definition, the relation is *reflexive*, *asymmetric*, and *transitive*. Two elements x, y of a partially ordered set are *comparable* provided $x \prec y$ or $y \prec x$. It should be noted that the definition of a partially ordered set permits the presence of non-comparable pairs of elements in the set.

A relation $\prec$ may be introduced in a lattice by defining $a \prec b$ to mean $ab = a$, or (what is equivalent) $a+b = b$. The reader is asked to verify that this relation is an inclusion relation.

The *complement* $C(A)$ of a subset A of S consists of all those elements of S that do not belong to A; that is, $x \in C(A)$ if and only if $x \in S$ and $x \bar{\in} A$, where $\bar{\in}$ is read 'is not an element of'. Obvious properties of set complement are $C(0) = S$, $C(S) = 0$, $C(C(A)) = A$ for each subset A of S, $A+C(A) = S$, and $AC(A) = 0$. If $A \subset B$ then $C(B) \subset C(A)$.

Slightly less obvious and quite useful properties of set complementation are exhibited by the so-called De Morgan formulae. To prove them in their most general form it is noted first that the definitions of set-sum and set-product given above may readily be extended to define the sum and the product of any collection or family of subsets of S. Thus if F is any family of subsets of S, the sum of the sets belonging to F, denoted by $\sum_{X \in F} X$, consists of all those elements of S belonging to at least one of the sets of the family, while the product $\prod_{X \in F} X$ of the sets belonging to F consists of all those elements of S which are common to all the members of the family. It is seen at once that both $\sum_{X \in F} X$ and $\prod_{X \in F} X$ are independent of any order or grouping of the sets constituting the family F, and hence set addition and multiplication are said to be *completely commutative* and *completely associative*.

The De Morgan formulae state that the complement of any sum of sets is the product of the complements of the summands, and the complement of any product of sets is the sum of the complements of the factors; that is,

$$C\left(\sum_{X \in F} X\right) = \prod_{X \in F} C(X), \tag{1}$$

$$C\left(\prod_{X \in F} X\right) = \sum_{X \in F} C(X). \tag{2}$$

† In all definitions 'provided' is used in the sense of 'if and only if'.

To prove (1) observe that x is an element of the left-hand side if and only if x is not an element of any of the sets X of F; that is, if and only if x *is* an element of each of the sets $C(X)$, $X \in F$. In a similar manner (2) is established.

EXERCISES

1. If in a lattice L, $a \prec b$ provided $ab = a$, prove that $ab \prec a$, $ab \prec b$, while if $c \in L$ such that $c \prec a$ and $c \prec b$ then $c \prec ab$ (that is, ab is the *greatest lower bound* of a, b). Show that $a+b$ is the *least upper bound* of a, b.
2. Show that the set of all subsets of a set S is partially ordered by set inclusion.
3. Prove that any relation of precedence can be interpreted as set inclusion.

3. Distributivity. Boolean algebra

The reader is familiar with the fact that the operations of addition and multiplication of numbers are connected by a distributive property; namely, the distributivity of multiplication with respect to addition. Thus if a, b, c are numbers, then $(a+b)c = ac+bc$; that is, the multiplication of the number $a+b$ by c may be *distributed* over the two summands a, b. On the other hand, addition of numbers is *not* distributive with respect to multiplication, for $ab+c \neq (a+c)(b+c)$.

Both kinds of distributivity are valid, however, for set addition and multiplication; that is, for each subset A of S and each family F of subsets of S,

$$A \sum_{X \in F} X = \sum_{X \in F} AX, \qquad (1)$$

$$A + \prod_{X \in F} X = \prod_{X \in F} (A+X). \qquad (2)$$

To prove (2), for example, it is seen that an element belongs to the left side of the equality if and only if it belongs either to A or to each set X of F; that is, if and only if it belongs to $A+X$ for each X of F and hence is an element of $\prod_{X \in F} (A+X)$.

DEFINITION 3.1. *A lattice L is called* distributive *provided $a, b, c \in L$ imply $(a+b)c = ac+bc$. It is* complemented *provided there corresponds to each element a of L an element a' of L such that*

$$a+a' = 1, \qquad aa' = 0,$$

where 0, 1 are elements of L with the property $0 \prec a \prec 1$ for each $a \in L$.

In the lattice of subsets of a set S, 0 is the null set, 1 is the whole set S, and $C(A)$ is the subset of S corresponding to A with $A+C(A) = 1$, $AC(A) = 0$ (§ 2).

DEFINITION 3.2. *A complemented, distributive lattice is called a* Boolean algebra.

The properties of set addition, multiplication, and complementation proved in the foregoing establish:

THEOREM 3.1. *The set of all subsets of a set S forms a Boolean algebra with respect to the operations of set sum, product, and complementation.*

EXERCISES

1. Prove that set multiplication is distributive with respect to addition. Consider generalizations of (1), (2) obtained by replacing the set A by a family of subsets of S.
2. Show that if a lattice L has elements 0, 1 they are unique.
3. Show that in a Boolean algebra each element has a unique complement. Prove that the De Morgan formulae are valid in every Boolean algebra.
4. Give an example of a non-distributive lattice.
5. If $a, b, c, d \in L$, a lattice, and $a \prec b, c \prec d$ show that $ac \prec bd$ and $a \mid c \prec b \mid d$.
6. Show that a partially ordered set that contains for each two of its elements a greatest lower bound and a least upper bound is a lattice.

4. Abstract spaces

Intuition endows a *space* with attributes that are not possessed by a mere collection of elements. For we think of objects being 'located' in a space, 'near' to some objects and 'far' from others, while our intuitive notion of a straight line suggests that each space, properly so called, should exhibit a similar continuity of structure. Hence in order that an abstract set may be spoken of as a space without sacrificing the innate significance of the term, it is necessary to give a set additional qualities that approximate in some measure the continuity and proximity properties that 'space' seems to have. The additional feature is called a *topology*, and so *an abstract space is an abstract set with a topology*.

There are several frequently used methods of assigning a topology to a set which, though based upon different primitive notions, define the same very general class of spaces. One of the earliest devices was that used by Fréchet in 1906 in defining an L-space or class (L); that is, a space in which the notion of 'limit of a sequence of elements' is primitive.

An abstract set forms an L-space provided for every infinite sequence $\{p_n\}$ of its elements, and each element p of the set, a convention is made which determines whether or not p is a 'limit' of the sequence (written $\lim p_n = p$). Any convention is permissible provided it satisfies the following three conditions:

1. If $p_n = p$, $n = 1, 2,...$, then $\lim p_n = p$.
2. If $\lim p_n = p$ and $\lim p_n = q$, then $p = q$.
3. If $\lim p_n = p$ and if $i_1, i_2,..., i_n,...$ is an infinite sequence of increasing indices, then $\lim p_{i_n} = p$.

This establishes a so-called *limit-topology* in the set.

The reader may show that the way in which a limit is ordinarily assigned to a sequence of numbers (real or complex) conforms to these three requirements and so the set of real numbers, for example, is an L-space with respect to the usual definition of limit.

Another means of introducing a topology (*open set topology*) in an abstract set is to assume a convention according to which certain subsets are called *open*. Such a convention is subjected to a few very simple restrictions as, for example, that the null set and the whole set be open, and that the sum of any collection of open sets be open. If it is further assumed that the product of any two open sets is open and that each two distinct elements are contained in open sets whose product is null, the resulting space K is *equivalent to the Hausdorff topological space* (Exercise 3). If, in a space K, $p = \lim p_n$ provided there corresponds to each open set U containing p an integer N such that $p_i \in U$ for $i > N$, the conditions for an L-space are satisfied (Exercise 1).

A set may be given a topology by associating with each of its subsets X a subset $\bar{X}$, called the *closure* of X. The association is subject to the requirements $\bar{X} \supset X$, $\bar{O} = O$, $\bar{\bar{X}} = \bar{X}$, and $\overline{X+Y} = \bar{X}+\bar{Y}$.

A development of the theory of any of the three spaces just defined is outside the purpose or scope of this book. In the next section we shall describe still another method of endowing an abstract set with a topology. That method is most intimately connected with our design and we shall study it in some detail.

EXERCISES

1. Prove that a space K is an L-space.
2. Give an example to show that the four axioms of closure do not imply that $\overline{(a)} = (a)$.
3. DEFINITION. An abstract set H forms a Hausdorff topological space provided certain subsets, called *neighbourhoods*, are selected to satisfy the following four conditions:
    I. There is associated with each element x of H at least one neighbourhood N_x; each neighbourhood N_x of x contains x.
    II. If N_x, N_x^* are two neighbourhoods of the same element x, there exists a neighbourhood M_x of x such that $M_x \subset N_x . N_x^*$.
    III. If $y \in N_x$ then a neighbourhood N_y of y exists with $N_y \subset N_x$.
    IV. If $x, y \in H$ ($x \neq y$), neighbourhoods N_x, N_y exist with $N_x . N_y = 0$.

    Show that if *neighbourhood of an element x* of a space K is defined as any open set containing x, the space K is a Hausdorff topological space. Prove also that if a subset G of a Hausdorff topological space be called *open* provided for each element x of G a neighbourhood $N_x \subset G$ exists, then the Hausdorff space is a space K. It follows easily that a space K and a Hausdorff topological space are equivalent; that is, the topology given to a set by 'neighbourhoods' satisfying I, II, III, IV is the same as that given by 'open' sets conforming to the five conditions for a space K.

5. Semimetric spaces

It was observed in the preceding section that the notion of proximity of elements seems inherent in our intuition of a space. But no one of the three methods for making a space out of an abstract set presented in that section seems to have any direct connexion with that notion, and indeed additional conditions must be imposed before any satisfactory basis for making measurements will exist in those spaces. Since the presence of a measure or *metric* is, from our point of view, a *sine qua non* for a space, it seems desirable to introduce it in an abstract set directly as a primitive notion. This leads to what we shall call a distance space.

In its most general aspects, a *distance space* arises from an abstract set by associating with each ordered pair p, q of its elements an element pq of a 'distance' set, the association being subjected to certain simple rules. The nature of these rules, as well as the character of the set from which the 'distances' are selected, depend upon the degree of generality that is sought. Those distance spaces which have been most studied have their distance sets subsets of numbers (real or complex) with by far the greatest amount of attention being given to the real case.

DEFINITION 5.1. *A distance space is called semimetric provided the distance set is a set of non-negative real numbers, and the association of distance pq with ordered element pairs p, q is any whatever, provided $pq = qp$ and $pq = 0$ if and only if $p = q$.*

Referring to the elements of a space as 'points', the distance of two points p, q of a semimetric space is a non-negative (unique) real number pq which is independent of any order of p, q (i.e. the number pq is assigned to the point-pair p, q) and which is positive when (and only when) the two points are distinct. Thus distances in a semimetric space are *symmetric* and *positive definite*.

It is with the class (S) of semimetric spaces and its important subclass (M) of metric spaces (to be defined later) that we are principally concerned.

Numerous examples of semimetric spaces (that satisfy an additional condition) will be given in a later section. The class contains nearly all of the most important spaces (euclidean, non-euclidean, spherical, etc.) but not, it may be observed, the directed straight line or the space-time continuum of relativity theory.

6. Topology of semimetric spaces

The notion of distance, primitive in a semimetric space, permits defining in such spaces various topological concepts such as 'limit of a sequence', 'accumulation point of a subset', etc. We shall see, however, that the distance-topology so obtained has certain features that must be regarded as undesirable.

DEFINITION 6.1. *An element p of a semimetric space S is called a limit of an infinite sequence $p_1, p_2, p_3, \ldots, p_n, \ldots$ of elements of S provided the limit of the sequence of non-negative numbers $p_1 p, p_2 p, \ldots, p_n p, \ldots$ is zero; that is, $\lim p_n = p$ if and only if $\lim p_n p = 0$.*

Thus the notion of limit of an infinite sequence of elements of a semimetric space is defined in terms of the well-known notion of convergence of an infinite sequence of real numbers to zero.

Clearly limit of a sequence of points, as just defined, satisfies condition (1) of a Fréchet L-space. The following example, however, shows that condition (2) may not be satisfied.

A semimetric space is formed by the points of the closed interval $[0, \frac{1}{2}]$ together with the point 1, by defining coincident points to have zero distance and

$$xy = |y-x|, \qquad x, y \in [0, \tfrac{1}{2}],$$
$$1y = 0y, \qquad y \in [0, \tfrac{1}{2}], \qquad y \neq 0,$$
$$10 = 01 = 1.$$

The sequence of points $\{1/n\}$, $n = 1, 2, \ldots$, has each of the distinct points 0 and 1 as limits.

Thus the assumptions giving a semimetric space are not strong enough to ensure the uniqueness of limits of sequences of points, and so a semimetric space is not necessarily an L-space.

DEFINITION 6.2. *A sequence $\{p_n\}$ of points of a semimetric space is called a Cauchy or convergent sequence provided $\lim_{i,j \to \infty} p_i p_j = 0$.*

In analysis it is shown that an infinite sequence $\{z_n\}$ of numbers (real or complex) has a limit if and only if $\lim_{i,j\to\infty} |z_i - z_j| = \lim_{i,j\to\infty} z_i z_j = 0$. This is Cauchy's criterion, asserting that an infinite sequence has a limit if and only if it is convergent.

But in a semimetric space, convergence of an infinite sequence of elements is neither a necessary nor a sufficient condition that it have a limit. That the condition is not sufficient is not surprising for the sufficiency would imply that the space had a structure similar to that of the line in being free from any gaps. But, as the reader may recall, the freedom from gaps enjoyed by the straight line was ensured by adjoining the limits of all convergent sequences! Thus the semimetric space obtained from the straight line by deleting the point 0, for example, and using the ordinary euclidean definition of distance, would contain the convergent sequence $\{1/n\}$, $n = 1, 2, 3,...$ which has no limit in the space.

More significant is the failure of the necessity of Cauchy's criterion to hold in semimetric spaces; that is, an infinite sequence of elements of a semimetric space might have a limit without being a convergent sequence. For the space consisting of the points 0, $\{1/n\}$, $n = 1, 2, 3,...$, with all distances euclidean *except*

$$\text{dist}(1/i, 1/j) = 1 \quad (i, j = 1, 2, 3, ..., i \neq j)$$

is semimetric. The sequence $\{1/n\}$ ($n = 1, 2,...$), has the limit 0, but it is evidently not a convergent sequence.

DEFINITION 6.3. *Let p, q be elements of a semimetric space S. If for any two sequences $\{p_n\}$, $\{q_n\}$ of elements of S, $\lim p_n = p$ and $\lim q_n = q$ imply $\lim p_n q_n = pq$, the distance function is said to be continuous at p, q. The distance function is continuous in S provided it is continuous at each point-pair of S.*

The semimetric space of all numbers between 0 and 1, inclusive, with $xy = (y-x)^2$ for x, y any two numbers of the set, has its distance function continuous. On the other hand, by defining the distance of points 0 and 1 of a line to be 2 and taking all other distances to be euclidean, a semimetric space is formed with a distance function that is evidently *discontinuous* at this point-pair 0, 1. For the sequences $\{1/n\}$ and $\{1-1/n\}$ have limits 0 and 1, respectively, but the sequence

$$\{\text{dist}(1/n, 1-1/n)\}$$

of corresponding distances has the number 1 as limit, which is *not* the distance of the two limit points 0, 1.

Hence the distance function in a semimetric space is not necessarily continuous.

Where a set E is referred to as a subset of a semimetric *space S* it is meant that E is a subset of the *set S* and *distance in E is defined as it is in the space S*.† Thus E is itself a semimetric space. In case distance in E is defined differently from distance in S then E is not called a subset of the space S even though E be a subset of the set S. For example, if

† The term 'subspace' which is more appropriate here than subset is reserved for a different notion.

distance of two points of a circle (curve) be the length of the chord joining them then the circle is a subset of the euclidean plane, but if the distance be defined as the length of the shorter arc of the circle joining the points, then the resulting semimetric space (called the *convex* circle) is *not* a subset of the euclidean plane.

DEFINITION 6.4. *Let E be a subset of a semimetric space S. An element p of S is called an accumulation point of E provided for each positive number ϵ there is a point q of E such that $0 < pq < \epsilon$.*

DEFINITION 6.5. *A subset E of a semimetric space is closed provided E contains each of its accumulation elements. E is open provided its complement $C(E)$ is closed.*

If E' denotes the subset of S that consists of all the accumulation elements of a subset E of S, then E is closed if and only if $E' \subset E$. The set E', called the *derived* set of E, may, of course, be the null set. The set $\bar{E} = E + E'$ is the *closure* of E.

If $p \in S$ and ρ a positive number, the subset $U(p; \rho)$ consisting of all those elements q of S with $pq < \rho$ is called the *spherical neighbourhood* of p with radius ρ.

In the euclidean plane, for example, $U(p; \rho)$ is simply the *interior* of the circle with centre p and radius ρ. It is easily seen to be an open set. *But it is not true that spherical neighbourhoods are open sets in every semimetric space.* In the semimetric space arising from the points of the straight line by defining distances as euclidean for each point-pair except 0, 2 and setting $02 = 20 = 1$, the spherical neighbourhood $U(0; \frac{3}{2})$ is not an open set. For since $U(0; \frac{3}{2})$ contains the point 2 which is clearly an accumulation point of the complement $CU(0; \frac{3}{2})$, this latter set is not closed and so $U(0; \frac{3}{2})$ is not open.

It may be noticed that the semimetric space just defined has its distance function discontinuous at the 'exceptional' point-pair 0, 2. That no example of the kind desired exists without a discontinuity in the distance function of the semimetric space is implied by the following theorem.

THEOREM 6.1. *If S is a semimetric space with continuous distance function then for each $p \in S$ and $\rho > 0$, $U(p; \rho)$ is an open set.*

Proof. Let q be an accumulation element of $CU(p; \rho)$. Then for each positive integer n there exists a point q_n of $CU(p; \rho)$ such that
$$0 < qq_n < 1/n.$$
The sequence $\{qq_n\}$ of non-negative numbers has limit 0 and hence

$\lim q_n = q$. Since $q_n \in CU(p; \rho)$, $pq_n \geqslant \rho$ $(n = 1, 2,...)$ and so, the distance function being continuous,

$$pq = \lim pq_n \geqslant \rho.$$

Hence $q \in CU(p; \rho)$, which is consequently a closed set. Then $U(p; \rho)$ is open and the theorem is proved.

It was remarked in the previous section that a topological space is an L-space. But since an example has been given of a semimetric space that is not an L-space, it follows that not every semimetric space is a topological space.

THEOREM 6.2. *A semimetric space S with continuous distance function is a Hausdorff topological space K.*

Proof. It follows from the definition of accumulation element that S and the null set are both closed, and hence both are open.

If F is any family of open subsets U of S we wish to show that the subset $\sum_{U \in F} U$ is open. Now by one of De Morgan's formulae

$$C\left(\sum_{U \in F} U\right) = \prod_{U \in F} C(U)$$

and since U is open, then $C(U)$ is closed, $U \in F$. Every accumulation element p of $\prod_{U \in F} C(U)$ is an accumulation element of each of the sets $C(U)$, $U \in F$; for if $\epsilon > 0$ there exists a point q belonging to the product set (and hence to each of the 'factors' $C(U)$, $U \in F$) with $0 < pq < \epsilon$. The sets $C(U)$ being closed, $p \in C(U)$, $U \in F$, and consequently $p \in \prod_{U \in F} C(U)$. Hence the product set is closed and so its complement $\sum_{U \in F} U$ is open. Thus the sum of any aggregate of open subsets of S is open.

We show next that the *product* of any two open subsets U, V of S is open. Use of a De Morgan formula gives

$$C(UV) = C(U) + C(V).$$

If we suppose an element p of S is *not* an accumulation element of $C(U)$, then there exists a positive number ϵ_0 such that no element of $C(U)$ has a positive distance from p less than ϵ_0. Assuming $p \in C'(UV)$, then $\epsilon > 0$ implies the existence of a point q of $C(U) + C(V)$ such that

$$0 < pq < \epsilon.$$

Hence for each $\epsilon < \epsilon_0$, $q \in C(V)$, and so $p \in C'(V)$. Consequently each element of $C'(UV)$ is an element of $C'(U)$ or of $C'(V)$; that is, an element of $C(UV)$ since $C(U)$, $C(V)$ are closed sets. Hence $C(UV)$ is closed and UV is open.

There remains to show, finally, that if $p, q \in S$, $p \neq q$, open subsets U, V of S exist such that $p \in U$, $q \in V$, and $UV = 0$. According to the previous theorem, the spherical neighbourhoods $U(p; \rho)$, $U(q; \rho)$ are open sets. If for each $\rho > 0$, these two open sets have a non-null product, there exists for each positive integer n an element r_n such that $pr_n < 1/n$, $qr_n < 1/n$, and so the infinite sequence $\{r_n\}$ has both p and q as limits. But from $\lim r_n = p$, $\lim r_n = q$ follows (by the continuity of the distance)

$$pq = \lim r_n r_n = 0,$$

which is impossible since $p \neq q$.

Hence there exists a $\rho > 0$ such that

$$p \in U(p; \rho), \qquad q \in U(q; \rho),$$
$$U(p; \rho) . U(q; \rho) = 0,$$

and the proof of the theorem is complete.

COROLLARY. *A semimetric space with continuous distance is an L-space.*

Thus the requirement that distance in a semimetric space be continuous eliminates some of the topological abnormalities in which general semimetric spaces abound. Many abnormalities of a different kind, however, remain, one of which will be exhibited in the next section.

7. An anomalous property of distance in semimetric spaces

Any positive definite, symmetric, distance shall be referred to as a *semimetric*, reserving the term *metric* for those distances that satisfy an additional requirement. We have seen that if a space has a continuous semimetric it is a Hausdorff topological space and is, consequently, rich in topological properties. It may be noted, moreover, that in such spaces a sequence with a limit is necessarily a convergent sequence.

Though a semimetric is positive definite in the sense that $pq \geq 0$, and $pq = 0$ if and only if $p = q$, two distinct points of a semimetric space may nevertheless be joined by an arc of zero length (extending the classical definition of arc length to semimetric spaces). This anomaly may be present even when the semimetric is continuous, as the following example shows.

Consider the semimetric space S obtained from the points of the unit interval $[0, 1]$ by defining the distance xy of points x, y of the set to be

$$xy = (x-y)^2.$$

This distance is easily seen to be *topologically equivalent* to the euclidean distance $|x-y|$ of points x, y, by which is meant that any point x of

[0, 1] which is the limit of a sequence of points of the interval when euclidean distance is used, will be the limit of the sequence when distance $(x-y)^2$ is employed, and conversely. It follows that the topological character of [0, 1] as a line segment in the euclidean distance is unchanged by the new semimetric, and so S is an arc.†

If a sequence of polygons P_n with vertices $0, 1/2^n, 2/2^n, ..., (2^n-1)/2^n, 1$ is inscribed in S ($n = 1, 2, ...$) then each pair of consecutive vertices has distance $1/2^{2n}$, and since the polygon has 2^n such pairs, the 'length' $l(P_n)$ of P_n is $1/2^n$. Now as $n \to \infty$, the distance of consecutive vertices approaches zero (that is, the sequence P_n is an allowable one according to the ordinary procedure for determining arc length). But

$$l(S) = \lim_{n \to \infty} l(P_n) = \lim_{n \to \infty} 1/2^n = 0.$$

Thus the classical method for defining length of arc leads to a paradoxical result when applied in semimetric spaces, even though the (positive definite) semimetric be continuous (indeed, uniformly continuous) in the space.

Though distance in a semimetric space has been given all those properties that one thinks of as most natural for 'distance' to possess, they are apparently not enough to prevent the occurrence of undesirable irregularities. Some additional, less superficial, property of distance must be required.

8. The triangle inequality. Metric spaces

It seems clear that the facility with which examples of semimetric spaces can be constructed is due to the complete independence of the distance function in the sense that the number assigned as distance to one pair of points does not depend in any way upon previously assigned distances. The great degree of freedom this circumstance permits allows one to form quite easily semimetric spaces with various kinds of bizarre properties. It also suggests that the more normal behaviour of the ordinary spaces is connected with properties of the distance which are much less obvious than those required of a semimetric. The problem, therefore, arises of seeking some kind of simple tie-up between the mutual distances of points which will eliminate the undesirable features which the weak assumptions on a semimetric permit, but which will not reduce too greatly the generality of the class of spaces.

It follows that we should first examine the possibilities afforded by sets of three points and ask whether there is any relation between the

---

† By definition an arc is a set with the same topological character as a segment of the straight line.

three mutual distances of three points which (when required of a semimetric to satisfy) will solve the problem just posed. Using the most important of all spaces—the euclidean—as a guide it is seen that the simplest relation satisfied by the mutual distances of each triple of points is the so-called *triangle inequality*, which asserts that the sum of any two sides of a triangle (proper or degenerate) is greater than or equal to the third side. Abstracting this fundamental property of euclidean triangles for application to a semimetric space yields the following important postulate.

TRIANGLE INEQUALITY POSTULATE. *If p, q, r are any three points of a semimetric space, the sum of any two of the distances pq, qr, pr is greater than or equal to the third.*

DEFINITION 8.1. *A semimetric space in which the Triangle Inequality Postulate is valid is called a metric space. The distance function of a metric space is called a metric.*

From our point of view the class (M) of metric spaces forms the most important class of abstract spaces. As shown by the examples given in the next section, the class is sufficiently general to embrace all of the best-known spaces, yet it is special enough to have an exceedingly rich topological and metric theory. It is remarkable that (as we shall soon see) the simple condition of the triangle inequality is strong enough to eliminate the 'pathological' aspects of semimetric space. No less remarkable is the fact that any more stringent demands on the mutual distances of three points, or restrictions on the mutual distances of *four points*, analogous to the triangle inequality, would too greatly specialize the class of spaces conforming to them. This will become clear later.

We have attempted to show how one might be lead to postulate the triangle inequality as a normalizing device from one elementary property of euclidean triangles. It is of interest to see how the same conclusion can be reached from quite different considerations; namely, from a desire to satisfy *in the simplest way* the demand that distance in a semimetric space be uniformly continuous.

If (a, b), (c, d) be any two pairs of ordered points in a semimetric space, let $ac+bd$ be defined as the distance of the pairs. Requiring distance to be uniformly continuous means that corresponding to an arbitrary positive ϵ there is a positive number $\delta(\epsilon)$ such that for all pairs (a, b), (c, d) of ordered points of the space, $ac+bd < \delta(\epsilon)$ shall imply

$$|ab-cd| < \epsilon.$$

Now this requirement will be most simply satisfied provided
$$|ab-cd| \leqslant ac+bd,$$
for then $\delta(\epsilon)$ may be taken equal to ϵ.

But the relation $|ab-cd| \leqslant ac+bd$ is equivalent to the triangle inequality. For if it holds for each two pairs (a,b), (c,d) of points then it holds, in particular, for the pair (a,b), (c,c), and we obtain $ab \leqslant ac+cb$. In a similar manner the other two inequalities are obtained.

Conversely, if the triangle inequality is valid then
$$|ab-cd| = |ab-bc+bc-cd| \leqslant |ab-bc|+|bc-cd| \leqslant ac+bd.$$

It is emphasized that uniform continuity of a semimetric does not imply the triangle inequality. The reader may easily show that the semimetric $xy = (x-y)^2$, $x, y \in [0,1]$, is uniformly continuous but *not a metric*. It is satisfying the demand for uniform continuity of distance *in the simplest manner* that gives rise to the triangle inequality. We have, incidentally, obtained the important result:

THEOREM 8.1. *The metric of a metric space is uniformly continuous.*

It is useful to observe that if a, b, c are elements of a metric space, then $|ab-bc| \leqslant ac$.

EXERCISE

1. Show that if distance xy is defined in a set so that $xy = 0$ if and only if $x = y$ and $xy+xz \geqslant yz$, the resulting space is metric.

9. Examples of metric spaces

It follows from Definition 8.1 that an abstract metric space is formed by attaching to each pair p, q of elements of an abstract set a non-negative real number pq, in accordance with the following agreements:

POSTULATE I. *If $p = q$ then $pq = 0$.*

POSTULATE II. *If $p \neq q$ then $pq > 0$.*

POSTULATE III. *The number attached to an element-pair is independent of any order of the elements in the pair; that is $pq = qp$.*

POSTULATE IV. *For each three elements p, q, r of the set, $pq+qr \geqslant pr$.*

The elements are spoken of (for suggestiveness) as 'points' and the number attached to a pair of points is the *distance* or *metric*. Thus the metric in a metric space is (*a*) *positive definite* (Postulates I, II imply $pq \geqslant 0$ and $pq = 0$ if and only if $p = q$), (*b*) *symmetric* (Postulate III), and (*c*) satisfies the *triangle inequality* (Postulate IV).

Fréchet defined a metric space in 1906, calling it a class (E) and referring to the distance pq as the *écart* of the points. The term *metric space* seems to have been used first by Hausdorff. It may be observed that any abstract set becomes a metric space upon defining $pq = 1$, $p \neq q$, and $pq = 0$, $p = q$, for each pair of elements p, q of the set. Hence the notion of an abstract set is no more general than that of metric space.

Example 1. The n-dimensional euclidean space E_n. The point set of the space is the set of all ordered n-tuples $(x_1, x_2,..., x_n)$ of real numbers. Distance is defined for each pair of elements $x = (x_1, x_2,..., x_n)$, $y = (y_1, y_2,..., y_n)$ by

$$xy = \Big[\sum_{i=1}^{n} (x_i - y_i)^2\Big]^{\frac{1}{2}}.$$

Defining $x = y$ if and only if $x_i = y_i$ ($i = 1, 2,..., n$), it is clear that Postulates I, II, III are satisfied. To show the triangle inequality valid let x, y, z be any three distinct elements of the space and put $a_i = x_i - y_i$, $b_i = y_i - z_i$ ($i = 1, 2,..., n$). Then $x_i - z_i = a_i + b_i$ ($i = 1, 2,..., n$). Now by the well-known Cauchy (or Schwarz) inequality,

$$(\sum a_i^2 \cdot \sum b_i^2)^{\frac{1}{2}} \geq \sum a_i b_i,$$

all summations going from 1 to n.

Multiplying both sides of the above inequality by 2 and adding $\sum a_i^2 + \sum b_i^2$ to both sides gives

$$[(\sum a_i^2)^{\frac{1}{2}} + (\sum b_i^2)^{\frac{1}{2}}]^2 \geq \sum (a_i + b_i)^2.$$

Whence
$$(\sum a_i^2)^{\frac{1}{2}} + (\sum b_i^2)^{\frac{1}{2}} \geq [\sum (a_i + b_i)^2]^{\frac{1}{2}};$$

that is,
$$xy + yz \geq xz,$$

and the n-dimensional euclidean space E_n is a metric space.

Example 2. The n-dimensional spherical space $S_{n,r}$ of radius r. The elements of the space are all the ordered $(n+1)$-tuples $(x_1, x_2,..., x_{n+1})$ of real numbers with $\sum_{i=1}^{n+1} x_i^2 = r^2$, $r > 0$. Distance is defined for each pair of elements x, y to be the smallest non-negative number xy such that

$$\cos(xy/r) = \Big(\sum_{i=1}^{n+1} x_i y_i\Big)\Big/r^2.$$

This space is the n-dimensional 'surface' or boundary of a sphere of radius r in E_{n+1} with geodesic (that is, shorter arc) metric. The $S_{n,r}$, referred to as the *convex n-sphere*, is not a subset of E_{n+1} since distance in $S_{n,r}$ is not euclidean (chord) distance.

To show $S_{n,r}$ is a metric space we make use of a generalization of the Cauchy inequality according to which the symmetric determinant

$$\begin{vmatrix} \sum x_i^2 & \sum x_i y_i & \sum x_i z_i \\ \sum y_i x_i & \sum y_i^2 & \sum y_i z_i \\ \sum z_i x_i & \sum z_i y_i & \sum z_i^2 \end{vmatrix}$$

is non-negative, together with all of its principal minors, for any three $(n+1)$-tuples $(x_1, x_2, ..., x_{n+1})$, $(y_1, y_2, ..., y_{n+1})$, $(z_1, z_2, ..., z_{n+1})$ of real numbers (all summations in the determinant being taken from 1 to $n+1$). The non-negativity of the second order principal minors is the ordinary Cauchy (or Schwarz) inequality; e.g.

$$\sum x_i^2 \cdot \sum y_i^2 \geqslant (\sum x_i y_i)^2.$$

Applying this inequality to two points

$$x = (x_1, x_2, ..., x_{n+1}), \qquad y = (y_1, y_2, ..., y_{n+1})$$

of $S_{n,r}$ it is seen that

$$-1 \leqslant (\sum x_i y_i)/r^2 \leqslant 1,$$

and it follows that xy is a real (non-negative) number, $0 \leqslant xy \leqslant \pi r$, $xy = 0$ if and only if $x = y$, and $xy = yx$. Hence $S_{n,r}$ is semimetric.

Using the generalized inequality for three distinct points $x^{(1)}$, $x^{(2)}$, $x^{(3)}$ of $S_{n,r}$ gives

$$\Delta_3 = \begin{vmatrix} 1 & \cos \alpha_{12} & \cos \alpha_{13} \\ \cos \alpha_{21} & 1 & \cos \alpha_{23} \\ \cos \alpha_{31} & \cos \alpha_{32} & 1 \end{vmatrix} \geqslant 0, \qquad (*)$$

where $\alpha_{ij} = x^{(i)} x^{(j)}/r$ $(i,j = 1, 2, 3)$. Expansion and obvious trigonometric substitutions yield

$$\Delta_3 = 4 \sin \tfrac{1}{2}(\alpha_{12}+\alpha_{23}+\alpha_{13}) \sin \tfrac{1}{2}(\alpha_{12}+\alpha_{23}-\alpha_{13}) \times$$
$$\times \sin \tfrac{1}{2}(\alpha_{12}-\alpha_{23}+\alpha_{13}) \sin \tfrac{1}{2}(-\alpha_{12}+\alpha_{23}+\alpha_{13}). \qquad (\dagger)$$

Now $\alpha_{12}+\alpha_{23}+\alpha_{13} \leqslant 2\pi$, for an assumption to the contrary, together with the fact that $0 \leqslant \alpha_{ij} \leqslant \pi$ $(i,j = 1, 2, 3)$, implies

$$2\pi < \alpha_{12}+\alpha_{23}+\alpha_{13} \leqslant 3\pi.$$

But then each of the angles

$$\tfrac{1}{2}(\alpha_{12}+\alpha_{23}-\alpha_{13}), \qquad \tfrac{1}{2}(\alpha_{12}-\alpha_{23}+\alpha_{13}), \qquad \tfrac{1}{2}(-\alpha_{12}+\alpha_{23}+\alpha_{13})$$

is seen to be positive and less than π. This results in Δ_3 being negative, contrary to $(*)$.

Hence $0 < \alpha_{12}+\alpha_{23}+\alpha_{13} \leqslant 2\pi$ and the first factor in $(\dagger)$ is non-negative. Clearly no two of the angles

$$\alpha_{12}+\alpha_{23}-\alpha_{13}, \qquad \alpha_{12}-\alpha_{23}+\alpha_{13}, \qquad -\alpha_{12}+\alpha_{23}+\alpha_{13}$$

can be negative, for this would imply that one of the angles α_{ij} ($i,j = 1, 2, 3$) is negative. If exactly one of them, say $\alpha_{12}+\alpha_{23}-\alpha_{13}$, is negative then (since $\Delta_3 \geqslant 0$) $\alpha_{12}+\alpha_{23}+\alpha_{13} = 2\pi$ and $\alpha_{13} > \pi$, which is impossible.

Therefore
$$\alpha_{12}+\alpha_{23} \geqslant \alpha_{13},$$
$$\alpha_{12}+\alpha_{13} \geqslant \alpha_{23},$$
$$\alpha_{23}+\alpha_{13} \geqslant \alpha_{12},$$
and the triangle inequality is satisfied by each three points of $S_{n,r}$.

Example 3. The n-dimensional elliptic space $\mathscr{E}_{n,r}$ of curvature $1/r^2$ ($r > 0$). This space may be obtained from the $S_{n,r}$ by identifying two points of $S_{n,r}$ if and only if the distance of the points is πr. Denoting distance of x, y in $S_{n,r}$ by (x, y), distance xy is defined in $\mathscr{E}_{n,r}$ by

$$xy = (x,y) \text{ if and only if } (x,y) \leqslant \tfrac{1}{2}\pi r,$$
$$xy = \pi r - (x,y) \text{ if } (x,y) > \tfrac{1}{2}\pi r.$$

Thus no two points of $\mathscr{E}_{n,r}$ have distance exceeding $\tfrac{1}{2}\pi r$.

$\mathscr{E}_{n,r}$ is semimetric, for $xy \geqslant 0$ (since no two points of $S_{n,r}$ have distance exceeding πr), $xy = yx$ (since $S_{n,r}$ is semimetric) and $xy = 0$ if and only if $x = y$ (since $xy = 0$ implies $(x, y) = 0$ or $(x, y) = \pi r$ and in either event $x = y$, and conversely).

Let $x^{(1)}$, $x^{(2)}$, $x^{(3)}$ be three distinct points of $\mathscr{E}_{n,r}$ and put
$$\beta_{ij} = x^{(i)}x^{(j)}/r \quad (i,j = 1, 2, 3).$$
Then $0 < \beta_{ij} \leqslant \tfrac{1}{2}\pi$ ($i,j = 1, 2, 3$; $i \neq j$). Since $x^{(i)}x^{(j)} = (x^{(i)}, x^{(j)})$ or $\pi r - (x^{(i)}, x^{(j)})$ then $\cos\beta_{ij} = \pm\cos(x^{(i)}, x^{(j)})/r = \pm\cos\alpha_{ij}$, and we have from (∗)

$$\begin{vmatrix} 1 & \epsilon_{12}\cos\beta_{12} & \epsilon_{13}\cos\beta_{13} \\ \epsilon_{21}\cos\beta_{21} & 1 & \epsilon_{23}\cos\beta_{23} \\ \epsilon_{31}\cos\beta_{31} & \epsilon_{32}\cos\beta_{32} & 1 \end{vmatrix} \geqslant 0$$

for at least one choice of $\epsilon_{ij} = \epsilon_{ji} = \pm 1$, $\epsilon_{ii} = 1$ ($i, j = 1, 2, 3$).

Then either
$$\Delta = \begin{vmatrix} 1 & \cos\beta_{12} & \cos\beta_{13} \\ \cos\beta_{21} & 1 & \cos\beta_{23} \\ \cos\beta_{31} & \cos\beta_{32} & 1 \end{vmatrix} \geqslant 0$$

or
$$\Delta^* = \begin{vmatrix} 1 & \cos\beta_{12} & \cos\beta_{13} \\ \cos\beta_{21} & 1 & -\cos\beta_{23} \\ \cos\beta_{31} & -\cos\beta_{32} & 1 \end{vmatrix} \geqslant 0.$$

But expansion of Δ and Δ^* shows that

$$\Delta = \Delta^* + 4\cos\beta_{12}\cos\beta_{23}\cos\beta_{13},$$

and since $\beta_{ij} \leq \frac{1}{2}\pi$ ($i,j = 1, 2, 3$) then $\Delta^* \geq 0$ implies $\Delta \geq 0$. Thus in any case $\Delta \geq 0$ and the argument used in Example 2 (simplified here since $\beta_{ij} \leq \frac{1}{2}\pi$ ($i,j = 1, 2, 3$)) is applied to show that the angles $\beta_{12}, \beta_{23}, \beta_{13}$ satisfy the triangle inequality. Hence $\mathscr{E}_{n,r}$ is a metric space.

Example 4. The n-**dimensional hyperbolic space** $\mathscr{H}_{n,r}$ **of curvature** $-1/r^2$ ($r > 0$). Let $K > 1$ be an arbitrary but fixed constant, and consider the set of all n-tuples $(x_1, x_2, ..., x_n)$ of real numbers satisfying the inequality

$$\sum_{i=1}^{n} x_i^2 < K.$$

If x, y are two elements of the set the number

$$xy = \tfrac{1}{2}r \ln[1+\sqrt{\{1-\gamma(x,y)\}}]/[1-\sqrt{\{1-\gamma(x,y)\}}] \quad (r > 0)$$

is attached to them as distance, where

$$\gamma(x,y) = \left(K - \sum_{i=1}^{n} x_i^2\right)\left(K - \sum_{i=1}^{n} y_i^2\right) \Big/ \left(K - \sum_{i=1}^{n} x_i y_i\right)^2,$$

and $\tanh(1/r) = 1/\sqrt{K}$.

Evidently $xy = yx$ and $xy = 0$ if $x = y$. We must show that xy is a non-negative real number, that $x = y$ if $xy = 0$, and that the triangle inequality holds.

Since

$$[\sqrt{(\sum x_i^2)} - \sqrt{(\sum y_i^2)}]^2 \geq 0,$$

and K is positive, then

$$-2K\sqrt{(\sum x_i^2 \cdot \sum y_i^2)} \geq -K[\sum x_i^2 + \sum y_i^2].$$

Adding $K^2 + \sum x_i^2 \cdot \sum y_i^2$ to both sides gives

$$[K - \sqrt{(\sum x_i^2 \cdot \sum y_i^2)}]^2 \geq (K - \sum x_i^2)(K - \sum y_i^2).$$

Now it follows from the Cauchy inequality that

$$K - \sum x_i y_i \geq K - \sqrt{(\sum x_i^2 \cdot \sum y_i^2)},$$

with both members of the inequality positive (since $\sum x_i^2 < K$) and the equality sign holding if and only if $x_i = \lambda y_i$ ($i = 1, 2, ..., n$). Hence

$$(K - \sum x_i y_i)^2 \geq [K - \sqrt{(\sum x_i^2 \cdot \sum y_i^2)}]^2 \geq (K - \sum x_i^2)(K - \sum y_i^2) > 0,$$

and so $0 < \gamma(x,y) \leq 1$, with the equality sign holding if and only if $x_i = y_i$ ($i = 1, 2, ..., n$). It follows that xy is a real, non-negative number and $xy = 0$ (i.e. $\gamma(x,y) = 1$) if and only if $x = y$.

To show $\mathscr{H}_{n,r}$ is metric, let x, y, z be three distinct points of $\mathscr{H}_{n,r}$ and consider the sphere in E_n whose equation in homogeneous cartesian coordinates is
$$X_1^2+X_2^2+\ldots+X_n^2-KX_{n+1}^2 = 0.$$
Interpreting the ordered n-tuple of numbers $x_1, x_2,\ldots, x_n$ forming the point x of $\mathscr{H}_{n,r}$ as non-homogeneous cartesian coordinates the point-set of $\mathscr{H}_{n,r}$ is represented by the set of all points of E_n interior to the above sphere. It is convenient to represent the elements of $\mathscr{H}_{n,r}$ in homogeneous form $(x_1, x_2,\ldots, x_n, x_{n+1})$ where x_{n+1} may be taken as 1.

The straight line joining the points x, y of $\mathscr{H}_{n,r}$ intersects the sphere in points u, v where the labelling is assumed so that the four points of the line are in the order v, x, y, u. Writing the homogeneous coordinates of u and v as $u_i = x_i+\lambda_1 y_i$, $v_i = x_i+\lambda_2 y_i$ ($i = 1, 2,\ldots, n+1$), it is seen that λ_1, λ_2 are roots of the quadratic equation
$$\lambda^2\Big(\sum_{i=1}^n y_i^2-Ky_{n+1}^2\Big)+2\lambda\Big(\sum_{i=1}^n x_i y_i-Kx_{n+1}y_{n+1}\Big)+\sum_{i=1}^n x_i^2-Kx_{n+1}^2 = 0.$$
It follows that (replacing x_{n+1}, y_{n+1} by 1)
$$\lambda_1/\lambda_2 = [1+\sqrt{\{1-\gamma(x,y)\}}]/[1-\sqrt{\{1-\gamma(x,y)\}}],$$
and since λ_1/λ_2 is the cross-ratio $(x,y; u,v)$ of the points x, y, $x+\lambda_1 y$, $x+\lambda_2 y$, then
$$xy = \tfrac{1}{2}r\ln(x,y; u,v) = \tfrac{1}{2}r\ln(xu.yv/yu.xv).$$

If three distinct points of $\mathscr{H}_{n,r}$ are on a chord of the sphere, say in the order x, y, z, then $(x,y; u,v).(y,z; u,v) = (x,z; u,v)$ and consequently $xy+yz = xz$. Then $xy+xz > yz$, $xz+yz > xy$, and the triangle inequality is satisfied.

If z is not on the chord joining x and y let the line $L(y,z)$ joining y, z intersect the sphere in u', v' and the line $L(x,z)$ joining x, z intersect it in u'', v'', so that
$$xy = \tfrac{1}{2}r\ln(x,y; u,v), \qquad yz = \tfrac{1}{2}r\ln(y,z; u',v'),$$
$$xz = \tfrac{1}{2}r\ln(x,z; u'',v'').$$
The lines $L(u',v'')$, $L(u'',v')$ intersect $L(x,y)$ in points x', y', respectively, and intersect one another in a point w.† If t denotes the intersection of $L(x,y)$ and $L(z,w)$ then x, z, u'', v'' are perspective from w to x, t, y', x' and y, z, u', v' are perspective from w to y, t, x', y'. Hence
$$(x,z; u'',v'') = (x,t; y',x'), \qquad (y,z; u',v') = (y,t; x',y'),$$
and so
$$(x,z; u'',v'')(y,z; u',v') = (x,t; y',x')(y,t; x',y')$$
$$= (x,y; y',x').$$

† The proof is modified in an obvious manner in case $L(u',v'')$, $L(u'',v')$ are parallel.

But since the points v, x', x, y, y', u are seen to be on a chord *in that order*, then
$$(x,y;\, y',x') > (x,y;\, u,v)$$
and so $\quad (x,z;\, u'',v'')(y,z;\, u',v') > (x,y;\, u,v).$

It follows at once that $\quad xz + yz > xy,$
and $\mathcal{H}_{n,r}$ is a metric space.†

A closed subset of E_n is *convex* (in the classical sense) provided it contains with each two of its points the segment joining them. A bounded, closed, and convex subset is a *convex body*, and the boundary of a convex body is a *convex surface*.

Example 5. The n-dimensional Minkowski space. Let Σ be a convex surface of E_n and $o \in E_n$ ($o \bar{\in} \Sigma$) such that every ray with initial point o intersects Σ in exactly one point. We suppose, moreover, that two oppositely directed rays beginning at o meet Σ in points equidistant from o; that is, Σ is symmetric with respect to o.

The n-dimensional Minkowski space is obtained by defining the distance $m(x,y)$ of elements x, y of the point-set of E_n in the following manner. If $x \neq y$ let the ray with initial point o which is parallel to the *directed* segment $\overrightarrow{xy}$ (and with the same sense) meet Σ at p. Then
$$m(x,y) = xy/op,$$
where xy and op are the euclidean distances of the points x, y and o, p, respectively. If $x = y$ put $m(x,y) = 0$.

It is clear that $m(x,y) = m(y,x) \geqslant 0$, and $m(x,y) = 0$ if and only if $x = y$ (the symmetry of the distance follows from the symmetry of Σ with respect to o).

To show the validity of the triangle inequality let x, y, z be non-collinear points of the space and p, q, r elements of Σ with ray $[o,p)$, ray $[o,q)$, ray $[o,r)$ parallel to (and with the same sense as) the directed segments $\overrightarrow{xz}, \overrightarrow{xy}, \overrightarrow{yz}$, respectively. It is seen that ray $[o,p)$ lies within the angle qor, and p is either within seg. $[q,r]$ or lies outside the triangle (o,q,r). Hence
$$A(o,p,q) + A(o,p,r) \geqslant A(o,q,r), \qquad (*)$$
where $A(s,t,u)$ denotes the area of the triangle with vertices s, t, u.

If $\phi = \angle qop, \psi = \angle por$, then
$$A(o,p,q) = \tfrac{1}{2} op \cdot oq \sin \phi,$$
$$A(o,p,r) = \tfrac{1}{2} op \cdot or \sin \psi,$$
$$A(o,q,r) = \tfrac{1}{2} oq \cdot or \sin(\phi + \psi).$$

† This method of proving hyperbolic space metric was chosen because of its applicability to more general spaces; see, for example, Exercise 2.

Now in the triangle (x, y, z) the angles at the vertices x, z, and y are ϕ, ψ, and $\pi-(\phi+\psi)$, respectively, and so
$$xy = C\sin\psi, \qquad yz = C\sin\phi, \qquad xz = C\sin(\phi+\psi).$$
Since
$$op = xz/m(x,z), \qquad oq = xy/m(x,y), \qquad or = yz/m(y,z),$$
then
$$A(o,p,q) = \frac{C^2 \sin\phi \sin\psi \sin(\phi+\psi)}{2m(x,y)m(x,z)},$$

$$A(o,p,r) = \frac{C^2 \sin\phi \sin\psi \sin(\phi+\psi)}{2m(x,z)m(y,z)},$$

$$A(o,q,r) = \frac{C^2 \sin\phi \sin\psi \sin(\phi+\psi)}{2m(x,y)m(y,z)}.$$

It follows directly from (∗) that
$$m(x,y)+m(y,z) \geqslant m(x,z),$$
and so the Minkowski space is metric.

The convex surface Σ is *strictly convex* provided it contains no segments. In this case the point p is always outside the triangle (o,q,r) (when x, y, z are not on a line) and consequently the equality sign in the triangle inequality must be excluded. Since for points of a euclidean line the change to the Minkowski metric amounts merely to a change of scale, three points are the vertices of a proper Minkowski triangle if and only if they are vertices of a proper euclidean triangle when Σ is strictly convex. If Σ is merely convex this is not the case, since three points not on a euclidean line may have the sum of two of the Minkowski distances equal to the third, while if Σ is not convex, the triangle inequality may not hold, for then p may fall *inside* the triangle (o,q,r).

Corresponding to the symmetric convex surface Σ is a function $f(x_1, x_2, ..., x_n)$ defined for each $x = (x_1, x_2, ..., x_n)$ of E_n as follows:
$$f(0,0,...,0) = 0,$$
$$f(x_1, x_2, ..., x_n) = t \quad (x_1, x_2, ..., x_n) \neq (0,0,...,0)$$
where t is the ratio of similitude of a homothetic transformation of E_n with centre o which carries Σ into a surface passing through x. Clearly $t = ox/op$ where p is the intersection of Σ with the ray $[o,x)$; that is
$$f(x_1, x_2, ..., x_n) = m(o, x).$$
It is easily seen that
$$f(x_1, x_2, ..., x_n) = f(-x_1, -x_2, ..., -x_n),$$
$$f(tx_1, tx_2, ..., tx_n) = tf(x_1, x_2, ..., x_n) \quad (t > 0),$$

and (from the validity of the triangle inequality)
$$f(x_1+y_1, x_2+y_2, \ldots, x_n+y_n) \leqslant f(x_1, x_2, \ldots, x_n) + f(y_1, y_2, \ldots, y_n).$$

Example 6. The Hilbert space. An infinite dimensional metric space of much importance has for elements the set of all infinite sequences $x_1, x_2, \ldots, x_n, \ldots$ of real numbers such that $\sum_{i=1}^{\infty} x_i^2$ converges. Distance xy is defined as the natural extension of the euclidean metric; that is,
$$xy = \left[\sum_{i=1}^{\infty}(x_i-y_i)^2\right]^{\frac{1}{2}}.$$
Since
$$\sum_{i=1}^{n}(|x_i|-|y_i|)^2 \geqslant 0,$$
then
$$\sum_{i=1}^{n}|x_i y_i| \leqslant \tfrac{1}{2}\left[\sum_{i=1}^{n} x_i^2 + \sum_{i=1}^{n} y_i^2\right],$$
and so the infinite series $\sum_{i=1}^{\infty}|x_i y_i|$ converges. Hence the series $\sum_{i=1}^{\infty} x_i y_i$ converges, and so does $\sum_{i=1}^{\infty}(x_i-y_i)^2$. Thus xy is defined for each pair of elements x, y of the set. Evidently $xy = yx$, $xy \geqslant 0$ and $xy = 0$ if and only if $x = y$ (where $x = y$ provided $x_i = y_i$ $(i = 1, 2, 3, \ldots)$).

To prove the triangle inequality, we have for each three distinct elements x, y, z
$$xz^2 = \sum (x_i-z_i)^2 = \sum [(x_i-y_i)+(y_i-z_i)]^2$$
$$= \sum (x_i-y_i)^2 + 2\sum (x_i-y_i)(y_i-z_i) + \sum (y_i-z_i)^2$$
$$\leqslant xy^2 + 2(xy)(yz) + yz^2,$$
since by the Schwarz inequality
$$[\sum (x_i-y_i)(y_i-z_i)]^2 \leqslant \sum (x_i-y_i)^2 \cdot \sum (y_i-z_i)^2,$$
(all summations being from 1 to infinity).

Hence $$xz^2 \leqslant (xy+yz)^2,$$
and so $xz \leqslant xy+yz$.

Example 7. The Hausdorff metric. A metric space M is *bounded* provided a positive number d exists such that $xy < d$ for each pair of elements x, y of M. Consider the set S of all closed, non-null, subsets of a bounded metric space M, and for $A, B \in S$ put
$$\overrightarrow{AB} = \underset{b\in B}{\text{l.u.b.}}\,(\underset{a\in A}{\text{g.l.b.}}\,ab),$$
$$\overrightarrow{BA} = \underset{a\in A}{\text{l.u.b.}}\,(\underset{b\in B}{\text{g.l.b.}}\,ab),$$

where l.u.b. and g.l.b. denote the least upper bound and greatest lower bound, respectively. S is a metric space if to each pair A, B of its elements the number
$$AB = \max[\overrightarrow{AB}, \overrightarrow{BA}]$$
is attached as distance.

Since M is bounded the set of numbers ab, $a \in A$, $b \in B$ is a bounded set of non-negative numbers, and hence the numbers $\overrightarrow{AB}$, $\overrightarrow{BA}$ exist and are non-negative. Then A, $B \in S$ implies AB exists and $AB \geqslant 0$.

If $A = B$ then obviously $\underset{a \in A}{\text{g.l.b.}} \, ab = 0$ for $b \in B$, and so $\overrightarrow{AB} = 0$; similarly $\overrightarrow{BA} = 0$, and consequently $AB = 0$. Conversely, if $AB = 0$ then $\overrightarrow{AB} = \overrightarrow{BA} = 0$. From $\overrightarrow{AB} = 0$ we have $\underset{a \in A}{\text{g.l.b.}} \, ab = 0$ for every b of B. It follows that $b \in A$ or $b \in A'$, and since A is closed, the second alternative implies the first. Thus $B \subset A$. In a similar manner from $\overrightarrow{BA} = 0$ we obtain $A \subset B$, and hence $A = B$.

Before proving the triangle inequality valid, it is useful to observe that if for each b of B there is an element a of A with $ab < \rho$ then $\underset{a \in A}{\text{g.l.b.}} \, ab < \rho$ for every element b of B and hence $\overrightarrow{AB} \leqslant \rho$. Also, if $\overrightarrow{AB} < \rho$ then clearly $\underset{a \in A}{\text{g.l.b.}} \, ab < \rho$ for b any element of B, and $b \in B$ implies the existence of $a \in A$ such that $ab < \rho$.

Let, now, A, B, C be three distinct elements of S and put $\overrightarrow{AB} = \rho$, $\overrightarrow{BC} = \sigma$. For each $\epsilon > 0$ $\overrightarrow{BC} < \sigma + \epsilon$ and hence, by the preceding remark, $c \in C$ implies the existence of $b \in B$ such that $bc < \sigma + \epsilon$. Similarly, from $\overrightarrow{AB} < \rho + \epsilon$ it follows that to each b of B there corresponds an element a of A with $ab < \rho + \epsilon$.

Then, since M is metric,
$$ac \leqslant ab + bc < \rho + \sigma + 2\epsilon,$$
and hence to each c of C and each positive ϵ there corresponds an element a of A with $ac < \rho + \sigma + 2\epsilon$. We may conclude from the above observation that
$$\overrightarrow{AC} \leqslant \rho + \sigma = \overrightarrow{AB} + \overrightarrow{BC}.$$

Now from
$$AB + BC \geqslant \overrightarrow{AB} + \overrightarrow{BC} \geqslant \overrightarrow{AC},$$
$$AB + BC \geqslant \overrightarrow{BA} + \overrightarrow{CB} \geqslant \overrightarrow{CA},$$
follows at once
$$AB + BC \geqslant AC,$$
and the triangle inequality is established.

Example 8. Metric lattices. The definition of a lattice is given in § 2. In some lattices a real, non-negative, function $|x|$, called the *norm*, can be defined for each element x of the lattice, such that (1) $x \prec y$ and $x \neq y$ implies $|x| < |y|$, and (2) for each pair of elements x, y
$$|x+y| + |xy| = |x| + |y|.†$$
Lattices in which a norm can be defined are called *normed lattices*.

A normed lattice L may be made into a metric space $D(L)$ by attaching to each pair of elements x, y of L the number
$$(x, y) = |x+y| - |xy|$$
as distance.

From the idempotent properties of lattice sum and product, $(x, y) = 0$ if $x = y$. Conversely, if $(x, y) = 0$ then $|xy| = |x+y|$ and since $xy \prec x+y$ property (1) of the norm allows us to conclude that $xy = x+y$. Then $xy \prec x$, $x \prec x+y = xy$ imply $xy = x$; similarly, it is seen that $xy = y$ and so $x = y$. Furthermore, $xy \prec x \prec x+y$ and consequently $|x+y| \geq |xy|$; that is, $(x, y) \geq 0$. The symmetry of the distance function, $(x, y) = (y, x)$, is an immediate consequence of the definition and the commutative properties of lattice sum and product.

It remains to show that for each three distinct elements x, y, z of L,
$$xy + yz \geq xz.$$

From
$$(x+z) + y = (x+y) + (y+z),$$
follows
$$(x+z)y \prec y \prec xz + y \prec (x+y)(y+z),$$
$$|x+z| + |y| = |(x+z)+y| + |(x+z)y|$$
$$\leq |(x+y)+(y+z)| + |(x+y)(y+z)|;$$
that is,
$$|x+z| + |y| \leq |x+y| + |y+z|. \qquad (*)$$
Since
$$(xz)y = (xy)(yz), \quad xy + yz \prec xz + y,$$
then
$$|xz| + |y| = |xz+y| + |(xz)y|$$
$$\geq |xy+yz| + |(xy)(yz)|.$$
Whence
$$|xz| + |y| \geq |xy| + |yz|. \qquad (**)$$
From $(*)$ and $(**)$
$$|x+z| - |xz| \leq |x+y| - |xy| + |y+z| - |yz|$$
or
$$(x, z) \leq (x, y) + (y, z).$$

The metric space $D(L)$ is called the metric space *associated with L*.

---

† In this formula (as in all lattice formulae that we shall write) xy denotes the *lattice product* of the elements x, y and *not* a distance.

EXERCISES

1. Using the law of cosines of hyperbolic trigonometry, prove that the determinant
$$|\cosh x^{(i)} x^{(j)}/r| \quad (i, j = 1, 2, 3)$$
is non-negative for any three points $x^{(i)}$ ($i = 1, 2, 3$) of $\mathcal{H}_{n,r}$, and show how the development of this determinant leads to another proof of the validity in $\mathcal{H}_{n,r}$ of the triangle inequality.

2. Let I_n denote the subset of E_n consisting of the interior points of a convex body K_n of E_n, and suppose $I_n \neq 0$. If $a, b \in I_n$, $a \neq b$, define
$$ab = |\ln(a, b; u, v)|,$$
where u, v are the points in which the line $L(a, b)$ intersects the boundary of K_n, and put $ab = 0$ if $a = b$. Apply the method of Example 4 to show that I_n is a metric space.

3. Attach to each two sequences $x = \{x_n\}$, $y = \{y_n\}$ of real numbers, the 'distance'
$$xy = \sum_{i=1}^{\infty} \frac{1}{i!} \frac{|x_i - y_i|}{1 + |x_i - y_i|}.$$
Show that the resulting space, denoted by E_ω, is metric.

4. Consider the set of all functions $f(x)$, continuous in the closed interval $[0, 1]$, and attach to each pair f, g the number $fg = \max|f(x) - g(x)|$, $0 \leq x \leq 1$. Show that the resulting space C is metric.

5. Let M be a metric space and $\phi(x)$ a monotone increasing concave function (x real) that vanishes for $x = 0$. If distance pq in M be redefined as $\phi(pq)$, prove that the new space is metric. Applying this for $\phi(x) = x/(1+x)$, we see that every metric space is homeomorphic to a bounded metric space.

6. Show that the space $M_n^{(p)}$ of ordered n-tuples of real numbers is metric when the distance of $x = (x_1, x_2, ..., x_n)$, $y = (y_1, y_2, ..., y_n)$ is $xy = \left[\sum_{i=1}^{n} |x_i - y_i|^p\right]^{1/p}$, $p \geq 1$.

7. Show that any set of non-negative numbers including zero is the distance set of some metric space.

10. Topology of metric spaces

The abstract metric space is the domain in which most topological investigations operate. A vast theory has been developed which it is not within the scope of this book to expose, since our chief interest is (as will appear later) with a certain subgroup of the group of topological transformations—the group of congruences. It is necessary, however, to define some topological notions and establish some topological properties that are needed in our study.

The intimate relationship existing between the continuity of the distance function and the triangle inequality was discussed in § 8. The importance of the continuity of the metric justifies another proof of that property of metric spaces.

THEOREM 10.1. *The metric of a metric space is continuous.*

Proof. Let $\{p_n\}$, $\{q_n\}$ be two infinite sequences of points of a metric space M, with $\lim p_n = p$, $\lim q_n = q$ $(p, q \in M)$. Clearly
$$|pq - p_n q_n| = |pq - pq_n + pq_n - p_n q_n| \leqslant |pq - pq_n| + |pq_n - p_n q_n|.$$
Applying the triangle inequality to the triples p, q, q_n and p, p_n, q_n yields
$$|pq - p_n q_n| \leqslant pp_n + qq_n.$$
Since $\lim p_n = p$, $\lim q_n = q$, to each $\epsilon > 0$ corresponds a positive integer $N(\epsilon)$ such that $pp_n < \tfrac{1}{2}\epsilon$, $qq_n < \tfrac{1}{2}\epsilon$ for $n > N(\epsilon)$. Hence for all such indices
$$|pq - p_n q_n| < \epsilon,$$
and consequently $pq = \lim p_n q_n$; that is, the metric is continuous at p, q.

It follows (Theorem 6.1) that in a metric space spherical neighbourhoods are open sets; that a metric space is a topological space (Theorem 6.2), and hence a class (L).

THEOREM 10.2. *Let M be a metric space and $E \subset M$. In order that E be open it is necessary and sufficient that each element p of E be the centre of a spherical neighbourhood contained in E.*

Proof. If $p \in E$ and for no $\rho > 0$, $U(p; \rho) \subset E$ then
$$U(p; 1/n) \cdot C(E) \neq 0,$$
for $n = 1, 2, \ldots$. Thus an infinite sequence $\{q_n\}$ exists such that (1) $p \neq q_n \in C(E)$ $(n = 1, 2, \ldots)$, and (2) $0 < pq_n < 1/n$ $(n = 1, 2, \ldots)$. Hence $p \in C'(E)$ and, since $p \bar\in C(E)$, then $C(E)$ is not a closed set and so E is not open. The necessity of the condition is thus established.

If, on the other hand, we suppose the condition satisfied then it is easily seen that
$$E = \sum_{p \in E} U(p; \rho(p))$$
and since each $U(p; \rho(p))$ is open, so is E.

COROLLARY 1. *If $p \in M$ and $E \subset M$, then $p \in E'$ if and only if every open set U containing p contains an element of E distinct from p.*

Proof. The sufficiency follows at once from the definition of accumulation point and the fact that $U(p; 1/n)$ is open for every $n = 1, 2, 3, \ldots$, while the necessity is an immediate consequence of the definition and the property of open sets proved in the theorem.

COROLLARY 2. *If $\lim p_n = p$ then corresponding to each open set U containing p an integer N exists such that $n > N$ implies $p_n \in U$.*

Proof. If $p \in U$ a positive ρ exists such that $U(p; \rho) \subset U$. Since $\lim pp_n = 0$ an integer N exists such that $pp_n < \rho$ for every index $n > N$. Hence $n > N$ implies $p_n \in U(p; \rho)$.

Thus $\lim p_n = p$ implies that for 'almost all' indices n (that is, all but a finite number) $p_n \in U$, where U is any open set containing p.

Theorem 10.3. *If $p \in E'$, $E \subset M$, there exists an infinite sequence $\{p_n\}$ of pairwise different elements of E, each distinct from p, with $\lim p_n = p$, and conversely.*

Proof. Let p_1 be an element of E with $0 < pp_1 < 1$, $p_2 \in E$ with $0 < pp_2 < \min[\tfrac{1}{2}, pp_1], \ldots,\ p_n \in E$ with $0 < pp_n < \min[1/n, pp_{n-1}]$. These n points being selected, p_{n+1} is any point of E with

$$0 < pp_{n+1} < \min[1/(n+1), pp_n].$$

The existence of the infinite sequence $\{p_n\}$ then follows by complete induction. Clearly $p_i \neq p_j$ $(i,j = 1, 2,\ldots,\ i \neq j)$, $p_n \neq p$ $(n = 1, 2,\ldots)$, and $\lim p_n = p$.

The converse is obvious.

Theorem 10.4. *If an infinite sequence of elements of a metric space M has a limit, the sequence is convergent.*

Proof. Let $\{p_n\}$ be an infinite sequence of elements of M and $p \in M$ with $\lim p_n = p$. For each pair of indices i, j

$$p_i p_j \leqslant pp_i + pp_j,$$

and consequently

$$0 \leqslant \lim_{i,j \to \infty} p_i p_j \leqslant \lim_{i \to \infty} pp_i + \lim_{j \to \infty} pp_j = 0,$$

since $\lim p_n = p$. Hence $\lim_{i,j \to \infty} p_i p_j = 0$ and the sequence $\{p_n\}$ is convergent.

By deleting a point from the straight line a metric space is obtained for which the converse of Theorem 10.4 is not valid. The condition that a sequence of elements of a metric space be convergent is (as previously remarked) a generalization of Cauchy's criterion which is necessary and sufficient for a sequence of real numbers to have a limit. A metric space for which the criterion is necessary and sufficient for an infinite sequence of elements to have a limit was said by Fréchet *to admit a generalization of Cauchy's theorem*. Such spaces were called by Hausdorff *complete*.

Definition 10.1. *A metric space is (metrically) complete provided for each infinite sequence $\{p_n\}$ of its elements $\lim_{i,j=\infty} p_i p_j = 0$ implies the existence of an element p of the space such that $\lim_{n \to \infty} pp_n = 0$.*

The first five metric spaces defined in the preceding section are easily shown to be complete.

The existence of an accumulation element for each bounded, infinite subset (which is an important property of euclidean space) is not valid in every metric space. It is certainly lacking in the metric space obtained from an infinite set by assigning to each pair of distinct points the distance 1. But it also fails to hold in some very important spaces as, for example, the Hilbert space. The infinite subset of Hilbert space formed by the points $x^{(i)} = (x_1, x_2, ..., x_i, ...)$, $x_i = 1$, $x_k = 0$ $(i \neq k)$ $(i = 1, 2, 3, ...)$, has distance $x^{(i)}x^{(j)} = \sqrt{2}$, $i \neq j$, and use of the triangle inequality shows that no point is an accumulation element of this set.

DEFINITION 10.2. *A subset E of a metric space is compact provided each of its infinite subsets has a non-null derived set.*

It is observed that every finite subset is compact, and that every subset of a compact set is compact. As the example just given shows, completeness does not imply compactness. A more familiar instance is that of the straight line which is complete by virtue of Cauchy's theorem, but not compact since the positive integers, for example, constitute an infinite subset without any accumulation element.

On the other hand, *a compact metric space is complete*. We leave the proof of this to the reader (Exercise 2). It may also be shown that a metric space which admits a generalization of the Weierstrass–Bolzano theorem (that is, in which *bounded* subsets are compact) is complete. Such spaces are sometimes called *finitely compact*.

The spherical space $S_{n,r}$, and elliptic space $\mathscr{E}_{n,r}$ are compact; the euclidean, hyperbolic, and Minkowski n-dimensional spaces are finitely compact, but not compact; and the Hilbert space is neither compact nor finitely compact.

Two important properties of compact subsets of metric spaces are given in the following two theorems.

THEOREM 10.5. *If E is a non-null compact subset of a metric space, then for each $\epsilon > 0$ a finite subset of E exists such that every element of E has distance less than ϵ from at least one of the elements of the finite subset.*

Proof. If the contrary be supposed then E must contain an infinite sequence $p_1, p_2, ..., p_n, ...$ with $p_i p_j \geqslant \epsilon$ $(i, j = 1, 2, ..., i \neq j)$. The point set represented by the terms of this infinite sequence is then an infinite subset of E which is obviously without any accumulation element, contradicting the compactness of E.

COROLLARY. *A compact subset of a metric space is bounded.*

A set is called *denumerable* or *countable* provided there exists a one-to-one correspondence between its elements and the set of all natural numbers.

THEOREM 10.6. *Every compact subset E of a metric space contains a subset F, countable at most, such that $E \subset \overline{F}$.*

Proof. From the preceding theorem, for each integer k a finite subset $p_1^k, p_2^k, \ldots, p_{n_k}^k$ of E exists such that each point of E has distance less than $1/k$ from at least one of the points p_j^k. Let F denote the set of all different terms of

$$p_1^1, p_2^1, \ldots, p_{n_1}^1, p_1^2, p_2^2, \ldots, p_{n_2}^2, p_1^3, p_2^3, \ldots, p_{n_3}^3, \ldots.$$

If $p \in E$ then for every integer m an element p_j^m ($j \leqslant n_m$) of F exists such that $pp_j^m < 1/m$. Hence p is an element or an accumulation element of F; that is $E \subset \overline{F}$, and the theorem is proved.

The set F is said to be *dense in E*.

COROLLARY. *If E is a compact metric space or a closed subset of a compact metric space then a subset F of E exists, countable at most, such that $E = \overline{F}$.*

Proof. Since E is closed in either case, $F \subset E$ implies $F' \subset E' \subset E$, and so $\overline{F} \subset E$.

DEFINITION 10.3. *A metric space is called separable provided it is the closure of a subset that is countable at most.*

From the corollary to the preceding theorem we have

THEOREM 10.7. *Each compact metric space is separable.*

DEFINITION 10.4. *Let P be a subset of a metric space M and Q a subset of a metric space M^* (which may coincide with M). A mapping of P into Q is any correspondence that associates with each element p of P exactly one element $f(p) = q$ of Q. The element $q = f(p)$ is called the image of p (by f). The function $f(p)$ defined by such an association is said to be on P to Q.*

It is important to observe that a mapping of P into Q does not imply that each element of Q is the image of some element of P. *If the set of images of elements of P exhausts the set Q*, we have a mapping of P onto Q, and we write $f(P) = Q$.† If, moreover, $p_1, p_2 \in P$ and $p_1 \neq p_2$ implies $f(p_1) \neq f(p_2)$, the mapping establishes a *one-to-one correspondence* between the elements of P and Q, and the function f so defined is called *biuniform*. The function ϕ on $f(P)$ to P, with $p = \phi(q)$ provided $q = f(p)$ is called the *inverse* of f.

DEFINITION 10.5. *A mapping f of a subset P of a metric space M onto a subset $f(P)$ of a metric space M^* is continuous at $p_0 \in P$ provided for*

† In this case P is called the *domain* of f and Q its *range*.

every infinite sequence $\{p_n\}$ of elements of P with $\lim p_n = p_0$, the infinite sequence $\{f(p_n)\}$ has limit $f(p_0)$. *The function f is continuous in P provided it is continuous at each element of P.*

If f is biuniform and continuous in P its inverse function ϕ may or may not be continuous in $f(P)$. If it is, then f is said to be *bicontinuous*, and each of the sets P and $Q = f(P)$ is a biuniform and bicontinuous *transform* of the other. Two such sets are said to be *homeomorphic*, and f is a *homeomorphism*.

DEFINITION 10.6. *Topology is the study of those properties of a set which are invariant under homeomorphisms.*

It is an important theorem in topology that every separable metric space is homeomorphic with a certain subset of Hilbert space, and hence every separable metric space is contained topologically in Hilbert space. Consequently the topology of (abstract) separable metric spaces might be completely investigated by studying the topology of a subset of the (concrete) Hilbert space. Though it is with the topology of separable metric spaces that many investigators are most concerned, there is not yet at hand any practicable means of exploiting this interesting result.

EXERCISES

1. If $\{p_n\}$ is an infinite sequence of elements of a closed, compact subset E of a metric space show that an element p of E exists such that $p = \lim p_{i_n}$, where $\{p_{i_n}\}$ is a sub-sequence of $\{p_n\}$.
2. Show that a compact metric space is complete.
3. Give an example of a continuous mapping f of a subset P of M onto a subset Q of M^* (M, M^* metric spaces) whose inverse function ϕ is not continuous at any point of Q.
4. Let $P = (r_1, r_2, ..., r_n, ...)$ be a countable (at most) subset of a metric space M with $\bar{P} = M$. Show that (1) if $p, q \in M$ ($p \neq q$) an index i exists such that $pr_i \neq qr_i$ and (2) $pq = \text{g.l.b.}_i(pr_i + qr_i)$.

REFERENCES

§§ 2–3. Birkhoff [1], Vaidyanathaswamy [1].

§ 4. Fréchet [1, 2], Hausdorff [1].

§ 5. The notion of a semimetric space is found in de Tilly [1]; the term 'halbmetrischer Raum' is due to Menger [1].

§ 8. Metric spaces were introduced in Fréchet [1] and called classes (E) (later, classes (D)); the term 'metrischer Raum' is due to Hausdorff.

§ 9. See Blumenthal [1] for additional examples of metric spaces.

§ 10. Chapters VI, VII of Sierpiński [1] are devoted to topological properties of metric spaces. In Whyburn [1], all spaces considered are metric.

SUPPLEMENTARY PAPERS

Wilson [1], Lindenbaum [1].

CHAPTER II

METRIC SEGMENTS AND LINES

11. Historical remarks

TOPOLOGY has been defined as the study of invariants of homeomorphisms.† The set of all homeomorphisms forms a group, and so topology fits the famous definition of a geometry, given by Klein, as the study of the invariants of figures under a selected group of transformations.

This book deals with a geometry that differs from topology not in the kind of spaces in which it operates, but in the group of transformations selected as basic. While the interest of topology is in all those transformations of sets which are biuniform and bicontinuous (homeomorphisms) our geometry is concerned with that subgroup of homeomorphisms for which *the distance of two points is an invariant*. The elements of that subgroup are called *congruences*, and the geometry is often referred to as *distance geometry* (though the terms *metric geometry* or *metric topology* are equally descriptive of its content).

Distance geometry may operate in any kind of space in which a notion of 'distance' is attached to each point-pair of the space. Distances need not be real or complex numbers but may be elements of an abstract group or of merely an ordered set. Whatever the underlying space might be, the objective of its distance geometry is to develop the theory of that space in terms of the fundamental notion of distance, and to express its properties as distance relations. The subordination of diverse fields of research to common ideas, which occurs when distances are associated with pairs of elements and the distance geometry of the resulting space is developed, is one of the features of the subject.

The notions of an abstract distance space and its geometry were undoubtedly held by the Belgian artillery officer de Tilly who (in a paper published in 1892) defined an abstract semimetric space and sought to show that the geometry is determined by specifying relations between the mutual distances of elements of certain finite subsets of the space. The idea of characterizing a space metrically by means of such *n-point relations* seems to have originated with de Tilly, and though his methods did not enable him to proceed very far in that direction, his introduction

† There is a tendency nowadays to broaden the term to include the study of transformations that are merely continuous mappings.

of the problem into geometry should be regarded as an important contribution.

Some of the questions raised by de Tilly about n-point relations were answered by Mansion in 1895, by Blichfeldt in 1902, and by de Donder in 1906 and 1907. Their investigations were concerned with special problems.† The first systematic development of abstract distance geometry was initiated by Karl Menger in 1928. His four *Untersuchungen über allgemeine Metrik* (the last one published in 1930) established the subject as a separate discipline, and they are the foundation upon which much of the subsequent development is based.

12. Betweenness in metric spaces. Congruence

Since distance geometry has 'distance' as its only primitive notion, those concepts and relations which axiomatic geometry defines only implicitly by means of postulates must here be defined explicitly in terms of distance. Among the important relations of axiomatic geometry is that of *betweenness*. Pasch axiomatized this notion in 1882. It plays a leading role in distance geometry, where it is a defined or derived notion.

DEFINITION 12.1. *The point q of a metric space is (metrically) between two points p, r of the space if and only if*

$$p \neq q \neq r, \qquad pq+qr = pr.$$

We shall symbolize this relation by writing pqr.

It is noted that pqr implies p, q, r are pairwise distinct, since $pr > 0$.

THEOREM 12.1. *In a metric space M the relation of betweenness has the following properties:*

(1) *symmetry of the outer points (pqr implies rqp),*

(2) *special inner point (if pqr then neither prq nor qpr),*

(3) *transitivity (pqr and prs are equivalent to pqs and qrs),*

(4) *closure (if $p, r \in M$, the set $\bar{B}(p,r) = (p)+(r)+B(p,r)$ is closed, where $B(p,r)$ denotes the set of all points of M between p and r),*

(5) *extension (if M is a metric space consisting of two distinct points p, r there exists a metric space M^* containing p, r, and a point q such that pqr holds).*

Proof. Parts (1) and (5) are obvious. To prove (2), pqr implies $p \neq q \neq r$, $pq+qr = pr$ and prq implies $p \neq r \neq q$, $pr+rq = pq$.

† For an account of these investigations as well as a fuller discussion of historical matters the reader is referred to Chapter I of the writer's *Distance Geometries*.

Addition of the two equalities gives $qr = 0$, which contradicts $q \neq r$. Similarly qpr cannot hold.

To prove (3) suppose pqr and prs hold. Then since each triple consists of pairwise distinct points, the points p, q, r, s are pairwise distinct except for, perhaps, the pair q, s. But $q \neq s$ for in the contrary case pqr and prq both hold, contradicting (2).

Addition of $pq+qr = pr$ and $pr+rs = ps$ gives

$$pq+qr+rs = ps \leqslant pq+qs,$$

and the triangle inequality applied to q, r, s shows that $qr+rs = qs$. This result substituted above gives $pq+qs = ps$. Since each two points are distinct, we have pqs and qrs. The converse is proved in a similar manner.

To prove (4) we show that if q is an accumulation element of $\bar{B}(p,r)$, with $p \neq q \neq r$, then $q \in B(p,r)$. From our assumption $q = \lim q_n$, $q_n \in B(p,r)$. Then

$$pq_n + q_n r = pr \quad (n = 1, 2, 3, \ldots)$$

and from the continuity of the metric,

$$pq + qr = pr;$$

that is, $q \in B(p,r)$ and the proof of the theorem is complete.

The relation of betweenness for triples on a straight line possesses all the properties of metric betweenness given in the above theorem. It has, moreover, properties that metric betweenness does not have; for example, (a) $p \neq q$, $q \neq r$, $r \neq p$ imply pqr or prq or qpr, and (b) pqr and qrs imply pqs. Though property (a) may be more appropriately regarded as a property of the line than of the betweenness relation, the *bona fide* property (b) is easily seen to fail for certain quadruples of the convex circle $S_{1,r}$.

It has been shown that properties (1), (2), (3) of Theorem 12.1, together with properties (a), (b), and (c) if pqr then $p \neq q$, $q \neq r$, $r \neq p$, *characterize* the notion of betweenness for triples on a line; that is, all the properties of the relation are consequences of those six properties. Thus, for example, the transitivity property pqr, ptr, qst imply psr, valid for each five points p, q, r, s, t of the line, follows from properties (1), (2), (3), (a), (b), (c). The reader is asked to prove this in an exercise. On the other hand, the above transitivity is *not* a property of metric betweenness (Exercise 3). We shall consider in the next section the problem of characterizing metric betweenness among all three-point relations that are definable in every metric space.

Another important primitive concept of axiomatic geometry is that of *congruence*. In distance geometry we have:

DEFINITION 12.2. *If $p, q \in M$, $p', q' \in M'$ (M, M' metric spaces) then p, q are congruent to p', q' if and only if $pq = p'q'$.*

Two subsets P, Q of the same or different metric spaces are congruent provided there exists a mapping f of P onto Q such that each point-pair of P is mapped onto a congruent point-pair of Q.

We write
$$P \underset{f}{\approx} Q$$
and call f a *congruent* (or *isometric*) *mapping of P onto Q*.

If f is a congruent mapping of P onto Q then f is biuniform, for $r, s \in P$ imply $rs = f(r)f(s)$ and so $f(r) = f(s)$ if and only if $r = s$. If ϕ denotes the inverse function of f, then clearly

$$P \underset{f}{\approx} Q \quad \text{implies} \quad Q \underset{\phi}{\approx} P,$$

and so congruence is a *symmetric* relation. It is also *reflexive* and *transitive*.

It is easily seen that a congruent mapping f of P onto Q is continuous in P and its inverse ϕ is continuous in Q. *Hence if two subsets of metric spaces are congruent they are homeomorphic.*†

Whenever it is unnecessary to indicate the mapping that establishes a congruence between two sets P and Q, we write merely

$$P \approx Q.$$

The notation
$$p_1, p_2, \ldots, p_n \approx p'_1, p'_2, \ldots, p'_n$$
signifies that for every index $i = 1, 2, \ldots, n$, p_i and p'_i are corresponding points in the congruence of the two n-tuples.

We note now a sixth property of metric betweenness; namely

(6) *if $p, q, r \approx p', q', r'$ then pqr implies $p'q'r'$.*

Thus metric betweenness is a *congruence invariant*.

EXERCISES

1. Give an example of a metric space in which pqr and qrs do not imply pqs.
2. Prove that the betweenness transitivity pqr, ptr, qst imply psr is a consequence of properties (1), (2), (3), (a), (b), (c).
3. Give an example of a metric space in which the transitivity of Exercise 2 is not valid.

† It is clear that the definition of congruence and the remarks following the definition are equally valid in semimetric spaces.

13. Characterization of metric betweenness

We now show that the six properties of metric betweenness established in the preceding section (namely, (1) symmetry of the outer points, (2) special inner point, (3) transitivity, (4) closure, (5) extension, (6) congruence invariance) *characterize* that relation. To do this it suffices to prove that if a relation $R(p,q,r)$ is defined for point-triples of *every* metric space and $R(p,q,r)$ *has the above six properties*, then points p, q, r are in the relation $R(p,q,r)$ if and only if pqr subsists. The following lemmas and theorems by which this is accomplished are due to Wald. We write $R(p,q,r)$ to denote that p, q, r are in that relation.

LEMMA 13.1. $R(p,q,r)$ *implies* $pq \neq pr \neq qr$.

Proof. If $pr = pq$ then $p, q, r \approx p, r, q$ and by (6) $R(p,r,q)$ holds which, by (2), contradicts $R(p,q,r)$. Similarly, $pr \neq qr$.

LEMMA 13.2. *If* $p, r \in E_2$ *(the euclidean plane)*, $p \neq r$, *two sequences* $\{x_i\}$, $\{y_i\}$ *of points of* E_2 *exist such that*

(i) $R(p, x_1, r)$, $R(p, x_{i+1}, x_i)$, $R(x_1, y_1, r)$, $R(y_i, y_{i+1}, r)$ $(i = 1, 2, ...)$;

(ii) *at least one of the sequences* $\{x_i\}$, $\{y_i\}$ *has an accumulation point.*

Proof. Consider the metric space M consisting of the two distinct points p, r. By (5) a metric space M^* exists containing p, r, and a point x with $R(p, x, r)$. The points p, x, r, being elements of a metric space, satisfy the triangle inequality and hence points $\bar{p}, \bar{x}, \bar{r}$ of E_2 exist with $p, x, r \approx \bar{p}, \bar{x}, \bar{r}$. Since $p, r \in E_2$ and $pr = \bar{p}\bar{r}$ it is clear that there is a point x_1 of E_2 with $px_1 = \bar{p}\bar{x}$ and $x_1 r = \bar{x}\bar{r}$; that is, $p, x_1, r \approx \bar{p}, \bar{x}, \bar{r}$. Then by (6) $R(p, x, r)$ implies $R(\bar{p}, \bar{x}, \bar{r})$, which implies $R(p, x_1, r)$. Similarly, the existence of $x_2 \in E_2$ and $y_1 \in E_2$ with $R(p, x_2, x_1)$, $R(x_1, y_1, r)$ is shown, and it is clear how the existence of the two sequences $\{x_i\}$, $\{y_i\}$ follows by complete induction.

Let n, m be two positive integers with $n > m$. Then $R(p, x_n, x_{n-1})$, $R(p, x_{n-1}, x_{n-2}), ..., R(p, x_{n-(n-m-2)}, x_{n-(n-m-1)})$, $R(p, x_{n-(n-m-1)}, x_m)$ hold. Using (3) on the last two relations gives $R(p, x_{n-(n-m-2)}, x_m)$, and by continued use of (3) we get $R(p, x_n, x_m)$. In a similar manner $R(y_m, y_n, r)$ is proved, and it follows from Lemma 13.1 that each of the sequences $\{x_i\}$, $\{y_i\}$ consists of pairwise different elements. Hence each of the sequences is an infinite subset of E_2.

To prove (ii) it suffices to show that at least one of the sets $\{x_i\}$, $\{y_i\}$ is bounded. If the contrary be assumed, a point x_k of $\{x_i\}$ exists such that $px_k > px_1$, $x_k r > x_1 r$, and a point y_k of $\{y_i\}$ exists such that $py_k > px_k$, $y_k r > x_k r$, and $y_k x_1 > x_k x_1$.

Let $\bar{Q} = (\bar{p}, \bar{r}, \bar{x}_1, \bar{x}_k, \bar{y}_k)$ be a quintuple of points with each distance equal to the corresponding distance in the quintuple $Q = (p, r, x_1, x_k, y_k)$ except that the distance $\bar{x}_k \bar{y}_k = x_1 y_k$. The quintuple $\bar{Q}$ is a metric space and hence by (6) $R(p, x_1, r)$, $R(p, x_k, x_1)$, $R(x_1, y_k, r)$ imply $R(\bar{p}, \bar{x}_1, \bar{r})$, $R(\bar{p}, \bar{x}_k, \bar{x}_1)$, $R(\bar{x}_1, \bar{y}_k, \bar{r})$. The last three relations imply (by (3)) $R(\bar{x}_k, \bar{x}_1, \bar{y}_k)$ which is impossible (Lemma 13.1) since $\bar{x}_k \bar{y}_k = x_1 y_k = \bar{x}_1 \bar{y}_k$.

LEMMA 13.3. *Let M be a metric space consisting of two points $\bar{p}, \bar{r}$. For an arbitrary $\epsilon > 0$ there exists a metric space M^* containing $\bar{p}, \bar{r}$ and a point $\bar{q}$ with $R(\bar{p}, \bar{q}, \bar{r})$ and $\bar{p}\bar{q} < \epsilon$.*

Proof. Let p, r be elements of E_2 with $pr = \bar{p}\bar{r}$. It suffices by (6) to prove the lemma for p, r. Let $\{x_i\}$, $\{y_i\}$ be two sequences of E_2 with $R(p, x_1, r)$, $R(p, x_{i+1}, x_i)$, $R(x_1, y_1, r)$, $R(y_i, y_{i+1}, r)$ $(i = 1, 2, \ldots)$. By Lemma 13.2 one of these sequences, say $\{x_i\}$, has an accumulation point. Then elements x_j, x_k of $\{x_i\}$ exist with $x_j x_k < \epsilon$. Assuming, as we may, that $j > k > 1$ then $R(p, x_j, x_k)$. There exists in E_2 a point q with $pq = x_j x_k < \epsilon$, $qx_k = x_j p$. Then $p, q, x_k \approx x_k, x_j, p$ and $R(p, q, x_k)$ follows. From $R(p, x_k, x_1)$ and $R(p, x_1, r)$ we have $R(p, x_k, r)$, which combined with $R(p, q, x_k)$ gives $R(p, q, r)$. A similar procedure proves the lemma in case $\{y_i\}$ has an accumulation point.

LEMMA 13.4. *If elements p, q, r of a metric space are in the relation $R(p, q, r)$ then $pq + qr = pr$ or $|pq - qr| = pr$.*

Proof. If $R(p, q, r)$ holds without either alternative, then (since p, q, r are elements of a metric space) positive numbers ϵ_1, ϵ_2 exist such that $pq + qr = pr + \epsilon_1$, and $|pq - qr| = pr - \epsilon_2$.

By Lemma 13.3 a metric space exists containing p, q and a point x with $R(p, x, q)$ and $px < \min(pr, \epsilon_1, \epsilon_2)$. Consider, now, the quadruple $\bar{Q} = (\bar{p}, \bar{q}, \bar{r}, \bar{x})$ with each distance equal to the corresponding distance in $Q = (p, q, r, x)$, except $\bar{r}\bar{x} = pr$. To show that $\bar{Q}$ is metric we observe first that triples $\bar{p}, \bar{q}, \bar{r}$ and $\bar{p}, \bar{q}, \bar{x}$ are metric since each is congruent with a metric triple, while triple $\bar{p}, \bar{r}, \bar{x}$ is 'isosceles' and hence metric since $\bar{p}\bar{r} + \bar{r}\bar{x} = 2pr > px = \bar{p}\bar{x}$. Finally, from the metricity of $\bar{p}, \bar{q}, \bar{x}$,

$$\bar{p}\bar{q} + \bar{p}\bar{x} \geq \bar{q}\bar{x} \geq |\bar{p}\bar{q} - \bar{p}\bar{x}|,$$

and it follows that

$$\bar{q}\bar{x} + \bar{q}\bar{r} \geq \bar{p}\bar{q} + \bar{q}\bar{r} - \bar{p}\bar{x} = pr + \epsilon_1 - px > pr = \bar{r}\bar{x},$$

$$|\bar{q}\bar{x} - \bar{q}\bar{r}| \leq |\bar{p}\bar{q} - \bar{q}\bar{r}| + \bar{p}\bar{x} = pr - \epsilon_2 + px < pr = \bar{r}\bar{x}.$$

Thus $\bar{q}\bar{x} + \bar{q}\bar{r} \geq \bar{r}\bar{x} \geq |\bar{q}\bar{x} - \bar{q}\bar{r}|$, and so $\bar{q}, \bar{r}, \bar{x}$ are metric.

Since $\bar{Q}$ is metric $R(\bar{p},\bar{x},\bar{q})$ and $R(\bar{p},\bar{q},\bar{r})$, valid by (6), imply $R(\bar{p},\bar{x},\bar{r})$ which (due to $\bar{r}\bar{x} = pr = \bar{p}\bar{r}$) contradicts a previous result and establishes the lemma.

COROLLARY. *If $R(p,q,r)$ holds for a metric triple p, q, r then one of the three distances pq, qr, pr is the sum of the other two, and the three points are congruent with three points of the straight line E_1.*

LEMMA 13.5. *If $p, r \in E_1$, $p \neq r$, then*

$$\bar{B}_R(p,r)_{E_1} = (p) + (r) + B_R(p,r)_{E_1},$$

where $B_R(p,r)_{E_1}$ denotes the set of all points x of E_1 with $R(p,x,r)$ holding, is a perfect set (that is $\bar{B}'_R(p,r)_{E_1} = \bar{B}_R(p,r)_{E_1}$).

Proof. By (5) and the preceding corollary, a point x of E_1 exists such that $R(p,x,r)$ holds and so the set $B_R(p,r)_{E_1}$ is not null. Let x be any element of $B_R(p,r)_{E_1}$. By Lemma 13.3 there corresponds to each $\epsilon > 0$ a metric space containing p, x, and a point $\bar{y}$ with $R(p,\bar{y},x)$ and

$$0 < x\bar{y} < \epsilon.$$

Hence an element y of E_1 exists with $p, y, x \approx p, \bar{y}, x$ and so $R(p,y,x)$ holds. Then $y \in B_R(p,r)_{E_1}$ by (3) and $0 < xy < \epsilon$. It follows that each element x of $B_R(p,r)_{E_1}$ is an accumulation element of the set, and in a similar manner it is seen that $p, r \in B'_R(p,r)_{E_1}$. Since by (4) $\bar{B}_R(p,r)_{E_1}$ is closed, the lemma is proved.

THEOREM 13.1. *If p, q, r are elements of a metric space, $R(p,q,r)$ implies pqr.*

Proof. By (2), $R(p,q,r)$ implies $p \neq q \neq r$, and from Lemma 13.4 we may conclude pqr when it is shown that $|pq-qr| = pr$ leads to a contradiction.

Assuming $|pq-qr| = pr$, we suppose the labelling of points p, r selected so that $pq - qr = pr$. By the now familiar process, a metric triple p, x, q exists with $R(p,x,q)$ and so $px + xq = pq$ or $|px - xq| = pq$. If the first alternative holds then $|pr - px| = |qr - qx|$. It is easy to show that the quadruple $\bar{Q} = (\bar{p}, \bar{q}, \bar{r}, \bar{x})$ is metric if each distance equals the corresponding distance in the quadruple $Q = (p,q,r,x)$, except for the distance $\bar{r}\bar{x}$. We define $\bar{r}\bar{x} = |pr - px| + \epsilon = |qr - qx| + \epsilon$, where $\epsilon > 0$ and is, moreover, so chosen that $\bar{r}\bar{x} < \min(pr + px, qr + qx)$. We may, in addition, suppose ϵ selected so that

$$\bar{p}\bar{x} + \bar{x}\bar{r} \neq \bar{p}\bar{r} \quad \text{and} \quad |\bar{p}\bar{x} - \bar{x}\bar{r}| \neq \bar{p}\bar{r}.$$

This contradicts $R(\bar{p},\bar{x},\bar{r})$, which follows from $R(\bar{p},\bar{q},\bar{r})$ and $R(\bar{p},\bar{x},\bar{q})$, and hence $px + xq = pq$ cannot hold. We have thus established that *if*

$R(p,q,r)$ implies $|pq-qr| = pr$ then $R(p,x,q)$ implies $|px-xq| = pq$ for each x such that p, x, q is a metric triple.

If $p, r \approx p', r'$, elements of E_1, there exists $q' \in E_1$ with
$$p, q, r \approx p', q', r'$$
and so $R(p',q',r')$ holds with $|p'q'-q'r'| = p'r'$. For each x of E_1 with $R(p',x,q')$ the relation $|p'x-xq'| = p'q'$ subsists.

ASSERTION. *If* $y \in E_1$ $(y \neq q')$ *and* $p'y - yq' = p'q'$, *then* $R(p',y,q')$ *holds*.

Let us assume the contrary. Then there exists a point y of E_1 with $p'y - yq' = p'q'$ but $y \bar{\in} B_R(p',q')_{E_1}$. Since q' is an accumulation element of $B_R(p',q')_{E_1}$, a point z of E_1. $B_R(p',q')_{E_1}$ exists with $q'z < \min(p'q', q'y)$ and $R(p',z,q')$. Whence, by the above, $|p'z-zq'| = p'q'$ and it follows that $p'z - zq' = p'q'$. Traversing the segment $[z,y]$ from z to y one encounters a *last* point z' of $[z,y]$. $\bar{B}_R(p',q')_{E_1}$ (since $\bar{B}_R(p',q')_{E_1}$ is closed); that is, $R(p',z',q')$, $p'z'-z'q' = p'q'$, and $q'z'y$ hold, while if $u \in E_1$ and $z'uy$ subsists, then $R(p',u,q')$ does not.

Now the point z' is an accumulation point of $\bar{B}_R(p',z')_{E_1}$ and so a point t of $B_R(p',z')_{E_1}$ exists with $tz' < \min(p'z', yz')$. We have $R(p',z',q')$ with $p'z'-z'q' = p'q'$, and $R(p',t,z')$. Consequently $|p't-tz'| = p'z'$ and hence $p't-tz' = p'z'$. But then $z'ty$ holds, and since $R(p',t,q')$ follows from $R(p',z',q')$ and $R(p',t,z')$, the 'last point' property of z' is contradicted and the assertion is established.

The proof of the theorem is now readily completed, for applying the assertion to the triple p', q', y' of E_1 with $p'y' = 2p'q', q'y' = p'q'$ gives $R(p',y',q')$. But since $p'q' = y'q'$ this contradicts Lemma 13.1, and the theorem is proved.

THEOREM 13.2. *Metric betweenness* pqr *implies* $R(p,q,r)$.

Proof. If pqr holds then points p', q', r' of E_1 exist with
$$p, q, r \approx p', q', r'.$$
Assuming $R(p',q',r')$ does not subsist then, since $\bar{B}_R(p',r')_{E_1}$ is a perfect subset of E_1, q' lies in the interior of a segment $[x_1, x_2]$ such that $R(p',x_1,r')$, $R(p',x_2,r')$ hold, while if $x \in E_1$ and $x_1 x x_2$ subsists then $R(p',x,r')$ does not. By Theorem 13.1, x_1 and x_2 are interior points of $[p',r']$, and we assume the labelling so that x_2 is between x_1 and r'.

Since $\bar{B}_R(x_1,r')_{E_1}$ is a perfect subset of E_1 a point y of E_1 exists with $R(x_1,y,r')$ holding and $x_1 y < x_1 x_2$. Then $x_1 y x_2$ subsists together with $R(p',y,r')$ which follows by transitivity. This contradiction establishes the theorem.

Thus metric betweenness is characterized by the six properties described above.

We have noted that betweenness in E_1 has a transitive property (pqr and qrs imply pqs and prs) not enjoyed by metric betweenness. It is interesting to observe that if this transitivity be substituted for (3) *there is no relation $R(p, q, r)$ definable for all triples in every metric space which has the new set of six properties.* We shall refer to this as Remark A.

Definition 12.1 is, of course, applicable in semimetric spaces also, and metric betweenness so defined has properties (1), (2), (5), and (6) in those spaces. In semimetric spaces with *continuous distance*, property (4) holds also, but the transitive property (3) does not. It may be replaced by the weaker property (3′) which asserts that pqr, prs, and pqs together imply qrs. The question then arises whether the six properties (1), (2), (3′), (4), (5), (6) characterize metric betweenness pqr among all relations $R(p, q, r)$ definable in every semimetric space with continuous distance. The answer is in the negative, for the relation

$$(pq+1)(qr+1) = pr+1, \qquad p \neq q \neq r,$$

which is different from metric betweenness, is easily shown to possess those six properties.

EXERCISES

1. Prove that the quintuple $\overline{Q}$ of Lemma 13.2 is metric.
2. Show that in a semimetric space with continuous distance, the relation

$$(pq+1)(qr+1) = pr+1$$

($p \neq q \neq r$), has properties (1), (2), (3′), (4), (5), (6).
3. Prove Remark A. (*Hint.* By (5) $R(p,q,r)$ subsists. Consider $\overline{Q} = (\bar{p}, \bar{q}, \bar{r}, \bar{s})$ with $\bar{p}, \bar{q}, \bar{r} \approx p, q, r$ and $\bar{q}, \bar{r}, \bar{s} \approx r, q, p$, while $\overline{ps} = pr$ in case $pq < 2pr$, but $\overline{ps} = pq$ if $pr < 2pq$).

14. Convexity in metric spaces. Metric segments

Though our extension to metric spaces of the notion of betweenness in elementary geometry is one which the obvious connexion of that notion with distance naturally suggests, it is much less evident how the important concept of convexity is to be extended to such spaces. That notion, as defined by Minkowski for closed subsets of euclidean space (a closed subset of E_n is convex provided it contains for each two of its points the segment joining them) seems bound up with the existence and uniqueness of segments joining each two points of E_n. It might seem, therefore, that an extension to metric spaces of the notion of segment must precede a definition of convexity in such spaces. This, however, is not the case, and convexity is definable (and useful) in metric spaces and subsets for which (1) not every pair of points is joined by a segment, or (2) if segments do exist, they are not unique. It will be shown, moreover, that *when applied to closed subsets of euclidean space, the metric notion of convexity coincides with the ordinary one.*

DEFINITION 14.1 (*Menger*). *A subset of a metric space is* (*metrically*) *convex provided it contains for each two of its points at least one betweenpoint.*

Clearly each subset of E_n that is Minkowski convex is metrically convex. Metric convexity is evidently a congruence invariant. But since the $S_{n,r}$ is metrically convex (for this reason we refer to it as the 'convex n-sphere') while the n-sphere $C_{n,r}$ with euclidean (chord) distance, which is homeomorphic to it, is not metrically convex, the property is not topological.

DEFINITION 14.2. *If p, q are elements of a subset of a metric space, then the subset is a metric segment with end-points p, q (denoted by $S_{p,q}$) provided it is congruent with a line segment of length pq.*

Since a congruence is a homeomorphism, a metric segment is an *arc* (that is, a homeomorph of a line segment).

It was proved by Menger that each two points of a closed and convex subset of a compact metric space are joined by a metric segment belonging to the subset. Using transfinite induction, he established this result also for complete metric spaces. The proof we shall give for this fundamental theorem, which avoids transfinite induction, is due to Aronszajn.

THEOREM 14.1. *Each two distinct points of a complete and convex metric space M are joined by a metric segment of the space.*

Proof. A real function f with domain $D(f)$ a subset of M and range $R(f)$ a subset of E_1 is called *isometric in M* provided $D(f) \approx R(f)$. A real function g is an *extension of f* (symbolized by $f \subset g$) whenever $D(f) \subset D(g) \subset M$ and $f(x) = g(x)$ for $x \in D(f)$.

If $\bar{p}, \bar{q} \in M$ with $\bar{p}\bar{q} = d$, consider the set $D_1 = (\bar{p}, \bar{q})$ and the real function f_1 with $f_1(\bar{p}) = 0, f_1(\bar{q}) = d$. Make now the inductive assumption that n real functions $f_1, f_2, \ldots, f_n$ have been defined isometric in M, whose ranges $R(f_i)$ are subsets of the line segment $[0, d]$, and

$$f_{i-1} \subset f_i \quad (i = 1, 2, \ldots, n),$$

where $f_0 \equiv f_1$. Denote by Σ_n the set of all functions f which are isometric in M, have range $R(f) \subset [0, d]$, and $f_n \subset f$. Let I_n denote the set of 2^n closed intervals obtained by dividing $[0, d]$ into equal, non-overlapping sub-intervals of length $d/2^n$. For each $f \in \Sigma_n$ the number of those intervals of I_n with points in common with $R(f)$ is denoted by $N_n(f)$. Since $N_n(f) \leqslant 2^n$ for each function f of Σ_n, we may select a function f_{n+1} of Σ_n such that $N_n(f_{n+1}) \geqslant N_n(f)$ for every $f \in \Sigma_n$.

Hence, by complete induction, an infinite sequence $f_1, f_2, \ldots, f_n, \ldots$ of functions is obtained, which uniquely determines a function f with

$$D(f) = \sum_{n=1}^{\infty} D(f_n) \subset M, \qquad R(f) = \sum_{n=1}^{\infty} R(f_n) \subset [0, d],$$

and $f_n \subset f$ ($n = 1, 2, \ldots$). Clearly $D(f) \underset{f}{\widetilde{\approx}} R(f)$, and so f is isometric in M. We show next that this congruence can be extended in a unique way to the closures of $D(f)$ and $R(f)$, and we thereby obtain a function g which is an extension of each f_n ($n = 1, 2, \ldots$).

If $p \in D'(f)$, an infinite sequence $\{p_n\}$ of pairwise different points of $D(f)$ exists with $\lim p_n = p$. Since $\{p_n\}$ has a limit (and M is metric) it is a Cauchy sequence, and consequently the sequence $\{f(p_n)\}$ of points of $[0, d]$ is also a Cauchy sequence (since $f(p_i)f(p_j) = p_i p_j$). Hence the E_1 contains a point p' such that $p' = \lim f(p_n)$. Clearly $p' \in R'(f)$ and since $R(f) \subset [0, d]$ then $p' \in [0, d]$.

It is easy to see that p' is uniquely determined by p; that is, p' is independent of the choice of the sequence $\{p_n\}$ with limit p. For let $\{\bar{p}_n\}$ be another sequence of points of $D(f)$ with $\lim \bar{p}_n = p$ and $\lim f(\bar{p}_n) = p''$. Then the 'staggered' sequence $p_1, \bar{p}_1, p_2, \bar{p}_2, \ldots, p_n, \bar{p}_n, \ldots$ has limit p and the sequence $f(p_1), f(\bar{p}_1), f(p_2), f(\bar{p}_2), \ldots, f(p_n), f(\bar{p}_n), \ldots$ has limit $p''' \in [0, d]$. But $\{f(p_n)\}$ and $\{f(\bar{p}_n)\}$ are each sub-sequences of the latter sequence and so $p' = p'' = p'''$. Hence to each $p \in D'(f)$ there corresponds a unique element $p' = g(p)$ of $R'(f)$, and it is evident that g is biuniform in $D'(f)$. Setting $g(p) = f(p)$, $p \in D(f)$, then

$$D(g) = D(f) + D'(f) = \bar{D}(f) \subset M,$$
$$R(g) = R(f) + R'(f) = \bar{R}(f) \subset [0, d].$$

If $x, y \in D(g)$ then $xy = \lim x_n y_n$, $(x_n, y_n \in D(f))$. Hence

$$xy = \lim f(x_n)f(y_n) = g(x)g(y),$$

by continuity of the metric, and so g is isometric in M with $f \subset g$. Consequently $f_n \subset g$, $n = 1, 2, \ldots$.

The reader might have observed that so far the convexity of M has not entered into consideration. We shall use it now in order to show that $R(g) = [0, d]$. This will complete the proof of the theorem, for from

$D(g) \approx R(g) = [0, d]$ it follows that $D(g)$ is a metric segment joining $\bar{p}$ and $\bar{q}$.

If we assume the existence of a point of $[0, d]$ not belonging to $R(g)$ then, since $R(g)$ is a closed subset of $[0, d]$ and $0, d \in R(g)$, there exists an open interval (a, c), $a < c$, which contains no points of $R(g)$ and whose end-points a, c belong to $R(g)$. Let r, t denote the elements of $D(g)$ such that $g(r) = a$ and $g(t) = c$. Then $rt = c-a$, making the labels of points of $[0, d]$ serve also as coordinates.

From the convexity of M, a point s of M exists with $r \neq s \neq t$, $rs+st = rt$. Hence a point b of (a, c) exists such that $rs = b-a$, $st = c-b$. Since $b \bar{\in} R(g)$ then $s \bar{\in} D(g)$. Let $D^* = D(g)+(s)$ and define $g^*(x) = g(x)$, $x \in D(g)$, $g^*(s) = b$. To show that g^* is isometric in M it suffices to show that $x \in D(g)$ implies

$$sx = g^*(s)g^*(x) = bg(x) = |b-g(x)|.$$

One of the three distinct points $a = g(r)$, $c = g(t)$, and $g(x)$ of E_1 is between the other two. Since g is isometric then one of the points r, x, t of $D(g)$ is between the other two. But $g(x) \bar{\in} (a, c)$ and so either xrt or rtx must hold. In the first case xrt and rst give xrs; that is, $xr+rs = sx$, or $sx = a-g(x)+b = b-g(x)$. In the second case, rtx and rst imply stx, and so $sx = c-b+g(x)-c = g(x)-b$. Thus in either event $sx = |b-g(x)|$ and so g^* is an isometric extension of f and every f_n, $n = 1, 2, \ldots$.

If n be chosen so that $d/2^n < \min(b-a, c-b)$, then at least one of the closed intervals of I_n contains b and is contained in (a, c). Then $R(g^*)$ has a point in common with at least one interval of I_n whose product with $R(f_{n+1})$ is null. Since $f_{n+1} \subset g^*$ then $R(f_{n+1}) \subset R(g^*)$ and so $R(g^*)$ surely has points in common with all those intervals of I_n that have non-empty products with $R(f_{n+1})$. Hence $N_n(g^*) \geq N_n(f_{n+1})+1$, which (since $g^* \in \Sigma_n$) contradicts the definition of f_{n+1} and establishes the theorem.

COROLLARY. *Metric convexity and Minkowski convexity (for closed subsets of E_n) are equivalent.*

Proof. We have already noted that Minkowski convexity implies metric convexity. On the other hand, if S is a closed, metrically convex subset of E_n then S is a complete convex metric space and so each two of its points are joined by a metric segment belonging to S. But a metric segment of E_n is obviously a straight line segment and so S is convex in the sense of Minkowski.

15. Two characteristic properties of metric segments

Since metric segments play an important role in the distance geometry of metric spaces, it is desirable to characterize them from many different standpoints. We begin by proving a useful lemma.

LEMMA 15.1. *If $S_{a,p}$, $S_{p,b}$ are metric segments, then $S_{a,p} + S_{p,b}$ is a metric segment with end-points a, b if and only if apb holds.*

Proof. The necessity of apb is obvious. Assuming apb, let
$$x \in S_{a,p} \cdot S_{p,b} \quad (x \neq p).$$
Then $ax + xp = ap$, $px + xb = pb$, whence $ax + xb + 2px = ap + pb = ab$. If $x \neq a$ then axp holds which (with apb) gives axb and so $ax + xb = ab$. This equality obviously is valid if $x = a$, and hence $2px = 0$ and $x = p$. Thus the metric segments $S_{a,p}$, $S_{p,b}$ have only the point p in common.

Let a', p', b' be points of E_1 with $a, p, b \approx a', p', b'$. Then
$$S_{a,p} \approx [a', p'], \qquad S_{p,b} \approx [p', b']$$
and since $S_{a,p} \cdot S_{p,b} = (p)$ these two congruences define a mapping of $S_{a,p} + S_{p,b}$ onto the line segment $[a', b']$.

To show the mapping is a congruence it suffices to prove that if $x, y \in S_{a,p} + S_{p,b}$ with $x \in S_{a,p}$, $y \in S_{p,b}$, $x \neq p \neq y$, then $xy = x'y'$, where x, y are mapped onto x', y', respectively. Now $x \in S_{a,p}$, $y \in S_{p,b}$ imply axp and pyb subsist provided $a \neq x$ and $b \neq y$. The relations axp, apb imply xpb, while xpy follows from xpb and pyb. If $x = a$ and $y = b$, xpy follows by hypothesis, and it is obtained by one application of the transitive property of betweenness in case $x = a$ or $y = b$. Hence in any event $xy = xp + py = x'p' + p'y' = x'y'$.

COROLLARY. *If p, q, r are points of a complete convex metric space M and pqr holds, then a metric segment $S_{p,r}$ of M exists containing q.*

Defining a point q to be a *middle-point* of p, r provided $pq = qr = \frac{1}{2}pr$, we use the lemma to give a new proof of a characterization theorem due to Menger.

THEOREM 15.1. *In a complete metric space M, a metric segment with end-points p, r is characterized among all closed subsets consisting of p, r and points between p, r by the property of containing for each pair of its points exactly one middle-point.*

Proof. From the congruence of $S_{p,r}$ with the line segment $[0, pr]$ it is clear that metric segments possess the property described in the theorem.

Let, now, F be any closed subset of M consisting of p, r and points between p, r, and having the above property. Then F is a complete convex metric space and, by Theorem 14.1, contains a metric segment $S_{p,r}$ with end-points p, r. We show $F = S_{p,r}$.

If $q \in F$ and $q \bar{\in} S_{p,r}$ then pqr implies the existence of a metric segment $S^*_{p,r}$ of F containing q (Corollary, Lemma 15.1) and $S^*_{p,r} \neq S_{p,r}$ since $q \bar{\in} S_{p,r}$. Traversing $S^*_{p,r}$ from q to p a first point s of $S_{p,r}$ is encountered, and similarly a first point t of $S_{p,r}$ is met when $S^*_{p,r}$ is traversed from q to r. (The reader is asked to prove this statement in an exercise.) Then the sub-segments of $S_{p,r}$ and $S^*_{p,r}$ with end-points s, t have only those points in common. But each of these sub-segments obviously contains a middle-point of s, t, and the existence of these two distinct middle-points for elements s, t of F is the contradiction that establishes the theorem.

THEOREM 15.2. *Among all arcs with end-points a, b of a metric space, a metric segment $S_{a,b}$ is characterized by the following property: if p, $q \in S_{a,b}$ (p, q each distinct from a, b) then apq or qpb or $p = q$.*

Proof. The property being a congruence invariant, its possession by metric segments follows at once from its obvious validity for line segments. On the other hand, suppose B (an arc with end-points a, b) has the property described in the theorem, and assume that B is not a metric segment. Then B contains at least one pair of points without containing a middle-point for the pair, for if this were not the case then B is convex, and since it is closed and compact (being an arc) it may be considered as a compact, convex metric space. Then by Theorem 14.1, B contains a metric segment (an arc) with end-points a, b and hence B *is* a metric segment contrary to assumption.

Further, it is clear that B contains at least one pair of points b_0, b_1, *each distinct from a, b* without containing a middle-point of the pair; for a denial of this assertion leads at once (because of the continuity of the metric and the compactness of B) to the conclusion that B contains a middle-point for each pair of its points.

Though B does not contain a middle-point for the pair b_0, b_1, that sub-arc of B joining b_0, b_1 does contain a point c such that

$$b_0 c = c b_1 > 0, \qquad b_0 c + c b_1 > b_0 b_1. \tag{*}$$

(The reader is asked to verify this in an exercise.)

Apply, now, the property stated in the theorem to the pairs of *distinct* points, (b_0, b_1), (c, b_0), (b_1, c) in turn. Then $ab_0 b_1$ or $b_1 b_0 b$ *and* acb_0 or $b_0 cb$ *and* $ab_1 c$ or $cb_1 b$ subsist. But an application of the transitive property of betweenness in metric spaces leads to the conclusion that one of the points b_0, c, b_1 is between the two others, which clearly violates the relations (∗). Hence B is a metric segment with end-points a, b.

EXERCISES

1. Prove the statement occurring in the proof of Theorem 15.1 that 'traversing $S_{p,r}^*$ from q to p a first point s of $S_{p,r}$ is encountered'.
2. Prove the validity of the relations in (∗) of Theorem 15.2.

16. Lemmas concerning closed and compact subsets. An equilateral triple theorem

The *diameter* $\delta(E)$ of a non-null subset E of a metric space is the l.u.b. pq for $p, q \in E$. Each bounded subset E is of finite diameter, though E may not contain a point-pair p, q with $pq = \delta(E)$.

LEMMA 16.1. *A closed and compact non-null subset E of a metric space contains a pair of points p, q such that $pq = \delta(E)$.*

Proof. Since E is compact, it is bounded and so $\delta(E)$ is finite. Corresponding to each positive integer n, points p_n, q_n of E exist with

$$\delta(E) - 1/n < p_n q_n < \delta(E) + 1/n,$$

and hence $\lim p_n q_n = \delta(E)$.

Now E contains a point p such that $p = \lim p_{i_n}$, where $\{p_{i_n}\}$ is a subsequence of $\{p_n\}$ (Exercise 1, § 10). Then $\lim p_{i_n} q_{i_n} = \delta(E)$ and E contains an element q such that $q = \lim q_{i_{j_n}}$ where $\{q_{i_{j_n}}\}$ is a sub-sequence of $\{q_{i_n}\}$. It follows that

$$pq = \lim p_{i_{j_n}} q_{i_{j_n}} = \delta(E).$$

If $E \subset M$, a metric space, and $p \in M$ then the *distance* $pE = \text{dist}(p, E)$ of p from E is the g.l.b. px, for x an element of E. Since $px \geq 0$, $x \in E$, pE exists for each subset E and element p of M. It is clear that $pE = 0$ if and only if $p \in \bar{E}$.

A point f_p of E such that $pf_p = pE$ is called a *foot* of p in E.

LEMMA 16.2. *If E is a closed and compact subset of a metric space M and $p \in M$, then E contains a foot f_p of p.*

Proof. If $\rho = pE$ then for each positive integer n a point p_n of E exists such that

$$\rho - 1/n < pp_n < \rho + 1/n,$$

and hence $\lim pp_n = \rho$. Since E contains a point q such that $q = \lim p_{i_n}$, where $\{p_{i_n}\}$ is a sub-sequence of $\{p_n\}$, then $pq = \lim pp_{i_n} = \lim pp_n = \rho$. Hence q is a foot f_p of p in E.

Three points p, q, r are said to form an *equilateral triple* provided $pq = qr = pr \neq 0$.

THEOREM 16.1. *Let $S_{a,b}$ be a metric segment with end-points a, b of a metric space M. If $p \in M$ with a foot f_p in $S_{a,b}$ such that*

$$\min[af_p, bf_p] > 3pf_p > 0,$$

then $S_{a,b}$ contains points q, r such that p, q, r form an equilateral triple.

Proof. Since $S_{a,b}$ is congruent with a line segment, we may introduce coordinates in $S_{a,b}$, with f_p as origin, in the obvious manner. Letters denoting points of $S_{a,b}$ being interpreted both as labels and numbers (coordinates), the distance xy of points x, y of $S_{a,b}$ is $|x-y|$. Put $pf_p = d$.

Denote the continuous function px ($0 \leqslant x \leqslant 3d$) by $S(x)$. The triangle inequality applied to x, p, f_p gives $|x - S(x)| \leqslant d$ and hence $y = x - S(x)$ is an element of $S_{a,b}$. The distance $py = T(x)$ is a continuous function of x. Use of the triangle inequality yields $S(x) \geqslant |x-d|$ and $d \leqslant T(x) \leqslant d + |x - S(x)| \leqslant 2d$.

Case 1. $T(0) = d$. Then the points p, $q = y(0) = -d$, $r = f_p$ form an equilateral triple, for $pq = py(0) = T(0) = d$, $pr = pf_p = d$, $qr = y(0)f_p = d$.

Case 2. $T(0) > d$. Let $R(x) = T(x) - S(x)$, a continuous function of x, $0 \leqslant x \leqslant 3d$. Then $R(0) = T(0) - S(0) = T(0) - d > 0$, and $R(3d) \leqslant 0$, since $T(3d) \leqslant 2d$ and $S(3d) \geqslant 2d$. Since

$$\min[af_p, bf_p] > 3pf_p = 3d,$$

the point with coordinate $3d$ belongs to $S_{a,b}$. Hence the sub-segment $S_{f_p,3d}$ contains a point q such that $R(q) = 0$; that is, $T(q) = S(q)$. Then p, q, and $y(q) = r$ form an equilateral triple, for $pq = S(q) > 0$, $qr = qy(q) = |q - S(q) - q| = S(q)$, and $pr = py(q) = T(q) = S(q)$.

The theorem has the following interesting corollary.

COROLLARY. *If a metric space M contains a subset L congruent to a straight line E_1, and a point p not belonging to L, then M contains an equilateral triple.*

This corollary is used later as a basis for a metric characterization of straight lines.

It is observed that the corollary is invalid if the metricity of M is relaxed to require that M be semimetric with continuous distance. The space formed by the spiral $r = 1 + e^{-\theta}$, $-\infty < \theta < \infty$, and the origin $r = 0$, with distance from the origin to points of the spiral euclidean, and distance for two points of the spiral the length of the arc joining them, is semimetric, with continuous distances. It contains a subset L congruent with E_1 (the spiral) and a point not belonging to L (the origin), but it does not contain an equilateral triple. For any such triple must contain the origin, and no two points of the spiral have the same distances from the origin.

17. An equilateral triple characterization of metric segments

This section is devoted to establishing a characterization of metric segments quite different from the two given in § 15.

THEOREM 17.1. *A metric segment is characterized among all closed compact and convex subsets* Σ *(with at least two points) of a metric space* M *by the property of not containing an equilateral triple.*

Proof. Since a metric segment is congruent with a line segment it evidently contains no equilateral triples.

Suppose, on the other hand, that Σ is a closed compact and convex subset of M (with at least two points) that does not contain an equilateral triple. Then there are two points p, q of Σ with maximum distance (Lemma 16.1) and Σ contains a metric segment $S_{p,q}$ with end-points p, q (Theorem 14.1). We assume that Σ has a point r not belonging to $S_{p,q}$, and show that this implies the existence of an equilateral triple in Σ.

If there exists a point s interior to $S_{p,q}$ (that is, $p \neq s \neq q$) such that a metric segment $S_{r,s} \subset \Sigma$ contains neither of the *sub-segments* $S_{p,s}$, $S_{s,q}$ of $S_{p,q}$, then $S_{r,s}$ has a point r^* with a foot f_{r^*} in $S_{p,q}$ ($p \neq f_{r^*} \neq q$) and $0 < 3r^*f_{r^*} < \min[pf_{r^*}, qf_{r^*}]$. Then by Theorem 16.1, $S_{p,q}$ contains two points forming with r^* an equilateral triple of Σ.

If the eventuality described above does not occur, then a point s of $S_{p,q}$ with *maximum distance d from r* is interior to $S_{p,q}$ and each metric segment of Σ with end-points r, s contains either $S_{p,s}$ or $S_{q,s}$ (sub-segments of $S_{p,q}$). Selecting the labelling so that $S_{p,s} \subset S_{r,s}$ then $rp+ps = rs = d$. It is easily seen that for each interior point t of $S_{q,s}$, any $S_{r,t} \subset \Sigma$ contains $S_{q,t}$, sub-segment of $S_{q,s}$. Letting t approach s, it follows that q is interior to a metric segment $S_{r,s}^*$ with end-points r, s.

Now $pq = ps+sq = d-pr+d-rq = 2d-(pr+rq)$. Since pq is a maximum distance in Σ, then $pq \geq d$ and so $d \geq pr+rq$. But

$$pq \leq pr+rq \leq d,$$

and hence $pq = d$ and prq holds.

The point p divides $S_{r,s}$ into two sub-segments denoted by $S_{r,p}$, $S_{p,s}$ and q divides $S_{r,s}^*$ into two sub-segments $S_{r,q}^*$, $S_{q,s}^*$.

ASSERTION. $S_{r,s} \cdot S_{r,s}^* = (S_{r,p}+S_{p,s})(S_{r,q}^*+S_{q,s}^*) = (r, s)$.

Case 1. If $x \in S_{r,p} \cdot S_{r,q}^*$, $x \neq r$, then $x \neq p$ (since otherwise $rp+pq = rq$ and $rq > pq$) and pxr holds. Similarly, $x \neq q$ and qxr subsists. But *prq* together with *pxr* implies *xrq*. This contradicts *qxr* and hence $S_{r,p} \cdot S^*_{r,q} = (r)$.

Case 2. If $x \in S_{r,p} \cdot S^*_{q,s}$ then $p \neq x \neq r$, $q \neq x \neq s$ and *pxr*, *qxs* subsist. But *rps* and *pxr* imply *xps*, while *psq* and *qxs* imply *psx*. Hence $S_{r,p} \cdot S^*_{q,s} = 0$.

The remaining two cases are similarly treated to yield $S_{p,s} \cdot S^*_{r,q} = 0$ and $S_{p,s} \cdot S^*_{q,s} = (s)$, and the assertion is established.

Let t and t^* be elements of $S_{r,s}$ and $S^*_{r,s}$, respectively, such that $rt = rt^* = \frac{2}{3}d$. Then $st = st^* = \frac{1}{3}d$, and $tt^* \leq st + st^* = \frac{2}{3}d$. The proof is concluded when it is shown that $tt^* = \frac{2}{3}d$.

If (a) $t^* \in S^*_{r,q}$, then one readily concludes that $t \in S_{p,s}$, while if (b) $t^* \in S^*_{q,s}$ then $t \in S_{p,s}$ gives at once $tt^* = \frac{2}{3}d$, and $t \in S_{r,p}$ differs from (a) only in labelling. Considering, then, $t^* \in S^*_{r,q}$, $t \in S_{p,s}$, we may suppose that each metric segment of Σ with end-points t, t^* contains either $S_{q,t}$ or $S_{p,t}$ (sub-segments of $S_{p,q}$ determined by t, an element of $S_{p,q}$) for otherwise the presence of an equilateral triple in Σ follows by the first part of the proof.

If such a metric segment contains $S_{q,t}$ then it contains s and so $tt^* = t^*s + st = \frac{2}{3}d$. If, finally, a metric segment S_{t,t^*} of Σ contains $S_{p,t}$, then $t^* \in S^*_{r,q}$, $t^* \neq q$, implies rt^*q which, together with *prq*, gives prt^* and $pt^* > pr$. (If $t^* = q$, $pt^* > pr$ follows from *prq*). Now $p \in S_{t,t^*}$ implies $tt^* = tp + pt^* = pt^* + \frac{2}{3}d - pr$ (for *pts* and *rps* imply *rpt*). Hence, $tt^* > \frac{2}{3}d$, which contradicts $tt^* \leq \frac{2}{3}d$ established above, and the proof of the theorem is complete.

18. Criteria for unique metric segments

Metric spaces exhibit great diversity of behaviour with regard to the number of metric segments that join a pair of points. In the chord n-sphere $C_{n,r}$, for example, no two points are joined by a metric segment. The convex n-sphere $S_{n,r}$ has exactly one metric segment with end-points p, q provided $pq \neq \pi r$, while if $pq = \pi r$, they are end-points of uncountably many metric segments. On the other hand, the euclidean space E_n has exactly one metric segment with end-points p, q for every pair of points p, q of the space. We present now some criteria for unique segments between two points.

THEOREM 18.1. *In order that each two points of a complete convex metric space M be end-points of exactly one metric segment it is necessary and sufficient that the common part of each two closed and convex subsets of the space be convex.*[†]

[†] We use 'convex' always in the sense of 'metrically convex'. From now on we shall refer to metric segments merely as 'segments'.

Proof. We prove the sufficiency by showing that if there is more than one segment between p, q then it is not the case that the common part of each two closed and convex subsets of the space is convex. For if $S_{p,q} \neq S^*_{p,q}$ exist and $S_{p,q} \cdot S^*_{p,q}$ is convex, then the product contains a segment $S^{**}_{p,q}$. Now $S^{**}_{p,q} \subset S_{p,q} \cdot S^*_{p,q}$ implies $S^{**}_{p,q} \subset S_{p,q}$ and $S^{**}_{p,q} \subset S^*_{p,q}$, from which we may conclude that

$$S_{p,q} = S^{**}_{p,q} = S^*_{p,q},$$

since an arc with end-points p, q does not contain, as a *proper* subset another arc between its end-points. It follows, then, that the product $S_{p,q} \cdot S^*_{p,q}$ of two closed and convex subsets of M is not convex.

Suppose now that each two points of M are end-points of exactly one segment, and let K_1, K_2 be two closed and convex subsets of M. If $p, q \in K_1 \cdot K_2$ ($p \neq q$) then $p, q \in K_1$ (closed and convex) implies the existence of $S_{p,q} \subset K_1$, and $p, q \in K_2$ implies $S^*_{p,q} \subset K_2$. Since $S_{p,q} = S^*_{p,q}$ then $S_{p,q} \subset K_1 \cdot K_2$ which is thus convex, and the theorem is proved.

THEOREM 18.2. *In order that two points a, b of a complete convex metric space M determine a unique segment between them it is necessary and sufficient that $\bar{B}(a,b)$ be an arc.*

Proof. It is clear that $\bar{B}(a,b)$ contains every segment $S_{a,b}$ and hence if $\bar{B}(a,b)$ is an arc then $S_{a,b} = \bar{B}(a,b)$.

Conversely, suppose $S_{a,b}$ is unique. Now $S_{a,b} \subset \bar{B}(a,b)$ and if

$$S_{a,b} \neq \bar{B}(a,b)$$

there exists a point p of $\bar{B}(a,b)$ and $p \bar{\in} S_{a,b}$. Then $a \neq p \neq b$ and $p \in B(a,b)$; that is, apb holds. By the corollary of Lemma 15.1, M contains a segment $S^*_{a,b}$ with $p \in S^*_{a,b}$. Then $S_{a,b} \neq S^*_{a,b}$, contradicting the uniqueness of $S_{a,b}$. Hence $S_{a,b} = \bar{B}(a,b)$, and $\bar{B}(a,b)$ is an arc.

THEOREM 18.3. *In order that two points a, b of a complete convex metric space M determine a unique segment between them it is necessary and sufficient that $p, q \in B(a,b)$ implies apq or qpb or $p = q$.*

Proof. If $S_{a,b}$ is unique then $S_{a,b} = \bar{B}(a,b)$ and the conclusion follows from Theorem 15.2.

If, on the other hand, a and b have at least two distinct segments $S_{a,b}, S^*_{a,b}$ between them, then it is seen (as in a previous theorem) that sub-segments $S_{r,s}, S^*_{r,s}$ of $S_{a,b}, S^*_{a,b}$ respectively, exist having only the points r, s in common. Then the middle-points p, q of $S_{r,s}, S^*_{r,s}$, respectively, are distinct, and $pr = ps = \frac{1}{2}rs$, $qr = qs = \frac{1}{2}rs$.

But then neither apq nor qpb exist, for evidently $ar + rp = ap$,

$ar+rq = aq$, so $ap = aq$. Hence apq does not hold. Similarly, it is seen that qpb is impossible.

19. Convex extension of a set

Defined first (for subsets of a euclidean space) by Minkowski in 1910, and slightly later by Carathéodory (1911), the notion of convex extension of a set has been found quite useful in many applications. We give a brief study of the notion for subsets of a convex metric space, utilizing results due to Menger. The term 'convex extension' (Dines) is used instead of the older 'convex hull' (konvexe Hülle).

DEFINITION 19.1. *A convex extension C_A of a subset A of a metric space is any subset of the space that is closed, convex, contains A, and is irreducible with respect to those properties (that is, no proper subset of C_A is closed, convex, and contains A).*

For each subset of E_n there exists one and only one convex extension. This statement is valid when E_n is replaced by any complete metric space in which each two distinct points determine exactly one segment between them. For if A is any subset of such a space, the product of all closed and convex subsets containing A is closed, convex (each 'factor' set of the product contains the one segment between two points of the product set), contains A, and is obviously irreducible with respect to those properties. Hence the product is the unique convex extension of A.

On the other hand, no subset of the chord n-sphere $C_{n,r}$ has a convex extension if it contains at least two points, while every subset A of the convex n-sphere $S_{n,r}$ has a convex extension which is unique if A contains at least three points. If $p, q \in S_{n,r}$, $pq = \pi r$, then every half great circle joining p, q is a convex extension of the set consisting of p and q. The last example shows also that the product of all closed and convex subsets of a compact, convex metric space, each of which contains a given subset A of the space, is not necessarily convex; since such a product, for $A = (p,q)$, $p, q \in S_{n,r}$, $pq = \pi r$, consists of the pair of points p, q which is not a convex set. Hence if segments are not unique, the simple device for finding a convex extension of a set used above is not valid.

Before proving the principal theorem of this section, we need to establish two lemmas.

LEMMA 19.1. *(Cantor Product Theorem.) If*

$$A_1 \supset A_2 \supset \ldots \supset A_n \supset A_{n+1} \ldots$$

is a monotone decreasing infinite sequence of non-null, closed and compact subsets of a metric space, then the product $\prod_{k=1}^{\infty} A_k \neq 0$.

Proof. Consider the infinite sequence $\{p_k\}, p_k \in A_k$ ($k = 1, 2,...$). Since $\{p_k\} \subset A_1$, closed and compact, an element p of A_1 and a subsequence $\{p_{i_k}\}$ of $\{p_k\}$ exist with $p = \lim p_{i_k}$. We may suppose that either $p_{i_k} = p$ ($k = 1, 2,...$) or $p_{i_k} \neq p_{i_j}$ ($k, j = 1, 2,...; k \neq j$). If the first alternative holds then clearly $p \in \prod_{k=1}^{\infty} A_k$. In the second case, p is seen to be an accumulation point of A_{i_j} (and hence a point of A_{i_j}) for every $j = 1, 2,...$; and again $p \in \prod_{k=1}^{\infty} A_k$.

LEMMA 19.2. *If*
$$A_1 \supset A_2 \supset ..., \qquad A_n \supset A_{n+1} ...$$
is a monotone decreasing infinite sequence of non-null, closed and convex subsets of a compact metric space, then $A = \prod_{k=1}^{\infty} A_k$ *is convex.*

Proof. Let p, q be points of A (if A does not contain two points, the theorem is proved) and denote by $D_1(p, q)$ the set of all points x of A_1 with $px = xq$. The set $M_1(p, q) = D_1(p, q) \cdot \bar{B}(p, q)$ of middle-points of p, q contained in A_1 is easily seen to be closed and (since A_1 contains a segment between $p, q \in A_1$) $M_1(p, q) \neq 0$.

Similarly, for each integer k, the non-null, closed set
$$M_k(p, q) = D_k(p, q) \cdot \bar{B}(p, q)$$
of middle-points of p, q contained in A_k is defined. Denoting by $M(p, q)$ the set of middle-points of p, q in the space, then $M_k = M(p, q) \cdot A_k$, and since $A_k \supset A_{k+1}$, then $M_k \supset M_{k+1}$. The space being compact, the sequence $\{M_k\}$ is seen to satisfy all the hypotheses of Lemma 19.1, and we conclude that $\prod_{k=1}^{\infty} M_k \neq 0$. But any element of this product set is a point of A_k for every integer k, and hence belongs to the set A. Then A contains for each two of its distinct points a between-point (indeed, a middle-point) and so A is convex.

THEOREM 19.1. *For each subset A of a compact convex metric space M there is at least one convex extension C_A.*

Proof. It follows from Theorem 10.5 that for every positive integer n there exists a finite number k_n of spherical neighbourhoods, each with radius $1/n$, whose sum 'covers' (that is, contains) M. Let $S_1, S_2,..., S_n,...$ denote the covering systems so obtained, and $\mu_n(E)$ the number of sets of the system S_n with points in common with a subset E of M, $n = 1, 2,...$. Clearly $1 \leqslant \mu_n(E) \leqslant k_n$.

Consider, now, the class of all closed, convex subsets of M containing A (since M is convex this class is not null) and let A_1 denote a member of this class for which $\mu_1(A_1)$ is a minimum. If sets $A_1, A_2,..., A_n$ have been defined, then A_{n+1} is any member of the class of all closed, convex subsets of A_n containing A for which $\mu_{n+1}(A_{n+1})$ is a minimum. The infinite sequence $\{A_k\}$ of sets so obtained satisfies the conditions of Lemma 19.2, and hence the product $\prod_{k=1}^{\infty} A_k$ is convex. It is also closed and contains A. Denoting the product by C, we show $C = C_A$ by proving that C is irreducible with respect to those properties.

Suppose that C^*, a proper subset of C, is closed, convex, and contains A. Then points p of C and q of C^* exist, with $p \bar{\in} C^*$ and $0 < pq \leqslant px$, for each element x of C^*. Let n be an integer such that

$$2/n > pq > 2/(n+1)$$

(replacing one of the inequalities by an equality in case pq is the reciprocal of an integer).

Since $C^* \subset C \subset A_{n+1}$ then $\mu_{n+1}(C^*) \leqslant \mu_{n+1}(A_{n+1})$. But $p \in A_{n+1}$ and there is a member of S_{n+1} that contains p but has no point in common with C^* (since pq is not less than the diameter $2/(n+1)$ of each member of S_{n+1}). Hence $\mu_{n+1}(C^*) < \mu_{n+1}(A_{n+1})$, contradicting the definition of the set A_{n+1} (since $C^* \subset A_n$) and the proof of the theorem is complete.

20. Passing points and terminal points

A point p of a metric space M is a *passing point* provided $a, b \in M$ exist such that apb holds. If P denotes the set of all passing points of a metric space, the set $T = C(P)$ consists of those points of M that are not between any two points of M. Such points are called *terminal* points.

Though the sum of a finite number of closed subsets of a metric space is a closed subset, the sum of denumerably many closed subsets is not necessarily closed. For example, if (r) denotes the set consisting of the rational point r, then (r) is a closed subset of E_1 but $\sum (r)$ (where the summation is extended over all the rational points of $[0, 1]$) is not a closed subset of E_1.

DEFINITION 20.1. *A sum of denumerably many closed subsets of a metric space is called an F_σ set. The complement of an F_σ set (that is, the product of denumerably many open subsets) is a G_δ set.*

There are obvious examples of compact metric spaces for which the set of passing points is (1) closed (the $S_{n,r}$), (2) not closed (the segment $[0, 1]$), and (3) not open (let the reader give an example).

THEOREM 20.1. *The set P of all passing points of a compact metric space M is an F_σ set.*

Proof. For each integer k let P_k denote the set of all passing points q of M for which points p, r of M exist with $pq \geqslant 1/k$, $qr \geqslant 1/k$ and pqr holds. The sets P_k are closed ($k = 1, 2,...$); for if $q^* \in P'_k$, let $\{q_n\}$ have limit q^*, $q_n \in P_k$ ($n = 1, 2,...$). Then sequences $\{p_n\}$, $\{r_n\}$ of M exist with $p_n q_n r_n$ holding and $p_n q_n \geqslant 1/k$, $q_n r_n \geqslant 1/k$ ($n = 1, 2,...$). From the compactness of M, the sub-sequences $\{p_{i_j}\}$, $\{r_{i_j}\}$ of $\{p_n\}$, $\{r_n\}$, respectively, and points p^*, r^* of M exist with $p^* = \lim p_{i_j}$, $r^* = \lim r_{i_j}$. Now

$$p_{i_j} q_{i_j} + q_{i_j} r_{i_j} = p_{i_j} r_{i_j}, \qquad p_{i_j} q_{i_j} \geqslant 1/k, \qquad q_{i_j} r_{i_j} \geqslant 1/k \quad (j = 1, 2,...)$$

and since $q^* = \lim q_{i_j}$, continuity of the metric gives

$$p^*q^* + q^*r^* = p^*r^*, \qquad p^*q^* \geqslant 1/k, \qquad q^*r^* \geqslant 1/k,$$

and hence $q^* \in P_k$.

Clearly $\sum_{k=1}^{\infty} P_k \subset P$. But also $P \subset \sum_{k=1}^{\infty} P_k$, for if $q \in P$ then points p, r of M exist with $pq + qr = pr$, $p \neq q \neq r$, and $q \in P_k$ for each integer k such that $pq \geqslant 1/k$, $qr \geqslant 1/k$.

COROLLARY. *The set T of all terminal points of a compact metric space is a G_δ.*

A metric subset is *segmentally connected* provided each two of its distinct points are joined by a segment belonging to the subset.

THEOREM 20.2. *In a complete convex metric space M containing at least two points, the set P of passing points is segmentally connected, dense in M, and dense-in-itself (that is, $P \subset P'$).*

The proof is left to the reader.

We have already seen that in any metric space M the set $\bar{B}(a,b)$, consisting of the points a, b and all points between a and b, is closed for each pair of elements a, b of M. If $\vec{B}(a,b)$ denotes the subset of M consisting of b and all points c of M for which abc holds, then it is easy to show that $\vec{B}(a,b)$ is closed, a, $b \in M$. Defining the subset $\overleftarrow{B}(a,b)$ in a similar manner, then the set

$$\overleftrightarrow{B}(a,b) = \bar{B}(a,b) + \vec{B}(a,b) + \overleftarrow{B}(a,b),$$

consisting of a, b and all points c of M for which acb or abc or cab holds, is closed, since it is the sum of three closed sets. Clearly $\overleftrightarrow{B}(a,b)$ is uniquely determined by the points a, b.

THEOREM 20.3. *The set $\overleftrightarrow{B}(a,b)$ determined by two distinct points a, b of a complete convex metric space is arc-wise connected (that is, each two distinct points of $\overleftrightarrow{B}(a,b)$ are end-points of an arc belonging to that set).*

The proof is left to the reader.

21. Metric lines

A metric space is *externally convex* provided it contains for each two of its distinct points p, q at least one point r such that pqr subsists.

LEMMA 21.1. *If a, b are distinct points of a complete, externally convex metric space M, the distances ap, $p \in \vec{B}(a,b)$ are unbounded.*

Proof. Assuming the contrary, let $\rho_0 = \text{l.u.b.} \, ap$, $p \in \vec{B}(a,b)$, and let p_1 be a point of $\vec{B}(a,b)$, $p_1 \neq b$, with $ap_1 > \rho_0 - 1$.

Since $\vec{B}(a,b) \supset \vec{B}(a,p_1)$ the distances ap, $p \in \vec{B}(a,p_1)$, are bounded also, and we write $\rho_1 = \text{l.u.b.} \, ap$, $p \in \vec{B}(a,p_1)$. Clearly $\rho_0 \geqslant \rho_1 > ab$.

Let p_2 be a point of $\vec{B}(a,p_1)$, $p_2 \neq p_1$ with $ap_2 > \rho_1 - \frac{1}{2}$. By complete induction an infinite sequence $\{p_n\}$ of pairwise distinct points of $\vec{B}(a,b)$, and a *bounded, monotone decreasing* sequence $\{\rho_n\}$ of positive numbers are obtained such that (setting $b = p_0$),

(i) $\rho_{n-1} = \text{l.u.b.} \, ap$, $p \in \vec{B}(a,p_{n-1})$,

(ii) $p_n \in \vec{B}(a,p_{n-1})$, $\rho_n > ap_n > \rho_{n-1} - 1/n$,

(iii) $\rho_0 \geqslant \rho_1 \geqslant \rho_2 \geqslant \ldots \geqslant \rho_n \geqslant \ldots > ab$,

(iv) $\vec{B}(a,b) \supset \vec{B}(a,p_1) \supset \vec{B}(a,p_2) \supset \ldots \supset \vec{B}(a,p_n) \supset \ldots$,

with (i), (ii) valid for every $n = 1, 2, \ldots$.

From (iii) the sequence $\{\rho_n\}$ has the limit ρ ($\rho \geqslant ab$), and from (ii) $\lim ap_n = \rho$. By (ii) also, $ap_n p_{n+1}$ subsists ($n = 1, 2, \ldots$), and induction readily shows that $ap_n p_{n+k}$ holds for every $k = 1, 2, \ldots$. Hence

$$ap_n + p_n p_{n+k} = ap_{n+k} \quad (n, k = 1, 2, \ldots), \qquad (*)$$

and consequently $\lim_{n,k \to \infty} p_n p_{n+k} = 0$.

Thus $\{p_n\}$ is a Cauchy sequence which (since M is complete) has the limit p, $p \in M$. Clearly $ap = \lim ap_n = \rho$.

Keeping n fixed in (*) and letting k approach infinity, we get $ap_n + p_n p = ap$, and since it is readily seen that p_n is distinct from a, p

then $ap_n p$ holds, $n = 1, 2, \ldots$. Let p' be any point of M such that app' subsists and choose n so that

$$\rho_n < \rho + \tfrac{1}{2} pp'. \qquad (**)$$

Now $ap_n p$ and app' imply $ap_n p'$, so that $p' \in \vec{B}(a, p_n)$, and consequently $ap' \leqslant \rho_n$. Hence $ap + pp' \leqslant \rho_n$; that is, $\rho + pp' \leqslant \rho_n$, which contradicts (**) and establishes the lemma.

DEFINITION 21.1. *A subset L of a metric space is a metric line provided L is congruent with the euclidean straight line E_1.*

THEOREM 21.1. *Each two distinct points of a complete, convex, externally convex metric space M are on a metric line of the space.*

Proof. Let a, b be distinct points of M. By Lemmas 15.1, 21.1 a segment $S_0 = S_{a,b}$ (which exists by Theorem 14.1) can be prolonged to a point p_1 of M, with $ap_1 > 2ab$. Denoting the new segment by S_1, it may be prolonged through a to a point p_2 with $p_1 p_2 > 2ap_1$, and the resulting segment S_2 can, in turn, be prolonged through p_1 to a point p_3 with $p_2 p_3 > 2 p_1 p_2$.

By induction, the infinite sequence $\{S_n\}$ of segments is obtained, with S_n a prolongation of S_{n-1} through more than twice its length. The set $L = \sum_{n=0}^{\infty} S_n$ is easily seen to be mapped congruently *onto* the euclidean line E_1 by associating with each point p of L the point p' of E_1 with $ap = a'p'$, $bp = b'p'$, where a', b' are two (fixed) points of E_1 with $ab = a'b'$. Hence L is a metric line and the proof is complete.

THEOREM 21.2. *A necessary and sufficient condition that a complete, convex, externally convex metric space M (containing at least two points) be a metric line is that M be free of equilateral triples.*

Proof. The necessity of the condition follows at once from the congruence of a metric line with E_1.

On the other hand, if M is not a metric line then (Theorem 21.1) it contains a metric line L and a point not on L. But then M contains an equilateral triple (Theorem 16.1, Corollary).

Three points of a metric space form a *linear triple* provided they are congruent with three points of E_1. If three pairwise distinct points are linear, then one is between the other two.

DEFINITION 21.2. *A metric space has the two-triple property provided for each of its quadruples of pairwise distinct points linearity of any two triples implies linearity of the remaining two triples.*

THEOREM 21.3. *Each two distinct points of a complete, convex, externally convex metric space M determine a unique line through them if and only if M has the two-triple property.*

Proof. By Theorem 21.1, each two distinct points of M are on at least one metric line. Suppose each two distinct points are on exactly one metric line, and let a, b, c, d be four points of M with a, b, c and a, b, d linear triples. Then the line $L(a, b)$ contains both c and d, and hence the triples a, c, d and b, c, d are linear (since $L(a, b) \approx E_1$). Thus the condition is necessary.

To show it also sufficient, assume that M has the two-triple property. If $a, b \in M$, $a \neq b$, such that metric lines $L(a,b)$, $L'(a,b)$ exist,

$$L(a,b) \neq L'(a,b),$$

a point c of one of these lines, say $L(a,b)$, exists and $c \bar{\in} L'(a,b)$. Two cases may be distinguished.

Case 1. The point c belongs to the metric segment $S_{a,b}$ determined on $L(a,b)$ by a, b.

Then $a \neq c \neq b$, and a point c' of $S'_{a,b}$ (the segment on $L'(a,b)$ determined by a, b) exists such that $ac' = ac$, $bc' = bc$. Now $c' \neq c$ and the points a, b, c, c' are pairwise distinct.

Since acb and $ac'b$ hold, we conclude from the two-triple property that a, c, c' and b, c, c' are each linear triples. But from $ac' = ac$ and $bc' = bc$, no one of the relations acc', $ac'c$, bcc', $bc'c$ can hold. It follows that cac' and cbc' must subsist.

Let d' be a point of $S'_{a,b}$ such that $ad'c'$ holds. This relation, together with $ac'b$, implies $d'c'b$ and $ad'b$ and the quadruple of distinct points b, c, c', d' has triples b, c, c' and b, c', d' linear (cbc' and $bc'd'$ hold). Hence triples c, c', d' and b, c, d' are linear.

Neither $cc'd'$ nor $c'cd'$ can hold, for the results of combining each with cac' contradict $ad'c'$. Hence $c'd'c$ subsists. Now $bd'c$ with $bc'd'$ gives $bc'c$, contradicting cbc'; bcd' and cad' (which holds as a consequence of cac' and $ad'c'$) yield bad', contrary to $ad'b$; while cbd' and $bc'd'$ imply $cc'd'$, contradicting $c'd'c$.

Hence Case 1 is impossible.

Case 2. The point c does not belong to the metric segment $S_{a,b}$.

We may assume the labelling so that abc holds, and denote by c' that point of $L'(a,b)$ such that abc' subsists and $ac = ac'$ (consequently $bc = bc'$). Then a, b, c, c' are pairwise distinct.

By the two-triple property, a, c, c' and b, c, c' are linear, and consequently cac' and cbc' must hold. Adding the two equalities
$$ab+bc = ac, \qquad ab+bc' = ac',$$
and taking into account cac' and cbc' gives $ab = 0$, which contradicts $a \neq b$, and the theorem is established.

THEOREM 21.4. *A necessary and sufficient condition that for each two distinct points a, b of a complete, convex, externally convex metric space $\overleftrightarrow{B}(a,b)$ be identical with $L(a,b)$ is that the space have the two-triple property.*

Proof. If for each $a, b \in M$, $a \neq b$, $\overleftrightarrow{B}(a,b)$ is identical with $L(a,b)$ then there is a unique line through each two distinct points of the space, which consequently has the two-triple property (Theorem 21.3).

On the other hand, if the space has the two-triple property then $L(a,b)$ is unique and is identical with $\overleftrightarrow{B}(a,b)$. For clearly $\overleftrightarrow{B}(a,b) \supset L(a,b)$, and if $p \in \overleftrightarrow{B}(a,b)$, $p \bar{\in} L(a,b)$ then $a \neq p \neq b$ and apb, abp, or pab exists. In any case there is a line through a, b, and p which is different from $L(a,b)$, contradicting the uniqueness of lines in the space.

REFERENCES

§ 11. Investigations of de Tilly, Mansion, Blichfeldt, and de Donder are discussed in Chapter I of Blumenthal [1].
§ 13. Wald [1].
§ 14. Menger [1], Aronszajn [1]; another proof of the existence of segments in complete convex metric spaces is sketched in Menger [2] and Milgram.
§ 16. Blumenthal [2] and Robinson.
§ 17. Blumenthal [3].

SUPPLEMENTARY PAPERS

Blanc [1]—a discussion of various convexity notions and relations between them. Huntington [1] and Kline, Huntington [2]—axiomatizations of betweenness on a line.
Bing [1]—it is shown that in each compact, locally connected continuum a topology-preserving convex metric may be defined.

CHAPTER III

CURVE THEORY

22. Arcs and arc length

A SUBSET A of a metric space is an *arc* provided it is a homeomorph of a line segment; that is, $A = f(I)$, where f is biuniform and bicontinuous in the line segment $I = [a,b]$, $a < b$. Since I is closed and compact, so is the arc A.

The points $\alpha = f(a)$, $\beta = f(b)$ are the end-points of A. Calling α the initial point and β the terminal point of A serves to orient the arc, and a finite subset $P = (\alpha_1, \alpha_2, ..., \alpha_n)$ of A is *normally ordered* provided these points are encountered in the order of their subscripts when the arc is traversed from α to β. Any homeomorphism of I with A for which α and β are first and last points, respectively, will leave the order of $\alpha_1, \alpha_2, ..., \alpha_n$ unchanged.

DEFINITION 22.1. *The length $l(A)$ of an arc A of a metric space M is the least upper bound of the numbers $l(P) = \sum_{i=1}^{n-1} \alpha_i \alpha_{i+1}$ for all (finite) normally ordered subsets $P = (\alpha_1, \alpha_2, ..., \alpha_n)$ of A. If $l(A)$ is finite, A is called rectifiable.*

It follows that arc length is an intrinsic property of the pointset which is the graph of the arc and is independent of the particular homeomorphism that maps I onto this pointset.

If, now, $P = (\alpha_1, \alpha_2, ..., \alpha_n)$, $P \subset A$, there corresponds to each permutation $i_1, i_2, ..., i_n$ of the indices $1, 2, ..., n$ a number $\sum_{j=1}^{n-1} \alpha_{i_j} \alpha_{i_{j+1}}$. Letting $\lambda(P)$ denote the smallest of the numbers so obtained, any permutation of the indices yielding this number selects a *shortest way* through the set P, and may be called a *geodesic ordering* of P. Clearly $\lambda(P) \leqslant l(P)$ for every P, and it is easy to construct an example for which $\lambda(P) < l(P)$.

It was shown by Menger, however, that if A is rectifiable then

$$\underset{P \subset A}{\text{l.u.b.}}\ \lambda(P) = \underset{P \subset A}{\text{l.u.b.}}\ l(P),$$

from which an interesting 'sharpening' of the definition of arc length is obtained. Before establishing this result we shall prove three lemmas of intrinsic value.

EXERCISES

1. Show that a metric segment is characterized among all arcs joining its endpoints by having the smallest length, namely, the distance of its end-points. (*Hint.* If an arc is not a segment it contains two points for which it has no middle-point.)
2. Give an example for which $\lambda(P) < l(P)$.

23. Homogeneous ϵ-chains and δ-density

A normally ordered subset $P = (\alpha_1, \alpha_2, ..., \alpha_n)$ of an arc A with end-points α, β is a *homogeneous ϵ-chain* provided

1. $\alpha\alpha_1 < \epsilon$, $\alpha_n\beta < \epsilon$,
2. $\alpha_i\alpha_j = \epsilon$, for $|i-j| = 1$ $(i,j = 1, 2, ..., n)$,
3. $\alpha_i\alpha_j \geqslant \epsilon$, for $|i-j| > 1$ $(i,j = 1, 2, ..., n)$.

LEMMA 23.1. *For each $\epsilon > 0$ there exists a homogeneous ϵ-chain in A.*

Proof. If we put $\alpha = \alpha_1$ then $P = (\alpha_1)$ is the desired ϵ-chain in case $\alpha_1\beta < \epsilon$, while $P = (\alpha_1, \alpha_2)$, where $\alpha_2 = \beta$, serves this purpose if

$$\alpha_1\beta = \epsilon.$$

In the event $\alpha_1\beta > \epsilon$ it follows from the compactness of A and the continuity of the metric that in traversing A from α_1 to β a *last point* α_2 is encountered such that $\alpha_1\alpha_2 = \epsilon$. If $\alpha_2\beta < \epsilon$ then $P = (\alpha_1, \alpha_2)$ is the desired chain; if $\alpha_2\beta = \epsilon$ then $P = (\alpha_1, \alpha_2, \alpha_3)$, $\alpha_3 = \beta$, serves that purpose.

Suppose $\alpha_2\beta > \epsilon$, and denote by α_3 the last point encountered, traversing A from α_2 to β, such that $\alpha_2\alpha_3 = \epsilon$. The set $P = (\alpha_1, \alpha_2, \alpha_3)$ has the properties: (1) α_{j+1} is the last point of A, from α_j to β, such that

$$\alpha_j\alpha_{j+1} = \epsilon \quad (j = 1, 2),$$

(2) $\alpha_1\beta > \epsilon$, $\alpha_2\beta > \epsilon$, and (3) each pair of the points α_1, α_2, α_3 has a distance at least ϵ. These properties are all obvious except, perhaps, that $\alpha_1\alpha_3 \geqslant \epsilon$. But if $\alpha_1\alpha_3 < \epsilon$ then, since $\alpha_1\beta > \epsilon$, continuity of the metric yields the existence of a point of the sub-arc of A with end-points α_3, β with distance from α_1 equal to ϵ. This contradicts the definition of α_2.

Make, now, the inductive assumption that the above procedure has yielded a set $P_k = (\alpha_1, \alpha_2, ..., \alpha_k)$ with properties (1), (2), (3), where now the index j of property (1) assumes the values $1, 2, ..., k-1$. If $\alpha_k\beta \leqslant \epsilon$, then ϵ-chains are obtained as before. If $\alpha_k\beta > \epsilon$ then

$$P_{k+1} = (\alpha_1, \alpha_2, ..., \alpha_k, \alpha_{k+1}),$$

where α_{k+1} is the last point of A (from α_k to β) with $\alpha_k\alpha_{k+1} = \epsilon$, is easily seen to have the properties:

1. $\alpha_i\alpha_j = \epsilon$, for $|i-j| = 1$ $(i,j = 1,2,...,k+1)$,
2. $\alpha_i\alpha_j \geqslant \epsilon$, for $|i-j| > 1$ $(i,j = 1,2,...,k+1)$.

Thus starting from a set P_k of k points with each distance at least ϵ, a set of $k+1$ points P_{k+1} can be constructed with the same property whenever $\alpha_k\beta > \epsilon$. It follows that an integer n exists such that the set P_n has $\alpha_n\beta \leqslant \epsilon$, for in the contrary case A contains an infinite sequence of points with each distance at least ϵ, contradicting the compactness of A.

If $P = (\alpha_1, \alpha_2,...,\alpha_n)$ is such that $\alpha\alpha_1 < \epsilon$, $\alpha_n\beta < \epsilon$, $\alpha_i\alpha_{i+1} \leqslant \epsilon$ $(i = 1,2,...,n-1)$, P is called an ϵ-chain.

A subset B of A is δ-*dense* in A provided each point of A has distance less than δ from B.

LEMMA 23.2. *For every $\delta > 0$ there exists an $\epsilon > 0$ such that every ϵ-chain in A is δ-dense in A.*

Proof. The homeomorphism f relating the arc A and segment I is uniformly continuous in I and its inverse f^{-1} is uniformly continuous in A. Hence corresponding to the given positive δ there is an $\eta > 0$ such that if s, $t \in I$, $st < \eta$ implies $f(s)f(t) < \delta$. Similarly, this positive η implies the existence of an $\epsilon > 0$ such that if σ, $\tau \in A$ with $\sigma\tau \leqslant \epsilon$, then $f^{-1}(\sigma)f^{-1}(\tau) < \eta$.

Thus, if P is an ϵ-chain in A then $f^{-1}(P)$ is an η-chain in I and hence is η-dense in I. To show that P is δ-dense in A let σ be any point of A and s its corresponding point in I. Then a point t of $f^{-1}(P)$ exists with $st < \eta$, and so $f(s)f(t) < \delta$; that is, a point $f(t)$ of P exists with $\sigma f(t) < \delta$, and hence P is δ-dense in A.

24. First sharpening of length concept

An additional lemma is needed before proceeding to the proof of the 'sharpening' theorem.

LEMMA 24.1. *Let $A = f(I)$ be a rectifiable arc of a metric space with length $l(A)$. Then $\eta > 0$ implies the existence of $\delta > 0$ such that for each normally ordered finite subset P of A which is δ-dense in A, $l(P) > l(A) - \eta$.*

Proof. We show first that $\eta > 0$ implies the existence of $\epsilon > 0$ such that if P is any normally ordered subset of A with $f^{-1}(P)$ ϵ-dense in I, then $l(P) > l(A) - \eta$.

By definition of $l(A)$, A contains a normally ordered subset $P^* = (\alpha_1, \alpha_2, ..., \alpha_n)$ with $l(P^*) > l(A) - \frac{1}{2}\eta$. Let $f^{-1}(P^*) = (a_1, a_2, ..., a_n)$. Since f is uniformly continuous in I, there corresponds to the number $\eta/4(n-1)$ a positive η' such that $s, t \in I$, $st < \eta'$ imply

$$f(s)f(t) < \eta/4(n-1).$$

If, now, $0 < \epsilon < \eta'$, let $E = (b_1, b_2, ..., b_m)$ be any subset of I that is ϵ-dense in I ($b_1 < b_2 < ... < b_m$) and put $P = (\beta_1, \beta_2, ..., \beta_m)$, where $\beta_i = f(b_i)$ ($i = 1, 2, ..., m$). For each a_j of $f^{-1}(P^*)$ there is a point b_{i_j} of E nearest a_j with $a_j b_{i_j} < \epsilon$ and hence $\alpha_j \beta_{i_j} < \eta/4(n-1)$. From the way in which b_{i_j} is selected it follows that $a_j < a_{j+1}$ implies $b_{i_j} \leqslant b_{i_{j+1}}$, and hence the subset $(\beta_{i_1}, \beta_{i_2}, ..., \beta_{i_n})$ of A is normally ordered.

Denote by B_j ($j = 1, 2, ..., n-1$), the points $b_{i_j}, b_{i_j+1}, ..., b_{i_{j+1}}$ of I. Then

$$l(P) = \sum_{i=1}^{m-1} \beta_i \beta_{i+1} \geqslant \sum_{j=1}^{n-1} l(f(B_j)),$$

the inequality being a consequence of the normal ordering of

$$(\beta_{i_1}, \beta_{i_2}, ..., \beta_{i_n}).$$

For each index j, $l(f(B_j)) \geqslant \beta_{i_j} \beta_{i_{j+1}}$, and by the triangle inequality

$$\alpha_j \alpha_{j+1} \leqslant \beta_{i_j} \alpha_{j+1} + \alpha_j \beta_{i_j}$$
$$\leqslant \beta_{i_j} \beta_{i_{j+1}} + \alpha_{j+1} \beta_{i_{j+1}} + \alpha_j \beta_{i_j}$$
$$\leqslant \beta_{i_j} \beta_{i_{j+1}} + \eta/2(n-1),$$

or

$$\beta_{i_j} \beta_{i_{j+1}} \geqslant \alpha_j \alpha_{j+1} - \eta/2(n-1).$$

Hence

$$l(P) \geqslant \sum_{j=1}^{n-1} \beta_{i_j} \beta_{i_{j+1}} \geqslant l(P^*) - \tfrac{1}{2}\eta,$$

and since $l(P^*) > l(A) - \frac{1}{2}\eta$, it follows that $l(P) > l(A) - \eta$, which was to be shown.

To complete the proof of the lemma, it suffices to observe that from the uniform continuity of f^{-1} on A, to each $\epsilon > 0$ there is a $\delta > 0$ such that each subset P of A which is δ-dense in A has $f^{-1}(P)$ ϵ-dense in I.

THEOREM 24.1. *Let A be a rectifiable arc of a metric space, with length $l(A)$. Then*

$$l(A) = \underset{P \subset A}{\text{l.u.b.}}\; \lambda(P).$$

Proof. Since $\lambda(P) \leqslant l(P)$, then clearly

$$\underset{P \subset A}{\text{l.u.b.}}\; \lambda(P) \leqslant l(A),$$

and the proof is complete when it is shown that corresponding to each arbitrary, positive η a normally ordered subset P of A exists with $\lambda(P) > l(A) - \eta$.

Now by the preceding lemma, $\eta > 0$ implies the existence of $\delta > 0$ such that for each P which is δ-dense in A, $l(P) > l(A) - \eta$. By Lemma 23.2, there corresponds to this δ a positive ϵ such that every ϵ-chain in A is δ-dense in A, and by Lemma 23.1 there exists for this ϵ a homogeneous ϵ-chain $P = (\alpha_1, \alpha_2, ..., \alpha_n)$ in A. Then P is δ-dense in A and $l(P) > l(A) - \eta$.

But $l(P) = \lambda(P)$, for since $\alpha_i \alpha_j = \epsilon$, $|i-j| = 1$ and

$$\alpha_i \alpha_j \geq \epsilon, \quad |i-j| > 1,$$

it follows that for any permutation $i_1, i_2, ..., i_n$ of the indices $1, 2, ..., n$,

$$\sum_{j=1}^{n-1} \alpha_{i_j} \alpha_{i_{j+1}} \geq (n-1)\epsilon = \sum_{i=1}^{n-1} \alpha_i \alpha_{i+1}.$$

Hence $\lambda(P) > l(A) - \eta$, and the theorem is proved.

EXERCISE

1. If A is a non-rectifiable arc of a metric space show that
$$\underset{P \subset A}{\text{l.u.b.}} \, l(P) = \underset{P \subset A}{\text{l.u.b.}} \, \lambda(P).$$

25. Second sharpening of length concept. Lower semi-continuity of arc length

Let E be any finite subset of a metric space and ρ any symmetric, non-reflexive, binary relation defined in E, with respect to which each two elements of E are comparable. The subset E is *connected with respect to* ρ provided, if $p, q \in E$, then E contains a subset $p_1, p_2, ..., p_n$ with $p = p_1$, $q = p_n$ and p_i, p_{i+1} are in the relation ρ for every $i = 1, 2, ..., n-1$. The *linear-content* $\mu_\rho(E)$ of E is the sum of the distances of those point-pairs of E that are in the relation ρ. (We put $\mu(E) = 0$ in case no two points of E are in the relation ρ.)

The foregoing may be thought of as an abstraction of the notion of a *linear graph* (the sum of a finite number of line segments whose endpoints form a subset E of a euclidean space). Here two points of E are in the relation ρ provided they are end-points of a line segment of the graph, while $\mu_\rho(E)$ is simply the sum of the lengths of these segments.

For A an arc of a metric space and $P = (\alpha_1, \alpha_2, ..., \alpha_n)$ a normally ordered subset of A, a relation ρ is defined in P by saying that α_i, α_j are in ρ relation if and only if $|i-j| = 1$. Clearly P is connected with respect to this ρ and $\mu_\rho(P) = l(P)$. A relation ρ^* may be defined in P by stating that α_i, α_j are in that relation provided they are consecutive in a given geodesic ordering of P. The set P is connected with respect to this relation ρ^* also and $\mu_{\rho^*}(P) = \lambda(P)$.

Let a linear graph be formed by S_{p_1,p_2}, S_{p_1,p_3}, S_{p_1,p_4}, three unit segments of the euclidean plane which make pairwise, at the common end-point p_1, angles of 120°. Considering the set P of the four points p_1, p_2, p_3, p_4 it is found that $l(P) = 1+2\sqrt{3}$, $\lambda(P) = 2+\sqrt{3}$, $\mu_\rho(P) = 3$ (the relation ρ being that induced by the given linear graph). Here

$$\mu_\rho(P) < \lambda(P) < l(P).$$

Denoting by $\mu(P)$ the smallest of the numbers $\mu_\rho(P)$ for all relations ρ with respect to which the finite, normally ordered, subset P of a metric arc A is connected, the above remarks show that there exist ρ relations for which $\mu_\rho(P) = l(P)$ and $\mu_\rho(P) = \lambda(P)$, while $\mu(P) \leqslant \lambda(P)$, with the inequality sign valid in certain cases. It would, therefore, be a further sharpening of the length concept for arcs if it were shown that

$$\underset{P \subset A}{\text{l.u.b.}}\ \mu(P) = l(A).$$

We shall establish this after proving the following lemma.

LEMMA 25.1. *If E is any metric set of n points ($n > 1$) which is connected with respect to a relation ρ then E contains at least $n-1$ pairs of points that are in the relation ρ.*

Proof. The lemma being clearly valid for $n = 2$, suppose it true for all sets of power less than or equal to $n-1$ and let p be any element of E. The set $E-(p)$ is the sum of a finite number of *components* $E_1, E_2,..., E_k$ each of which is either connected with respect to relation ρ but ceases to be so after the adjunction of any point of $E-(p)$ not belonging to it, or is a single point. It follows that $E_i . E_j = 0$ ($i,j = 1,2,...,k; i \neq j$).

If E_i consists of n_i points then evidently $n_i \leqslant n-1$ ($i = 1,2,...,k$). By the inductive hypothesis each set E_i contains at least n_i-1 pairs of points in ρ relation, and hence the set $E-(p)$ contains at least

$$\sum_{i=1}^{k}(n_i-1) = \sum_{i=1}^{k} n_i - k = n-1-k$$

pairs of points in ρ relation. Since E is ρ connected, each of the components $E_1, E_2,..., E_k$ of $E-(p)$ contains at least one element which is in the relation ρ with p. Thus besides the (at least) $n-1-k$ pairs of ρ related points of $E-(p)$, the set E contains k additional pairs, and so E has at least $n-1$ pairs of ρ related points.

THEOREM 25.1. *For each rectifiable arc A of a metric space*

$$\underset{P \subset A}{\text{l.u.b.}}\ \mu(P) = l(A).$$

Proof. Since $\mu(P) \leqslant l(P)$, $P \subset A$, then $\underset{P \subset A}{\text{l.u.b.}}\ \mu(P) \leqslant l(A)$.

To show that l.u.b.$_{P \subset A}$ $\mu(P) \geqslant l(A)$, the procedure used in Theorem 24.1 shows that corresponding to each $\eta > 0$ a homogeneous ϵ-chain P exists in A with $l(P) > l(A)-\eta$. But for each homogeneous ϵ-chain P, $l(P) = \mu(P)$, for (by the preceding lemma) if P consists of n points, then at least $n-1$ pairs of its points are ρ related for any relation ρ with respect to which P is connected. Hence for each homogeneous ϵ-chain P, $\mu(P) \geqslant (n-1)\epsilon$ since each two points of such a chain have a distance at least ϵ.

The normal ordering of P yields, as seen above, a ρ relation with respect to which P is connected, and $\mu_\rho(P) = l(P) = (n-1)\epsilon$. It follows that $\mu(P) = l(P)$ for each homogeneous ϵ-chain P in A, and hence corresponding to each $\eta > 0$ a subset P of A exists with $\mu(P) > l(A)-\eta$; that is, l.u.b.$_{P \subset A}$ $\mu(P) \geqslant l(A)$, and the theorem is established.

If A, A^* are two rectifiable arcs of a metric space with Hausdorff distance AA^* (Example 7, § 9) easy examples show that the lengths of A and A^* may differ greatly even though the distance AA^* is very small. On the other hand, if A is a fixed rectifiable arc of a metric space then corresponding to each $\epsilon > 0$ there is a $\delta > 0$ such that if A^* is any arc of the space with $AA^* < \delta$, then $l(A^*) > l(A)-\epsilon$. This important property is expressed by saying that arc length in a metric space is a *lower semi-continuous* function of arcs. It is proved in the following theorem.

THEOREM 25.2. *Arc length in a metric space is a lower semi-continuous function of arcs.*

Proof. Let A be a rectifiable arc of a metric space M. By the preceding theorem $l(A) = $ l.u.b.$_{P \subset A}$ $\mu(P)$, and hence for each $\epsilon > 0$ a normally ordered subset $P = (\alpha_1, \alpha_2, ..., \alpha_n)$ of A exists such that

$$\mu(P) > l(A)-\tfrac{1}{2}\epsilon.$$

We shall show that if A^* is any arc of M with $AA^* < \delta = \epsilon/2n(n-1)$, then $l(A^*) > l(A)-\epsilon$.

Since $AA^* < \delta$, then to each of the n points of $P \subset A$ there is at least one point of A^* whose distance from it is less than δ. Denote by P^* the n (not necessarily pairwise distinct) points of A^* so obtained. Consider $\mu(P^*)$ and let ρ^* denote any ρ relation in P^* with respect to which $\mu(P^*)$ is the linear-content of P^*. Then P^* is connected with respect to ρ^*.

Define, now, a relation ρ in the set P by agreeing that α_i, α_j ($i \neq j$)

are in the relation ρ if and only if their corresponding points in P^* are either coincident or in the relation ρ^*. Then P is connected with respect to this ρ relation. Denote by $\mu_\rho(P)$ the linear content of P with respect to this ρ.

Since $\alpha_i \alpha_i^* < \delta$, $\alpha_j \alpha_j^* < \delta$, then
$$|\alpha_i^* \alpha_j^* - \alpha_i \alpha_j^* + \alpha_i \alpha_j^* - \alpha_i \alpha_j| \leqslant |\alpha_i^* \alpha_j^* - \alpha_i \alpha_j^*| + |\alpha_i \alpha_j^* - \alpha_i \alpha_j|,$$
or
$$|\alpha_i^* \alpha_j^* - \alpha_i \alpha_j| \leqslant \alpha_i \alpha_i^* + \alpha_j \alpha_j^* < 2\delta.$$

Now P contains n points and hence at most $\tfrac{1}{2}n(n-1)$ pairs in the relation ρ. Since the distance of each such pair differs from the distance of its corresponding pair in P^* by less than $2\delta = \epsilon/n(n-1)$, $\mu_\rho(P)$ differs from $\mu(P^*)$ by less than
$$\tfrac{1}{2}n(n-1) \cdot \epsilon/n(n-1) = \tfrac{1}{2}\epsilon.$$
Hence $|\mu_\rho(P) - \mu(P^*)| < \tfrac{1}{2}\epsilon$; that is,
$$\mu(P^*) > \mu_\rho(P) - \tfrac{1}{2}\epsilon.$$
Since P is connected with respect to ρ, then $\mu_\rho(P) \geqslant \mu(P)$, and so $\mu(P^*) > \mu(P) - \tfrac{1}{2}\epsilon$. This, combined with $\mu(P) > l(A) - \tfrac{1}{2}\epsilon$, gives
$$\mu(P^*) > l(A) - \epsilon.$$
But $\mu(P^*) \leqslant l(A^*)$ and thus we obtain
$$l(A^*) > l(A) - \epsilon,$$
and the theorem is proved.†

If E consists of the four vertices of a unit square of the euclidean plane, it is easily seen that $\mu(E) = 3$. Adjoining to E the centre p of the square gives a set $E^* = E + (p)$ with $\mu(E^*) = 2\sqrt{2} < \mu(E)$. Hence the linear content μ of a set E may be *decreased* by adjoining points to E.

Let, now, E be any finite subset of a metric space M and define
$$\nu(E) = \underset{E \subset E^* \subset M}{\text{g.l.b.}} \mu(E^*).$$
Then $\nu(E) \leqslant \mu(E)$, with the inequality sign valid in certain cases. The question arises whether for each arc A of a metric space,
$$\underset{P \subset A}{\text{l.u.b.}} \nu(P) = l(A).$$
If this question, which is an open one, were answered in the affirmative then a still further sharpening of the notion of arc length is possible.

† It is clear that the concept of lower semi-continuity of a function is relative to the definition of distance used for the elements constituting the domain of the function. In our definition of the notion, the domain (of arcs) is metrized by the Hausdorff metric and Theorem 25.2 is proved on that basis. Its validity for Fréchet distance (customarily used in this connexion) follows at once, however, from the fact that the Hausdorff distance of two arcs is less than or equal to the Fréchet distance.

For arcs A of a *euclidean space* an affirmative answer to the above question was established by Mimura (*Ergebnisse eines mathematischen Kolloquiums*, Wien, Heft 4 (1932), 20–22).

EXERCISE

Give an example to show that arc length is not an upper semi-continuous function (and hence not a continuous function) of arcs; that is, $\lim A_n A = 0$ does not imply that for all sufficiently large indices n, $l(A_n) < l(A) + \epsilon$, for each positive ϵ.

26. A relaxing of the length concept

In the two preceding sections it was shown that the set Σ of numbers (arising from the extension to arcs of a metric space of the classical definition of arc length) whose least upper bound is the arc length can be replaced by sets of numbers which are, in general, smaller than the elements of Σ but which, it turns out, have the same least upper bounds as Σ. This section deals with a departure from the classical procedure of a different nature. Here the set Σ is replaced by a set of numbers which are, in general, *larger* than those of Σ, and it is proved that the least upper bound of the new set equals that of Σ.

Let M be a metric space, each two points of which are joined by at least one arc of finite length. If $p, q \in M$ ($p \neq q$), denote by $\gamma(p, q)$ the greatest lower bound of the lengths of all arcs of M joining p and q.

The number $\gamma(p, q)$ thus attached to each pair of distinct points of M is called the *geodesic* distance of p, q. The question whether there exists an arc of M joining p, q with length equal to $\gamma(p, q)$ will be discussed later. The existence of such an arc is, of course, not implied by the definition of $\gamma(p, q)$.

It is clear that if p, q are distinct points of M then $\gamma(p, q) \geqslant pq$, for if the contrary be assumed an arc of M, with end-points p, q, exists whose length is less than pq, which (using the triangle inequality) is seen from the definition of arc length to be impossible.

Now let A be an arc of a metric space and $P = \{\alpha_1, \alpha_2, ..., \alpha_n\}$ a normally ordered subset of A. Writing

$$\gamma(P) = \sum_{i=1}^{n-1} \gamma(\alpha_i, \alpha_{i+1}),$$

then $\gamma(P) \geqslant l(P)$, with the inequality sign valid in general. Let

$$\gamma(A) = \underset{P \subset A}{\text{l.u.b.}}\, \gamma(P).$$

THEOREM 26.1. *For each arc A of a metric space, $l(A) = \gamma(A)$.*

Proof. Since for each normally ordered subset P of A, $\gamma(P) \geq l(P)$, then $\gamma(A) \geq l(A)$.

On the other hand, if $P = \{\alpha_1, \alpha_2, ..., \alpha_n\}$ is any normally ordered subset of A, then clearly $l(A)]_{\alpha_i}^{\alpha_{i+1}} \geq \gamma(\alpha_i, \alpha_{i+1})$, where the symbol in the left-hand member of the inequality denotes the length of that sub-arc of A with end-points α_i, α_{i+1}. Hence $l(A) \geq \sum_{i=1}^{n-1} \gamma(\alpha_i, \alpha_{i+1})$ for each P, and therefore
$$l(A) \geq \underset{P \subset A}{\text{l.u.b.}}\, \gamma(P) = \gamma(A),$$
completing the proof of the theorem.

EXERCISES†

Spread of a transformation. Associated with a homeomorphism f on the interval $I = [a, b]$ to the metric arc A is a *real* function
$$f^*(t) = \lim_{x,y \to t} f(x)f(y)/xy,$$
$x, y, t \in I$, defined for all points t of I at which the limit exists (Wilson [4]). This real function is called the *spread of f at t*.
1. What is the spread of a real function defined on I?
2. Prove that if f^* exists at each point of I the distance quotient $f(x)f(y)/xy$ is bounded on I, and the arc $A = f[I]$ is rectifiable.
3. Prove that if f^* exists at each point of a subset E of I then f^* is continuous in E.
4. Prove that if the set E of Exercise 3 is closed, the distance quotient converges uniformly to $f^*(t)$ for $t \in E$.

Length of arc in any metric space as a Riemann integral
5. Show that if $f^*(t)$ exists at each point of $I = [a, b]$ then the length $l(A)$ of the arc $A = f[I]$ is given by $l(A) = \int_a^b f^*(t)\, dt$.
6. If $f^*(t)$ exists at each point t of I, then at each point at which $f^*(t) \neq 0$,
$$\lim_{x,y \to t} f(x)f(y)/l(f(x), f(y)) = 1,$$
where $l(f(x), f(y))$ denotes the length of that part of A with end-points $f(x), f(y)$.

27. Continuous curves

Some of the theorems proved in the preceding sections for metric arcs are valid also for *continuous curves*, that is, for subsets of a metric space which are continuous images of a line segment I.

If f denotes a continuous mapping of I onto a subset C of a metric space M, then f is uniformly continuous over I; that is, for each $\epsilon > 0$ there is a $\delta > 0$ such that if $r, r' \in I$ and $rr' < \delta$, then $f(r)f(r') < \epsilon$. We write $C = f(I)$.

† These exercises are results obtained in the Missouri Doctoral Dissertation of J. W. Gaddum.

Let $E = \{r_1, r_2, ..., r_n\}$ be a normally ordered subset of I and put

$$l[f(E)] = \sum_{i=1}^{n-1} f(r_i)f(r_{i+1}),$$

$$l_f(C) = \underset{E \subset I}{\text{l.u.b.}}\, l[f(E)].$$

The number $l_f(C)$, finite or infinite, is the path-length of C corresponding to the function f. It obviously depends not only on the point-set C, but on the particular way of 'running through' the points of C prescribed by the function f.

THEOREM 27.1. *If $C = f(I)$, f continuous over I, then for each $\eta > 0$ there is a $\delta > 0$ such that if E is any normally ordered subset of I that is δ-dense in I, then*

$$l[f(E)] > l_f(C) - \eta.$$

Proof. The proof is the same as that given for the first part of Lemma 24.1.

We leave it to the reader to show that for each $\epsilon > 0$, a homogeneous ϵ-chain exists in $C = f(I)$. The argument is quite similar to the one used to establish Lemma 23.1. On the other hand, Lemma 23.2 is not valid for continuous curves, nor can Theorem 24.1 be extended to continuous images of segments.

The following theorem, comparing path-lengths of a subset of a metric space with the length of an arc of the subset is of interest.

THEOREM 27.2. *Let S be any subset of a metric space which is a continuous image of a line segment I, and let A be an arc contained in S. Then for each function f, continuous over I, with respect to which S is a continuous curve $C = f(I)$, $l_f(C) \geqslant l(A)$.*

Proof. If a function f exists such that $l_f(C) < l(A)$ then a finite subset P of A exists with $\lambda(P) > l_f(C)$. Let E denote a (normally ordered) deleted counter-image of P; that is, E is a subset of I such that corresponding to each point p_i of P, E contains *exactly* one point whose image by f is p_i.

Then $l[f(E)] \geqslant \lambda(P) > l_f(C)$, which contradicts the definition of $l_f(C)$ and establishes the theorem.

Thus if A is an arc of any continuous image S of a segment, then however one passes continuously through the points of S a path-length is obtained which is at least equal to the length of A.

EXERCISES

1. Show that for each $\epsilon > 0$ a homogeneous ϵ-chain exists in $C = f(I)$.
2. Show by an example that Theorem 24.1 is not valid for every continuous curve.

28. Existence of geodesic arcs

A rectifiable arc A is called a *geodesic* provided its length equals the greatest lower bound of the lengths of all arcs joining its end-points; that is, A is a geodesic if and only if $l(A) = \gamma(p,q)$, where p, q are the end-points of A. Recalling that a finitely compact metric space is one in which bounded subsets are compact, the following existence theorem for geodesic arcs is proved.

THEOREM 28.1. *In a finitely compact metric space M, each two points that can be joined by a rectifiable arc can be joined by a geodesic arc.*

Proof. Let p, q be distinct points of M joinable by a rectifiable arc. If only a finite number of such arcs exist, then clearly one of them has length $\gamma = \gamma(p,q)$, and the theorem is proved.

In the contrary case, an infinite sequence $\{A_n^*\}$ of rectifiable arcs joining p, q exists with

$$\gamma \leqslant l(A_n^*) \leqslant \gamma + 1/n \quad (n = 1, 2, \ldots).$$

We may suppose that for each index n, A_n^* is a homeomorph of the interval $I = [0, 1]$.

Let $D = (d_1, d_2, \ldots, d_n, \ldots)$ be any denumerable subset of I such that $I = \bar{D}$. Our first objective is to define a certain uniformly continuous mapping of D onto a subset of M.

For each positive integer n, a point $f_n^*(d_1)$ of A_n^* exists *dividing that arc in the ratio d_1*; that is, $l(p, f_n^*(d_1)) = d_1 \cdot l(A_n^*)$, where $l(p, f_n^*(d_1))$ denotes the length of the sub-arc of A_n^* joining p and $f_n^*(d_1)$. (If $d_1 = 0$, $f_n^*(d_1) = p$.) The point-set of the sequence $\{f_n^*(d_1)\}$ is easily seen to be bounded, and hence a point $f(d_1)$ of M exists with $f(d_1) = \lim f_{n;d_1}^*(d_1)$, where $\{f_{n;d_1}^*(d_1)\}$ is a sub-sequence of $\{f_n^*(d_1)\}$. Denote the corresponding sub-sequence of $\{A_n^*\}$ by $\{A_{n;d_1}^*\}$.

Make now the inductive hypothesis that for the m-tuple $d_1, d_2, \ldots, d_m$ the corresponding m-tuple $f(d_1), f(d_2), \ldots, f(d_m)$ of M has been obtained by selecting a sub-sequence $\{A_{n;d_1,d_2,\ldots,d_m}^*\}$ of $\{A_{n;d_1,d_2,\ldots,d_{m-1}}^*\}$ such that
(1) $f_{n;d_1,d_2,\ldots,d_m}^*(d_i) \in A_{n;d_1,d_2,\ldots,d_m}^*$ with

$$l(p, f_{n;d_1,d_2,\ldots,d_m}^*(d_i)) = d_i \cdot l(A_{n;d_1,d_2,\ldots,d_m}^*)$$

and (2) $\lim f_{n;d_1,d_2,\ldots,d_m}^*(d_i) = f(d_i)$ $(i = 1, 2, \ldots, m)$. Each of the arcs $A_{n;d_1,d_2,\ldots,d_m}^*$ contains a point $f_{n;d_1,d_2,\ldots,d_m}^*(d_{m+1})$ dividing it in the ratio d_{m+1}, and a point $f(d_{m+1})$ of M exists which is the limit of a sub-sequence $\{f_{n;d_1,d_2,\ldots,d_m,d_{m+1}}^*(d_{m+1})\}$ of the infinite sequence of these points. The mapping $d_i \to f(d_i)$ is thus defined for every $i = 1, 2, \ldots$.

Ch. III, § 28 CURVE THEORY 71

The above procedure shows that for each positive integer m a subsequence $\{A_k^m\}$ of $\{A_n^*\}$ exists with $\gamma \leq l(A_k^m) \leq \gamma+1/k$ and for each k, A_k^m contains points $f_k^m(d_s)$ dividing it in the ratio d_s, with

$$f(d_s) = \lim f_k^m(d_s) \quad (s = 1, 2, \ldots, m).$$

To show that the mapping is uniformly continuous in D, let $\epsilon > 0$ be arbitrarily assigned and let d_i, d_j be any elements of D with distance $d_i d_j = |d_i - d_j| < 5\epsilon/8\gamma$. Then, considering the sequence $\{A_k^m\}$ described above, for $m = \max[i,j]$,

$$f(d_i)f(d_j) \leq f(d_i)f_k^m(d_i) + f_k^m(d_i)f_k^m(d_j) + f_k^m(d_j)f(d_j).$$

Choosing k so that $f(d_i)f_k^m(d_i)$ and $f_k^m(d_j)f(d_j)$ are each less than $\tfrac{1}{8}\epsilon$, and so that $l(A_k^m) < \gamma + \tfrac{1}{8}\epsilon$, then

$$f(d_i)f(d_j) < \tfrac{1}{4}\epsilon + |d_i - d_j| \cdot (\gamma + \tfrac{1}{8}\epsilon),$$

since $\quad f_k^m(d_i)f_k^m(d_j) \leq l(f_k^m(d_i), f_k^m(d_j)) = |d_i - d_j| \cdot l(A_k^m).$

But $|d_i - d_j| \leq 1$, and so $f(d_i)f(d_j) < \epsilon$.

It is clear from this that if r is the limit of a sequence of D then the sequence of corresponding points of M is a Cauchy sequence and hence has a limit $f(r)$ in M (§ 10). It is easy to show that $f(r)$ is independent of the choice of the sequence of elements of D with limit r, and so the mapping is extended to a mapping f of I onto a subset of M. Since f is obviously continuous over I, it defines a continuous curve $C = f(I)$ in M with $f(0) = p$, $f(1) = q$.

ASSERTION. *The path-length* $l_f(C) \leq \gamma$. Assume the contrary and write $l_f(C) = \gamma + \epsilon$, $\epsilon > 0$. Then (Theorem 27.1) corresponding to this ϵ there is a $\delta > 0$ such that if E is any normally ordered subset of I which is δ-dense in I, then

$$l[f(E)] > l_f(C) - \tfrac{1}{2}\epsilon = \gamma + \tfrac{1}{2}\epsilon.$$

We shall obtain the desired contradiction by showing that for each $\delta > 0$ a normally ordered subset E of I exists which is δ-dense in I and $l[f(E)] < \gamma + \tfrac{1}{2}\epsilon$.

Let a positive integer s be chosen so that $1/2^s < \delta$, and decompose I into a sum of 2^s sub-segments $\sigma_1, \sigma_2, \ldots, \sigma_{2^s}$ ordered as encountered in traversing I from 0 to 1, each of length 2^{-s}. For $j = 1, 2, \ldots, 2^s$, points d_{i_j} of D exist such that d_{i_j} is an interior point of σ_j and $d_{i_j}d_{i_{j+1}} < 1/2^s$ ($j = 1, 2, \ldots, 2^s - 1$). The set $E = (d_{i_1}, d_{i_2}, \ldots, d_{i_{2^s}})$ is a normally ordered subset of I which is clearly δ-dense in I.

If $m = \max[i_1, i_2, ..., i_{2^s}]$ consider the sub-sequence of arcs $\{A_k^m\}$ and select k so that

$$f(d_{i_j})f_k^m(d_{i_j}) < \epsilon/2^{s+3} \quad (j = 1, 2, ..., 2^s),$$

$$l(A_k^m) < \gamma + \tfrac{1}{4}\epsilon.$$

Then it is readily seen that

$$f(d_{i_j})f(d_{i_{j+1}}) < \gamma/2^s + \epsilon/2^{s+1} \quad (j = 1, 2, ..., 2^s-1).$$

Hence

$$l[f(E)] = \sum_{j=1}^{2^s-1} f(d_{i_j})f(d_{i_{j+1}}) < (2^s-1)[\gamma/2^s + \epsilon/2^{s+1}];$$

and so

$$l[f(E)] < \gamma + \tfrac{1}{2}\epsilon.$$

To complete the proof of the theorem it is observed that since each continuous image of a segment contains for each two of its points an arc joining them, C contains an arc A joining p and q. Since $l_f(C) \leqslant \gamma$ and (Theorem 27.2) $l(A) \leqslant l_f(C)$, then $l(A) \leqslant \gamma$. But obviously $l(A) \geqslant \gamma$; that is, $l(A) = \gamma$ and the theorem is proved.

The proof we have given is a slight modification of one due to Menger which is based on a procedure due to Hilbert. More recently it has been proved by S. B. Myers that it suffices to suppose that the metric space M is *locally compact* (that is, each point of M is contained in a spherical neighbourhood which is compact) and *almost complete* (that is, if $\{p_n\}$ is an infinite sequence of points of M and for each $\epsilon > 0$ an integer N exists such that $i, j > N$ imply that p_i, p_j are end-points of an arc of length less than ϵ, then $\{p_n\}$ has a limit in M). In any such space, each two points that are joined by a rectifiable arc are joined by a geodesic arc.

EXERCISES

Let M be a finitely compact metric space in which each two points are joined by at least one rectifiable arc. Denote by M_γ the space whose points are the points of M and in which distance $\widehat{pq}$ is defined as $\widehat{pq} = 0$ if $p = q$ and $\widehat{pq} = \gamma(p, q)$ if $p \neq q$.
1. Show that M_γ is a convex metric space.
2. Prove that an arc has the same length in M_γ that it has in M, and hence

$$(M_\gamma)_\gamma = M_\gamma.$$

3. Does compactness of M_γ follow from that of M; compactness of M from that of M_γ?
4. Let M have the additional property that $\epsilon > 0$ implies the existence of $\delta > 0$ such that if $p, q \in M$, $pq < \delta$ then $\gamma(p, q) < \epsilon$. Show that then M and M_γ are homeomorphic, and hence M_γ is a *convexification* of M.

It suffices to assume in Exercise 4 that for each $\epsilon > 0$ and each $p \in M$ a $\delta = \delta(p, \epsilon) > 0$ exists such that if $pq < \delta$ then p and q are joined by an arc of length less than ϵ. (The condition of Exercise 4 is a uniformization of this one.)

29. The n-lattice theorem

Let $C = f(I)$ be a continuous curve of a metric space M, where $I = [0, 1]$ and $f_0 = f(0) \neq f(1) = f_1$. An $(n+1)$-tuple $p_0, p_1, ..., p_n$ of C forms an n-lattice in C provided

$$p_0 = f_0, \quad p_n = f_1, \quad \text{and} \quad p_0 p_1 = p_1 p_2 = ... = p_{n-1} p_n.$$

The question whether for each such continuous curve C, and each positive integer n, an n-lattice exists in C, is both intrinsically interesting and important for later developments. It seems strange that even for arcs of euclidean space no answer to this question was available until 1935 when Alt and Beer obtained an affirmative result. The problem was solved for continuous curves C of a semimetric space with continuous metric (and hence for continuous curves of a metric space) by Schoenberg in 1940. We give in this section his proof of the n-lattice theorem.

THEOREM 29.1. *Let $C = f(I)$ be a continuous curve of a semimetric space with continuous metric. Denoting by p_t the point of C corresponding to the point t of the parameter interval I, $0 \leqslant t \leqslant 1$, it is assumed that $p_0 \neq p_1$. For each positive integer n and each set of $n+1$ positive numbers $r_0, r_1, ..., r_n$ there exist n points $p_{t_1}, p_{t_2}, ..., p_{t_n}$ of C $(0 < t_1 < t_2 < ... < t_n < 1)$ such that*

$$p_0 p_{t_1}/r_0 = p_{t_1} p_{t_2}/r_1 = p_{t_2} p_{t_3}/r_2 = ... = p_{t_n} p_1/r_n.$$

Proof. Clearly the function $F(s, t) = p_s p_t$ is continuous for

$$0 \leqslant s \leqslant t \leqslant 1,$$

with $F(s, t) \geqslant 0$, $F(s, s) = 0$, and $F(0, 1) > 0$. Consider now the set Π of all partitions π of the segment $[0, 1]$ into a sum of $n+1$ non-overlapping segments $\sigma_0, \sigma_1, ..., \sigma_n$ (ordered as encountered when traversing $[0, 1]$ from 0 to 1) with lengths $x_0, x_1, ..., x_n$, respectively. Interpreting the ordered $(n+1)$-tuple $(x_0, x_1, ..., x_n)$ as the cartesian coordinates of a point in euclidean $(n+1)$-space, the set Π is in one-to-one correspondence with the *interior* Σ_n of the closed simplex $\overline{\Sigma}_n$ defined by the inequalities

$$x_0 \geqslant 0, \quad x_1 \geqslant 0, \quad ..., \quad x_n \geqslant 0, \quad x_0 + x_1 + ... + x_n = 1.$$

We now map $\overline{\Sigma}_n$ into itself by the transformation

$$x_0' = F(0, x_0)/N,$$

$$x_1' = F(x_0, x_0+x_1)/N, \quad ..., \quad x_n' = F(x_0+x_1+...+x_{n-1}, 1)/N,$$

where

$$N = F(0, x_0) + F(x_0, x_0+x_1) + ... + F(x_0+x_1+...+x_{n-1}, 1).$$

Since $N = 0$ if and only if
$$F(0, x_0) = F(x_0, x_1+x_2) = \ldots = F(x_0+x_1+\ldots+x_{n-1}, 1) = 0$$
(that is, if and only if
$$p_0 p_{x_0} = p_{x_0} p_{x_0+x_1} = \ldots = p_{x_0+x_1+\ldots+x_{n-1}} p_1 = 0)$$
from which $p_0 = p_1$, contrary to hypothesis, it follows that $N > 0$ and the transformation is continuous in $\overline{\Sigma}_n$. Clearly $x'_i \geqslant 0$ ($i = 0, 1,\ldots, n$) and $x'_0+x'_1+\ldots+x'_n = 1$. The reader may easily show that the transformation maps each k-dimensional 'face' $\overline{\Sigma}_k$ of $\overline{\Sigma}_n$ into itself ($k = 1, 2,\ldots, n$) and leaves each of the $n+1$ vertices fixed. We may conclude from a theorem of topology that *each point of $\overline{\Sigma}_n$ is an image point of the mapping* (see Exercise below).

Applying the above observation to the point $(r_0/r, r_1/r,\ldots, r_n/r)$, $r = r_0+r_1+\ldots+r_n$, of Σ_n, a point $(X_0, X_1,\ldots, X_n)$ of $\overline{\Sigma}_n$ exists such that
$$r_0/r = F(0, X_0)/N,$$
$$r_1/r = F(X_0, X_0+X_1)/N, \quad \ldots, \quad r_n/r = F(X_0+X_1+\ldots+X_{n-1}, 1)/N,$$
and hence
$$\frac{F(0, X_0)}{r_0} = \frac{F(X_0, X_0+X_1)}{r_1} = \ldots = \frac{F(X_0+X_1+\ldots+X_{n-1}, 1)}{r_n}.$$
Setting $X_0 = t_1$, $X_0+X_1 = t_2$, ..., $X_0+X_1+\ldots+X_{n-1} = t_n$ then it easily follows that $0 < t_1 < t_2 < \ldots < t_n < 1$ (that is,
$$(X_0, X_1,\ldots, X_n) \in \Sigma_n)$$
and
$$\frac{p_0 p_{t_1}}{r_0} = \frac{p_{t_1} p_{t_2}}{r_1} = \ldots = \frac{p_{t_n} p_1}{r_n},$$
and the theorem is proved.

For $r_0 = r_1 = \ldots = r_n$, the existence of an $(n+1)$-lattice of C is proved.

EXERCISE

Show that if a point of $\overline{\Sigma}_n$ exists which is not an image point of the mapping, then to each point of $\overline{\Sigma}_n$ a vector may be attached such that (i) the resulting vector field is continuous, (ii) no vector is null, and (iii) the vectors attached to boundary points of $\overline{\Sigma}_n$ all point towards the interior of $\overline{\Sigma}_n$. Hence derive the desired contradiction.

30. Metric definitions of curvature

The preceding sections of this chapter are devoted to topics in the *integral* geometry of curves. The first notion of the differential geometry of curves to be metrized is that of *curvature*. A purely metric definition of this concept was given in 1930 by Menger (for arbitrary continua of a metric space). A modification of Menger's definition by Alt yields a more general concept. Recently an expression for curvature due to Finsler (*Dissertation*, Basle, 1918) has been applied by Haantjes to the study of rectifiable arcs in any metric space.

The programme of studying the *local properties* of a geometrical configuration metrically—that is, of developing a differential geometry without the use of coordinates—is still in its introductory phase. The theorems obtained by its methods are freed of many of the numerous geometrically unessential restrictions (for example, differentiability conditions) with which classical differential geometry abounds, and which are imposed merely to make the problems amenable to the methods of analysis. Thus besides being more general in scope than those heretofore employed (since they operate in very general spaces), metric methods often yield better results even when applied to the spaces of classical geometry. These advantages are, however, obtained by arguments which, though more direct (geometrically) than the customary ones, seem more complicated, since the elaborate but well-known machinery of the calculus is not employed to smooth the way.

If q, r, s are any three pairwise distinct points of a metric space M, the expression

$$K(q,r,s) = \frac{\sqrt{\{(qr+rs+sq)(qr+rs-sq)(qr-rs+sq)(-qr+rs+sq)\}}}{qr.rs.sq}$$

is called by Menger the *curvature of the three points*. This is motivated by the fact that if q, r, s are points of the euclidean plane E_2, then the expression for $K(q,r,s)$ is, according to elementary geometry, the reciprocal of the radius $\rho(q,r,s)$ of the circle passing through q, r, s. We call $\rho(q,r,s)$ the *radius of curvature* of q, r, s. Clearly for each point-triple q, r, s of a metric space, $K(q,r,s)$ is real, non-negative, and vanishes if and only if q, r, s are congruently contained in E_1.

DEFINITION 30.1. *Menger curvature. A metric space M has, at an accumulation point p, a curvature $K_M(p)$ provided that, corresponding to each $\epsilon > 0$, there is a $\delta > 0$ such that for every triple q, r, s of points of M with pq, pr, ps each less than δ,*

$$|K_M(p) - K(q,r,s)| < \epsilon.$$

It is noted that $K_M(p)$ is finite and non-negative whenever it exists, and $K_M(p) = 0$ for each point p of the E_1.

The above definition demands that the limit of $K(q,r,s)$ exist as q, r, s approach p in any manner whatever. This severe restriction may be relaxed by supposing that one of the points q, r, s is the point p. We are led in this way to the more general notion of Alt curvature.

DEFINITION 30.2. *Alt curvature.* *A metric space M has, at an accumulation point p, a curvature $K_A(p) \geqslant 0$ provided that*
$$K_A(p) = \lim_{q,r \to p} K(p,q,r).$$
Obviously the existence of $K_M(p)$ implies that of $K_A(p)$, and
$$K_M(p) = K_A(p).$$
On the other hand, $K_A(p)$ might exist when $K_M(p)$ does not. For if $[p,x]$, $[p,y]$, $[p,z]$ are three line segments of E_2 which have, pairwise, only the end-point p in common, the set $[p,x]+[p,y]+[p,z]$ forms a metric space when the distance ab of points a, b is defined as euclidean provided both a, b belong to one of the segments, and $ab = pa+pb$ otherwise. It is easily seen that $K_A(p) = 0$, while $K_M(p)$ does not exist. Thus the Alt curvature is a more general notion than the Menger curvature.

Both the Menger and Alt definitions of curvature are open to the objection that since the basis of their formulations is a function of distances which, in *euclidean space*, gives the desired result, they impose a euclidean notion of curvature upon all metric spaces. The inappropriateness of this euclidean notion in certain metric spaces leads to paradoxical results.

The extension of Finsler's curvature to *rectifiable arcs* of any metric space, made by Haantjes, and his study of it in that environment, seems to justify calling it by the latter's name.

If q, s are points of a rectifiable arc A of a metric space M, let $l(q,s)$ denote the length of the sub-arc of A joining q and s.

DEFINITION 30.3. *Haantjes curvature.* *The arc A has at a point p a curvature $K_H(p) \geqslant 0$ provided that, corresponding to each $\epsilon > 0$, a $\delta > 0$ exists such that $q, s \in A . U(p; \delta)$ implies*
$$\left| 4! \left(\frac{l(q,s)-qs}{l^3(q,s)} \right) - K_H^2(p) \right| < \epsilon.$$
It is convenient to put $K_H^2(q,s) = 4! [l(q,s)-qs]/l^3(q,s)$.

The reader will observe that while the notions of Menger and Alt curvatures are applicable to any metric space, that of Haantjes is restricted to rectifiable arcs. But the greater generality of the first two curvature concepts is more apparent than real, for it has been shown by Pauc that (1) *if $K_M(p)$ exists at a point p of a metric continuum M, then*

M is a rectifiable arc in a neighbourhood of p, while (2) if $K_A(p)$ exists then M is, in a neighbourhood of p, either (a) the sum of finitely many arcs which have pairwise only the end-point p in common, or (b) the sum of countably many such arcs with diameters converging to zero. Hence the restriction of Haantjes curvature to rectifiable arcs is quite unessential when comparing the notion with Menger curvature.

EXERCISES

1. Show that if $q, r, s \in E_2$, then $\rho(q,r,s) = qr.qs.rs/4A(q,r,s)$, where $A(q,r,s)$ denotes the area of the triangle with vertices q, r, s.
2. Prove that if $K_M(p)$ is defined for each point p of F, a subset of a metric space M, then $K_M(p)$ is continuous in F.
3. Prove that if $K_H(p)$ is defined for each point p of F, a subset of a rectifiable metric arc A, then $K_H(p)$ is continuous in F.
4. Show that $K_A(p)$ is not a continuous function of p.
5. Using the result of Pauc referred to above, show that a metric arc is rectifiable if the Menger curvature exists at each of its points.

31. Some properties of K_M, K_A, and K_H

It has been observed that if $K_M(p)$ exists then so does $K_A(p)$, and the two are equal. The following theorem, due to Haantjes, relates K_H with the other two curvatures.

THEOREM 31.1. *If p is a point of a rectifiable metric arc A and both $K_A(p)$ (or $K_M(p)$) and $K_H(p)$ exist, then these curvatures are equal.*

Proof. Let $K(p)$ denote $K_A(p)$ or $K_M(p)$, depending upon which exists, and let q, r be two points of A, on the same side of p, with
$$pq = qr = d, \quad pr = a.$$
Then
$$K^2(p) = \lim_{d \to 0} a^2(2d+a)(2d-a)/a^2 d^4$$
$$= \lim_{d \to 0} \frac{4}{d^2}(1-a/2d)(1+a/2d).$$

Since $K^2(p)$ exists, $\lim_{d \to 0} a/2d = 1$, and consequently
$$K^2(p) = 4\lim_{d \to 0}(2d-a)/d^3.$$

On the other hand, since $K_H(p)$ exists, then
$$\frac{1}{4!}K_H^2(p) = \lim_{d \to 0}\frac{l(p,q)-d}{l^3(p,q)} = \lim_{d \to 0}\frac{l(q,r)-d}{l^3(q,r)} = \lim_{d \to 0}\frac{l(p,q)+l(q,r)-a}{[l(p,q)+l(q,r)]^3}.$$

Putting $l(p,q) = l_1$, $l(q,r) = l_2$, we have
$$\frac{l_1+l_2-a}{(l_1+l_2)^3} = \frac{l_1-d}{l_1^3}\frac{l_1^3}{(l_1+l_2)^3}+\frac{l_2-d}{l_2^3}\frac{l_2^3}{(l_1+l_2)^3}+\frac{2d-a}{d^3}\frac{d^3}{(l_1+l_2)^3},$$

and since l_1/d and l_2/d approach 1 as $d \to 0$ (why ?) then an obvious computation yields
$$K_H^2(p) = 4\lim_{d \to 0}(2d-a)/d^3,$$
and the theorem is proved.

Since $K_H(p)$ is a continuous function and $K_A(p)$ is not, it is clear that the existence of Alt curvature does not imply the existence of $K_H(p)$. This is exemplified by the plane curve $y = x^4 \sin(1/x)$, $x \neq 0$; $y = 0$, $x = 0$, which possesses an Alt but not a Haantjes curvature at the origin. It may be shown, however, that *the existence of $K_M(p)$ implies that of $K_H(p)$, but not conversely*. Indeed, $K_H(p)$ may exist at a point p of a rectifiable metric arc A without $K_M(p)$ or even $K_A(p)$ existing. For if the euclidean distance $t = |x-y|$ of *distinct* points x, y of the line segment $[0, 1]$ is redefined by putting

$$xy = t - \frac{1}{3!}t^3 + \frac{1}{4!}t^4 \sin\frac{1}{t}, \qquad xy = 0 \quad \text{if} \quad t = 0,$$

then the resulting space may be shown to be a rectifiable metric arc with $K_H(p)$ existing at each point p, but $K_A(p)$, $K_M(p)$ exist at no point. The reader is asked to supply some of the details in an exercise.

We observe, finally, that for arcs in E_n, K_M and K_H are equivalent.

EXERCISES

1. Prove that $y = x^4 \sin(1/x), x \neq 0; y = 0, x = 0$ has an Alt but not a Haantjes curvature at the origin.
2. Prove that the *metric transform* of the unit segment $[0, 1]$ defined above is a rectifiable metric arc with constant Haantjes curvature 2. (*Hint.* To show the arc rectifiable, compute the spread of the transformation.)
3. If p is a regular point of an analytic space curve C, show that
$$\frac{1}{R} = \lim_{q \to p} K_H(p, q),$$
where $1/R$ is the classical curvature of C at p, and q is a variable point of C. (*Hint.* Use the *canonical representation* of C in the neighbourhood of p.) (See, for example, Graustein, *Differential Geometry*, Macmillan (1935), 39.)

32. Ptolemaic spaces. A lemma

Since for each three points q, r, s of a metric segment, one is between the other two, $K(q, r, s) = 0$. It follows then, from the definition of $K_M(p)$, that each metric space M which is locally a segment (that is, for each element p of M a neighbourhood of p exists which is a metric segment) has $K_M(p) = 0$ for every element p of M. But such spaces are, of course, not necessarily metric segments—indeed, they may not even be arcs, as the convex circle $S_{1,r}$ and the E_1 show. If, moreover, $A_{1,r}$

is an arc of $S_{1,r}$ of length at least $\frac{4}{3}\pi r$, then $K_M(p) = 0$, $p \in A_{1,r}$, but $A_{1,r}$ is not a metric segment since it contains an equilateral triple. Thus not all 'uncurved' arcs of a metric space are metrically 'straight' (that is, congruent with a straight line segment whose length is the distance of the end-points), and the problem arises of determining what additional metric properties suffice to ensure that each arc with everywhere vanishing curvature K_M be metrically straight.

Menger proved that it is sufficient to assume that the space M has the *euclidean four-point property*; that is, that *each four points of M be congruent with a subset of E_3*. This property of M ensures the validity of a conclusion that Menger used in proving that each arc of M with $K_M(p) = 0$ is a metric segment. It was shown later by Schoenberg that the desired conclusion is valid under a somewhat weaker restriction.

DEFINITION 32.1. *A space is called ptolemaic provided for each quadruple p, q, r, s of its points, the three products $pq.rs, pr.qs, ps.qr$ of 'opposite' distances satisfy the triangle inequality.*

We shall refer to the inequality

$$pq.rs + pr.qs \geq ps.qr$$

as the *ptolemaic inequality*.

The reader may readily show by easy examples that a metric space is not necessarily ptolemaic and a semimetric ptolemaic space is not necessarily metric.

LEMMA 32.1. *Let $p_0, p_1, ..., p_n$ be a semimetric ptolemaic $(n+1)$-tuple, and let $p'_0, p'_1, ..., p'_n$ be an $(n+1)$-tuple of the euclidean plane E_2 such that*

(1) $p_0 p_1 = p_1 p_2 = ... = p_{n-1} p_n$
    $= p'_0 p'_1 = p'_1 p'_2 = ... = p'_{n-1} p'_n > 0$,

(2) $p_0 p_2 \geq p'_0 p'_2$, $p_1 p_3 \geq p'_1 p'_3$, ..., $p_{n-2} p_n \geq p'_{n-2} p'_n$,

(3) *the polygon $p'_0 p'_1 ... p'_n p'_0$ is convex and inscribed in a circle C.*

Then $p_0 p_n \geq p'_0 p'_n$.

Proof. Putting $p'_0 p'_k = s_k$ ($k = 1, 2, ..., n$), it is easily seen that

$$p'_i p'_{i+k} = s_k \quad (0 \leq i < i+k \leq n).$$

Since each of the convex quadruples $p'_0, p'_{\nu-2}, p'_{\nu-1}, p'_\nu$ ($\nu = 3, 4, ..., n$) is inscribed in a circle, it follows from elementary geometry (Ptolemy's equality) that $s_1 s_{\nu-2} + s_1 s_\nu = s_2 s_{\nu-1}$, or

$$s_{\nu-2} - a s_{\nu-1} + s_\nu = 0 \quad (\nu = 3, 4, ..., n), \qquad (*)$$

where $a = s_2/s_1$.

The ptolemaic inequality applied to each of the quadruples

$$p_0, p_{\nu-2}, p_{\nu-1}, p_\nu \quad (\nu = 3, 4, ..., n)$$

yields

$$p_0 p_{\nu-2} \cdot p_{\nu-1} p_\nu + p_0 p_\nu \cdot p_{\nu-2} p_{\nu-1} \geqslant p_0 p_{\nu-1} \cdot p_{\nu-2} p_\nu,$$

and writing $x_k = p_0 p_k$ $(k = 0, 1, 2, ..., n)$, we have

$$x_{\nu-2} s_1 + x_\nu s_1 \geqslant x_{\nu-1} \cdot p_{\nu-2} p_\nu \geqslant x_{\nu-1} s_2 = x_{\nu-1} a s_1.$$

Hence

$$x_{\nu-2} - a x_{\nu-1} + x_\nu \geqslant 0 \quad (\nu = 3, 4, ..., n). \tag{$\dagger$}$$

Using the relations (∗) it follows that

$$s_{n-1}(x_0 - a x_1 + x_2) + s_{n-2}(x_1 - a x_2 + x_3) + ... + s_1(x_{n-2} - a x_{n-1} + x_n)$$
$$= s_1 x_n + (s_{n-2} - a s_{n-1}) s_1 = s_1(x_n - s_n),$$

and since $x_0 - a x_1 + x_2 = p_0 p_2 - a s_1 \geqslant s_2 - a s_1 = 0$, relations (†) give $x_n \geqslant s_n$, and the lemma is proved.

EXERCISES

1. Show that a semimetric quadruple p_1, p_2, p_3, p_4 is ptolemaic if and only if the determinant
$$C(p_1, p_2, p_3, p_4) = |p_i p_j^2| \quad (i, j = 1, 2, 3, 4)$$
is negative or zero.
2. Prove that the three-dimensional euclidean space E_3 is ptolemaic.
3. Prove that if p, q, r, s are a metric ptolemaic set, then pqr and qps imply rps and rqs.
4. Prove that if p, q, r, s, t are five points of a metric ptolemaic space, then psr, ptr, sqt imply pqr.
5. Show that the convex n-sphere $S_{n,r}$ is not even locally ptolemaic. Suggest an analogue of the ptolemaic inequality valid throughout the $S_{n,r}$.
6. Is the hyperbolic space $\mathcal{H}_{n,r}$ ptolemaic?
7. Show that the space $M_n^{(p)}$ of Exercise 6, § 9 is ptolemaic if and only if $p = 2$ (that is, if and only if it is euclidean).

33. Segments characterized as arcs with vanishing Menger curvature

It has been observed that a metric segment has $K_M(p) = 0$ at each point p, but that metric arcs exist with $K_M(p) = 0$ at each point which are not metric segments. The object of this section is to show that in metric *ptolemaic* spaces an arc is a metric segment if and only if it has zero Menger curvature at each point. We consider first the case of rectifiable arcs.

THEOREM 33.1. *In order that a rectifiable arc A of a metric ptolemaic space be a metric segment it is necessary and sufficient that for each element x of A, $K_M(x) = 0$.*

Ch. III, § 33   CURVE THEORY   81

Proof. The necessity being obvious, we suppose A is a metric, ptolemaic arc of finite length $l(A)$, and $p \in A$ implies $K_M(p) = 0$.

For each positive integer $n \geqslant 2$, an n-lattice

$$a = b_0^n, b_1^n, b_2^n, \ldots, b_{n-1}^n, b_n^n = b \quad (a, b \text{ end-points of } A)$$

exists in A (Theorem 29.1). Let $b_i^n b_{i+1}^n = l(n)$ $(i = 0, 1, \ldots, n-1)$.

If $\quad \rho(n) = \min \rho(b_i^n, b_{i+1}^n, b_{i+2}^n) = \min \dfrac{l^2(n)}{\sqrt{\{4l^2(n) - (b_i^n b_{i+2}^n)^2\}}}$ (*)

$(i = 0, 1, \ldots, n-2)$, then $2\rho(n) > l(n)$ and so $n+1$ points $p_0^n, p_1^n, \ldots, p_n^n$ may be marked off on a circle C_n of E_2 with radius $\rho(n)$, such that

$$p_i^n p_{i+1}^n = b_i^n b_{i+1}^n = l(n) \quad (i = 0, 1, \ldots, n-1).$$

Then for each consecutive three of these points

$$\rho(n) = \rho(p_i^n, p_{i+1}^n, p_{i+2}^n) \leqslant \rho(b_i^n, b_{i+1}^n, b_{i+2}^n),$$

and it follows that

$$p_i^n p_{i+2}^n \leqslant b_i^n b_{i+2}^n \quad (i = 0, 1, \ldots, n-2).$$

Denoting by p the centre of the circle C, then

$$\angle p_0^n p p_n^n = 2n \arcsin(l(n)/2\rho(n)),$$

and consequently

$$p_0^n p_n^n = 2\rho(n) |\sin[n \arcsin(l(n)/2\rho(n))]|. \quad (**)$$

The reader will observe that the two sets of points $b_0^n, b_1^n, \ldots, b_n^n$ and $p_0^n, p_1^n, \ldots, p_n^n$ satisfy all the conditions of Lemma 32.1 except, possibly, the demand that the polygon $p_0^n p_1^n p_2^n \ldots p_n^n p_0^n$ be convex. Nor is this polygon necessarily convex for *every* integer n. It suffices for our purpose to establish that for all sufficiently large positive integers n, the polygon *is* convex. In order to do this we prove the following statement.

ASSERTION. *The radius $\rho(n)$ approaches infinity with n.*

If the assertion is false, then a real number r and an infinite subsequence $\{k_n\}$ of the sequence of positive integers exist such that

$$\rho(k_n) \leqslant r \quad (n = 1, 2, \ldots).$$

From the definition of $\rho(n)$ there exists in the k_n-lattice of A a triple T_{k_n} of consecutive points for which the radius of curvature equals $\rho(k_n)$ and hence does not exceed r. Clearly $\delta(T_{k_n}) \leqslant 2l(k_n)$.

Now the length $l(L_{k_n})$ of each k_n-lattice L_{k_n} of A is $k_n \cdot l(k_n)$, and hence for every integer n,

$$k_n \cdot l(k_n) \leqslant l(A).$$

Consequently $\lim_{n\to\infty} l(k_n) = 0$, and so $\delta(T_{k_n}) \to 0$. From the compactness of A it follows that a point t of A exists such that each neighbourhood of t contains a member of the sequence $\{T_{k_n}\}$; that is, in every neighbourhood of t three points of A may be found whose radius of curvature does not exceed r, and consequently, whose curvature is greater than or equal to $1/r$. But then $K_M(t) > 0$, contrary to hypothesis, and the assertion is established.

Now $p_0^n p_n^n \leqslant p_0^n p_1^n + p_1^n p_2^n + \dots + p_{n-1}^n p_n^n = n \cdot l(n) \leqslant l(A)$ and so $p_0^n p_n^n$ does *not* approach infinity with n. Then (∗∗) implies that

$$\sin\left[\lim_{n\to\infty} n \arcsin(l(n)/2\rho(n))\right] = 0,$$

and so

$$\lim_{n\to\infty} \arcsin(l(n)/2\rho(n)) = 0.$$

But

$$\lim_{n\to\infty} n \arcsin(l(n)/2\rho(n)) = \lim_{n\to\infty} n \frac{\arcsin(l(n)/2\rho(n))}{l(n)/2\rho(n)} \frac{l(n)}{2\rho(n)}$$

$$= \lim_{n\to\infty} n \cdot l(n)/2\rho(n)$$

$$= 0.$$

Thus a positive integer N exists such that $n > N$ implies

$$n \arcsin(l(n)/2\rho(n)) < \tfrac{1}{2}\pi,$$

that is, $\angle p_0^n p p_n^n = 2n \arcsin(l(n)/2\rho(n)) < \pi$ and $p_0^n p_1^n p_2^n \dots p_n^n p_0^n$ is a convex polygon for every $n > N$.

Applying, now, Lemma 32.1 to the $(n+1)$-tuples $b_0^n, b_1^n, \dots, b_n^n$ and $p_0^n, p_1^n, \dots, p_n^n$, we conclude that for every $n > N$,

$$ab = b_0^n b_n^n \geqslant p_0^n p_n^n = 2\rho(n)\sin[n \arcsin(l(n)/2\rho(n))].$$

But it is readily seen from the foregoing that

$$\lim_{n\to\infty} 2\rho(n)\sin[n \arcsin(l(n)/2\rho(n))] = \lim_{n\to\infty} n \cdot l(n) = l(A),$$

and hence $ab \geqslant l(A)$. The arc A being metric, $ab \leqslant l(A)$ and consequently $ab = l(A)$. It follows (Exercise 1, § 22) that the arc A is a metric segment and the theorem is proved.

It was assumed in Theorem 33.1 that arc A is rectifiable, and this hypothesis was made use of several times in the proof of the theorem. We show next that the hypothesis of rectifiability of A may, however, be suppressed for it is implied by the everywhere vanishing curvature of the arc.

THEOREM 33.2. *If for each point x of a metric, ptolemaic arc A, $K_M(x) = 0$, then A is rectifiable.*†

Proof. The procedure adopted in the proof of the preceding theorem is applied here also and we get, as before,

$$p_0^n p_n^n = 2\rho(n)|\sin[n\arcsin(l(n)/2\rho(n))]|.$$

If, now, $n\arcsin(l(n)/2\rho(n)) \leqslant \tfrac{1}{2}\pi$, define $m_n = n$, while in case

$$\nu\arcsin(l(n)/2\rho(n)) \leqslant \tfrac{1}{2}\pi < (\nu+1)\arcsin(l(n)/2\rho(n)),$$

for $\nu < n$, put $m_n = \nu$. For each positive integer $n \geqslant 2$, consider the point $p_{m_n}^n$ of the circle C and the corresponding point $b_{m_n}^n$ of arc A. Then the polygon $p_0^n p_1^n \ldots p_{m_n}^n p_0^n$ is convex, and Lemma 32.1 gives

$$b_0^n b_{m_n}^n \geqslant p_0^n p_{m_n}^n.$$

Assuming that A is not rectifiable, we shall obtain a contradiction by showing that then points $b_{m_n}^n$ of A exist with distances $b_0^n b_{m_n}^n$ *unbounded*, which violates the compactness of A. To accomplish this, it suffices, by the preceding inequality, to prove $\lim_{n \to \infty} p_0^n p_{m_n}^n = \infty$.

We show first that $\lim_{n \to \infty} l(n) = 0$. For in the contrary case a positive number λ exists such that for any positive integer N there is a positive integer $n > N$ for which the n-lattice $b_0^n, b_1^n, \ldots, b_n^n$ has

$$b_i^n b_{i+1}^n = l(n) \geqslant \lambda \quad (i = 0, 1, \ldots, n-1).$$

Supposing A to be a homeomorph of the line segment $[0, 1]$, let $f^{-1}(b_0^n), f^{-1}(b_1^n), \ldots, f^{-1}(b_n^n)$ be the counter-image of $b_0^n, b_1^n, \ldots, b_n^n$ respectively. Evidently at least one pair of consecutive points in this counter-image has distance less than $1/(n-1)$. Now since the mapping f of $[0, 1]$ onto A is uniformly continuous, there corresponds to $\lambda > 0$ an $\eta > 0$ such that if $u, v \in [0, 1]$ with $uv < \eta$, then $f(u)f(v) < \lambda$. Choosing n so that $1/(n-1) < \eta$ and letting $f^{-1}(b_j^n), f^{-1}(b_{j+1}^n)$ be a pair of consecutive points in the counter-image of the n-lattice $b_0^n, b_1^n, \ldots, b_n^n$ (with $l(n) \geqslant \lambda$) then $b_j^n b_{j+1}^n = l(n) < \lambda$, and so $\lim_{n \to \infty} l(n) = 0$.

Since $\lim_{n \to \infty} l(n) = 0$, the same argument used in the preceding theorem to show $\lim_{n \to \infty} \rho(n) = \infty$ (the everywhere zero curvature of A enters at this point) is also valid here, and so in this case too we have

$$\lim_{n \to \infty} \rho(n) = \infty.$$

---

† Since the more general result of Exercise 5, § 30 is based on a theorem of Pauc that is not proved in this book, it seems desirable to give the argument for this special case of that result.

Consider now
$$\lim_{n\to\infty} p_0^n p_{m_n}^n = \lim_{n\to\infty} 2\rho(n)\sin[m_n \arcsin(l(n)/2\rho(n))].$$
Since $m_n \arcsin(l(n)/2\rho(n)) \leqslant \tfrac{1}{2}\pi$ for every m_n, clearly $\lim_{n\to\infty} p_0^n p_{m_n}^n = \infty$ in the event that $\sin[m_n \arcsin(l(n)/2\rho(n))]$ has no limit as $n \to \infty$ or has a limit different from zero.

On the other hand, if
$$\lim_{n\to\infty} \sin[m_n \arcsin(l(n)/2\rho(n))] = 0,$$
then
$$\lim_{n\to\infty} m_n \arcsin(l(n)/2\rho(n)) = 0,$$
$$\lim_{n\to\infty} \arcsin(l(n)/2\rho(n)) = 0,$$
and we have
$$\lim_{n\to\infty} p_0^n p_{m_n}^n = \lim_{n\to\infty} 2\rho(n) m_n \arcsin(l(n)/2\rho(n))$$
$$= \lim_{n\to\infty} m_n . l(n).$$

Since $m_n \arcsin(l(n)/2\rho(n)) \to 0$, an integer N exists such that $n > N$ implies $m_n \arcsin(l(n)/2\rho(n)) < \tfrac{1}{8}\pi$. Then $m_n = n$, for in the contrary case
$$m_n \arcsin(l(n)/2\rho(n)) \leqslant \tfrac{1}{2}\pi < (m_n+1)\arcsin(l(n)/2\rho(n)),$$
which is clearly impossible.

Hence for all sufficiently large values of n, $m_n = n$ and so
$$\lim_{n\to\infty} p_0^n p_{m_n}^n = \lim_{n\to\infty} p_0^n p_n^n = \lim_{n\to\infty} n . l(n) = l(A) = \infty,$$
and the theorem is proved.

Combining the two preceding theorems, we obtain

THEOREM 33.3. *A metric, ptolemaic arc is a metric segment if and only if its Menger curvature vanishes at each point.*

Since euclidean space is metric and ptolemaic, this theorem characterizes line segments as arcs with zero curvature at each point.

Remark. A very simple proof that a metric, ptolemaic arc with everywhere vanishing Haantjes curvature is a segment has recently been given (Haantjes [2]), and since $K_H(p)$ exists and equals $K_M(p)$ whenever the latter exists (§ 31) an easier proof of Theorem 33.3 is available. It is, however, rather tedious to prove that the existence of $K_M(p)$, p a point of an arc, implies that of $K_H(p)$.

34. Metrization of torsion

A solution to the problem of metrizing the notion of curvature is quite naturally suggested by the existence at a point p, of a sufficiently smooth curve C, of a circle (the osculating circle) the reciprocal of whose radius is the curvature of C at p. For the radius of the circle through three points of E_n is easily expressible in terms of the three mutual distances of the points, and taking the limit of that expression as the points approach independently a fourth point p gives the radius of the limiting circle. All of this is applicable to a general metric space, and the metric curvature K_M results.

Though the notions of torsion and curvature are quite similar (torsion being, apart from sign, the rate of change in direction of the *binormal* rather than the *tangent*) nothing analogous to a centre of curvature exists for torsion, and consequently its metrization does not lie quite so close to hand. To obtain a *suggestion* for a solution of the problem we shall first seek to metrize the notion of torsion for skew curves C of euclidean three-space, of the kind usually considered in classical differential geometry.

Let $x = x(t)$, $y = y(t)$, $z = z(t)$, $t_0 \leqslant t \leqslant t_1$ be the equations of such a curve C, p a point of C with coordinates $(x(t), y(t), z(t))$, and p_i neighbouring points of p with coordinates

$$(x(t+h_i), y(t+h_i), z(t+h_i)) \quad (i = 1, 2, 3, 4)$$

with no three of the four points linear. We shall call such quadruples *quasi-independent*.† Using the Taylor series expansions of $x(t+h_i)$, $y(t+h_i)$, $z(t+h_i)$, and *neglecting powers of h_i exceeding 3*, the numerical value $V(p_1, p_2, p_3, p_4)$ of the volume of the tetrahedron with vertices p_1, p_2, p_3, p_4 is readily found to be given by

$$72 V(p_1, p_2, p_3, p_4) = |\Delta Z(h_1, h_2, h_3, h_4)|, \qquad (*)$$

where
$$\Delta = \begin{vmatrix} x' & y' & z' \\ x'' & y'' & z'' \\ x''' & y''' & z''' \end{vmatrix}$$

and
$$Z(h_1, h_2, h_3, h_4) = \prod_{\substack{i,j=1 \\ i<j}}^{4} (h_i - h_j).$$

If the parameter t of C is the arc length, then approximately

$$(x_i - x_j)^2 + (y_i - y_j)^2 + (z_i - z_j)^2 = (h_i - h_j)^2 (x'^2 + y'^2 + z'^2)$$
$$= (h_i - h_j)^2,$$

and hence
$$|Z(h_1, h_2, h_3, h_4)| = \prod_{\substack{i,j=1 \\ i<j}}^{4} p_i p_j. \qquad (**)$$

† More generally, any *metric* quadruple with no triple linear is called quasi-independent.

By the classical formula for torsion $1/T$ of C at p,

$$\Delta = -1/T \cdot 1/R^2, \qquad (***)$$

where $1/R$ is the curvature of C at p. Replacing $1/R^2$ by

$$[R(p_1,p_2,p_3)R(p_1,p_2,p_4)R(p_1,p_3,p_4)R(p_2,p_3,p_4)]^{-\frac{1}{2}},$$

and recalling that

$$1/R(q,r,s) = 4A(q,r,s)/qr \cdot qs \cdot rs$$

for $q, r, s \in E_2$, substitution of (**) and (***) in (*) suggests that the torsion of four points p_1, p_2, p_3, p_4 of E_3, no three of which are linear, be defined as

$$t(p_1,p_2,p_3,p_4) = \frac{\frac{9}{2}V(p_1,p_2,p_3,p_4)}{\sqrt{\left\{\prod_{i_1,i_2,i_3} A(p_{i_1},p_{i_2},p_{i_3})\right\}}},$$

the product being extended over the four combinations of the indices 1, 2, 3, 4 taken three at a time.

Since both $V(p_1,p_2,p_3,p_4)$ and $A(p_{i_1},p_{i_2},p_{i_3})$ are meaningless for points of a general metric space, it is necessary to express them both in terms of distances alone before the formula for torsion of four points just obtained can be applied in each metric space. The well-known formulae†

$$288 V^2(p_1,p_2,p_3,p_4) = D(p_1,p_2,p_3,p_4),$$

$$-16 A^2(p_1,p_2,p_3) = D(p_1,p_2,p_3),$$

where
$$D(p_1,p_2,p_3,p_4) = \begin{vmatrix} 0 & 1 & 1 & 1 & 1 \\ 1 & 0 & p_1p_2^2 & p_1p_3^2 & p_1p_4^2 \\ 1 & p_1p_2^2 & 0 & p_2p_3^2 & p_2p_4^2 \\ 1 & p_1p_3^2 & p_2p_3^2 & 0 & p_3p_4^2 \\ 1 & p_1p_4^2 & p_2p_4^2 & p_3p_4^2 & 0 \end{vmatrix},$$

and $D(p_1,p_2,p_3)$ is the co-factor of the element in the last row and column of this determinant, furnish the necessary expressions, and we make the following definitions.

DEFINITION 34.1. *The torsion* $t(p_1,p_2,p_3,p_4)$ *of four quasi-independent points* p_1, p_2, p_3, p_4 *of a metric space* M *is the positive square root of*

$$t^2(p_1,p_2,p_3,p_4) = \frac{18|D(p_1,p_2,p_3,p_4)|}{\left\{\prod_{i_1,i_2,i_3} D(p_{i_1},p_{i_2},p_{i_3})\right\}^{\frac{1}{2}}}.$$

† § 40.

DEFINITION 34.2. *A metric space M has torsion $t(p)$ at a point p (in each neighbourhood of which a quasi-independent quadruple exists) provided for each positive ϵ there exists a positive δ such that for each quasi-independent quadruple p_1, p_2, p_3, p_4 of M with $pp_i < \delta$ ($i = 1, 2, 3, 4$), $|t(p) - t(p_1, p_2, p_3, p_4)| < \epsilon$; that is, $t(p) = \lim_{Q \to p} t(Q)$, where Q is a quasi-independent quadruple of M.*

The reader will note that (1) $t(p) \geq 0$, and if E denotes the set of all points p of a metric space at which $t(p)$ is defined, then $t(p)$ is continuous in E, (2) $t(p)$ is not defined at any point of a metric segment, (3) $t(p) = 0$ at each point of a subset of the E_2 whose torsion is defined at p, (4) $t(p_1, p_2, p_3, p_4)$ is a function of the *quadruple* Q and is independent of any ordering of the points of the quadruple.

It is observed, moreover, that (5) if

$$t^{*2}(p) = \lim_{p_1, p_2, p_3 \to p} \frac{18|D(p, p_1, p_2, p_3)|}{[D(p, p_1, p_2)D(p, p_1, p_3)D(p, p_2, p_3)D(p_1, p_2, p_3)]^{\frac{1}{2}}}$$

with p, p_1, p_2, p_3 a quasi-independent quadruple, then the existence of $t(p)$ does *not* imply the existence of $t^*(p)$ since the space may not contain any quasi-independent quadruple with p a member, though such quadruples *without* p might be found in every neighbourhood of p. Let the reader give an example of this.

It can be shown that $t(p)$ gives, within sign, the classical torsion of sufficiently smooth curves of E_3.

EXERCISES

1. Prove the five properties of $t(p)$ stated above.
2. Show that

$$K_M^2(p_1, p_2, p_3) = \frac{8|D(p_1, p_2, p_3)|}{D(p_1, p_2)D(p_2, p_3)D(p_1, p_3)},$$

and note the analogy between this expression and the definition of $t^2(p_1, p_2, p_3, p_4)$. Justify the square root in the denominator of the latter by a 'dimensional' argument.

3. Give an example of a metric arc with everywhere vanishing torsion which is not congruent with a planar subset.†
4. If p is a regular point of an analytic curve C of E_3, show that the classical torsion $1/T$ of C at p is given, within sign, by $3 \lim_{q \to p} (\sin \Delta \psi)/pq$, where $\Delta \psi$ denotes the angle between the osculating plane of C at p and the plane determined by q and

† In the Missouri Doctoral Dissertation of J. W. Sawyer it is shown that if A is an arc of euclidean three-space with everywhere positive Menger curvature and everywhere vanishing metric torsion $t(p)$, then A is planar.

the tangent to C at p. Considering the osculating plane of C at p as the limit plane of $P(p, r_1, r_2)$ as r_1, r_2 approach p along C, then

$$|1/T| = 3 \lim_{q \to p} \left[\lim_{r_1, r_2 \to p} \frac{\sin \angle (P(p, r_1, r_2), P(q, r_1, r_2))}{pq} \right].$$

Since

$D(p, q, r_1, r_2)$
$\quad = 8[r_1 p \cdot r_1 q \cdot r_1 r_2 \cdot \sin \angle p r_1 r_2 \cdot \sin \angle q r_1 r_2 \cdot \sin \angle (P(p, r_1, r_2), P(q, r_1, r_2))]^2,$

show that

$$|1/T| = \lim_{q \to p} \left[\lim_{r_1, r_2 \to p} \frac{r_1 r_2}{pq} \sqrt{\left(\frac{18 |D(p, q, r_1, r_2)|}{D(p, r_1, r_2) D(q, r_1, r_2)} \right)} \right].$$

5. In Alexits [1] the iterated limit in Exercise 4 is taken as the *definition* of the torsion of a metric space at the point p, the points p, q, r_1, r_2 being taken in all orders. Show that this definition is equivalent to Definition 34.2 at all points p of a euclidean arc with $K_M(p) \neq 0$. (Find the 'euclidean form' of each definition.)
6. Give an example of an arc C of E_3 for which the two definitions give different torsions at a point.
7. What can be said of a metric continuum in the neighbourhood of a point p at which $t(p)$ exists?

35. Wald's metrization of Gauss curvature

The metrizations of curvature and torsion presented in the preceding sections give a basis for developing the properties of a curve in a neighbourhood of a point by purely metric methods and furnish, accordingly, the foundation for a *metric* differential geometry of curves in which coordinates are not employed.† In order that the differential geometry of surfaces might be included in such a study, it is, first of all, necessary to devise a means for bringing the important notion of Gauss curvature within the range of metric methods. This was achieved by Wald.

The metrization of the Gauss or total curvature of a surface at a point is based on the notion of an *imbedding curvature of a quadruple of points*. A quadruple Q has an imbedding curvature $k(Q)$, positive, negative, or zero, according as Q is congruent with a quadruple of the convex two-sphere, the hyperbolic plane, or the euclidean plane of Gauss curvature k, respectively. Linear quadruples being excluded, it may be shown that a quadruple Q has at most two imbedding curvatures, and there exist non-linear quadruples which are (1) congruently contained in two convex spheres of different radii, and (2) imbeddable congruently in both an $S_{2,r}$ and the euclidean plane. Indeed, it is easy to show that in every neighbourhood of every point the $S_{2,r}$ contains a non-linear quadruple that is congruent with a quadruple of E_2.

† The fundamental theorem of curve theory, for example, according to which two arcs of E_3 are congruent if there exists a correspondence between their points preserving arc length, curvature, and torsion has been given a purely metric proof (J. W. Gaddum, Missouri Dissertation). It is assumed that $K_M(p) \neq 0$, $t(p) \neq 0$ for each p of the arc.

DEFINITION 35.1. *The metric space M has at an accumulation point p the Wald curvature $k(p)$ provided* (1) *no neighbourhood of p is linear, and* (2) *corresponding to each $\epsilon > 0$ there is a $\delta > 0$ such that each quadruple p_1, p_2, p_3, p_4 of M with $pp_i < \delta$ ($i = 1, 2, 3, 4$) has an imbedding curvature $k(p_1, p_2, p_3, p_4)$ with*

$$|k(p) - k(p_1, p_2, p_3, p_4)| < \epsilon.$$

Wald proved that if M is a surface of the kind considered in classical differential geometry, then the curvature $k(p)$, $p \in M$, is the Gauss (total) curvature of M at p. Bounded portions of such surfaces are, moreover, characterized among all compact and convex metric spaces by the property of having at each point p a Wald curvature $k(p)$. The detailed proofs of these remarkable results would occupy more space than we can give to them in this book. A brief sketch of the procedure is to be found in the author's *Distance Geometries*, while the reader is referred to Wald's original paper for the details.

EXERCISES

1. Prove that in every neighbourhood of every point p of $S_{2,r}$ a non-linear quadruple exists that is congruent with four points of E_2.
2. Show that if a non-linear metric quadruple Q has a linear triple then Q has at most one imbedding curvature.

REFERENCES

§§ 22–28. Modified versions of proofs appearing in Menger [3].
§ 29. Schoenberg [1].
§ 30. Menger [3], Alt [1], Haantjes [1], Pauc [1].
§ 31. Haantjes [1].
§ 32. Blumenthal [4], Schoenberg [1].
§ 33. Menger [3].
§ 35. Wald [2], Robinson [2].

SUPPLEMENTARY PAPERS

Haantjes [2]—characterizations of geodesics (arcs that are locally segments) by everywhere vanishing metric curvature simply obtained in a wide class of spaces.
Milgram [2]—proof of the n-lattice theorem for a metrized simple closed curve.

PART II

EUCLIDEAN AND HILBERT SPACES

CHAPTER IV

CONGRUENT IMBEDDING IN EUCLIDEAN SPACE

36. Two fundamental problems of distance geometry

WHENEVER a given class S of distance spaces is thought of as a proper sub-class of a class Σ of such spaces, a fundamental problem of distance geometry is that of determining those metric properties which serve to distinguish the members of S from among all other spaces of Σ.

DEFINITION 36.1. *A sub-class of a class of distance spaces is characterized metrically among the spaces of the class whenever conditions (expressed wholly and explicitly in terms of the metric) are obtained which are necessary and sufficient to ensure that any member of the class satisfying them be congruent with a member of the sub-class.*

Obviously, the more highly differentiated the sub-class is from the class, the more complicated is the metric characterization of the sub-class with respect to the class. The simple distance relation expressed by the triangle inequality, for example, suffices to characterize metrically the sub-class of metric spaces with respect to the class of semimetric spaces, but the characterization, with respect to the same *comparison* class, of a greatly restricted sub-class (the euclidean spaces, for instance) is not so simply expressed.

In many characterization problems the sub-class to be characterized consists of only a single space, and it was in this form that the characterization problem first arose. In 1928 Menger characterized the euclidean n-space E_n among all semimetric spaces. His investigations were apparently initiated by a remark of Biedermann to the effect that a connected metric space with each triple linear is *homeomorphic* with a segment, ray, or line. Biedermann then inquired concerning the existence of distance relations which imply the homeomorphism of a metric space with a euclidean subset of given dimension.

Replacing 'homeomorphism' by 'congruence', Biedermann's inquiry gives rise to a special case of a problem—the subset problem—that should be clearly distinguished from the characterization or *space problem* just discussed. The subset problem is an *isometric imbedding* problem; it seeks necessary and sufficient metric conditions that an arbitrary distance space of a specified class must satisfy in order that it may be congruent with a subset of a member of a prescribed class of spaces. It is clear that the subset problem is more general than the space problem—indeed, it contains the space problem as a special case. It should also be remarked that a complete solution of the space problem may contribute little towards a solution of the subset problem, since in characterizing a space one frequently imposes from the outset certain obvious necessary conditions which are extraneous to the imbedding problem. On the other hand, if one has obtained necessary and sufficient conditions for congruent imbedding in a given space of an arbitrary member of a class of spaces, the characterization of the given space with respect to that class is reduced to characterizing the space *among its subsets*.

For the characterization and imbedding problems to be discussed in this book, *the comparison class is usually the class of semimetric spaces*. Now a semimetric space of m distinct elements $p_1, p_2, ..., p_m$ is completely specified by a matrix $(p_i p_j)$ $(i,j = 1, 2, ..., m)$, where $p_i p_j$ denotes, as usual, the distance of p_i, p_j. The semimetric character of the space is reflected in the *structure* of this matrix. It is a symmetric matrix whose principal diagonal consists exclusively of zeros and the remaining elements of which are positive real numbers. If such a matrix be formed for m points of a particular semimetric space $(m > 2)$, it is clear that the individuality of that space will give rise to additional properties (minor relations) of the matrix, and it is natural to seek to study the well-known spaces in terms of the structure of such matrices.

There are many ways in which the structure of a distance matrix may be described. One may, for example, base the description upon the signs of principal minors of certain determinants formed from the elements of the matrix, the character (definite, semi-definite, etc.) of certain quadratic forms associated with the matrix, etc. We shall be especially concerned with the first method of description.

37. Congruence indices

We define now a property of a space R relative to a class $\{S\}$ of spaces which is of much importance in the metric study of the space.

DEFINITION 37.1 (*Congruence indices*). *A space R has congruence indices (n, k) with respect to a class $\{S\}$ of spaces provided any space S of $\{S\}$, containing more than $n+k$ pairwise distinct points, is congruently imbeddable* in R whenever each n of its points (not necessarily pairwise distinct) has that property. The symbol (n, k) is called a congruence symbol of R with respect to $\{S\}$.

We shall write $S \subsetsim R$ to denote that S is congruently imbeddable in R.

The cases of congruence indices $(n, 0)$ and $(n, 1)$ are of special interest. We say that R has *congruence order n* or *quasi-congruence order n* with respect to $\{S\}$ according as R has congruence indices $(n, 0)$ or $(n, 1)$, respectively, with respect to $\{S\}$. It is clear that R has congruence order n if and only if $S \subsetsim R$ provided each n-tuple of S has that property, and R has quasi-congruence order n if and only if each element S of $\{S\}$, *with more than $n+1$ distinct points*, is congruently contained in R provided each n-tuple of its points is imbeddable in R.

Let $(n, k) \to (n', k')$ symbolize the proposition, 'If R has congruence indices (n, k) with respect to $\{S\}$, then R has congruence indices (n', k') with respect to $\{S\}$'. The reader may readily show that $(n, 1) \to (n+1, 0)$; that is, quasi-congruence order n implies congruence order $n+1$. Also $(n, 1) \to (n, 2) \to (n+1, 1)$, which interpolates a property between quasi-congruence order n and quasi-congruence order $n+1$. As a direct consequence of Definition 37.1, it is seen that if $n \leqslant n'$ and $n+k \leqslant n'+k'$, then $(n, k) \to (n', k')$.

Since indices (n, k) are in terms of the congruent imbedding of n points, each of the symbols (n, k) $(k = 0, 1, 2, \ldots)$ is to be 'preferred' to the symbol $(n+1, 0)$, though the implication $(n, k) \to (n+1, 0)$ is valid only for $k = 0, 1$. Though the symbol (n, k) describes a property of R with respect to spaces S containing *more than $n+k$* points, this restriction in the class of comparison spaces is less important than the advantage gained by keeping the first index as small as possible.

This suggests a preferential ordering of the symbols (n, k) according to which (n, k) precedes (n', k') if and only if $n < n'$ or, in case $n = n'$, then $k < k'$. The ordered set so obtained has a first element which is called the *best* congruence symbol of R with respect to $\{S\}$. A *complete set* of congruence symbols for R with respect to $\{S\}$ is a set which logically implies all the congruence symbols of R with respect to $\{S\}$.

The indices n, k forming a congruence symbol are, of course, non-negative integers. It is occasionally useful, however, to have the notion of congruence order η for η transfinite. Intermediate is the concept of *unbounded congruence order* which is a property of a space R provided $S \subsetsim R$, $S \in \{S\}$, whenever each finite subset of S is congruently contained in R. These notions will be discussed further in a later section.

EXERCISES

1. Find the complete set of congruence symbols for a semimetric space of exactly n points, the class $\{S\}$ being the set of all semimetric spaces.
2. Show by an example that $(n+1, 0)$ does not imply $(n, 1)$.
3. Show that if the best congruence symbol of R has its second index 0 or 1, the complete set of congruence symbols consists of only one symbol.

38. Congruence order of the E_n

Menger showed in 1928 that any semimetric space S is congruently imbeddable in the n-dimensional euclidean space E_n if and only if each $(n+3)$-tuple of S has that property; that is, the E_n has congruence indices $(n+3, 0)$ with respect to the class $\{S\}$ of semimetric spaces. This important result reduces the problem of imbedding an arbitrary semimetric space S into E_n to the 'finite' problem of imbedding each $(n+3)$-tuple into E_n. We derive this fundamental metric property of the E_n as a corollary of a theorem whose hypotheses do not explicitly require the imbedding in E_n of *every* $(n+3)$-tuple of S. Thus there are present in S certain $(n+3)$-tuples which are not assumed to be congruently contained in E_n. These $(n+3)$-tuples are called *free*.

DEFINITION 38.1. *A congruent mapping of a metric space onto itself is called a motion.*

DEFINITION 38.2. *Two subsets of a metric space are superposable provided a motion exists that maps one onto the other.*

DEFINITION 38.3. *A $(k+1)$-tuple of a euclidean space is called independent provided it is not a subset of an E_{k-1}. A $(k+1)$-tuple that is not independent is called dependent.*

In our study of the E_n we shall make use of some of its elementary properties which are either well known to the reader or the proofs of which he can readily supply. Of particular importance for this chapter are:

PROPERTY I. *Two congruent $(k+1)$-tuples of E_n are either both independent or both dependent.*

PROPERTY II. *Each two k-dimensional euclidean spaces are congruent.*

Any two k-dimensional subspaces of E_n are, in fact, superposable.

PROPERTY III. *There is at most one point of E_k with given distances from the points of an independent $(k+1)$-tuple of E_k.*

PROPERTY IV. *Any congruence between any two subsets of E_n can be extended to a motion.*

Thus, if P, Q are any two subsets of E_n and f is any congruent mapping of P onto Q ($f(P) = Q$), there exists a congruent mapping g of E_n onto itself such that $g = f$ over P. This strong property of the E_n simplifies greatly the problems of characterizing metrically the E_n and its subsets. We shall see later how much more difficult such problems are for spaces (for example, the elliptic spaces) that do not possess this property.

PROPERTY V. *There exists one and only one motion g of the E_k such that if $p_0, p_1, \ldots, p_k \approx q_0, q_1, \ldots, q_k$; $p_i, q_i \in E_k$ ($i = 0, 1, \ldots, k$); and $p_0, p_1, \ldots, p_k$ are independent, then $g(q_i) = p_i$ ($i = 0, 1, \ldots, k$).*

This property is an immediate consequence of Properties III and IV.

PROPERTY VI. *A subset P of E_k is irreducibly contained in E_k (that is, $P \subset E_k$ but $P \not\subset E_m \subset E_k$, $m < k$) if and only if P contains an independent $(k+1)$-tuple.*

THEOREM 38.1. *A non-null semimetric space $S \Subset E_n$ if and only if there exists an integer r ($0 \leqslant r \leqslant n$), such that (a) S contains $r+1$ points $p_0, p_1, \ldots, p_r$ congruent with an independent $(r+1)$-tuple of an E_r, and (b) each $(r+3)$-tuple of S containing $p_0, p_1, \ldots, p_r$ is congruently imbeddable in an E_r. Then S is congruently contained in an r-dimensional (but in no lower dimensional) subset of E_n.*

Proof. The necessity of the conditions is obvious, with r the dimension of the lowest dimensional subspace of E_n containing a subset S' of E_n congruent to S (Property VI).

To prove the sufficiency, we observe first that by (a) and Property II, an r-dimensional subspace E_r^* of E_n exists, containing an *independent* $(r+1)$-tuple $p_0', p_1', \ldots, p_r'$ with

$$p_0, p_1, \ldots, p_r \approx p_0', p_1', \ldots, p_r'.$$

If, now, $x \in S$, then by (b) and Property II points $\bar{p}_0, \bar{p}_1, \ldots, \bar{p}_r, \bar{x}$ of E_r^* exist with
$$p_0, p_1, \ldots, p_r, x \approx \bar{p}_0, \bar{p}_1, \ldots, \bar{p}_r, \bar{x}.$$
It follows that $\quad p_0', p_1', \ldots, p_r' \approx \bar{p}_0, \bar{p}_1, \ldots, \bar{p}_r,$
and by Property V a unique motion g of E_r^* exists with $g(\bar{p}_i) = p_i'$ ($i = 0, 1, \ldots, r$). Denoting $g(\bar{x})$ by x', a *mapping of S into E_r^** is established (Property III).

To show that the mapping is a congruence suppose $s, t \in S$, and let s', t' be their respective images under the above mapping. By (b) and Property II, points $p_0'', p_1'', \ldots, p_r'', s'', t''$ of E_r^* exist with

$$p_0, p_1, \ldots, p_r, s, t \approx p_0'', p_1'', \ldots, p_r'', s'', t''$$
$$\approx p_0', p_1', \ldots, p_r', s', t',$$

by Property V and the definitions of s', t'. Hence $st = s't'$, and S is mapped congruently into E_r^*. Since $p_0, p_1, ..., p_r$ are congruent with an independent $(r+1)$-tuple of E_r, then by Property I, $p_0, p_1, ..., p_r$ are not congruently contained in a euclidean subspace of dimension less than r, and the proof of the theorem is complete.

COROLLARY. *A semimetric space* $S \subseteqq E_n$ *if and only if each set of* $n+3$ *points of* S *is congruently imbeddable in* E_n; *that is*, E_n *has the congruence symbol* $(n+3, 0)$.

For then S evidently contains a set of $r+1$ points, $0 \leqslant r \leqslant n$, satisfying the conditions of the theorem.

EXERCISES
1. Prove Properties III and IV for the plane.
2. Give examples of congruent sets which are not superposable.
3. Show that Theorem 38.1 is not valid if hypothesis (*b*) be weakened by replacing $r+3$ by $r+2$.
4. May 'independent' be deleted from (*a*) of Theorem 38.1 if we wish to conclude merely that $S \subseteqq E_r$?

39. A set of sufficient conditions for congruence indices $(m+3, k)$

We have shown in the preceding section that E_n has congruence order $n+3$ (that is, E_n has congruence indices $(n+3, 0)$) with respect to the class $\{S\}$ of semimetric spaces. This property does not characterize E_n since other spaces which are metrically and even topologically quite different from E_n (for example, the $\mathscr{H}_{n,r}$ and $S_{n,r}$) possess it also. It would be desirable to know just what metric properties of a space are necessary and sufficient to endow it with congruence indices (m, k) with respect to a given class of spaces. Such a general characterization of the notion of congruence indices has not yet been obtained even for the case of congruence order. We shall, however, establish a set of sufficient conditions for congruence indices of a space R with respect to $\{S\}$, applicable to some of the more important spaces.

DEFINITION 39.1. *A subset B of a semimetric space S is a metric basis of S provided each point of S is uniquely determined by its distances from the points of B.*

Clearly $n+1$ independent points of E_n form a metric basis of E_n.

DEFINITION 39.2. *A metric basis B of a semimetric space S is a complete metric basis provided for each subset B^* of S with $B \approx B^*$, and each subset T^* of S containing B^*, a subset T of S exists such that the congruence of B with B^* can be extended to $T \approx T^*$. The extension is obviously unique.*

THEOREM 39.1. *Let R be any semimetric space with the properties*:
(1) *There is defined in R a system $\{U\}$ of subsets containing a member U^* in which every member of the system may be congruently imbedded.*
(2) *The subset U^* has congruence indices $(m+2, k)$ with respect to $\{S\}$.*
(3) *If $m+2$ points of R are such that corresponding to each $m+1$ of the points there is a member of $\{U\}$ containing them, then the $m+2$ points lie in some member of $\{U\}$.*
(4) *Each subset of $m+1$ points of R is either contained in a member of $\{U\}$ or is congruent with a complete metric basis of R.*

Then R has congruence indices $(m+3, k)$ with respect to $\{S\}$.

Proof. Let S be any member of $\{S\}$ containing more than $m+k+3$ points, with each $(m+3)$-tuple congruently imbeddable in R. We wish to prove that $S \subseteqq R$.

If each $(m+2)$-tuple of S is congruently contained in a member of $\{U\}$, then by (1) and (2) $S \subseteqq U^* \subset R$. Let $p_0, p_1, \ldots, p_m, p_{m+1}$ be an $(m+2)$-tuple of S and $p'_0, p'_1, \ldots, p'_m, p'_{m+1}$ points of R, not in any member of $\{U\}$, with

$$p_0, p_1, \ldots, p_m, p_{m+1} \approx p'_0, p'_1, \ldots, p'_m, p'_{m+1}.$$

By (3) an $(m+1)$-tuple of the primed points exists, say $p'_0, p'_1, \ldots, p'_m$, which is not contained in any member of $\{U\}$. Property (4) ensures that $p'_0, p'_1, \ldots, p'_m$ are congruent with a complete metric basis $p''_0, p''_1, \ldots, p''_m$ of R.

Now let x be any element of S, different from $p_0, p_1, \ldots, p_m$ or not. Then $p_0, p_1, \ldots, p_m, x \approx \bar{p}_0, \bar{p}_1, \ldots, \bar{p}_m, \bar{x}$ of R, and since

$$\bar{p}_0, \bar{p}_1, \ldots, \bar{p}_m \approx p_0, p_1, \ldots, p_m \approx p''_0, p''_1, \ldots, p''_m,$$

a unique point x'' of R exists such that

$$p_0, p_1, \ldots, p_m, x \approx p''_0, p''_1, \ldots, p''_m, x''.$$

We have thus defined a mapping of S into R.

To show the mapping is a congruence, we have for each pair of elements x, y of S (Definition 39.2)

$$p_0, p_1, \ldots, p_m, x, y \approx p''_0, p''_1, \ldots, p''_m, x^*, y^*,$$

and since

$$p''_0, p''_1, \ldots, p''_m, x^* \approx p''_0, p''_1, \ldots, p''_m, x'',$$

$$p''_0, p''_1, \ldots, p''_m, y^* \approx p''_0, p''_1, \ldots, p''_m, y'',$$

and $p''_0, p''_1, \ldots, p''_m$ is a metric basis, then $x^* = x''$, $y^* = y''$. Hence

$$xy = x^*y^* = x''y'',$$

and the proof is complete.

To apply this theorem to the E_n, the subsets forming the system $\{U\}$ are taken as the $(n-1)$-dimensional hyperplanes E_{n-1}. The smallest values of m, k for which E_n has (with respect to that choice of the system $\{U\}$) the properties (1), (2), (3), (4) are $m = n$, and $k = 0$. Then indices $(3, 0)$ for E_0 imply (by the theorem) indices $(4, 0)$ for E_1 and, by complete induction, indices $(n+3, 0)$ for E_n, for every non-negative integer n.

Applying the theorem to the convex n-sphere $S_{n,r}$ the sets $\{U\}$ are great hyperspheres $S_{n-1,r}$ of dimension $n-1$, with $S_{0,r}$ consisting of a *pair of diametral points*. It is easily seen that the $S_{n,r}$ has the four properties required in Theorem 39.1, with the system $\{U\}$ so defined and $m = n$, $k = 0$.

THEOREM 39.2. *The convex n-sphere $S_{n,r}$ has congruence indices $(n+3, 0)$ with respect to the class of all semimetric spaces.*

The reader is asked to establish the same result for the hyperbolic space $\mathscr{H}_{n,r}$ (Exercise 5). On the other hand, the elliptic space $\mathscr{E}_{n,r}$ does *not* have property (4), with $\{U\}$ *taken as the hyperplanes $\{\mathscr{E}_{n-1,r}\}$ and $m = n$, $k = 0$*; for the vertices of an equilateral triangle of $\mathscr{E}_{2,r}$, for example, with side $\tfrac{1}{2}\pi r$ are neither contained in an $\mathscr{E}_{1,r}$ nor are they congruent with a complete metric basis of $\mathscr{E}_{2,r}$ since no three points of $\mathscr{E}_{2,r}$ form a metric basis for the space. We shall study the difficult problem of obtaining congruence indices for the $\mathscr{E}_{n,r}$ in a later section.

EXERCISES

1. A metric basis B is *minimal* provided no proper subset of B is a metric basis. What subsets of the space formed by the euclidean segment $[0, 1]$ are minimal metric bases; minimal complete metric bases?
2. Show that every separable metric space has a denumerable metric basis. Does this statement remain valid when 'separable metric space' is replaced by 'separable semimetric space'?
3. Give examples of metric spaces in which proper subsets forming metric bases exist, but no proper subset is a complete metric basis.
4. Prove that for $n = 1, 2$ the $S_{n,r}$ has Properties (1)–(4) of Theorem 39.1, with $m = n$ and $k = 0$.
5. Use Theorem 39.1 to show that $\mathscr{H}_{n,r}$ has congruence indices $(n+3, 0)$ with respect to $\{S\}$.
6. Show that $(1, 0)$, $(0, 1)$, $(-1, 0)$, and $(0, -1)$ of the space $M_2^{(1)}$ (Exercise 6, § 9) form a maximal equilateral set but not a metric basis for the space.

40. The Cayley–Menger determinant

Theorem 38.1 leads to a solution of the problem of imbedding an arbitrary semimetric space congruently into E_n when necessary and sufficient metric conditions are obtained in order that (1) a semimetric $(r+1)$-tuple may be congruent with an independent $(r+1)$-tuple of E_r, and (2) a semimetric $(r+3)$-tuple containing such an $(r+1)$-tuple may be imbedded in the E_r. To secure necessary conditions we seek in this section relations among the mutual distances of $k+1$ points $q_0, q_1, ..., q_k$ of a euclidean space.

Now $k+1$ points of a euclidean space are always contained in some E_k, in which we may establish a rectangular coordinate system. Letting the coordinates of q_i be $(x_i^{(1)}, x_i^{(2)}, ..., x_i^{(k)})$ $(i = 0, 1, ..., k)$, and denoting by $V(q_0, q_1, ..., q_k)$ the volume of the simplex (perhaps degenerate) with vertices $q_0, q_1, ..., q_k$, a formula of elementary analytic geometry gives

$$V(q_0, q_1, ..., q_k) = \frac{1}{k!} \begin{vmatrix} x_0^{(1)} & x_0^{(2)} & . & . & x_0^{(k)} & 1 \\ x_1^{(1)} & x_1^{(2)} & . & . & x_1^{(k)} & 1 \\ . & . & . & . & . & . \\ x_k^{(1)} & x_k^{(2)} & . & . & x_k^{(k)} & 1 \end{vmatrix}.$$

The determinant is unaltered in value by bordering it with a $(k+2)$th row and column, with 'intersecting' element 1 and the remaining elements zero. Multiplying this bordered determinant by the transpose of the determinant obtained from it by interchanging the last two rows and last two columns, we have

$$V^2(q_0, q_1, ..., q_k) = \frac{-1}{(k!)^2} \begin{vmatrix} (q_0, q_0) & (q_0, q_1) & . & . & (q_0, q_k) & 1 \\ (q_1, q_0) & (q_1, q_1) & . & . & (q_1, q_k) & 1 \\ . & . & . & . & . & . \\ (q_k, q_0) & (q_k, q_1) & . & . & (q_k, q_k) & 1 \\ 1 & 1 & . & . & 1 & 0 \end{vmatrix}, \quad (*)$$

where $(q_i, q_j) = x_i^{(1)} x_j^{(1)} + x_i^{(2)} x_j^{(2)} + ... + x_i^{(k)} x_j^{(k)}$ $(i, j = 0, 1, ..., k)$.

If we substitute

$$(q_i, q_j) = \tfrac{1}{2}[(q_i, q_i) + (q_j, q_j) - q_i q_j^2] \quad (i, j = 0, 1, ..., k)$$

in the determinant in (∗), subtract from the ith row (column) the product of the last row (column) by

$$\tfrac{1}{2}(q_{i-1}, q_{i-1}) \quad (i = 1, 2, ..., k+1),$$

we obtain after easy reductions,

$$V^2(q_0, q_1, ..., q_k) = \frac{(-1)^{k+1}}{2^k (k!)^2} \begin{vmatrix} 0 & 1 & 1 & . & . & 1 \\ 1 & 0 & q_0 q_1^2 & . & . & q_0 q_k^2 \\ 1 & q_1 q_0^2 & 0 & . & . & q_1 q_k^2 \\ . & . & . & . & . & . \\ 1 & q_k q_0^2 & q_k q_1^2 & . & . & 0 \end{vmatrix}. \quad (**)$$

The determinant in (∗∗) is evidently a *congruence invariant*. Denoting it by $D(q_0, q_1, ..., q_k)$, we have (since $V^2(q_0, q_1, ..., q_k) \geqslant 0$ for every euclidean

$(k+1)$-tuple, and $V^2(q_0, q_1,..., q_k) > 0$ if and only if $q_0, q_1,..., q_k$ form an independent $(k+1)$-tuple):

THEOREM 40.1. *A necessary condition that a semimetric $(r+1)$-tuple $p_0, p_1,..., p_r$ be congruently contained in a euclidean space E_n is that for every $k = 1, 2,..., r$, the determinant $D(p_0, p_1,..., p_k)$ either vanish or have the sign of $(-1)^{k+1}$. If $n < r$, then $D(p_0, p_1,..., p_k) = 0$ ($n < k \leqslant r$).*

We shall write the condition of the preceding theorem in the form

$$\operatorname{sgn} D(p_0, p_1,..., p_k) = (-1)^{k+1} \text{ or } 0 \quad (k = 1, 2,..., r).$$

Since every subset of an independent euclidean $(r+1)$-tuple is an independent set, another consequence of (**) is

THEOREM 40.2. *A necessary condition that a semimetric $(r+1)$-tuple $p_0, p_1,..., p_r$ be congruent with an independent euclidean $(r+1)$-tuple is that*

$$\operatorname{sgn} D(p_0, p_1,..., p_k) = (-1)^{k+1} \quad (k = 1, 2,..., r).$$

In the first paper published by Cayley he showed that the mutual distances of five points $q_0, q_1,..., q_4$ in E_3 satisfy the relation

$$D(q_0, q_1,..., q_4) = 0.$$

Since Menger used the determinants D to express *sufficient* conditions for the congruent imbedding of semimetric m-tuples in euclidean n-space, it seems appropriate to call these determinants Cayley–Menger determinants.

EXERCISES

1. Let p_1, p_2, p_3 be any three pairwise distinct points of E_1 and p any point of E_2 not collinear with p_1, p_2, p_3. Apply the law of cosines to triangles p_1, p_2, p and p_2, p_3, p to show the existence of constants a, b, c, d (functions of $p_1 p_2, p_2 p_3, p_1 p_3$) such that

$$a \cdot p_1 p^2 + b \cdot p_2 p^2 + c \cdot p_3 p^2 + d = 0, \quad a+b+c = 0.$$

Show that $D(p_1, p_2, p_3) = 0$ by letting p approach p_1, p_2, p_3 in turn.

2. Use Exercise 1 to obtain $D(p_1, p_2, p_3, p_4) = 0$, $D(p_1, p_2,..., p_5) = 0$ for quadruples and quintuples of E_2 and E_3, respectively.

3. Show that

$$D(p_1, p_2, p_3) = -(a+b+c)(a+b-c)(a-b+c)(-a+b+c),$$

where $a = p_1 p_2$, $b = p_2 p_3$, $c = p_1 p_3$, and hence prove that a semimetric triple p_1, p_2, p_3 is metric if and only if $D(p_1, p_2, p_3) \leqslant 0$.

4. If p_1, p_2, p_3, p_4 are non-collinear points of E_2, find and interpret geometrically the signs of *all* fourth-order minors of $D(p_1, p_2, p_3, p_4)$. Extend to $n+2$ points of E_n with no $(n+1)$-tuple in a hyperplane. See Blumenthal [28] and Gillam.

41. Imbedding a semimetric $(r+1)$-tuple irreducibly in E_r

A semimetric $(r+1)$ tuple is said to be *irreducibly imbeddable* in E_r provided it is congruent with an independent $(r+1)$-tuple of E_r.

THEOREM 41.1. *A necessary and sufficient condition that a semimetric $(r+1)$-tuple $p_0, p_1, ..., p_r$ be irreducibly imbeddable in E_r is that*

$$\operatorname{sgn} D(p_0, p_1, ..., p_k) = (-1)^{k+1} \quad (k = 1, 2, ..., r).$$

Proof. The necessity follows from Theorem 40.2.

To prove the sufficiency, make the inductive hypothesis (obviously valid for $r = 2$) that $\operatorname{sgn} D(p_0, p_1, ..., p_k) = (-1)^{k+1}$ ($k = 1, 2, ..., r-1$), is necessary and sufficient for $p_0, p_1, ..., p_{r-1}$ to be congruent with an independent r-tuple of E_{r-1}.

ASSERTION. *Each r-tuple of $p_0, p_1, ..., p_r$ is irreducibly imbeddable in E_{r-1}.* Consider the r-tuples

$$p_0, p_1, ..., p_{i-1}, p_{i+1}, ..., p_{r-1}, p_r \quad (i = 0, 1, ..., r-1).$$

Denoting by $[i, r]$ the cofactor of the element $p_i p_r^2$ in $D(p_0, p_1, ..., p_r)$, a theorem of determinants gives

$$D(p_0, p_1, ..., p_{r-1}) D(p_0, p_1, ..., p_{i-1}, p_{i+1}, ..., p_r) - [i, r]^2$$
$$= D(p_0, p_1, ..., p_{i-1}, p_{i+1}, ..., p_{r-1}) D(p_0, p_1, ..., p_r) \quad (i = 0, 1, ..., r-1).$$

By the hypothesis of the theorem and the inductive hypothesis, $p_0, p_1, ..., p_{r-1}$ are congruent with an independent r-tuple of E_{r-1} and so (Theorem 40.2)

$$\operatorname{sgn} D(p_0, p_1, ..., p_{i-1}, p_{i+1}, ..., p_{r-1}) = (-1)^{r-1}.$$

Then the above determinant relation gives

$$\operatorname{sgn} D(p_0, p_1, ..., p_{i-1}, p_{i+1}, ..., p_r) = (-1)^r \quad (i = 0, 1, ..., r-1),$$

and the validity of the assertion now follows from the inductive hypothesis and Theorem 40.2 applied to $p_0, p_1, ..., p_{r-1}$.

Let, now, E_{r-1}^* be an $(r-1)$-dimensional subspace of E_r. We have, by the foregoing,

$$p_0, p_1, ..., p_{r-2}, p_{r-1} \approx p'_0, p'_1, ..., p'_{r-2}, p'_{r-1}, \tag{*}$$
$$p_0, p_1, ..., p_{r-2}, p_r \approx p''_0, p''_1, ..., p''_{r-2}, p''_r,$$

with the two r-tuples on the right contained in E_{r-1}^*. Then

$$p''_0, p''_1, ..., p''_{r-2} \approx p'_0, p'_1, ..., p'_{r-2},$$

and a motion of E_{r-1}^* exists mapping p''_i onto p'_i ($i = 0, 1, ..., r-2$). Since $p'_0, p'_1, ..., p'_{r-2}$ is a subset of an independent r-tuple of E_{r-1}^*, these points do not lie in any E_{r-3} and hence the above motion sends p''_r into one of two distinct points $\bar{p}_r, p_r^*$ which are reflections of one another in the $(r-2)$-dimensional subspace of E_{r-1}^* determined by $p'_0, p'_1, ..., p'_{r-2}$. Thus

$$p_0, p_1, ..., p_{r-2}, p_r \approx p'_0, p'_1, ..., p'_{r-2}, \bar{p}_r$$
$$\approx p'_0, p'_1, ..., p'_{r-2}, p_r^*. \tag{**}$$

Denote by $D(p_0, p_1,..., p_r; x)$ the function obtained from $D(p_0, p_1,..., p_r)$ upon replacing $(p_{r-1}p_r)^2$ and $(p_r p_{r-1})^2$ by x. Clearly
$$D(p_0, p_1,..., p_r; x) = Ax^2 + Bx + C,$$
with $A = -D(p_0, p_1,..., p_{r-2})$. Hence $\operatorname{sgn} A = (-1)^r$.

From congruences (∗) and (∗∗),
$$D(p_0, p_1,..., p_{r-1}, p_r; x) \equiv D(p'_0, p'_1,..., p'_{r-1}, \bar{p}_r; x)$$
$$\equiv D(p'_0, p'_1,..., p'_{r-1}, p^*_r; x),$$
and it follows from Theorem 40.1 that the two roots of
$$D(p_0, p_1,..., p_r; x) = 0$$
are $x = (p'_{r-1}\bar{p}_r)^2$ and $x = (p'_{r-1}p^*_r)^2$. These roots are *distinct*, since otherwise p'_{r-1} lies in the E_{r-2} determined by $p'_0, p'_1,..., p'_{r-2}$ (the locus of points of E^*_{r-1} equidistant from $\bar{p}_r$ and p^*_r), which contradicts the independence of $p'_0, p'_1,..., p'_{r-2}, p'_{r-1}$.

Assume the labelling so that
$$p'_{r-1}\bar{p}_r < p'_{r-1}p^*_r.$$
Since $\operatorname{sgn} D(p_0, p_1,..., p_{r-1}, p_r) = (-1)^{r+1}$, it follows that
$$p'_{r-1}\bar{p}_r < p_{r-1}p_r < p'_{r-1}p^*_r.$$
If, now, the E^*_{r-1} be revolved 180° about the $E_{r-2}(p'_0, p'_1,..., p'_{r-2})$ into the E_r, the point $\bar{p}_r$ describes a semicircle Γ on the segment with endpoints $\bar{p}_r, p^*_r$. The distance $p'_{r-1}y$ ($y \in \Gamma$) is a continuous function of y and hence it assumes every value between its minimum $p'_{r-1}\bar{p}_r$ and its maximum $p'_{r-1}p^*_r$. Thus a point p'_r of Γ exists ($\bar{p}_r \neq p'_r \neq p^*_r$) with $p'_{r-1}p'_r = p_{r-1}p_r$. Since $p'_i p'_r = p'_i \bar{p}_r = p_i p_r$ ($i = 0, 1,..., r-2$), then
$$p_0, p_1,..., p_{r-1}, p_r \approx p'_0, p'_1,..., p'_{r-1}, p'_r,$$
with the 'primed' points in E_r, not in any E_{r-1}, and the theorem is proved.

COROLLARY. *If $p_0, p_1,..., p_{r-1}$ is a semimetric r-tuple irreducibly imbeddable in E_{r-1}, then the semimetric (r+1)-tuple $p_0, p_1,..., p_{r-1}, p_r$ is congruent with an independent (r+1)-tuple of E_r if and only if*
$$\operatorname{sgn} D(p_0, p_1,..., p_r) = (-1)^{r+1}.$$

Thus the inequality demanded of $D(p_0, p_1,..., p_r)$ is strong enough to force the selection of the r distances of p_r from $p_0, p_1,..., p_{r-1}$ so that the r-tuples $p_0, p_1,..., p_{i-1}, p_{i+1},..., p_{r-1}, p_r$ ($i = 0, 1,..., r-1$) are euclidean.

42. A solution of the euclidean imbedding problem

To have a complete solution of the problem of finding necessary and sufficient conditions, expressed wholly and explicitly in terms of the metric, in order that an arbitrary semimetric space may be congruent with a euclidean subset, it remains (according to Theorem 38.1) to obtain metric conditions under which a semimetric $(r+3)$-tuple may be congruently imbedded in E_r, when $r+1$ of its points have this property irreducibly. We establish first a lemma.

LEMMA 42.1. *If p_0, p_1,..., p_r are irreducibly imbeddable in E_r, then the semimetric $(r+2)$-tuple p_0, p_1,..., p_r, p_{r+1} is congruent with a subset of E_r if and only if* $$D(p_0, p_1,..., p_r, p_{r+1}) = 0.$$

Proof. The necessity follows from Theorem 40.1. Suppose, now, $D(p_0, p_1,..., p_r, p_{r+1}) = 0$ and make the inductive hypothesis of the theorem's validity when stated for r points $p_0, p_1,..., p_{r-1}$. The induction may be anchored for $r = 1$, for if p_0 is congruently imbeddable in E_0 (which consists of a single point) and not in E_{-1} (which is null) then the semimetric pair p_0, p_1, is congruently imbeddable in E_0 if and only if $D(p_0, p_1) = 2(p_0 p_1)^2 = 0$; that is, if and only if $p_0 = p_1$.

Applying to $D(p_0, p_1,..., p_r, p_{r+1})$ the determinant theorem used in the proof of Theorem 41.1 gives

$$D(p_0, p_1,..., p_r)D(p_0, p_1,..., p_{i-1}, p_{i+1},..., p_r, p_{r+1}) - [i, r+1]^2 = 0$$

$(i = 0, 1,..., r)$, and since $p_0, p_1,..., p_r$ are irreducibly imbeddable in E_r, then
$$\operatorname{sgn} D(p_0, p_1,..., p_r) = (-1)^{r+1}.$$
It follows that
$$\operatorname{sgn} D(p_0, p_1,..., p_{i-1}, p_{i+1},..., p_r, p_{r+1}) = (-1)^{r+1} \text{ or } 0.$$

In the first case, $p_0, p_1,..., p_{i-1}, p_{i+1},..., p_r, p_{r+1}$ are congruently imbeddable in E_r (Corollary, Theorem 41.1), while in the second case these $r+1$ points are congruent with a subset of E_{r-1} (inductive hypothesis). Hence every $(r+1)$-tuple of $p_0, p_1,..., p_r, p_{r+1}$ is congruently imbeddable in E_r.

The method of Theorem 41.1 is again employed. We have

$$p_0, p_1,..., p_{r-1}, p_r \approx p'_0, p'_1,..., p'_{r-1}, p'_r,$$
$$p_0, p_1,..., p_{r-1}, p_{r+1} \approx p'_0, p'_1,..., p'_{r-1}, \bar{p}_{r+1}$$
$$\approx p'_0, p'_1,..., p'_{r-1}, p^*_{r+1},$$

where all points on the right-hand sides of these congruences lie in an E_r, $p'_0, p'_1,..., p'_{r-1}, p'_r$ form an independent $(r+1)$-tuple, and $\bar{p}_{r+1}$, p^*_{r+1} are reflections of each other in the $(r-1)$-dimensional subspace of E_r determined by $p'_0, p'_1,..., p'_{r-1}$. In contrast with the situation existing in the preceding theorem, the points $\bar{p}_{r+1}, p^*_{r+1}$ are not necessarily distinct since $D(p_0, p_1,..., p_{r-1}, p_{r+1})$ may vanish.

Replacing $(p_r p_{r+1})^2$ and $(p_{r+1} p_r)^2$ in $D(p_0, p_1,..., p_r, p_{r+1})$ by x, we have

$$D(p_0, p_1,..., p_r, p_{r+1}; x) \equiv D(p'_0, p'_1,..., p'_r, \bar{p}'_{r+1}; x)$$
$$\equiv D(p'_0, p'_1,..., p'_r, p'^*_{r+1}; x)$$
$$\equiv Ax^2 + Bx + C,$$

with $A = -D(p_0, p_1,..., p_{r-1}) \neq 0$.

Case 1. The points $\bar{p}'_{r+1}, p'^*_{r+1}$ are distinct. Then

$$D(p_0, p_1,..., p_r, p_{r+1}; x) = 0$$

has, by Theorem 40.1, the two distinct roots $x = (p'_r \bar{p}'_{r+1})^2$ and $x = (p'_r p'^*_{r+1})^2$. But since $D(p_0, p_1,..., p_r, p_{r+1}) = 0$, then $(p_r p_{r+1})^2$ is also a root of the quadratic equation. It follows that either $p_r p_{r+1} = p'_r \bar{p}'_{r+1}$ or $p_r p_{r+1} = p'_r p'^*_{r+1}$, and in any event $p_0, p_1,..., p_r, p_{r+1}$ are congruent with an $(r+2)$-tuple of E_r.

Case 2. The points $\bar{p}'_{r+1}, p'^*_{r+1}$ coincide. Then $D(p_0, p_1,..., p_{r-1}, p_{r+1}) = 0$ and $D(p_0, p_1,..., p_r, p_{r+1}; x) = -[r+2, r+3]^2 / D(p_0, p_1,..., p_{r-1})$ where $[r+2, r+3]$ denotes the cofactor of the element appearing in the $(r+2)$th row and $(r+3)$th column of $D(p_0, p_1,..., p_r, p_{r+1}; x)$. It follows that $D(p_0, p_1,..., p_r, p_{r+1}; x)$ is the square of a linear function of x, which vanishes for $x = p'_r p'_{r+1}$, where $\bar{p}'_{r+1} = p_{r+1} = p'^*_{r+1}$. Then

$$p_r p_{r+1} = p'_r p'_{r+1} \quad \text{and} \quad p_0, p_1,..., p_r, p_{r+1} \approx p'_0, p'_1,..., p'_r, p'_{r+1}$$

(points of E_r).

THEOREM 42.1. *If $p_0, p_1,..., p_r$ are congruent with an independent $(r+1)$-tuple of E_r, then the semimetric $(r+3)$-tuple $p_0, p_1,..., p_{r+1}, p_{r+2}$ is congruently contained in E_r if and only if*

$$D(p_0, p_1,..., p_r, p_{r+1}) = D(p_0, p_1,..., p_r, p_{r+2})$$
$$= D(p_0, p_1,..., p_r, p_{r+1}, p_{r+2})$$
$$= 0.$$

Proof. By the preceding lemma (and properties of E_r)

$$p_0, p_1,..., p_r, p_{r+1} \approx p'_0, p'_1,..., p'_r, p'_{r+1},$$
$$p_0, p_1,..., p_r, p_{r+2} \approx p'_0, p'_1,..., p'_r, p'_{r+2},$$

with the 'primed' points in E_r and $p'_0, p'_1,..., p'_r$ an independent $(r+1)$-tuple.

Since $D(p_0, p_1,..., p_r, p_{r+1}) = 0$, the quadratic function

$$D(p_0, p_1,..., p_r, p_{r+1}, p_{r+2}; x)$$

is the square of a linear function. It vanishes, by hypothesis, for

$x = (p_{r+1}p_{r+2})^2$, and, according to Theorem 40.1, for $x = (p'_{r+1}p'_{r+2})^2$. Hence $p_{r+1}p_{r+2} = p'_{r+1}p'_{r+2}$ and

$$p_0, p_1, \ldots, p_r, p_{r+1}, p_{r+2} \approx p'_0, p'_1, \ldots, p'_r, p'_{r+1}, p'_{r+2}.$$

Summarizing the results of previous sections, the following *complete solution of the problem of congruently imbedding a semimetric space in a euclidean space* is obtained.

THEOREM 42.2. *An arbitrary semimetric space S is congruently imbeddable in a euclidean n-dimensional space E_n if and only if* (i) *S contains an $(r+1)$-tuple $p_0, p_1, \ldots, p_r$ ($r \leqslant n$) such that*

$$\operatorname{sgn} D(p_0, p_1, \ldots, p_k) = (-1)^{k+1} \quad (k = 1, 2, \ldots, r),$$

(ii) *for every pair x, y of points of S the determinants $D(p_0, p_1, \ldots, p_r, x)$, $D(p_0, p_1, \ldots, p_r, y)$, $D(p_0, p_1, \ldots, p_r, x, y)$ vanish. Then S is irreducibly imbeddable in E_r.*

Later developments will make use of the following immediate corollary of this theorem.

COROLLARY. *A necessary and sufficient condition that a semimetric space S be congruently imbeddable in E_n ($n > 0$), is that if any two distinct points of S be labelled p_0, p_1 an $(r+1)$-tuple $p_0, p_1, \ldots, p_r$ of S exists ($1 \leqslant r \leqslant n$) such that* (i) $\operatorname{sgn} D(p_0, p_1, \ldots, p_k) = (-1)^{k+1}$ ($k = 1, 2, \ldots, r$), *and* (ii) *if $x, y \in S$, then*

$$D(p_0, p_1, \ldots, p_r, x) = D(p_0, p_1, \ldots, p_r, y) = D(p_0, p_1, \ldots, p_r, x, y) = 0.$$

By Theorem 42.2 the condition is surely sufficient. It is also necessary, for if $S \subsetneq E_n$, a positive integer r exists ($r \leqslant n$) such that $S \approx S'$ ($S' \subset E_r$, S' not contained in any E_{r-1}). Clearly *any two distinct points of S' are contained in an independent $(r+1)$-tuple of S'*, and the corollary is established.

For semimetric spaces containing only a finite number of points the following formulation of the preceding theorem is useful.

THEOREM 42.3. *A necessary and sufficient condition that a semimetric $(n+1)$-tuple may be irreducibly congruently imbeddable in E_r ($r \leqslant n$) is that a labelling $p_0, p_1, \ldots, p_n$ of the points exist so that* (i) *the bordered principal minors $D(p_0, p_1, \ldots, p_k)$ of $D(p_0, p_1, \ldots, p_n)$ have the sign of $(-1)^{k+1}$ ($1 \leqslant k \leqslant r$), and* (ii) *the rank of $D(p_0, p_1, \ldots, p_n)$ is $r+2$.*

Proof. If the $(n+1)$-tuple is irreducibly congruently imbeddable in E_r, then the points may be labelled so that

$$p_0, p_1, \ldots, p_r, \ldots, p_n \approx p'_0, p'_1, \ldots, p'_r, \ldots, p'_n,$$

with the 'primed' points in E_r and $p'_0, p'_1, ..., p'_r$ an independent $(r+1)$-tuple. Then by Theorem 40.2

$$\operatorname{sgn} D(p_0, ..., p_k) = (-1)^{k+1} \quad (1 \leqslant k \leqslant r),$$

while $\quad D(p_0, p_1, ..., p_r, p_{r+i}) = D(p_0, p_1, ..., p_r, p_{r+i}, p_{r+j}) = 0$

$(i, j = 1, 2, ..., n-r; i \neq j)$. It follows that the rank of $D(p_0, p_1, ..., p_n)$ is $r+2$.

Conversely, if a labelling $p_0, p_1, ..., p_n$ of the points exists satisfying (i) and (ii), then $D(p_0, ..., p_r) \neq 0$, while each principal minor of $D(p_0, ..., p_n)$ formed by annexing one row or two rows to $D(p_0, ..., p_r)$ vanishes. The irreducible congruent imbedding of $p_0, p_1, ..., p_n$ in E_r is then ensured by Theorem 42.2.

COROLLARY. *A semimetric $(n+1)$-tuple is congruently imbeddable in E_n if and only if the determinant D of each $(k+1)$-tuple of its points either vanishes or has the sign of $(-1)^{k+1}$ ($k = 1, 2, ..., n$).*

EXERCISE

From the corollary of Theorem 41.1 and Lemma 42.1 it follows that if $p_0, p_1, ..., p_{r-1}$ are irreducibly imbeddable in E_{r-1}, then the semimetric $(r+1)$-tuple $p_0, p_1, ..., p_{r-1}, p_r$ is imbeddable in E_r if and only if

$$\operatorname{sgn} D(p_0, p_1, ..., p_{r-1}, p_r) \neq (-1)^r.$$

Give an example to show that this is not valid if the restriction that $p_0, p_1, ..., p_{r-1}$ be not imbeddable in E_{r-2} is dropped.

43. Additional criteria for imbedding finite semimetric spaces in E_r

In the preceding sections we have shown that any semimetric space S is congruently imbeddable in E_r whenever each $(r+3)$-tuple of S has that property, and criteria for imbedding an $(r+3)$-tuple in E_r were developed in terms of the signs of certain principal minors of a bordered determinant whose elements (apart from the bordering) are the squares of the distances involved in the $(r+3)$-tuple. Other expressions of these criteria are available. In 1935 Schoenberg obtained a necessary and sufficient condition for the irreducible imbedding of a finite semimetric space in E_r. His condition, obtained quite independently of previous results, is phrased in terms of quadratic form theory. We show that his theorem is, in fact, an immediate consequence of Theorem 42.3.

Theorem 43.1 (*Schoenberg*). *A necessary and sufficient condition that a semimetric* $(n+1)$-*tuple* $p_0, p_1, \ldots, p_n$ *be irreducibly congruently imbeddable in* E_r $(r \leqslant n)$ *is that the quadratic form*

$$F(x_1, x_2, \ldots, x_n) = \tfrac{1}{2} \sum_{i,j=1}^{n} [p_0 p_i^2 + p_0 p_j^2 - p_i p_j^2] x_i x_j$$

be positive semi-definite (that is, non-negative) and of rank r.

Proof. By despoiling $D(p_0, p_1, \ldots, p_k)$ of its bordering, we obtain

$$D(p_0, p_1, \ldots, p_k) = (-1)^{k+1} 2^k |p_{ij}| \quad (i,j = 1, 2, \ldots, k), \qquad (*)$$

where $p_{ij} = \tfrac{1}{2}(p_0 p_i^2 + p_0 p_j^2 - p_i p_j^2)$. Now by Theorem 42.3, the $(n+1)$-tuple $p_0, p_1, \ldots, p_n$ is irreducibly imbeddable in E_r if and only if for $r+1$ of its points, say $p_0, p_1, \ldots, p_r$ $(r \leqslant n)$, $\operatorname{sgn} D(p_0, p_1, \ldots, p_k) = (-1)^{k+1}$ $(k = 1, 2, \ldots, r)$, while in case $r < n$, all $(r+3)$th and $(r+4)$th-order principal minors of $D(p_0, p_1, \ldots, p_n)$ containing $D(p_0, p_1, \ldots, p_r)$ vanish. By (*) this is equivalent to $|p_{ij}| > 0$ $(i,j = 1, 2, \ldots, r)$ while all $(r+1)$th- and $(r+2)$th-order principal minors of $|p_{ij}|$ $(i,j = 1, 2, \ldots, n)$, containing $|p_{ij}|$ $(i,j = 1, 2, \ldots, r)$, vanish. But these are precisely necessary and sufficient conditions that the quadratic form $\sum_{i,j=1}^{n} p_{ij} x_i x_j$ be positive semi-definite of rank r, and the theorem is proved.

It is occasionally useful to express imbedding conditions in terms of properties of the quadratic form $\sum_{i,j=0}^{n} p_i p_j^2 . x_i x_j$, whose coefficients are the squares of the mutual distances of the $n+1$ points $p_0, p_1, \ldots, p_n$, rather than in terms of the form $\sum_{i,j=1}^{n} p_{ij} x_i x_j$ with more complicated coefficients. Clearly

$$\sum_{i,j=0}^{n} p_i p_j^2 . x_i x_j = 2 x_0 \sum_{i=1}^{n} p_0 p_i^2 . x_i + \sum_{i,j=1}^{n} p_i p_j^2 . x_i x_j,$$

and for all values of $x_0, x_1, \ldots, x_n$ such that $x_0 + x_1 + \ldots + x_n = 0$, we have

$$\sum_{i,j=0}^{n} p_i p_j^2 . x_i x_j = -2\Big(\sum_{j=1}^{n} x_j\Big)\Big(\sum_{i=1}^{n} p_0 p_i^2 . x_i\Big) + \sum_{i,j=1}^{n} p_i p_j^2 . x_i x_j$$

$$= -2 \sum_{i,j=1}^{n} p_0 p_i^2 . x_i x_j + \sum_{i,j=1}^{n} p_i p_j^2 . x_i x_j$$

$$= -\sum_{i,j=1}^{n} p_0 p_i^2 . x_i x_j - \sum_{i,j=1}^{n} p_0 p_j^2 . x_i x_j + \sum_{i,j=1}^{n} p_i p_j^2 . x_i x_j$$

$$= -\sum_{i,j=1}^{n} [p_0 p_i^2 + p_0 p_j^2 - p_i p_j^2] . x_i x_j.$$

Thus $\sum_{i,j=0}^{n} p_i p_j^2 . x_i x_j$

is negative semi-definite of rank r on the hyperplane $x_0+x_1+\ldots+x_n = 0$ if and only if
$$\sum_{i,j=1}^{n} p_{ij} x_i x_j$$
is positive semi-definite of rank r, and the preceding theorem has the following

COROLLARY. *A semimetric $(n+1)$-tuple $p_0, p_1, \ldots, p_n$ is irreducibly imbeddable in E_r if and only if the quadratic form $\sum_{i,j=0}^{n} p_i p_j^2 \cdot x_i x_j$ is negative semi-definite of rank r on the hyperplane $\sum_{i=0}^{n} x_i = 0$.*

Another criterion for the congruent imbedding in a euclidean space of a semimetric $(n+1)$-tuple is obtained with the aid of a lemma.

LEMMA 43.1. *The determinant $\Delta_3 = |\cos \alpha_{ij}|$, $0 \leq \alpha_{ij} = \alpha_{ji} \leq \pi$, $\alpha_{ii} = 0$ $(i,j = 1, 2, 3)$, is non-negative if and only if $\alpha_{12}, \alpha_{23}, \alpha_{13}$ satisfy the triangle inequality, and have a sum less than or equal to 2π.*

Proof. The necessity is shown in Example 2, § 9, and the sufficiency follows at once from the factored form (†) of Δ_3 given there.

Subtracting from the second and third columns of Δ_3 the products of the first column by $\cos \alpha_{12}$ and $\cos \alpha_{13}$, respectively, yields

$$\Delta_3 = \begin{vmatrix} \sin^2 \alpha_{12} & \cos \alpha_{23} - \cos \alpha_{12} \cos \alpha_{13} \\ \cos \alpha_{23} - \cos \alpha_{12} \cos \alpha_{13} & \sin^2 \alpha_{13} \end{vmatrix}.$$

If, now, angle α_{213} (non-negative and at most π, *if real*) is defined by
$$\sin \alpha_{12} \sin \alpha_{13} \cos \alpha_{213} = \cos \alpha_{23} - \cos \alpha_{12} \cos \alpha_{13},$$
then
$$\Delta_3 = \sin^2 \alpha_{12} \sin^2 \alpha_{13} \sin^2 \alpha_{213},$$
and it follows easily from the lemma and the definition of α_{213} that $\Delta_3 \geq 0$ *if and only if α_{213} is a real angle*.

We observe, further, that if p_0, p_i, p_j are semimetric ($p_0 \neq p_i, p_j$) and the angle $p_0 \colon p_i, p_j$ (non-negative and at most π if real) is defined by the cosine law

$$2 p_0 p_i \cdot p_0 p_j \cos p_0 \colon p_i, p_j = p_0 p_i^2 + p_0 p_j^2 - p_i p_j^2, \qquad (**)$$

then $D(p_0, p_i, p_j) \leq 0$ (that is, p_0, p_i, p_j are *metric*) if and only if $p_0 \colon p_i, p_j$ is a real angle.

THEOREM 43.2. *A semimetric quadruple of distinct points p_0, p_1, p_2, p_3 is congruently imbeddable in E_3 if and only if the angles*
$$p_0 \colon p_i, p_j \quad (i,j = 1, 2, 3)$$
defined by (**) are real, satisfy the triangle inequality, and
$$p_0 \colon p_1, p_2 + p_0 \colon p_2, p_3 + p_0 \colon p_1, p_3 \leq 2\pi.$$

Proof. The necessity is clear from elementary geometry.

To prove the sufficiency, it is observed that the hypotheses on the angles $p_0: p_i, p_j$ $(i,j = 1, 2, 3)$ imply that

$$D(p_0, p_i, p_j) \leq 0 \quad \text{and} \quad D(p_0, p_1, p_2, p_3) \geq 0$$

(Lemma 43.1) since

$$D(p_0, p_1, p_2, p_3) = 8 \prod_{i=1}^{3} p_0 p_i^2 |\cos p_0: p_s, p_t| \quad (s, t = 1, 2, 3).$$

If at least one of the angles $p_0: p_i, p_j$ $(i,j = 1, 2, 3)$ is different from 0 or π then $D(p_0, p_i, p_j) < 0$ for at least one pair of indices i, j and the imbedding of p_0, p_1, p_2, p_3 in E_3 follows from Theorem 41.1 or Lemma 42.1. If, on the other hand, each of the angles $p_0: p_i, p_j$ $(i,j = 1, 2, 3)$ is 0 or π, then at least one of the angles is 0, say $p_0: p_1, p_2$. Since the three angles satisfy the triangle inequality, $p_0: p_1, p_3 = p_0: p_2, p_3$ and the determinant $|\cos p_0: p_i, p_j|$ $(i, j = 1, 2, 3)$ vanishes. Hence $D(p_0, p_1) > 0$ but

$$D(p_0, p_1, p_2) = D(p_0, p_1, p_3) = D(p_0, p_1, p_2, p_3) = 0,$$

and p_0, p_1, p_2, p_3 are imbeddable in E_3 by Theorem 42.1.

Theorem 43.2 may be extended to semimetric $(n+1)$-tuples $p_0, p_1, \ldots, p_n$ by defining angles $p_0: p_i, p_j$ $(i,j = 1, 2, \ldots, n)$ by the cosine law (✻✻) and higher dimensional angles $p_0 p_1 \ldots p_k: p_i, p_j$ by a spherical law of cosines

$$\sin(p_0 p_1 \ldots p_{k-1}: p_k, p_i) \sin(p_0 p_1 \ldots p_{k-1}: p_k, p_j) \cos(p_0 p_1 \ldots p_{k-1} p_k: p_i, p_j)$$
$$= \cos(p_0 p_1 \ldots p_{k-1}: p_i, p_j) - \cos(p_0 p_1 \ldots p_{k-1}: p_k, p_i) \cos(p_0 p_1 \ldots p_{k-1}: p_k, p_j)$$

$(k = 1, 2, \ldots, n-2)$, in case angles $p_0 p_1 \ldots p_{k-1}: p_k, p_i$ and $p_0 p_1 \ldots p_{k-1}: p_k, p_j$ are different from zero or π.

In terms of these higher dimensional angles, taken *non-negative and at most π if real*, the determinant $D(p_0, p_1, \ldots, p_k)$ of $k+1$ points of the $(n+1)$-tuple is factorable, and we have

$$D(p_0, p_1, \ldots, p_k) = (-1)^{k+1} 2^k \prod_{i=1}^{k} p_0 p_i^2 |\cos p_0: p_s, p_t| \quad (s, t = 1, 2, \ldots, k),$$
(†)

with $\quad |\cos p_0: p_i, p_j| = \prod_{i=2}^{k} \sin^2(p_0: p_1, p_i) \prod_{i=3}^{k} \sin^2(p_0 p_1: p_2, p_i) \ldots \times$

$$\times \prod_{i=k-1}^{k} \sin^2(p_0 p_1 \ldots p_{k-3}: p_{k-2}, p_i) \sin^2(p_0 p_1 \ldots p_{k-2}: p_{k-1}, p_k), \quad (\dagger\dagger)$$

provided the introduction of higher dimensional angles can be carried out to the end.

Postulate, now, for all angles introduced the following:

I. The angles $p_0 \colon p_i, p_j$ are real $(i,j = 1,2,...,n)$.

II. $p_0 p_1 \cdots p_k \colon p_i, p_j + p_0 p_1 \cdots p_k \colon p_j, p_l + p_0 p_1 \cdots p_k \colon p_i, p_l \leqslant 2\pi$.

III. $|p_0 p_1 \cdots p_k \colon p_i, p_j - p_0 p_1 \cdots p_k \colon p_j, p_l| \leqslant p_0 p_1 \cdots p_k \colon p_i, p_l$
$\leqslant p_0 p_1 \cdots p_k \colon p_i, p_j + p_0 p_1 \cdots p_k \colon p_j, p_l$.

It is easy to show that (1) *all higher dimensional angles are real*, and (2) *if $p_0 p_1 \cdots p_k \colon p_{k+1}, p_i$ is zero or π then the angles $p_0 p_1 \cdots p_k \colon p_{k+1}, p_j$ and $p_0 p_1 \cdots p_k \colon p_i, p_j$ are equal or supplementary for every index j.* In either event the determinant $|\cos p_0 p_1 \cdots p_k \colon p_i, p_j|$ $(i,j = k+1, k+2,..., n)$ is zero, and the introduction of angles of still higher dimensions is unnecessary.

It follows that Postulates I, II, III imply that

$$\operatorname{sgn} D(p_0, p_1,..., p_k) = (-1)^{k+1} \text{ or } 0,$$

and we are thus led to another imbedding criterion.

THEOREM 43.3. *A semimetric $(n+1)$-tuple is imbeddable in E_n unless its points can be labelled $p_0, p_1,..., p_n$ so that* (i) *either at least one of the angles $p_0 \colon p_i, p_j$ $(i,j = 1,2,...,n)$ is not real, or* (ii) *higher dimensional angles*

$$p_0 p_1 \cdots p_k \colon p_i, p_j, \qquad p_0 p_1 \cdots p_k \colon p_j, p_l, \qquad p_0 p_1 \cdots p_k \colon p_i, p_l$$

exist such that their sum exceeds 2π or for which the triangle inequality is not valid.

Essentially this theorem was obtained by W. A. Wilson in 1935 as a generalization of Theorem 43.2 due to the author.

EXERCISES

1. Obtain formulae (†) and (††).
2. Supply the details in the proof of Theorem 43.3.

44. An example of a pseudo-E_n space

It was proved in § 38 that E_n has congruence indices $(n+3, 0)$ with respect to the class of semimetric spaces. In examining the question of the existence of a 'better' congruence symbol for E_n, we shall first of all show that it cannot be replaced by the symbol $(n+2, 0)$.

Let $p'_0, p'_1,..., p'_n$ be the vertices of an equilateral simplex of E_n, and p'_{n+1}, the centre of the circumscribing sphere; that is,

$$p'_0 p'_{n+1} = p'_1 p'_{n+1} = \cdots = p'_n p'_{n+1}.$$

Denote by q'_i the reflection of p'_{n+1} in the $(n-1)$-dimensional 'face' of the simplex determined by $p'_0, p'_1, \ldots, p'_{i-1}, p'_{i+1}, \ldots, p'_n$ ($i = 0, 1, \ldots, n$), and let $p_0, p_1, \ldots, p_{n+1}, p_{n+2}$ be a semimetric $(n+3)$-tuple with

$$p_0, p_1, \ldots, p_n, p_{n+1} \approx p'_0, p'_1, \ldots, p'_n, p'_{n+1},$$

$$p_0, p_1, \ldots, p_n, p_{n+2} \approx p'_0, p'_1, \ldots, p'_n, p'_{n+2},$$

where $$p'_{n+2} = p'_{n+1},$$

and $$p_{n+1}p_{n+2} = p'_{n+1}q'_0 = p'_{n+1}q'_1 = \ldots = p'_{n+1}q'_n.$$

It may be observed that p'_{n+2} is the circumcentre of the $(n+1)$-tuple $q'_0, q'_1, \ldots, q'_n$, and $p_{n+1}p_{n+2} = r$, the circumradius of these points.

Now, by the foregoing, evidently

$$p_0, p_1, \ldots, p_{i-1}, p_{i+1}, \ldots, p_{n+1}, p_{n+2} \approx p'_0, p'_1, \ldots, p'_{i-1}, p'_{i+1}, \ldots, p'_{n+1}, q'_i$$

($i = 0, 1, \ldots, n$), and so each $n+2$ of the points $p_0, p_1, \ldots, p_{n+1}, p_{n+2}$ is congruent with an $(n+2)$-tuple of E_n. But $p_0, p_1, \ldots, p_{n+1}, p_{n+2}$ are not congruently imbeddable in E_n (or, indeed, in any euclidean space) for in the contrary case *distinct* points $\bar{p}_{n+1}, \bar{p}_{n+2}$ of E_n exist, each of which is equidistant from the points of an *independent* $(n+1)$-tuple $\bar{p}_0, \bar{p}_1, \ldots, \bar{p}_n$ of E_n, which is clearly impossible.

DEFINITION 44.1. *A semimetric space S is called pseudo-E_n provided each subset of $n+2$ points of S (not necessarily pairwise distinct) is congruently imbeddable in E_n, but S is not congruent with a subset of E_n.*

The semimetric space consisting of the points $p_0, p_1, \ldots, p_{n+1}, p_{n+2}$ defined above is, then, pseudo-E_n and we have shown

THEOREM 44.1. *For each positive integer n, pseudo-E_n spaces exist.*

Thus for no euclidean space E_n, $n > 0$, can the congruence symbol $(n+3, 0)$ be replaced by $(n+2, 0)$.

EXERCISE

Prove that the pseudo-E_n $(n+3)$-tuple $p_0, p_1, \ldots, p_{n+1}, p_{n+2}$ defined in this section cannot be imbedded congruently in any euclidean space by showing that $\operatorname{sgn} D(p_0, p_1, \ldots, p_{n+2}) = (-1)^n$.

45. The structure of pseudo-E_n $(n+3)$-tuples

The example of a pseudo-E_n space constructed in the preceding section gives rise to two important problems: (1) *to determine the structure of the most general pseudo-E_n $(n+3)$-tuple*, and (2) *to ascertain whether or not pseudo-E_n spaces of more than $n+3$ points exist.* We shall find that the results obtained here in solving the first problem are needed when we turn to the consideration of the second one.

LEMMA 45.1. *Each pseudo-E_n $(n+3)$-tuple P contains at least one independent $(n+1)$-tuple (that is, an $(n+1)$-tuple whose congruent image in E_n is an independent set).*

Proof. By definition of P, each of its $(n+2)$-tuples is imbeddable in E_n, and if the lemma is false, then each of these $(n+2)$-tuples is imbeddable in E_{n-1}. Since E_{n-1} has congruence order $n-1+3 = n+2$, it follows that P is imbeddable in E_{n-1} (and hence in E_n), contrary to its definition.

LEMMA 45.2. *If $p_0, p_1, \ldots, p_{n+2}$ is a pseudo-E_n $(n+3)$-tuple with $p_0, p_1, \ldots, p_n$ independent, points $p'_0, p'_1, \ldots, p'_{n+2}$ of E_n exist such that*

$$p_0, p_1, \ldots, p_n, p_{n+1} \approx p'_0, p'_1, \ldots, p'_n, p'_{n+1},$$

$$p_0, p_1, \ldots, p_n, p_{n+2} \approx p'_0, p'_1, \ldots, p'_n, p'_{n+2},$$

$$p_{n+1} p_{n+2} \neq p'_{n+1} p'_{n+2}$$

and p'_{n+1}, p'_{n+2} are uniquely determined by $p'_0, p'_1, \ldots, p'_n$.

Proof. We have

$$p_0, p_1, \ldots, p_n \approx p'_0, p'_1, \ldots, p'_n,$$

with the primed points an independent subset of E_n.
Then

$$p_0, p_1, \ldots, p_n, p_{n+1} \approx \bar{p}_1, \bar{p}_2, \ldots, \bar{p}_n, \bar{p}_{n+1} \approx p'_0, p'_1, \ldots, p'_n, p'_{n+1},$$

$$p_0, p_1, \ldots, p_n, p_{n+2} \approx p^*_1, p^*_2, \ldots, p^*_n, p^*_{n+2} \approx p'_0, p'_1, \ldots, p'_n, p'_{n+2},$$

with the 'barred' and 'starred' points in E_n, and the last congruences in the above sets, together with the uniqueness of p'_{n+1}, p'_{n+2}, consesequences of Property V, § 38.

Clearly $p_{n+1} p_{n+2} \neq p'_{n+1} p'_{n+2}$, since otherwise

$$p_0, p_1, \ldots, p_{n+1}, p_{n+2} \subseteq E_n.$$

We say that $p_0, p_1, \ldots, p_{n+2}$ and $p'_0, p'_1, \ldots, p'_{n+2}$ are *almost congruent*, and shall denote this relation by writing

$$p_0, p_1, \ldots, p_{n+1}, p_{n+2} \sim p'_0, p'_1, \ldots, p'_{n+1}, p'_{n+2}.$$

THEOREM 45.1. *If $p_0, p_1, \ldots, p_{n+2}$ is a pseudo-E_n $(n+3)$-tuple, then*

$$\operatorname{sgn} D(p_0, p_1, \ldots, p_{n+2}) = (-1)^n.$$

Proof. Assuming the labelling so that $p_0, p_1, \ldots, p_n$ is an independent $(n+1)$-tuple, let $p'_0, p'_1, \ldots, p'_{n+2}$ be points of E_n with

$$p_0, p_1, \ldots, p_{n+1}, p_{n+2} \sim p'_0, p'_1, \ldots, p'_{n+1}, p'_{n+2}.$$

Denote by $D(p_0, p_1,..., p_{n+2}; x)$ the quadratic function obtained from $D(p_0, p_1,..., p_{n+2})$ by substituting x for $(p_{n+1}p_{n+2})^2$ and $(p_{n+2}p_{n+1})^2$. Then $x = (p'_{n+1}p'_{n+2})^2$ is a root of $D(p_0, p_1,..., p_{n+2}; x) = 0$, since

$$D(p_0, p_1,..., p_{n+2}; (p'_{n+1}p'_{n+2})^2) = D(p'_0, p'_1,..., p'_{n+1}, p'_{n+2}) = 0.$$

Now by an expansion used in § 41,

$$D(p_0, p_1,..., p_{n+2}; x) = \frac{D(p_0, p_1,..., p_{n+1}) D(p_0, p_1,..., p_n, p_{n+2}) - (Ex+F)^2}{D(p_0, p_1,..., p_n)},$$

where $E = D(p_0, p_1,..., p_n)$ and F is independent of x.

Since $\qquad D(p_0, p_1,..., p_{n+1}) = D(p'_0, p'_1,..., p'_{n+1}) = 0,$

and $\qquad \operatorname{sgn} D(p_0, p_1,..., p_n) = (-1)^{n+1}$

(due to the independence of $p_0, p_1,..., p_n$) we conclude that

$$D(p_0, p_1,..., p_{n+2}; x)$$

has $x = (p'_{n+1}p'_{n+2})^2$ as its *only* root. But $p_{n+1}p_{n+2} \neq p'_{n+1}p'_{n+2}$ and so $(p_{n+1}p_{n+2})^2 E + F \neq 0$. Hence

$$\operatorname{sgn} D(p_0, p_1,..., p_{n+2}) = \operatorname{sgn} \frac{-[(p_{n+1}p_{n+2})^2 E + F]^2}{D(p_0, p_1,..., p_n)}$$
$$= (-1)^n.$$

COROLLARY. *A pseudo-E_n $(n+3)$-tuple is not congruently imbeddable in any euclidean space.*

Let, now, $p_0, p_1,..., p_{n+2}$ be a pseudo-E_n $(n+3)$-tuple with $p_0, p_1,..., p_n$ independent and let $p'_0, p'_1,..., p'_{n+2}$ be points of E_n with

$$p_0, p_1,..., p_n, p_{n+1} \approx p'_0, p'_1,..., p'_n, p'_{n+1},$$
$$p_0, p_1,..., p_n, p_{n+2} \approx p'_0, p'_1,..., p'_n, p'_{n+2}, \qquad (\dagger)$$
$$p_{n+1}p_{n+2} \neq p'_{n+1}p'_{n+2}.$$

For each index $i = 0, 1, 2,..., n$, the $(n+2)$-tuple

$$p_0, p_1,..., p_{i-1}, p_{i+1},..., p_n, p_{n+1}, p_{n+2}$$

is congruent with a subset of E_n, say

$$p_0, p_1,..., p_{i-1}, p_{i+1},..., p_n, p_{n+1}, p_{n+2}$$
$$\approx \bar{p}_0, \bar{p}_1,..., \bar{p}_{i-1}, \bar{p}_{i+1},..., \bar{p}_n, \bar{p}_{n+1}, \bar{p}_{n+2}.$$

Then

$$\bar{p}_0, \bar{p}_1,..., \bar{p}_{i-1}, \bar{p}_{i+1},..., \bar{p}_n, \bar{p}_{n+1} \approx p_0, p_1,..., p_{i-1}, p_{i+1},..., p_n, p_{n+1}$$
$$\approx p'_0, p'_1,..., p'_{i-1}, p'_{i+1},..., p'_n, p'_{n+1},$$

with $p'_0, p'_1,..., p'_{i-1}, p'_{i+1},..., p'_n$ an independent subset of E_n (since $p'_0, p'_1,..., p'_n$ is an independent $(n+1)$-tuple), and there exists a motion of E_n carrying the barred points into the corresponding primed ones. If q'_i denotes the point into which $\bar{p}_{n+2}$ is sent by one such motion, then

$$p_0, p_1, \ldots, p_{i-1}, p_{i+1}, \ldots, p_n, p_{n+1}, p_{n+2}$$
$$\approx p'_0, p'_1, \ldots, p'_{i-1}, p'_{i+1}, \ldots, p'_n, p'_{n+1}, q'_i, \quad (*)$$

and since $p'_{n+1} q'_i = p_{n+1} p_{n+2} \neq p'_{n+1} p'_{n+2}$, $q'_i \neq p'_{n+2}$. But

$$p'_{n+2} p'_j = p_{n+2} p_j \quad (j = 0, 1, \ldots, n)$$

by (†) and $p_{n+2} p_j = q'_i p'_j$ $(j = 0, 1, \ldots, i-1, i+1, \ldots, n)$ by (∗). Hence $p'_{n+2} p'_j = q'_i p'_j$ $(j = 0, 1, \ldots, i-1, i+1, \ldots, n)$, and so q'_i is the reflection of p'_{n+2} in that $(n-1)$-dimensional face of the simplex $(p'_0, p'_1, \ldots, p'_n)$ whose vertices are $p'_0, p'_1, \ldots, p'_{i-1}, p'_{i+1}, \ldots, p'_n$ $(i = 0, 1, \ldots, n)$. It follows that

$$p_0, p_1, \ldots, p_{i-1}, p_{i+1}, \ldots, p_n, p_{n+2} \quad (i = 0, 1, \ldots, n)$$

are independent $(n+1)$-tuples.

We obtain from (∗) also that

$$p'_{n+1} q'_i = p_{n+1} p_{n+2} \quad (i = 0, 1, \ldots, n);$$

that is, p'_{n+1} is the centre of the sphere through $q'_0, q'_1, \ldots, q'_n$, and $p_{n+1} p_{n+2}$ equals the radius of this sphere. (It is easily shown that if $q'_0, q'_1, \ldots, q'_n$ lie in an E_{n-1}, no point is equidistant from them (Exercise 4).)

Two points p'_{n+1}, p'_{n+2} that are related in this manner, with respect to a non-degenerate simplex $\{p'_0, p'_1, \ldots, p'_n\}$ of E_n are said to be *isogonal conjugates* with respect to the simplex. (It is well known that the relation is a reciprocal one.) Calling the circumsphere of $q'_0, q'_1, \ldots, q'_n$ the *isogonal sphere* of p'_{n+2}, we have the following characterization theorem.

THEOREM 45.2. *A semimetric $(n+3)$-tuple P is a pseudo-E_n set if and only if a labelling $p_0, p_1, \ldots, p_{n+2}$ of the points of P exists so that*

$$p_0, p_1, \ldots, p_{n+2} \sim p'_0, p'_1, \ldots, p'_{n+2} \in E_n,$$

with p'_{n+1}, p'_{n+2} isogonal conjugates with respect to the non-degenerate simplex of E_n with vertices $p'_0, p'_1, \ldots, p'_n$, and $p_{n+1} p_{n+2}$ equals the radius of the isogonal sphere of p'_{n+2}.

Proof. The necessity is established by the preceding argument. To prove the sufficiency, it is first of all clear that a set P with the properties stated in the theorem has each of its $(n+2)$-tuples congruently imbeddable in E_n. But that P is not congruent with a subset of E_n follows from the fact that $p'_{n+1} p'_{n+2} \neq r = p_{n+1} p_{n+2}$. For if

$$p'_{n+1} p'_{n+2} = r = p'_{n+1} q'_i \quad (i = 0, 1, \ldots, n),$$

then p'_{n+1} must lie in every 'face' of the simplex $(p'_0, p'_1, \ldots, p'_n)$ (since $q'_0, q'_1, \ldots, q'_n$ are pairwise distinct) which is impossible.

It will be noticed that the example of a pseudo-E_n $(n+3)$-tuple given in § 44 is a very special case of the general such set described in Theorem 45.2. It should be observed, further, that *any* independent $(n+1)$-tuple of a pseudo-E_n $(n+3)$-tuple may be labelled $p_0, p_1,..., p_n$ and the euclidean subset $p'_0, p'_1,..., p'_n, p'_{n+1}, p'_{n+2}$ almost congruent to $p_0, p_1,..., p_{n+1}, p_{n+2}$ is then uniquely determined (apart from position in E_n).

THEOREM 45.3. *Every subset of $n+1$ points of a pseudo-E_n $(n+3)$-tuple P is an independent $(n+1)$-tuple.*

Proof. If the points of P are labelled $p_0, p_1,..., p_n, p_{n+1}, p_{n+2}$, with $p_0, p_1,..., p_n$ the independent $(n+1)$-tuple guaranteed to exist by Lemma 45.1, then we have seen that for each index $i = 0, 1,..., n$ the sets

$$p_0, p_1,..., p_{i-1}, p_{i+1},..., p_n, p_{n+1} \quad \text{and} \quad p_0, p_1,..., p_{i-1}, p_{i+1},..., p_n, p_{n+2}$$

are independent $(n+1)$-tuples. In a similar manner, starting from the independence of an $(n+1)$-tuple $p_0, p_1,..., p_{i-1}, p_{i+1},..., p_n, p_{n+1}$, it is proved that all of the $(n+1)$-tuples

$$p_0, p_1,..., p_{i-1}, p_{i+1},..., p_{j-1}, p_{j+1},..., p_n, p_{n+1}, p_{n+2}$$

$(i,j = 0, 1,..., n;\ i \neq j)$ are independent sets. Since obviously no repetitions are involved, this gives

$$1+n+1+n+1+\tfrac{1}{2}n(n+1) = \tfrac{1}{2}(n+2)(n+3)$$

pairwise distinct independent $(n+1)$-tuples, which are all that the $n+3$ points contain.

THEOREM 45.4. *A semimetric $(n+3)$-tuple P is a pseudo-E_n set if and only if for any labelling whatever, say $p_0, p_1,..., p_{n+2}$, of the points of P, an $(n+3)$-tuple $p'_0, p'_1,..., p'_{n+2}$ of E_n exists such that*

$$p_0, p_1,..., p_{n+2} \sim p'_0, p'_1,..., p'_{n+2},$$

with p'_{n+1}, p'_{n+2} isogonal conjugates with respect to the non-degenerate simplex of E_n with vertices $p'_0, p'_1,..., p'_n$, and $p_{n+1}p_{n+2}$ equals the radius of the isogonal sphere of p'_{n+2}.

Proof. The proof follows at once from the preceding theorem and the argument establishing Theorem 45.2.

EXERCISES

1. Show that the points of any pseudo-E_1 (pseudo-linear) quadruple may be labelled p_0, p_1, p_2, p_3 so that

$$p_0p_1 = p_2p_3, \quad p_1p_2 = p_0p_3, \quad p_0p_2 = p_0p_1 + p_1p_2 = p_1p_3,$$

and conversely. Show that p_0, p_1, p_2, p_3 (metric and distinct) are pseudo-linear provided $p_0p_1p_2,\ p_1p_2p_3,\ p_2p_3p_0,\ p_3p_0p_1$ exist under appropriate labelling.

2. Compute $D(p_0, p_1, p_2, p_3)$ for a pseudo-linear quadruple.
3. Let $q' = f(q)$ denote the isogonal conjugate of q with respect to the vertices a, b, c of a proper triangle T of E_2 ($q, q' \in E_2$). Show that (1) f is involutory, (2) the fixed points are the centres of the inscribed and escribed circles of T, (3) each point of the lines determined by a, b, c has the opposite vertex as isogonal conjugate, and (4) if $q \neq a, b, c$ is a point of the circumcircle of T, then q' does not exist, and conversely.
4. Show that if no two of the $n+1$ points obtained by reflecting a point in the $n+1$ faces of a non-degenerate simplex of E_n coincide, then no point of E_n is equidistant from them if they lie in an E_{n-1}.
5. What is the locus of points of E_n with no isogonal conjugates with respect to the vertices of a non-degenerate simplex of E_n?

46. Quasi-congruence order of E_n

Turning now to the second of the problems posed in the preceding section, we shall establish here the important result that *pseudo-E_n spaces of more than $n+3$ points do not exist*.

LEMMA 46.1. *If P and Q are two pseudo-E_n $(n+3)$-tuples with $n+2$ points of P congruent with $n+2$ points of Q, then $P \approx Q$.*

Proof. If the points of P and Q be labelled $p_0, p_1, ..., p_{n+2}$ and $q_0, q_1, ..., q_{n+2}$, respectively, with
$$p_0, p_1, ..., p_{n+1} \approx q_0, q_1, ..., q_{n+1},$$
then by Lemma 45.2 and Theorem 45.4 it follows that points
$$p'_0, p'_1, ..., p'_{n+2}$$
of E_n exist such that
$$p_0, p_1, ..., p_{n+2} \sim p'_0, p'_1, ..., p'_{n+2} \sim q_0, q_1, ..., q_{n+2}.$$
It follows that
$$p_{n+2} p_j = p'_{n+2} p'_j = q_{n+2} q_j \quad (j = 0, 1, ..., n),$$
and
$$p_{n+1} p_{n+2} = r = q_{n+1} q_{n+2},$$
where r is the radius of the isogonal sphere of p'_{n+2}; that is, $P \approx Q$.

LEMMA 46.2. *A pseudo-E_n $(n+4)$-tuple P contains at least three pseudo-E_n $(n+3)$-tuples.*

Proof. Since E_n has congruence order $n+3$, P surely contains at least one pseudo-E_n $(n+3)$-tuple, say $p_0, p_1, ..., p_{n+2}$. Assume the lemma false and let p_{n+3} denote the remaining point of P. Then at least two $(n+3)$-tuples, say $p_0, p_1, ..., p_{n+1}, p_{n+3}$ and $p_0, p_1, ..., p_n, p_{n+2}, p_{n+3}$, are congruent with subsets of E_n.

Let $p'_0, p'_1, ..., p'_{n+2}$ be points of E_n with
$$p_0, p_1, ..., p_{n+1}, p_{n+2} \sim p'_0, p'_1, ..., p'_{n+1}, p'_{n+2}.$$
Then
$$p_0, p_1, ..., p_n, p_{n+1}, p_{n+3} \approx p'_0, p'_1, ..., p'_n, p'_{n+1}, p'_{n+3},$$
$$p_0, p_1, ..., p_n, p_{n+2}, p_{n+3} \approx p'_0, p'_1, ..., p'_n, p'_{n+2}, p'_{n+3}, \qquad (*)$$

with p'_{n+3} unique since (Theorem 45.3) $p_0, p_1,..., p_n$ is an independent $(n+1)$-tuple. It follows that $p_{n+3}p_i = p'_{n+3}p'_i$ ($i = 0, 1,..., n+2$), and hence $p'_{n+3} \neq p'_i$ ($i = 0, 1,..., n$). Since p'_{n+3} is, therefore, not a vertex of the non-degenerate simplex $\{p'_0, p'_1,..., p'_n\}$ of E_n, there are at least two 'faces' of this simplex, say

$$\{p'_0, p'_1,..., p'_{i-1}, p'_{i+1},..., p'_n\} \quad \text{and} \quad \{p'_0, p'_1,..., p'_{j-1}, p'_{j+1},..., p'_n\}$$

(i, j different integers between 0 and n) not containing p'_{n+3}.

We assert that

$$p_0, p_1,..., p_{i-1}, p_{i+1},..., p_n, p_{n+1}, p_{n+2}, p_{n+3},$$
$$p_0, p_1,..., p_{j-1}, p_{j+1},..., p_n, p_{n+1}, p_{n+2}, p_{n+3},$$

are pseudo-E_n $(n+3)$-tuples. Since all $(n+2)$-tuples of P are congruently imbeddable in E_n, it suffices to show that neither of these $(n+3)$-tuples is congruent with a subset of E_n. Suppose

$$p_0, p_1,..., p_{i-1}, p_{i+1},..., p_n, p_{n+1}, p_{n+2}, p_{n+3}$$
$$\approx \bar{p}_0, \bar{p}_1,..., \bar{p}_{i-1}, \bar{p}_{i+1},..., \bar{p}_n, \bar{p}_{n+1}, \bar{p}_{n+2}, \bar{p}_{n+3},$$

with the barred points in E_n. By the first of congruences (∗) a motion of E_n exists mapping

onto $\quad \bar{p}_0, \bar{p}_1,..., \bar{p}_{i-1}, \bar{p}_{i+1},..., \bar{p}_n, \bar{p}_{n+1}, \bar{p}_{n+3}$
$\quad\quad p'_0, p'_1,..., p'_{i-1}, p'_{i+1},..., p'_n, p'_{n+1}, p'_{n+3},$

and since the $(n+1)$-tuple $p'_0, p'_1,..., p'_{i-1}, p'_{i+1},..., p'_n, p'_{n+3}$ is independent, the motion is unique. By the second of congruences (∗), this motion carries $\bar{p}_{n+2}$ into p'_{n+2} and we have

$$p_0, p_1,..., p_{i-1}, p_{i+1},..., p_n, p_{n+1}, p_{n+2}, p_{n+3}$$
$$\approx p'_0, p'_1,..., p'_{i-1}, p'_{i+1},..., p'_n, p'_{n+1}, p'_{n+2}, p'_{n+3}.$$

But this gives $p_{n+1}p_{n+2} = p'_{n+1}p'_{n+2}$, contrary to

$$p_0, p_1,..., p_{n+1}, p_{n+2} \sim p'_0, p'_1,..., p'_{n+1}, p'_{n+2}.$$

Similarly it is shown that

$p_0, p_1,..., p_{j-1}, p_{j+1},..., p_n, p_{n+1}, p_{n+2}, p_{n+3}$ is pseudo-E_n, and the lemma is proved.

LEMMA 46.3. *Each* $(n+3)$-*tuple of a pseudo-*E_n $(n+4)$-*tuple* P *is pseudo-*E_n.

Proof. Let

$$p_0, p_1,..., p_n, p_{n+1}, p_{n+2},$$
$$p_0, p_1,..., p_n, p_{n+1}, p_{n+3},$$
$$p_0, p_1,..., p_n, p_{n+2}, p_{n+3}.$$

denote three pseudo-E_n ($n+3$)-tuples of P. By Lemma 46.1 these three ($n+3$)-tuples are pairwise congruent and hence

$$p_{n+1}p_j = p_{n+2}p_j = p_{n+3}p_j \quad (j = 0,1,...,n),$$

$$p_{n+1}p_{n+2} = p_{n+2}p_{n+3} = p_{n+1}p_{n+3}.$$

If, now, $p_0, p_1,..., p_{i-1}, p_{i+1},..., p_n, p_{n+1}, p_{n+2}, p_{n+3}$ is assumed imbeddable in E_n, for i any one of the numbers $0, 1,..., n$, then the E_n contains $n+3$ points

$$p'_0, p'_1,..., p'_{i-1}, p'_{i+1},..., p'_n, p'_{n+1}, p'_{n+2}, p'_{n+3}$$

with $\quad p'_{n+1}p'_j = p'_{n+2}p'_j = p'_{n+3}p'_j \quad (j = 0,1,...,i-1,i+1,...,n),$

$$p'_{n+1}p'_{n+2} = p'_{n+2}p'_{n+3} = p'_{n+1}p'_{n+3}.$$

It follows easily (Exercise 2) that the points

$$p'_0, p'_1,..., p'_{i-1}, p'_{i+1},..., p'_n$$

form a *dependent* n-tuple and hence the points $p_0, p_1, p_{i-1}, p_{i+1},..., p_n$ are dependent. But this is impossible, since they are part of the ($n+1$)-tuple $p_0, p_1,..., p_n$ which is independent (being an ($n+1$)-tuple of a pseudo-E_n ($n+3$)-tuple). This contradiction yields the lemma.

THEOREM 46.1. *If a semimetric space S of more than $n+3$ distinct points has each of its ($n+2$)-tuples congruently contained in E_n, then S is congruent with a subset of E_n.*

Proof. If S is not congruently contained in E_n, then S has at least one ($n+3$)-tuple with this property, and any ($n+4$)-tuple containing it is a pseudo-E_n set. Let $P = (p_0, p_1,..., p_{n+3})$ be such an ($n+4$)-tuple and consider the $n+2$ points $p_0, p_1,..., p_{n+1}$. We shall obtain the desired contradiction by showing that $p_0, p_1,..., p_{n+1}$ is an equilateral set and hence is not congruently imbeddable in E_n.

Let i, j, r, s be pairwise distinct indices selected from $0, 1,..., n+1$. The two ($n+3$)-tuples

$$p_0, p_1,..., p_{i-1}, p_{i+1},..., p_{r-1}, p_{r+1},..., p_{n+1}, p_{n+2}, p_{n+3}, p_i,$$

$$p_0, p_1,..., p_{i-1}, p_{i+1},..., p_{r-1}, p_{r+1},..., p_{n+1}, p_{n+2}, p_{n+3}, p_r$$

are pseudo-E_n sets (Lemma 46.3) and from Lemma 46.1 we have $p_i p_j = p_r p_j$.

From the two pseudo-E_n ($n+3$)-tuples

$$p_0, p_1,..., p_{j-1}, p_{j+1},..., p_{s-1}, p_{s+1},..., p_{n+1}, p_{n+2}, p_{n+3}, p_j,$$

$$p_0, p_1,..., p_{j-1}, p_{j+1},..., p_{s-1}, p_{s+1},..., p_{n+1}, p_{n+2}, p_{n+3}, p_s,$$

we obtain $p_j p_r = p_s p_r$. Hence $p_i p_j = p_r p_s$, $p_0, p_1,..., p_{n+1}$ has all distances equal and is therefore not imbeddable in E_n.

Thus, in Menger's terminology, *the E_n has quasi-congruence order $n+2$ (with respect to the class of semimetric spaces).*

COROLLARY. *For every positive integer n, the n-dimensional euclidean space E_n has the best congruence indices $(n+2, 1)$ with respect to the class of semimetric spaces.*

Proof. The theorem shows that E_n has indices $(n+2, 1)$ and since E_n contains an equilateral $(n+1)$-tuple, the space does not have indices $(n+1, k)$ for any $k = 0, 1, 2,\ldots$. Indices $(n+2, 0)$ are invalid by § 44.

EXERCISES

1. Give a purely algebraic proof that E_1 has indices $(3, 1)$. (No entirely algebraic proof of Theorem 46.1 has yet been obtained.)
2. Prove the property of E_n used in the proof of Lemma 46.3.
3. Prove that the E_1 is characterized among all complete, convex, externally convex metric spaces containing at least two points by having indices $(3, 1)$ with respect to the class of semimetric spaces. (*Hint.* Apply Theorem 21.2.)
4. Prove that a semimetric $(n+2)$-tuple is congruently imbeddable in E_n if and only if each of its $(n+1)$-tuples has that property, and the determinant D of the $n+2$ points is zero.

47. Free $(n+2)$-tuples

In proving that E_n has congruence order $n+3$ with respect to the class of semimetric spaces (§ 38) we found it unnecessary to assume that *every* $(n+3)$-tuple of a semimetric space S is imbeddable in E_n. It sufficed to require that every $(n+3)$-tuple *containing an independent $(n+1)$-tuple* be congruent with a subset of E_n. This allows the presence in S of $(n+3)$-tuples which are not assumed congruently imbeddable in E_n. Such unrestricted subsets of S are said to be *free*.

The question arises whether Theorem 46.1, which establishes that E_n has quasi-congruence order $n+2$, can be strengthened in a similar manner by permitting the presence in S of free $(n+2)$-tuples. If so, how many of the $(n+2)$-tuples might be allowed to be free?

In answering these questions we observe first of all that it does not suffice for the imbedding of S in E_n to assume merely that each $(n+2)$-tuple of S containing an independent $(n+1)$-tuple be imbeddable in E_n, even if the power of S is that of the continuum. For $n = 1$, for example, this requires only that every triple containing a given pair of distinct points be imbeddable in E_1—a condition satisfied by the convex circle upon taking the pair of distinct points to be diametral.

A valid analogue of the previous type of weakened hypothesis is given in the following theorem.

THEOREM 47.1. *Let S be a semimetric space of at least $n+4$ distinct points (n, a non-negative integer). S is congruently contained in E_n, not in E_{n-1}, if and only if (a) S contains a set Q of $n+2$ distinct points of which at least one $(n+1)$-tuple is not imbeddable in E_{n-1}, and (b) each $(n+2)$-tuple of S with at least n points in common with Q is imbeddable in E_n.*

Proof. Let $p_1, p_2, ..., p_{n+1}, p_{n+2}$ denote the points of Q, with $p_1, p_2, ..., p_{n+1}$ not imbeddable in E_{n-1}. From (b) these $n+1$ points are imbeddable in E_n, and hence form an independent $(n+1)$-tuple. If r, s are any two distinct elements of S, each different from $p_1, p_2, ..., p_{n+2}$, it follows from (b) and Theorem 46.1 that

$$p_1, p_2, ..., p_{n+1}, p_{n+2}, r, s \cong E_n;$$

that is, every $(n+3)$-tuple of S containing the independent $(n+1)$-tuple $p_1, p_2, ..., p_{n+1}$ is imbeddable in E_n. Hence (Theorem 38.1) $S \cong E_n$.

An example shows that this theorem is no longer valid if the assumption that at least one $(n+1)$-tuple of Q is not imbeddable in E_{n-1} is not made (suppressing at the same time the assertion that S is not imbeddable in E_{n-1}). For let p_1, p_2, p_3 be the vertices of an equilateral triangle of E_2, with circumcentre labelled both p_4 and p_5, and consider the semimetric space S whose points are p_3, p_4, p_5 together with the points of the straight line joining p_1 and p_2, with all distances euclidean *except the distance* $p_4 p_5$, which is defined to be the circumradius of p_1, p_2, p_3. If x, y are any two distinct points of the line (other than p_1, p_2) the four distinct points $Q = (p_1, p_2, x, y)$ are such that every quadruple of S with two points in common with Q is imbeddable in E_2. But the space S is not imbeddable in E_2 since p_1, p_2, p_3, p_4, p_5 form a pseudo-E_2 quintuple.

A more interesting way of allowing free $(n+2)$-tuples is given by the following theorems.

THEOREM 47.2. *Let S be semimetric with at least $n+k+4$ distinct points (n, k non-negative integers). Then $S \cong E_n$ if and only if S contains at most k free $(n+2)$-tuples.*

Proof. The necessity is obvious. To prove the sufficiency note that if $k = 0$ the theorem reduces to Theorem 46.1. Assuming $k > 0$, let $p_4, p_5, ..., p_{n+5}$ be any free $(n+2)$-tuple. Then an $(n+5)$-tuple of S exists, say $p_1, p_2, ..., p_{n+5}$, that has $p_4, p_5, ..., p_{n+5}$ as its only free $(n+2)$-tuple. We shall prove the theorem by showing that $p_4, p_5, ..., p_{n+5}$ (and hence each $(n+2)$-tuple of S) is imbeddable in E_n.

By Theorem 46.1 each of the $(n+4)$-tuples

$$p_1, p_2, p_3, p_4, ..., p_{i-1}, p_{i+1}, ..., p_{n+2}, p_{n+3}, p_{n+4}, p_{n+5} \quad (i = 4, 5, ..., n+5)$$

is imbeddable in E_n. If each is imbeddable in E_{n-1}, then every $(n+1)$-tuple of $p_1, p_2, \ldots, p_{n+5}$ is imbeddable in E_{n-1} and consequently so is the $(n+5)$-tuple. If, on the other hand, one of these $(n+4)$-tuples is not imbeddable in E_{n-1}, we may assume the labelling so that

$$p_1, p_2, p_3, \ldots, p_{n+3}, p_{n+4} \approx p'_1, p'_2, p'_3, \ldots, p'_{n+3}, p'_{n+4}, \quad (*)$$

with the primed points in E_n but not in any E_{n-1}. It follows easily that p'_1, p'_2 are contained in an *independent* $(n+1)$-tuple I'_{n+1} of these points. If I_{n+1} denotes the corresponding (independent) $(n+1)$-tuple of $p_1, p_2, \ldots, p_{n+4}$ congruent to I'_{n+1} by $(*)$ then $D(I_{n+1}) \neq 0$, while every principal minor of $D(p_1, p_2, \ldots, p_{n+5})$ obtained by annexing one row or two rows to $D(I_{n+1})$ vanishes.

Hence the rank of $D(p_1, p_2, \ldots, p_{n+5})$ is $n+2$ and so

$$D(p_4, p_5, \ldots, p_{n+5}) = 0.$$

Since each $(n+1)$-tuple of $p_4, p_5, \ldots, p_{n+5}$ is imbeddable in E_n, then clearly $p_4, p_5, \ldots, p_{n+5} \subseteq E_n$, and the theorem is proved.

The same kind of argument yields the next theorem.

THEOREM 47.3. *Let S be semimetric with power exceeding $\aleph_0$. Then S is congruently contained in E_n if and only if S has at most $\aleph_0$ free $(n+2)$-tuples.*

Theorem 47.2 is sharp in the sense that it allows the maximum number of free $(n+2)$-tuples in semimetric spaces S with no more than $n+k+4$ distinct points. This is seen, for example, in the case of $n = 1$, $k = 0$, by showing that a semimetric quintuple may contain exactly one non-linear triple, and is, consequently, not imbeddable in E_1. Such a semimetric set is formed by p_1, p_2, p_3, p_4, p_5 with

$$p_1 p_2 = p_1 p_3 = p_1 p_4 = p_2 p_5 = p_3 p_5 = p_4 p_5 = a > 0,$$
$$p_1 p_5 = p_2 p_3 = p_2 p_4 = p_3 p_4 = 2a.$$

It will be found that all triples of this quintuple are linear except p_2, p_3, p_4. Since the only non-linear triple is metric, the quintuple is a metric quintuple. Other such examples, as well as an extended study of imbedding theorems under weakened hypotheses, are due to P. H. Pepper and B. J. Topel.

REFERENCES

§ 37. Congruence indices were introduced in Blumenthal [5].

§ 38. Theorem 38.1 (and its corollary) are strengthened forms (under weaker hypotheses) of the congruence order theorem for E_n given in Menger [1].

§§ 41–42. Blumenthal [6]. The original forms of these theorems (with stronger hypotheses) are in Menger [1].

§ 43. Schoenberg [2], Blumenthal [7], Wilson [2].
§ 45. The algebraic proof of Theorem 45.1, Blumenthal [8], is much simpler than the geometric proof offered in Menger [1].
§ 46. Theorem 46.1 was proved originally in Menger [1, 4]. The greatly simplified proof of the text appears in Blumenthal [9].
§ 47. Blumenthal [10], Pepper [1], and Topel.

SUPPLEMENTARY PAPERS

Pepper [2, 3]—an imbedding method, and spaces containing congruently pseudo-E_n sets.

CHAPTER V

METRIC AND VECTOR CHARACTERIZATIONS OF EUCLIDEAN AND HILBERT SPACES. IMBEDDING IN HILBERT SPACE. NORMED LINEAR SPACES

48. First metric characterization of the E_n

WE recall that a given space R is characterized metrically among a given class of spaces when necessary and sufficient conditions, expressed wholly and explicitly in terms of the metric, are given in order that an arbitrary space of the class be congruent with R. The first approach to such a characterization of the E_n, due to Menger, was through the imbedding problem. In terms of it we may easily prove

THEOREM 48.1. *A necessary and sufficient condition that a semimetric space S be congruent with the E_n is that S be complete, metrically convex and externally convex, and irreducibly imbeddable in E_n.*

Proof. The necessity is obvious. To prove the sufficiency, let S' be a subset of E_n congruent with S. Then S' contains an independent $(n+1)$-tuple and (from the ordinary convexity and external convexity of S' (Corollary, Theorem 14.1)) it contains the $(n-1)$-dimensional hyperplanes determined by the points of the $(n+1)$-tuple. If q' is any element of E_n, a line through q' will intersect two of these hyperplanes in points p', r' of S'. The convexity properties of S' imply $q' \in S'$, and hence $S' = E_n$.

Combining this theorem with Theorem 42.2 gives

THEOREM 48.2. *A semimetric space S is congruent with the E_n if and only if (i) S is complete, metrically convex and externally convex, (ii) S contains an $(n+1)$-tuple $p_0, p_1,..., p_n$ such that $\operatorname{sgn} D(p_0, p_1,..., p_k) = (-1)^{k+1}$ ($k = 1, 2,..., n$), and (iii) for every pair x, y of points of S the determinants $D(p_0, p_1,..., p_n, x)$, $D(p_0, p_1,..., p_n, y)$, $D(p_0, p_1,..., p_n, x, y)$ vanish.*

The way in which the above theorem was obtained suggests possible redundancies in the hypotheses. Thus while hypotheses (ii), (iii) can scarcely be weakened in the situation for which they were developed (namely the irreducible imbedding in E_n of an *arbitrary* semimetric space) they might be susceptible of considerable abbreviation when the space S to be imbedded is assumed complete, convex and externally convex. This was shown to be the case by W. A. Wilson. We shall in the next two sections obtain his principal results by methods which (unlike those used by him) are entirely independent of the imbedding theorems of the preceding sections, and permit a significant weakening of his hypotheses.

49. The weak euclidean four-point property. Some lemmas

A semimetric space that has each three of its points congruently contained in E_2 (and hence is a metric space) clearly approximates more closely a euclidean space than one without that property. We define now a property that *a priori* seems only slightly stronger than metricity, but which, it turns out, has exceedingly restrictive consequences.

DEFINITION 49.1. *A semimetric space S has the weak euclidean four-point property provided that each quadruple of pairwise distinct points of S containing a linear triple is congruently imbeddable in E_2.*

Remark. A metrically convex semimetric space S with the weak euclidean four-point property is metric. For if p, q, r are three non-linear points of S, they are contained in a quadruple of distinct points p, q, r, s of S with psq holding, and hence are imbeddable in E_2.

Remark. A semimetric space S with the weak euclidean four-point property has the two-triple property (§ 21). The proof is clear.

LEMMA 49.1. *Each two distinct points p, q of a complete, convex, externally convex semimetric space S with the weak euclidean four-point property are on a unique line $L(p, q)$ of S.*

Proof. The proof follows at once from the above two remarks and Theorem 21.3.

In the lemmas and theorems of this and the next section we shall mean by 'a space S' a space satisfying the hypotheses of Lemma 49.1.

LEMMA 49.2. *If p_0, p_1, p_2 are elements of a space S ($p_1 \neq p_2$), and $p'_0, p'_1, p'_2 \in E_2$ with $p_0, p_1, p_2 \approx p'_0, p'_1, p'_2$, this congruence has a unique extension to*
$$(p_0) + L(p_1, p_2) \approx (p'_0) + E_1(p'_1, p'_2). \qquad (*)$$

Proof. Clearly the congruence $p_1, p_2 \approx p'_1, p'_2$ has the unique extension $L(p_1, p_2) \approx E_1(p'_1, p'_2)$, and so the lemma is proved if $p_0 \in L(p_1, p_2)$. In the contrary case, that congruence, together with $p_0 \leftrightarrow p'_0$, defines a one-to-one correspondence between the points of $(p_0) + L(p_1, p_2)$ and $(p'_0) + E_1(p'_1, p'_2)$. In this correspondence let x, y, elements of the first set, correspond to x', y', respectively, of the second set. If both
$$x, y \in L(p_1, p_2),$$
then $xy = x'y'$ by the congruence of the two lines.

Suppose, now, that $x = p_0$ and hence $x' = p_0'$. We have
$$p_0, p_1, p_2, y \approx \bar{p}_0, \bar{p}_1, \bar{p}_2, \bar{y}$$
of E_2 and since
$$\bar{p}_0, \bar{p}_1, \bar{p}_2 \approx p_0, p_1, p_2 \approx p_0', p_1', p_2',$$
it follows that $p_0, p_1, p_2, y \approx p_0', p_1', p_2', y^*$, with y^* uniquely determined. But $p_1', p_2', y^* \approx p_1, p_2, y \approx p_1', p_2', y'$, and hence $y^* = y'$. Whence
$$p_0, p_1, p_2, y \approx p_0', p_1', p_2', y' \quad \text{and} \quad p_0 y = p_0' y'.$$
The uniqueness of the extension is obvious.

LEMMA 49.3. *If p_0, p_1, p_2 are pairwise distinct elements of a space S and p_0', p_1', p_2' are elements of E_2 with $p_0, p_1, p_2 \approx p_0', p_1', p_2'$, this congruence has a unique extension to*
$$L(p_0, p_1) + L(p_1, p_2) \approx E_1(p_0', p_1') + E_1(p_1', p_2').$$

Proof. A mapping of $L(p_0, p_1) + L(p_1, p_2)$ onto $E_1(p_0', p_1') + E_1(p_1', p_2')$ is provided by the congruences
$$L(p_0, p_1) \approx E_1(p_0', p_1'), \tag{1}$$
$$L(p_1, p_2) \approx E_1(p_1', p_2'), \tag{2}$$
which are unique extensions of $p_0, p_1 \approx p_0', p_1'$; $p_1, p_2 \approx p_1', p_2'$ respectively.

If $x, y \in L(p_0, p_1) + L(p_1, p_2)$ and x', y' are their corresponding points in the mapping we may suppose $x \in L(p_0, p_1)$, $y \in L(p_1, p_2)$. By the previous lemma $p_0, p_1, p_2, y \approx p_0', p_1', p_2', y'$. Now $p_0, p_1, x, y \approx \bar{p}_0, \bar{p}_1, \bar{x}, \bar{y}$ of E_2 and since $\bar{p}_0, \bar{p}_1, \bar{y} \approx p_0, p_1, y \approx p_0', p_1', y'$ we have
$$p_0, p_1, x, y \approx p_0', p_1', x^*, y',$$
with x^* a uniquely determined element of $E_1(p_0', p_1')$. But
$$p_0', p_1', x' \approx p_0, p_1, x \approx p_0', p_1', x^*,$$
and so $x' = x^*$. Hence $p_0, p_1, x, y \approx p_0', p_1', x', y'$ and $xy = x'y'$. The uniqueness of the extension of the congruence is obvious.

LEMMA 49.4. *If p_0 is any point and L any line of a space S, there exists a unique foot $f(p_0)$ of p_0 on L (that is, a unique point $f(p_0)$ of L such that $p_0 f(p_0) \leqslant p_0 x$, $x \in L$).*

Proof. If $p_0 \in L$, then $f(p_0) = p_0$. In the contrary case, let p_1, p_2 be distinct points of L and p_0', p_1', p_2' points of E_2 with
$$p_0, p_1, p_2 \approx p_0', p_1', p_2'.$$
Then $(p_0) + L(p_1, p_2) \approx (p_0') + E_1(p_1', p_2')$. If $f(p_0')$ denotes that point of

$E_1(p_1', p_2')$ nearest p_0' and $f(p_0)$ corresponds to $f(p_0')$ in the above congruence, then $p_0 f(p_0) = p_0' f(p_0') \leqslant p_0' x' = p_0 x$. The uniqueness of $f(p_0)$ follows at once from that of $f(p_0')$.

LEMMA 49.5. *The congruence*

$$(p) + L \approx (p') + E_1$$

is valid if and only if $\mathrm{dist}(p, L) = \mathrm{dist}(p', E_1)$.

Proof. The necessity is obvious, as is also the sufficiency in case $\mathrm{dist}(p, L) = 0$. Suppose, then, $\mathrm{dist}(p, L) = \mathrm{dist}(p', E_1) \neq 0$, and let $f(p), f(p')$ be the feet of p, p' on L, E_1, respectively. Then $pf(p) = p'f(p')$. Select a point q on L and a point q' on E_1 with

$$qf(p) = q'f(p') \neq 0.$$

Since $p, q, f(p)$ are congruently contained in E_2, distinct points $p^\mathrm{I}, p^\mathrm{II}$ of that E_2 exist (reflections of each other in $E_1(q', f(p'))$) such that

$$p, q, f(p) \approx p^\mathrm{I}, q', f(p') \approx p^\mathrm{II}, q', f(p').$$

By Lemma 49.2, these congruences extend uniquely to

$$(p) + L(q, f(p)) \approx (p^\mathrm{I}) + E_1(q', f(p')) \approx (p^\mathrm{II}) + E_1(q', f(p')),$$

and $f(p')$ is the foot of p^I and of p^II on $E_1(q', f(p'))$. Since

$$p^\mathrm{I} f(p') = pf(p) = p'f(p'),$$

it follows that either p^I or p^II coincides with p'. Hence

$$(p) + L \approx (p') + E_1,$$

and the lemma is proved.

We shall refer to the lines L of a space S as *one-dimensional subspaces*, and denote them by $\bar{L}_1$.

Defining a zero-dimensional subspace $\bar{L}_0$ of a space S as a single point it follows from the preceding lemmas that, for $k = 1$, (i) a k-dimensional subspace $\bar{L}_k$ of S is the *closure* of the set of all points of S on a line with a point of $\bar{L}_{k-1}$ and a point not belonging to it, (ii) if $p_0, p_1, \ldots, p_k$ are not elements of a $(k-1)$-dimensional subspace $\bar{L}_{k-1}$, there is one and only one subspace $\bar{L}_k$ of S of dimension k containing them, (iii) $\bar{L}_k$ is a linear space (that is, it contains with each two of its distinct points the line joining them), and (iv) $\bar{L}_k$ is congruent with the k-dimensional euclidean space E_k.

If, for a given non-negative integer k, a space S contains a subspace $\bar{L}_k$ and a point p not belonging to it, a subspace $\bar{L}_{k+1}$, sometimes denoted by $\{p; \bar{L}_k\}$ is defined as the *closure* of the set L_{k+1} of all points of S on a line with p and a point of $\bar{L}_k$.

We make the inductive hypothesis that all properties of $\bar{L}_1$ listed above, as well as all those proved in the foregoing, are valid for every $\bar{L}_k$ ($k = 1, 2, ..., n$), and shall establish those properties (which are pertinent to our purpose) for ($n+1$)-dimensional subspaces $\bar{L}_{n+1}$ of S.

Consider, now, a point p and an n-dimensional subspace $\bar{L}_n$ of S. Let p' be a point of a euclidean space E_{n+1}, and E_n one of its n-dimensional subspaces with $\text{dist}(p', E_n) = \text{dist}(p, \bar{L}_n)$. Denoting by $f(p)$, $f(p')$ the feet of p, p' on $\bar{L}_n$, E_n, respectively, let q, q' be corresponding points in a congruence $\bar{L}_n \approx E_n$ (inductive hypothesis) which associates $f(p)$ and $f(p')$. To show that $pq = p'q'$ suppose $q \neq f(p)$ (trivial otherwise) and note that $f(p)$ a foot of p on $\bar{L}_n$ implies $f(p)$ a foot of p on $\bar{L}_1(q, f(p))$ which, by linearity of $\bar{L}_n$ (inductive hypothesis), is contained in $\bar{L}_n$. Hence $\text{dist}[p, \bar{L}_1(q, f(p))] = \text{dist}[p', E_1(q', f(p'))]$ and, by Lemma 49.5,

$$(p) + \bar{L}_1(q, f(p)) \approx (p') + E_1(q', f(p')).$$

Thus $pq = p'q'$ and an extension of that lemma is obtained.

LEMMA 49.6. *The congruence*

$$(p) + \bar{L}_n \approx (p') + E_n$$

is valid if and only if $\text{dist}(p, \bar{L}_n) = \text{dist}(p', E_n)$.

50. Second metric characterization of the E_n

We are now in a position to prove a principal result.

THEOREM 50.1. *Each k-dimensional subspace $\bar{L}_k$ of a space S is congruent with E_k ($k = 0, 1, 2, ...$).*

Proof. From the foregoing it suffices to prove the assertion for $k = n+1$ ($n > 0$). If $\bar{L}_{n+1} = \{p_0; \bar{L}_n\}$, let E_n be an n-dimensional subspace of E_{n+1} and p'_0 a point of E_{n+1} with $\text{dist}(p_0, \bar{L}_n) = \text{dist}(p'_0, E_n)$. By Lemma 49.6,
$$(p_0) + \bar{L}_n \approx (p'_0) + E_n. \qquad (*)$$

If $x \in L_{n+1}$ ($x \neq p_0$), let p be the point of $\bar{L}_n$ such that p_0, p, x are on a line and p' the point of E_n corresponding to p by congruence $(*)$. Let

$$\bar{L}_1(p_0, p) \approx E_1(p'_0, p') \qquad (\dagger)$$

contain the congruence $p_0, p \approx p'_0, p'$, and let x' be the (unique) point of $E_1(p'_0, p')$ corresponding to x, an element of $\bar{L}_1(p_0, p)$, by $(\dagger)$. We wish to show that this mapping of $L_{n+1} - (p_0)$ onto the subset of E_{n+1} obtained by deleting the n-dimensional subspace through p'_0 which is parallel to E_n is a congruent one.

Let x, x' and y, y' be two pairs of corresponding points in this mapping, with $p_0, p, x \approx p'_0, p', x'$; $p_0, q, y \approx p'_0, q', y'$ (all triples lying on lines and

q, q' corresponding points of $\bar{L}_n, E_n$ by (∗)). Since $p_0, p, q \approx p_0', p', q'$, Lemma 49.3 gives
$$\bar{L}_1(p_0, p) + \bar{L}_1(p_0, q) \approx E_1(p_0', p') + E_1(p_0', q'),$$
and since x and x', y and y' are seen to be corresponding points in this congruence we have $xy = x'y'$. The mapping is thus a congruence.

Since S and E_{n+1} are complete, the mapping may be extended in the usual way to the *closures* of the two sets (see, for example, the proof of Theorem 14.1) and we obtain $\bar{L}_{n+1} \approx E_{n+1}$. This completes the proof of the theorem.

COROLLARY. *Each k-dimensional subspace $\bar{L}_k$ of S is a linear space* ($k = 0, 1, 2, ...$).

THEOREM 50.2. *If k is any positive integer and $p_0, p_1, ..., p_k$ are $k+1$ points of the space S that are not contained in any $(k-1)$-dimensional subspace, they lie in one and only one k-dimensional subspace of S.*

Proof. The theorem is established when it is proved for $k = n+1$.

Now the points $p_1, p_2, ..., p_{n+1}$ are not in any $\bar{L}_{k-1}$ of S ($k = 1, 2, ..., n$), for otherwise, since p_0 would clearly not belong to such a subspace, this point and subspace generate a subspace $\bar{L}_k$ ($k \leq n$) which contains $p_0, p_1, ..., p_{n+1}$, contrary to hypothesis. By the inductive hypothesis it follows that $p_1, p_2, ..., p_{n+1}$ are in one and only one $\bar{L}_n$, and $p_0, \bar{L}_n$ generate a subspace $\bar{L}_{n+1}$ containing $p_0, p_1, ..., p_{n+1}$.

If, now, $\bar{L}_{n+1}^*$ is another $(n+1)$-dimensional subspace containing $p_0, p_1, ..., p_{n+1}$, then, by the linearity of all subspaces, it contains the unique $\bar{L}_n$ containing the points $p_1, p_2, ..., p_{n+1}$, and it follows that $\bar{L}_{n+1}^* \supset \bar{L}_{n+1}$. On the other hand, if $\bar{L}_{n+1}^*$ contained a point q which does not belong to $\bar{L}_{n+1}$, then $\bar{L}_{n+1}^*$ would contain the $\bar{L}_{n+2}$ generated by q and $\bar{L}_{n+1}$. This is impossible since $\bar{L}_{n+1}^* \approx E_{n+1}$, $\bar{L}_{n+2} \approx E_{n+2}$, and $\bar{L}_{n+1}^* \supset \bar{L}_{n+2}$ would imply that a subset of E_{n+1} is congruent with E_{n+2}. Hence $\bar{L}_{n+1}^* \equiv \bar{L}_{n+1}$.

DEFINITION 50.1. *A semimetric space has the euclidean $(k+1)$-point property (k, a fixed, non-negative integer) provided each $(k+1)$-tuple of pairwise distinct points of the space is congruently contained in E_k.*

THEOREM 50.3. *A complete, convex, externally convex semimetric space S with the weak euclidean four-point property has the euclidean $(n+1)$-point property for every non-negative integer n.*

Proof. Let $p_0, p_1, ..., p_n$ be any subset of $n+1$ points of S. By Theorem 50.2 either these $n+1$ points are contained in some k-dimensional subspace $\bar{L}_k$ of S for some integer $k \leq n-1$, or they lie in a unique

n-dimensional subspace $\bar{L}_n$ of S. In either event their congruence with a subset of E_n follows from Theorem 50.1.

This was the principal theorem obtained by W. A. Wilson (in a quite different way) in the work referred to in the concluding remarks of §48, except that our hypothesis of *weak* euclidean four-point property replaces his assumption of euclidean four-point property. It should be observed that the procedure we use here to establish Theorem 50.3 is entirely independent of results obtained in solving the problem of imbedding semimetric spaces in E_n. The hypothesis of external convexity can be suppressed by a slight alteration of our procedure, but since it is needed for the characterization theorem (which is, after all, our main objective) and facilitates the developments leading to Theorem 50.3, its retention seems justified.

THEOREM 50.4. *A semimetric space S is congruent with the n-dimensional euclidean space E_n (n, a given non-negative integer) if and only if (i) S is complete, convex, externally convex, (ii) S has the weak euclidean four-point property, (iii) the determinant D of each $n+2$ points of S vanishes, and (iv) n is the smallest integer for which (iii) is valid.*

Proof. The necessity is clear. To prove the sufficiency we observe that from hypothesis (iv) (and Theorem 50.3) S contains an $(n+1)$-tuple which does not lie in any $\bar{L}_{n-1}$ and hence is contained in a unique $\bar{L}_n$. By (iii) S does not contain any $\bar{L}_{n+1}$, and consequently each element of S lies in $\bar{L}_n$; that is, $S = \bar{L}_n$. But, by Theorem 50.1, $\bar{L}_n \approx E_n$, and our theorem is proved.

The reader will notice that the metric characterization theorem just proved is superior to characterization Theorem 48.2 in that it dispenses with the chain of determinant conditions of the latter, which are equivalent to euclidean $(k+1)$-point properties for $k = 4, 5,..., n$. The surprising strength of the weak euclidean four-point property could hardly have been anticipated. Its effect in restricting what seemed, *a priori*, a wide class of spaces, to euclidean spaces of finite or infinite dimensions is remarkable.

If the completeness condition be replaced by the somewhat stronger one of 'finitely compact', we obtain the interesting result:

THEOREM 50.5. *A finitely compact, convex, externally convex semimetric space S with the weak euclidean four-point property is congruent with a euclidean space of finite dimension.*

Proof. Since S is finitely compact it is easily seen that an integer n exists such that $S = \bar{L}_n \approx E_n$.

EXERCISE

If M, N are metric spaces, the cartesian product space $M \times N$ consists of all ordered pairs (m, n) with $m \in M$, $n \in N$ and

$$\text{dist}[(m,n), (m',n')] = [(mm')^2 + (nn')^2]^{\frac{1}{2}}.$$

Show that if M and N have the euclidean k-point property for a fixed integer k, so does $M \times N$.

51. The pythagorean property

An easy application of Theorem 50.4 shows that in a wide class of spaces the validity of the pythagorean theorem ensures the euclidean character of the metric and hence the basic role that this theorem plays in euclidean geometry is seen to be fully justified.

If a metric space M contains a line L and a point p not on L, a foot $f(p)$ of p on L exists; that is, L contains at least one point $f(p)$ such that $0 < pf(p) \leqslant px$, for every $x \in L$. If $q \in L$ and a, b are points of L on opposite sides of q with $qa = qb = 2pq$, then by Lemma 16.2 the closed segment $[a, b]$ contains at least one foot of p on it, and any such point $f(p)$ is a foot of p on L. For $0 < pf(p) \leqslant px$, $x \in [a, b]$, and if $x \in L$, $x \bar{\in} [a, b]$, then $qa + ax = qx$ or $qb + bx = qx$. In the first case

$$px + pq \geqslant qx = qa + ax = 2pq + ax$$

and hence $\quad px \geqslant pq + ax > pq \geqslant pf(p).$

The second case is treated similarly.

We say that the pythagorean theorem is valid in the space M provided, for each element x of L, $pp_0^2 + p_0 x^2 = px^2$, where p_0 is any foot of p on L.

Remark. If the pythagorean theorem is valid in M, then p_0 is unique.

THEOREM 51.1. *A finitely compact, convex, externally convex metric space M in which the pythagorean theorem is valid is congruent with a euclidean space.*

Proof. By Theorem 50.5, it suffices to show that M has the weak euclidean four-point property. If p, q, r, $s \in M$ with q, r, $s \in L$, a line of M, and p_0 the foot of p on L, let q', r', s', p_0' be points of an E_1 congruent with q, r, s, p_0. In the plane E_2 formed by this E_1 and a line perpendicular to it at p_0', let p' denote a point on that perpendicular with $p'p_0' = pp_0$. From the validity of the pythagorean theorem in M and E_2 it follows that the distances of p' from q', r', s' equal, respectively, the distances of p from q, r, s, and hence $p, q, r, s \approx p', q', r', s'$.

Thus M has the weak euclidean four-point property, and the theorem is proved.

It might be remarked that though the proof of Theorem 51.1 is trivial when we know that it suffices to show that the space has the *weak* euclidean four-point property, the demonstration that *every* four points of the space are imbeddable in E_3 does not lie nearly so close at hand. The proof serves, therefore, as an example of the usefulness of the weaker property.

We observe, moreover, that the condition of finite compactness performs for us the single service of ensuring the finite dimensionality of the space while preserving completeness. It does not enter into our methods. We shall see later that adjoining the weaker property of separability (and eliminating finite compactness) gives stronger theorems characterizing infinite dimensional 'euclidean' spaces. In that respect the use of finite compactness (when not necessitated by the *methods*) would weaken the results.

52. Metric transforms and the euclidean four-point property

The results of the preceding sections of this chapter focus attention on the euclidean four-point property. We show in this section that *any* metric space M can be given the euclidean four-point property by defining in M a new metric (a function of the old one) which is topologically equivalent to the original metric. Whence, *any metric space is homeomorphic to one with each four of its points congruently imbeddable in E_3.*

DEFINITION 52.1. *Let S be a semimetric space and $\phi(x)$ a real function defined for every value of $x = pq$, where $p, q \in S$. A space $\phi(S)$ is the metric transform of S by ϕ provided* (1) *the points of S and $\phi(S)$ are in a one-to-one correspondence, and* (2) *if points p', q' of $\phi(S)$ correspond, respectively, to points p, q of S, then $p'q' = \phi(pq)$.*

In many applications of the notion just defined, the space $\phi(S)$ has the same point-set as S, and the biuniform correspondence is the identity; that is, $\phi(S)$ arises by redefining the distance pq of points p, q to be $\phi(pq)$. This is the case in Exercise 5, §9, where the reader was asked to prove (in the new terminology) that the *metric transform of any metric space M by any monotone increasing concave function $\phi(x)$, vanishing at the origin, is a metric space (that is, has the euclidean three-point property).*

The concavity of $\phi(x)$ implies that for $x_1 < x < x_2$, the determinant

$$\begin{vmatrix} \phi(x_1) & x_1 & 1 \\ \phi(x) & x & 1 \\ \phi(x_2) & x_2 & 1 \end{vmatrix} > 0,$$

and it is easy to show that $\phi(p_1p_2)+\phi(p_2p_3) > \phi(p_1p_3)$, for each three points p_1, p_2, p_3. Hence no three points of $\phi(M)$ are linear, and so $\phi(M)$ is *not convex* if it contains more than one point. Applying this to the metric transform of a metric triple (linear or not) by

$$\phi(x) = x^\alpha \quad (0 \leqslant \alpha < 1),$$

we see that p'_1, p'_2, p'_3, with $p'_i p'_j = (p_i p_j)^\alpha$, form a non-linear metric triple, and *for $0 \leqslant \alpha < \tfrac{1}{2}$, each angle (defined by the cosine law) of the triangle p'_1, p'_2, p'_3 is acute.*

THEOREM 52.1. *The metric transform $\phi(M)$ of any metric space M by $\phi(x) = x^\alpha$, $0 \leqslant \alpha \leqslant \tfrac{1}{2}$, has the euclidean four-point property.*

Proof. Consider first the case $\alpha = \tfrac{1}{2}$. Since M is metric so is $\phi(M)$, and it suffices to show that if p'_1, p'_2, p'_3, p'_4 are any four points of $\phi(M)$, then

$$D(p'_1, p'_2, p'_3, p'_4) = \begin{vmatrix} 0 & 1 \\ 1 & (p'_i p'_j)^2 \end{vmatrix}_{(i,j=1,2,3,4)} = \begin{vmatrix} 0 & 1 \\ 1 & p_i p_j \end{vmatrix}_{(i,j=1,2,3,4)}$$

is not negative.

Now

$$D(\alpha) = \begin{vmatrix} 0 & 1 \\ 1 & (p_i p_j)^{2\alpha} \end{vmatrix}_{(i,j=1,2,3,4)}$$

is a continuous function of α which is positive for $\alpha = 0$. If we suppose $D(\tfrac{1}{2}) < 0$, then a number α_1 exists, $0 < \alpha_1 < \tfrac{1}{2}$, with $D(\alpha_1) = 0$. This implies the existence of four points p_1^*, p_2^*, p_3^*, p_4^* of E_2 with

$$p_i^* p_j^* = (p_i p_j)^{\alpha_1} \quad (i,j = 1, 2, 3, 4).$$

But this is impossible, for (by the remarks preceding this theorem) each of the twelve angles determined by the points p_i^* $(i = 1, 2, 3, 4)$ is acute (since $0 < \alpha_1 < \tfrac{1}{2}$), which cannot be the case for four points of a plane. Hence $D(\tfrac{1}{2}) \geqslant 0$ and the theorem is proved for that value of α.

We have actually proved, however, that $D(\alpha)$ cannot vanish for any value of α between zero and $\tfrac{1}{2}$, and it follows that $D(\alpha)$ must be *positive* for $0 \leqslant \alpha < \tfrac{1}{2}$. Hence the elements p'_i $(i = 1, 2, 3, 4)$ of $\phi(M)$, with $p'_i p'_j = (p_i p_j)^\alpha$ $(i,j = 1, 2, 3, 4)$, $0 \leqslant \alpha \leqslant \tfrac{1}{2}$, are congruent with four points of E_3, and the theorem is proved.

It is easily seen that $\alpha = \tfrac{1}{2}$ *is the greatest exponent for which the above theorem is valid*; for if the points p_i $(i = 1, 2, 3, 4)$, forming a pseudo-linear quadruple with

$$p_1 p_2 = p_2 p_3 = p_3 p_4 = p_1 p_4 = 1, \qquad p_1 p_3 = p_2 p_4 = 2,$$

be metrically transformed by $\phi(x) = x^{\frac{1}{2}(1+\epsilon)}$, $0 \leqslant \epsilon < 1$, we find that

$$D(p'_1, p'_2, p'_3, p'_4) = -32 \cdot 2^{2\epsilon}(2^\epsilon - 1),$$

which vanishes for $\epsilon = 0$ and is negative for $\epsilon > 0$. Hence p'_1, p'_2, p'_3, p'_4 are congruent with four points of a plane for $\epsilon = 0$, while if $\epsilon > 0$, the four points are not congruently imbeddable in any euclidean space.

The theorem is sharp in another respect also, for Ville has shown that a metric *quintuple* exists whose metric transform by $\phi(x) = x^{\frac{1}{2}}$ is not congruently imbeddable in a euclidean space.

EXERCISES

1. Prove that not every one of the twelve angles determined by four points of a plane is acute. Show that the minimum of the maximum angle determined by quadruples of E_2 is $\frac{1}{2}\pi$.
2. Prove that for $p = 3, 4, 5, 6$ the minimum of the maximum angle determined by planar subsets of p points is $(p-2)\pi/p$. Show that this formula fails for $p = 7$. (The problem of determining the desired minimax is unsolved for $p > 6$.)
3. What can you say concerning the minimum of the maximum of the thirty angles determined by five points of E_3?

53. Congruent imbedding in Hilbert space

The Hilbert space $\mathscr{H}$ has been defined and proved metric in §9. It is clear from the definition that the subset of $\mathscr{H}$ consisting of all elements $(x_1, x_2, x_3,...)$ which, for a given non-negative integer n, have $x_i = 0$ ($i = n+1, n+2,...$), form an n-dimensional euclidean space E_n. Thus for every non-negative integer n, the E_n is contained in $\mathscr{H}$. It follows that each *finite* subset of $\mathscr{H}$ is contained in a *euclidean space* that is part of $\mathscr{H}$.

LEMMA 53.1. *A denumerable semimetric space P is congruently imbeddable in $\mathscr{H}$ if and only if, for every positive integer k, every $(k+1)$-tuple of P is imbeddable in E_k.*

Proof. The necessity follows from the remark preceding this lemma. To prove the sufficiency, we observe that if P is not imbeddable in some euclidean subspace of $\mathscr{H}$, then a serialization $p_0, p_1,..., p_n,...$ of the points of P exists such that $p_0, p_1,..., p_k$ are irreducibly imbeddable in E_k for every positive integer k. For a fixed value of k, points

$$x^{(i)} = (x_1^{(i)}, x_2^{(i)},..., x_i^{(i)}, 0, 0,...) \quad (i = 0, 1,..., k)$$

of $\mathscr{H}$ are easily seen to exist such that

$$p_0, p_1,..., p_k \approx x^{(0)}, x^{(1)},..., x^{(k)}$$

with $x_i^{(i)} > 0$ ($i = 1, 2,..., k$). These points lie in the E_{k+1} formed by those elements $(x_1, x_2, x_3,...)$ of $\mathscr{H}$ with $x_i = 0$ ($i > k+1$), and this E_{k+1} contains exactly one point $x^{(k+1)}$ with $x_{k+1}^{(k+1)} > 0$ and

$$p_0, p_1,..., p_k, p_{k+1} \approx x^{(0)}, x^{(1)},..., x^{(k)}, x^{(k+1)}.$$

Hence $P \subseteqq \mathscr{H}$.

COROLLARY. *A denumerable set* $P \cong \mathscr{H}$ *if and only if the determinant D of each* $(k+1)$-*tuple of P either vanishes or has the sign of* $(-1)^{k+1}$, *for every positive integer k.*

THEOREM 53.1. *A separable semimetric space S is congruently imbeddable in* $\mathscr{H}$ *if and only if, for every positive integer k, the determinant D of each* $(k+1)$-*tuple of S either vanishes or has the sign of* $(-1)^{k+1}$.

Proof. The necessity is clear from the above. The sufficiency follows by applying the corollary of the lemma to a denumerable subset P of S, with $\bar{P} = S$ (which exists, since S is separable).† Then $P \approx Q \subset \mathscr{H}$, and since $\mathscr{H}$ is complete, this congruence may be extended to a congruence of $\bar{P}$ with a subset of $\mathscr{H}$.

COROLLARY. *No pseudo-*E_n *set is imbeddable in Hilbert space.*

Though in the euclidean imbedding theorems separability of S is not explicitly assumed (it follows, of course, from the other hypotheses) this requirement may not be suppressed in Theorem 53.1, since a space with power exceeding $\mathfrak{c}$, the power of Hilbert space, in which $pq = a > 0$ for $p \neq q$ and $pp = 0$, has every $k+1$ of its points satisfying the determinant conditions of the theorem without being imbeddable in $\mathscr{H}$.

Quite a different formulation of the conditions for congruent imbedding in Hilbert space is due to Schoenberg. A real continuous even function $g(t)$, defined for all values $t = \pm pq$ ($p, q \in S$) is called *positive definite in a semimetric space S* provided for every positive integer n and every $(n+1)$-tuple $p_0, p_1, \ldots, p_n$ of S,

$$\sum_{i,j=0}^{n} g(p_i p_j) x_i x_j \geq 0,$$

for x_i ($i = 0, 1, \ldots, n$) real.

THEOREM 53.2. *Let S be a separable semimetric space. If the family of functions* $e^{-\lambda t^2}$ *be positive definite in S for a set of positive values of* λ *which has zero as an accumulation point, then S is congruently imbeddable in* $\mathscr{H}$.

Proof. It suffices to show that for every positive integer n, each $(n+1)$-tuple $p_0, p_1, \ldots, p_n$ of S is congruently contained in E_n. By hypothesis

$$\sum_{i,j=0}^{n} \exp[-\lambda \cdot p_i p_j^2] x_i x_j \geq 0$$

for all real values of $x_0, x_1, \ldots, x_n$.

† If S is finite, the sufficiency follows from the corollary of Theorem 42.3.

Now

$$\sum_{i,j=0}^{n} \exp[-\lambda \cdot p_i p_j^2] x_i x_j = \sum_{i,j=0}^{n} [1 - \lambda \cdot p_i p_j^2 + \tfrac{1}{2}\lambda^2 \cdot p_i p_j^4 + \ldots] x_i x_j$$

$$= \Big(\sum_{i=0}^{n} x_i\Big)^2 - \lambda \sum_{i,j=0}^{n} p_i p_j^2 \cdot x_i x_j +$$

$$+ \tfrac{1}{2}\lambda^2 \sum_{i,j=0}^{n} p_i p_j^4 \cdot x_i x_j + \ldots$$

$$= -\lambda \sum_{i,j=0}^{n} p_i p_j^2 \cdot x_i x_j + \tfrac{1}{2}\lambda^2 \sum_{i,j=0}^{n} p_i p_j^4 \cdot x_i x_j + \ldots$$

on the hyperplane $x_0 + x_1 + \ldots + x_n = 0$.

Since the quadratic form is non-negative for all real values of $x_0, x_1, \ldots, x_n$, the right-hand side of the above is non-negative, which for small values of λ implies that

$$\sum_{i,j=0}^{n} p_i p_j^2 \cdot x_i x_j \leqslant 0$$

for $x_0 + x_1 + \ldots + x_n = 0$. Then, by the corollary of Theorem 43.1, the points $p_0, p_1, \ldots, p_n$ are congruently imbeddable in E_n, and the theorem is proved.

It is not difficult to show that for each $\lambda > 0$, $e^{-\lambda t^2}$ is positive definite in every euclidean space and hence is positive definite in Hilbert space. Then $e^{-\lambda t^2}$ ($\lambda > 0$) is evidently positive definite in any semimetric space that is congruently imbeddable in Hilbert space, so *the sufficient condition of Theorem 53.2 is also necessary.*

EXERCISES

1. Prove that the metric transform of the line segment $[0, 1]$ by $\phi(x) = x^{\frac{1}{2}}$ is congruent to a subset Γ of Hilbert space which is not contained in any finite dimensional subspace (that is, in any E_k).
2. Show that the set Γ of Exercise 1 is (i) an arc, (ii) has each three of its points the vertices of a right triangle, and (iii) possesses neither a right nor left (unilateral) tangent at any of its points (Blumenthal [27]).
3. Though Γ is a continuous curve that has no tangent (even a unilateral one) at any point, show that it has equations $x_i = x_i(t)$ ($0 \leqslant t \leqslant 1$), with $x_i(t)$ *analytic* ($i = 1, 2, \ldots$) and hence admitting derivatives of all orders!

54. Metric transforms of euclidean and Hilbert spaces

We saw in a preceding section that the metric transform of an arbitrary metric quadruple by $\phi(x) = x^{\frac{1}{2}}$ is imbeddable in E_3, and that $\tfrac{1}{2}$ is the greatest exponent and four is the maximum number of points for which this is true. If, however, the space being metrically transformed is specialized, both the exponent and the number of points may be increased. Thus, for example, the reader was asked to show that the metric transform of $[0, 1]$ by $\phi(x) = x^{\frac{1}{2}}$ is imbeddable in Hilbert space, and hence that metric transform of each finite subset of $[0, 1]$ is imbeddable in a euclidean space.

Extensive investigations concerning metric transforms of euclidean and Hilbert spaces, particularly by $\phi(x) = x^\alpha$, $0 < \alpha < 1$, have been made by Schoenberg, some of whose results we present in this section.

It was remarked in § 53 that $e^{-\lambda t^2}$ ($\lambda > 0$) is positive definite in Hilbert space. Now, for any $t > 0$ and $0 < \alpha < 2$, each of the integrals

$$\int_0^\infty (1-e^{-s^2 t^2}) s^{-1-\alpha}\, ds$$

and

$$\int_0^\infty (1-e^{-s^2}) s^{-1-\alpha}\, ds \quad (s\ \text{real})$$

exists, and the second integral is *positive*.

Now substituting s/t for s in the first integral

$$\int_0^\infty (1-e^{-s^2 t^2}) s^{-1-\alpha}\, ds = t^\alpha \int_0^\infty (1-e^{-s^2}) s^{-1-\alpha}\, ds,$$

or

$$t^\alpha = c(\alpha) \int_0^\infty (1-e^{-s^2 t^2}) s^{-1-\alpha}\, ds,$$

where $c(\alpha)$ denotes the reciprocal of the second of the above two integrals.

Let $p_0, p_1, \ldots, p_n$ be $n+1$ points of any semimetric space S which is imbeddable in Hilbert space. We have, for $0 < \alpha < 2$,

$$(p_i p_k)^\alpha = c(\alpha) \int_0^\infty [1-\exp(-s^2 \cdot p_i p_k^2)] s^{-1-\alpha}\, ds$$

and if $x_0, x_1, \ldots, x_n$ are any real numbers with $x_0 + x_1 + \ldots + x_n = 0$, then

$$\sum_{i,k=0}^n (p_i p_k)^\alpha x_i x_k = -c(\alpha) \int_0^\infty \left\{\sum_{i,k=0}^n x_i x_k \cdot \exp(-s^2 \cdot p_i p_k^2)\right\} s^{-1-\alpha}\, ds$$
$$\leqslant 0,$$

for since the function $e^{-s^2 t^2} = e^{-\lambda t^2}$ ($\lambda > 0$) is positive definite in $\mathscr{H}$, it is also positive definite in S ($S \subseteq \mathscr{H}$) and so the integrand is non-negative.

Putting $\alpha = 2\beta$, then the quadratic form

$$\sum_{i,k=0}^n (p_i p_k^\beta)^2 \cdot x_i x_j$$

is negative or zero for all real numbers $x_0, x_1, \ldots, x_n$ with $\sum_{i=0}^n x_i = 0$. Hence (corollary, Theorem 43.1) the metric transform of the $(n+1)$-tuple $p_0, p_1, \ldots, p_n$ by $\phi(x) = x^\beta$, $0 < \beta < 1$, is imbeddable in E_n.

Thus for every positive integer k, the metric transform of each $(k+1)$-tuple of S by $\phi(x) = x^\beta$, $0 < \beta < 1$, is imbeddable in E_k. Since $S \Subset \mathscr{H}$, then S is separable and we have established the interesting result:

THEOREM 54.1. *If S is congruently imbeddable in Hilbert space, then the metric transform of S by $\phi(x) = x^\beta$ is congruently imbeddable in Hilbert space for every exponent β such that $0 < \beta \leqslant 1$.*

Applying the theorem for (i) $S = \mathscr{H}$, and (ii) $S = E_n$ (n, an arbitrary positive integer) two important corollaries are obtained.

COROLLARY 1. *The metric transform of Hilbert space by $\phi(x) = x^\beta$, $0 < \beta \leqslant 1$, is congruently imbeddable in Hilbert space.*

COROLLARY 2. *For each positive integer n, the metric transform of the euclidean space E_n by $\phi(x) = x^\beta$, $0 < \beta \leqslant 1$, is congruently imbeddable in Hilbert space.*

Since for $\beta < 1$ the transformed spaces are not convex, they are congruent with a *proper* subset of Hilbert space. We note, without proof, that the metric transform of Hilbert space by $\phi(x) = x/(1+x)$ *is congruent with a bounded subset of that space*.

55. Metric characterizations of Hilbert space

We saw in § 50 that a complete, convex, externally convex semimetric space, with the weak euclidean four-point property, has the euclidean k-point property for every positive integer k, and the k-dimensional subspace generated by any $k+1$ points with non-vanishing determinant D is congruent with E_k. If such a space is also separable, it is congruent with a subset of Hilbert space (Theorem 53.1). If, further, it contains for each positive integer k, a $(k+1)$-tuple that is not imbeddable in E_{k-1}, then the space is *not* congruent with any *finite* dimensional subspace of $\mathscr{H}$ (that is, with any E_n, n a positive integer), and we readily obtain the following characterization theorem.

THEOREM 55.1. *A semimetric space S is congruent with Hilbert space if and only if S is* (i) *separable, complete, convex, externally convex,* (ii) *has the weak euclidean four-point property, and* (iii) *contains for each positive integer k a $(k+1)$-tuple with non-vanishing determinant D.*

It is worth remarking that the correspondence

$$(x_1, x_2, x_3, \ldots) \longleftrightarrow (0, x_1, x_2, x_3, \ldots)$$

defines a biuniform and distance-preserving mapping of the whole

Hilbert space onto its 'hyperplane' $X_1 = 0$. Thus $\mathscr{H}$ is congruent with a proper part of itself. Denoting by K the subset of Hilbert space which is the sum of this hyperplane and a point not on it, say $(1, 0, 0,...)$, we have $K \subsetneq \mathscr{H}$, $\mathscr{H} \subsetneq K$, but K and $\mathscr{H}$ are surely not congruent (since $\mathscr{H}$ is convex and K is not). Thus each of two sets may be congruently contained in the other without being congruent to each other. The reader is asked to give additional (simpler) instances of this.

It was shown in § 51 that a complete, convex, externally convex metric space with the pythagorean property has the weak euclidean four-point property. *Hence Theorem 55.1 remains true for metric spaces if hypothesis (ii) is replaced by the assumption that the theorem of Pythagoras is valid*

EXERCISE

Supply the details to complete the proof of Theorem 55.1.

56. Normed linear spaces, inner product, and quasi inner product space

A space S is a (real) linear space provided (1) for each element x of S and each real number λ, S contains a unique element $\lambda.x$ (called the *scalar multiple* of x by λ), and (2) S contains for each pair of elements x, y a uniquely defined element $x+y$ (the *sum* of x and y) with the operations of scalar multiplication and addition satisfying the following conditions:

1. $x+y = y+x$,
2. $x+(y+z) = (x+y)+z$,
3. $x+y = x+z$ implies $y = z$,
4. $\lambda.(x+y) = \lambda.x+\lambda.y$,
5. $(\lambda_1+\lambda_2).x = \lambda_1.x+\lambda_2.x$ $\quad(\lambda_1, \lambda_2$ real$)$,
6. $\lambda_1(\lambda_2.x) = (\lambda_1\lambda_2).x$ $\quad(\lambda_1, \lambda_2$ real$)$,
7. $1.x = x$.

A linear space is *normed* provided there is attached to each element x of the space a *non-negative* real number $||x||$, called the norm of x, such that:

1. $||x|| = 0$ if and only if $x = \theta$, where θ denotes the unique element of S with the property $x+\theta = x$, for each element x of S,
2. $||\lambda.x|| = |\lambda|.||x||$ $\quad(\lambda$ real$)$,
3. $||x+y|| \leqslant ||x||+||y||$ $\quad(x, y \in S)$.

Normed linear spaces, which were introduced almost simultaneously by Banach, Hahn, and Wiener, have been intensively studied, for in addition to their manifold applications in mathematics they play an important role in modern physics.

If the norm $||x-y|| = ||x+(-1).y||$ be attached to elements x, y as *distance*, it is seen at once that the space is metric. It is, moreover, convex and externally convex, for if (for example) $x, y \in S$ ($x \neq y$), it is easily verified that for each λ ($0 < \lambda < 1$) the element $\lambda x + (1-\lambda)y$ is metrically between x and y. A *complete* normed linear space is usually called a Banach space. It follows from earlier theorems of this chapter that *a separable Banach space is congruent with Hilbert or euclidean space provided it has the weak euclidean four-point property.*

If P is any abstract set in which for each element x and each real number λ a unique element $\lambda.x$ is defined, and if to each pair of elements x, y of P there is attached a unique element $x+y$ of P, then a real function $((x,y))$, defined over the set of element pairs of P, is a (real) *inner product* provided:

1. $((x,y)) = ((y,x))$,
2. $((x,x)) \geqslant 0$,
3. $((x,x)) = 0$ if and only if $x = \theta$, where θ is a neutral element of P with respect to addition,
4. $((\lambda.x, y)) = \lambda((x,y))$ (λ real),
5. $((x+y,z)) = ((x,z)) + ((y,z))$ ($x, y, z \in P$).

DEFINITION 56.1. *A normed linear space in which an inner product is defined so that for each element x, $||x|| = ((x,x))^{\frac{1}{2}}$, is called a generalized euclidean space.*

Various necessary and sufficient conditions in order that a normed linear space be generalized euclidean are known. Thus, for example, von Neumann and Jordan require each element-pair x, y to satisfy the norm equality $||x+y||^2 + ||x-y||^2 = 2(||x||^2 + ||y||^2)$. (The equality sign may, in fact, be replaced by $\geqslant$ or by $\leqslant$). This is equivalent to demanding that $x, y, z, w \in E_2$ whenever z is the (algebraic) middle-element of x, y and hence is similar to the weak euclidean four-point property. In a very recent note, Schoenberg has shown that it suffices to require that the normed linear space be ptolemaic.

We study here the reverse problem. Starting with an abstract set in which a 'quasi inner product' is defined, we seek conditions, expressed wholly and explicitly in terms of it, in order that the space be normed linear (generalized euclidean).

Postulates for a quasi inner product

To each pair of elements x, y of an abstract set Σ there is attached a real number (x, y), called a *quasi inner product*, in conformity with the following conventions.

Q_1. (*Symmetry.*) If $x, y \in \Sigma$, then $(x, y) = (y, x)$.

Q_2. (*Definiteness.*) For each element x of Σ $(x, x) \geqslant 0$.

Q_3. (*Identification.*) If $x, y \in \Sigma$ and $(x, x) = (x, y) = (y, y)$, then $x = y$.

The set Σ may then be referred to as a quasi inner product space.

Schwarz postulates

If $x_1, x_2, \ldots, x_n \in \Sigma$, denote by $G(x_1, x_2, \ldots, x_n)$ the Gram determinant $|(x_i, x_j)|$ $(i, j = 1, 2, \ldots, n)$.

S_1. If $x_1, x_2 \in \Sigma$, then $G(x_1, x_2) \geqslant 0$.

S_2. If $x_1, x_2, x_3 \in \Sigma$ and $G(x_1, x_2) = 0$, then $G(x_1, x_2, x_3) \geqslant 0$.

S_3. If $x_1, x_2, x_3, x_4 \in \Sigma$ and $G(x_1, x_2, x_3) = 0$, then $G(x_1, x_2, x_3, x_4) \geqslant 0$.

Existence postulates

If $x_1, x_2, \ldots, x_n \in \Sigma$, denote by $B(x_1, x_2, \ldots, x_n)$ the symmetric determinant obtained by bordering $G(x_1, x_2, \ldots, x_n)$ with a row and column of 1's with intersection 0.

E_1. There exists at least one element θ of Σ such that $(\theta, x) = 0$ for each element x of Σ.

E_2. For each element x of Σ and each real number λ there exists at least one element y of Σ such that $G(x, y) = 0$ and $(x, y) = \lambda(x, x)$.

E_3. If $x, z \in \Sigma$ and $G(x, z) \neq 0$, there exists at least one element y of Σ such that $B(x, y, z) = G(x, y, z) = 0$ and $G(x, y) = G(y, z)$.

57. Quasi inner product space satisfying Schwarz inequalities

We consider first the system $\{\Sigma: Q_1, Q_2, Q_3, S_2, E_1\}$.

Theorem 57.1. *There is exactly one element θ of Σ with $(\theta, x) = 0$ for each element x of Σ.*

Proof. If θ, θ' are two such elements, then $(\theta, \theta) = (\theta, \theta') = (\theta', \theta') = 0$, and so $\theta = \theta'$ by Q_3. The theorem now follows from E_1.

Theorem 57.2. *The quasi inner product (x, x) vanishes if and only if $x = \theta$.*

Proof. By E_1 $(\theta, \theta) = 0$, and if $(x, x) = 0$, then $(x, x) = (\theta, x) = (\theta, \theta)$, which implies $x = \theta$ by Q_3.

THEOREM 57.3. *Corresponding to each element x of Σ ($x \neq \theta$), and each real number λ, there is at most one element y of Σ such that*

$$(x, y) = \lambda(x, x) \quad \text{and} \quad G(x, y) = 0.$$

Proof. If $y, y^* \in \Sigma$ satisfying those two relations, then from the second one we get $(y, y) = (y^*, y^*) = \lambda^2 \cdot (x, x)$. By computation

$$G(x, y, y^*) = -(x, x)[\lambda^2 \cdot (x, x) - (y, y^*)]^2 \leqslant 0$$

by Q_2 and Theorem 57.2. Use of S_2 gives $(y, y^*) = \lambda^2(x, x)$, and hence $(y, y) = (y^*, y^*) = (y, y^*)$, from which $y = y^*$ follows by Q_3.

DEFINITION 57.1. *If $x \in \Sigma$ ($x \neq \theta$), and λ is a real number, an element y of Σ such that $(x, y) = \lambda(x, x)$, $G(x, y) = 0$ is called a scalar multiple of x by λ and is denoted by $\lambda \cdot x$. If $x = \theta$, define $\lambda \cdot x = \theta$.*

THEOREM 57.4. *For each x and each real λ, Σ contains at most one element $\lambda \cdot x$.*

Remark. For each element x of Σ, $0 \cdot x = \theta$.

THEOREM 57.5. *If $x, \lambda \cdot x, y \in \Sigma$, $(\lambda \cdot x, y) = \lambda(x, y)$.*

Proof. The theorem being obvious for $x = \theta$, suppose $x \neq \theta$.

Now $\qquad G(x, \lambda \cdot x, y) = -(x, x)[(\lambda \cdot x, y) - \lambda \cdot (x, y)]^2 \leqslant 0,$

and using S_2 gives $(\lambda \cdot x, y) = \lambda(x, y)$.

THEOREM 57.6. *If $x, 1 \cdot x \in \Sigma$ then $1 \cdot x = x$.*

Proof. Use of Theorem 57.5 gives $(1 \cdot x, 1 \cdot x) = (1 \cdot x, x) = (x, x)$ and $x = 1 \cdot x$ by Q_3.

THEOREM 57.7. *If $x, \lambda_2 \cdot x, \lambda_1 \cdot (\lambda_2 \cdot x), (\lambda_1 \lambda_2) \cdot x \in \Sigma$ (λ_1, λ_2, real), then $\lambda_1 \cdot (\lambda_2 \cdot x) = (\lambda_1 \lambda_2) \cdot x$.*

Proof. The proof follows immediately from Q_2, Theorem 57.5, and Q_3.

THEOREM 57.8. *If $x, y \in \Sigma$ ($x \neq \theta$), then y is a scalar multiple of x if and only if $G(x, y) = 0$.*

Proof. If $y = \lambda \cdot x$, then $G(x, y) = 0$ by Definition 57.1. If $G(x, y) = 0$, put $(x, y)/(x, x) = \lambda$. Then $(x, y) = \lambda(x, x)$ which, together with $G(x, y) = 0$, gives $y = \lambda \cdot x$.

THEOREM 57.9. *If $x, y, z \in \Sigma$ ($y \neq \theta$), and $G(x, y) = G(y, z) = 0$, then $G(x, z) = 0$.*

Proof. The theorem is trivial if $x = \theta$. In the contrary case, $y = \lambda_1 \cdot x$ and $z = \lambda_2 \cdot y$ by Theorem 57.8. Then Theorem 57.7 gives

$$z = \lambda_2(\lambda_1 \cdot x) = (\lambda_1 \lambda_2) \cdot x \quad \text{and} \quad G(x, z) = 0.$$

We now introduce a distance in $\{\Sigma\colon Q_1, Q_2, Q_3, S_1, S_2, E_1\}$. Attach to each two elements x, y of Σ the number

$$xy = [(x,x)+(y,y)-2(x,y)]^{\frac{1}{2}} \qquad (*)$$

as distance. Clearly $xy = yx$ and $xx = 0$. Now

$$(x,x)+(y,y)-2(x,y) = [(x,x)^{\frac{1}{2}}-(y,y)^{\frac{1}{2}}]^2 + 2[(x,x)^{\frac{1}{2}}(y,y)^{\frac{1}{2}}-(x,y)]$$
$$\geqslant 0 \qquad (\dagger)$$

by S_1, and so xy is real and non-negative. Moreover, if $xy = 0$, then each summand in the right-hand member of $(\dagger)$ vanishes. This gives

$$(x,x) = (x,y) = (y,y)$$

and consequently $x = y$. Thus the distance xy defined by $(*)$ makes Σ a *semimetric* space.

We note that $\theta x = (x,x)^{\frac{1}{2}}$ and $(x,y) = \frac{1}{2}(\theta x^2 + \theta y^2 - xy^2)$.

DEFINITION 57.2. *A distance space is metric about one of its elements t provided t, u, v are congruent with three points of the euclidean plane for every pair of elements u, v of the space.*

THEOREM 57.10. *The space Σ is metric about θ.*

Proof. Since Σ is semimetric, it suffices to show that the determinant $D(\theta, x, y)$ is negative or zero for every x, $y \in \Sigma$. An easy computation shows that $D(\theta, x, y) = -4G(x,y)$, and hence $D(\theta, x, y) \leqslant 0$ by S_1.

Remark. The space Σ is not necessarily metric. For if Σ consists of elements θ, x_1, x_2, x_3 with

$$(\theta, x_i) = (x_i, \theta) = 0, \qquad (x_i, x_i) = 16 \quad (i = 1, 2, 3),$$

and
$$(x_1, x_2) = (x_2, x_1) = (x_1, x_3) = (x_3, x_1) = \tfrac{1}{2}(31),$$
$$(x_2, x_3) = (x_3, x_2) = \tfrac{1}{2}(23),$$

then Postulates Q_1, Q_2, Q_3, S_1, S_2, E_1 are seen to be satisfied. But

$$x_1 x_2 = x_1 x_3 = 1, \qquad x_2 x_3 = 3,$$

and consequently Σ is not metric.

DEFINITION 57.3. *Three elements x, y, z of Σ are algebraically linear provided $B(x,y,z) = G(x,y,z) = 0$.*

Remark. Since $B(x,y,z) = \frac{1}{4}D(x,y,z)$, the elements x, y, z are *metrically linear* (that is, the sum of two of the distances xy, yz, and xz equals the third) if and only if $B(x,y,z) = 0$. This, however, does not imply that $G(x,y,z) = 0$. Hence algebraic linearity implies metric linearity, but not conversely—a well-known phenomenon in normed linear spaces.

DEFINITION 57.4. *If $x, z \in \Sigma$ ($x \neq z$), an element y of Σ is an algebraic middle element of x, z provided*

(a) $G(x,z) \neq 0$, x, y, z are algebraically linear, and $G(x,y) = G(y,z)$, or

(b) $G(x,z) = 0$ and $y = \frac{1}{2}(1+\lambda).x$, where $z = \lambda.x$ (with the roles of x and z reversed in the two preceding relations in case $x = \theta$).

THEOREM 57.11. *If y is an algebraic middle element of x, z, then y is a metric middle element of x and z.*

Proof. From $D(x,y,z) = 4B(x,y,z) = 0$, it follows that x, y, z are metrically linear. It remains to show that $xy = yz$.

Case 1. $G(x,z) \neq 0$. Since $G(x,y,z) = 0$, the easily established relation $D(\theta,x,y,z) = 8G(x,y,z)$ gives $D(\theta,x,y,z) = 0$, with θ, x, y, z a metric quadruple (due to Theorem 57.10 and the linearity of x, y, z). Hence points θ', x', y', z' of the euclidean plane exist such that the congruence (∗∗) $\theta, x, y, z \approx \theta', x', y', z'$, holds and x', y', z' lie on a straight line. Now $G(x,y) = -\frac{1}{4}D(\theta,x,y) = -\frac{1}{4}D(\theta',x',y') = 4\Delta^2(\theta',x',y')$ and, similarly, $G(y,z) = 4\Delta^2(\theta',y',z')$, where $\Delta(\ ,\ ,\ )$ denotes the area of the triangle whose vertices appear within the parentheses.

Then $\Delta(\theta',x',y') = \Delta(\theta',y',z')$, which follows from $G(x,y) = G(y,z)$, gives (using the linearity of x', y', z') $x'y' = y'z'$ and so $xy = yz$ from congruence (∗∗).

Case 2. $G(x,z) = 0$. If $x \neq \theta$, then $y = \frac{1}{2}(1+\lambda).x$ and

$$xy = \{(x,x) + [\tfrac{1}{2}(1+\lambda)]^2(x,x) - (\lambda+1)(x,x)\}^{\frac{1}{2}};$$

that is, $xy = \frac{1}{2}|1-\lambda|(x,x)^{\frac{1}{2}}$. A similar computation shows that yz has the same value. If, finally, $x = \theta$, then $y = \frac{1}{2}.z$ and we get

$$xy = \theta y = \tfrac{1}{2}(z,z)^{\frac{1}{2}} = yz.$$

Though the determinant D formed for a finite subset of Σ is evidently a congruence invariant, the Gram determinants G do not have this property. If, however, $x_1, x_2,..., x_n$ and $y_1, y_2,..., y_n$ are two n-tuples of Σ such that $x_1, x_2,..., x_n \approx y_1, y_2,..., y_n$ and $\theta x_i = \theta y_i$ ($i = 1, 2,..., n$), then $G(x_1, x_2,..., x_n) = G(y_1, y_2,..., y_n)$, since each is equal to

$$(-1)^{n+1} 2^{-n} D(\theta, x_1, x_2,..., x_n).$$

LEMMA 57.1. *If y is an algebraic middle element of x and z, then $\theta y = \frac{1}{2}[2(\theta x^2 + \theta z^2) - xz^2]^{\frac{1}{2}}$.*

Proof. Developing $D(\theta,x,y,z) = 8G(x,y,z) = 0$, and using

$$xy = yz = \tfrac{1}{2}xz \quad \text{(Theorem 57.11)}$$

the expression for θy given in the lemma is obtained.

Adjoining the third Schwarz postulate to the preceding system, we prove the following theorem.

THEOREM 57.12. *There is at most one algebraic middle element of two elements of Σ.*

Proof. If y, y^* are middle elements of x and z, Theorem 57.11 and Lemma 57.1 give $xy = yz = \tfrac{1}{2}xz = xy^* = y^*z$, $\theta y = \theta y^*$.

Case 1. $G(x, z) \neq 0$. From $D(\theta, x, y, z, y^*) = -16G(x, y, z, y^*) \leqslant 0$ (S_3), we obtain $-D(\theta, x, z)(yy^*)^4 \leqslant 0$.

But $G(x, z) \neq 0$ gives $D(\theta, x, z) = -4G(x, z) < 0$ (by S_1). Hence $D(\theta, x, z).yy^* = 0$ and so $yy^* = 0$. Since Σ is semimetric, this implies $y = y^*$, and the theorem is proved in this case.

Case 2. $G(x, z) = 0$. If $x \neq \theta$, then $y = \tfrac{1}{2}(1+\lambda).x$, where $z = \lambda.x$, and the uniqueness of y follows from Theorem 57.4. A similar argument is employed in case $x = \theta$ (and consequently $z \neq \theta$), and the proof is complete.

DEFINITION 57.5. *Let x, z be elements of Σ with algebraic middle element y. An element $2.y$ is called a sum of x and z (written, $2.y = x+z$).*

If $2.y = x+z$ then $\tfrac{1}{2}(2.y) = \tfrac{1}{2}.(x+z)$, or (Theorem 57.7) $y = \tfrac{1}{2}(x+z)$.

THEOREM 57.13. *Two elements x, y of Σ have at most one sum $x+y$.*

LEMMA 57.2. *If x, y, $x+y \in \Sigma$, then*

$$(x+y, x) = (x, x)+(x, y), \qquad (x+y, x+y) = (x, x)+(y, y)+2(x, y).$$

Proof. Writing $2.w = x+y$, we have

$$(x+y, x) = (2.w, x) = 2(w, x) = \theta w^2 + \theta x^2 - xw^2.$$

Since $xw^2 = \tfrac{1}{4}xy^2$ and (Lemma 57.1)

$$\theta w^2 = \tfrac{1}{2}(\theta x^2 + \theta y^2) - \tfrac{1}{4}xy^2,$$

substitution yields

$$(x+y, x) = \theta x^2 + \tfrac{1}{2}(\theta x^2 + \theta y^2 - xy^2)$$
$$= (x, x) + (x, y).$$

Also,

$$(x+y, x+y) = (2.w, 2.w) = 4(w, w) = 4\theta w^2$$
$$= 2(\theta x^2 + \theta y^2) - xy^2$$
$$= 2(\theta x^2 + \theta y^2) - (\theta x^2 + \theta y^2 - 2(x, y))$$
$$= \theta x^2 + \theta y^2 + 2(x, y)$$
$$= (x, x) + (y, y) + 2(x, y).$$

THEOREM 57.14. *If x, y, $x+y$, $z \in \Sigma$, then*

$$(x+y, z) = (x, z)+(y, z).$$

Proof. Suppose, first, that $G(x, y) \neq 0$. Since
$$G(x, y, x+y) = 4G(x, y, \tfrac{1}{2}(x+y)) = 0,$$
S_3 yields $G(x, y, x+y, z) \geqslant 0$.

Use of Lemma 57.2 gives
$$G(x, y, x+y, z) = -[(x+y, z)-(x, z)-(y, z)]^2 G(x, y) \leqslant 0.$$
Consequently $G(x, y, x+y, z) = 0$ and $(x+y, z) = (x, z)+(y, z)$.

If $G(x, y) = 0$, then (assuming $x \neq \theta$) $\tfrac{1}{2}(x+y) = [\tfrac{1}{2}(1+\lambda)]x$ with $y = \lambda.x$, and so $x+y = (1+\lambda).x$ (Theorem 57.7). Whence
$$(x+y, z) = ((1+\lambda).x, z) = (1+\lambda)(x, z)$$
$$= (x, z)+(\lambda.x, z)$$
$$= (x, z)+(y, z).$$
The desired result is obtained in a similar manner in case $x = \theta$ (and consequently $y \neq \theta$).

58. Generalized euclidean space in terms of a quasi inner product

We adjoin now the two existence postulates E_2 and E_3 to our system.

THEOREM 58.1. *For each element x of Σ and each real number λ, Σ contains exactly one element $\lambda.x$. If $x, y \in \Sigma$ there is exactly one element $x+y$ of Σ.*

Proof. The proof of the first statement of the theorem is an immediate consequence of Postulate E_2, Definition 57.1, and Theorem 57.4, while the proof of the second statement follows from the first one and Theorem 57.12 when it is shown that at least one algebraic middle element exists for every pair of distinct elements of Σ.

If $x, z \in \Sigma$ ($x \neq z$), Postulate E_3 ensures the existence of at least one middle element of x, z whenever $G(x, z) \neq 0$. If $x, z \in \Sigma$ with $G(x, z) = 0$, then $z = \lambda.x$ (assuming $x \neq \theta$) and the element $[\tfrac{1}{2}(1+\lambda)]x$, which exists by the first part of the theorem, is the desired algebraic middle element of x and z. If, finally, $x = \theta$, then the element $\tfrac{1}{2}.z$ is an algebraic middle element of x and z.

THEOREM 58.2. *Addition and scalar multiplication in Σ have the following properties*:

(i) $x+y = y+x$,
(ii) $x+(y+z) = (x+y)+z$,
(iii) $x+y = x+z$ *implies* $y = z$,
(iv) $\lambda.(x+y) = \lambda.x+\lambda.y$,
(v) $(\lambda_1+\lambda_2).x = \lambda_1.x+\lambda_2.x$.

Proof. Properties (i), (ii), (iv), (v) follow at once from Postulates Q_1, Q_3 and Theorems 57.5, 57.14. To establish (iii), we have $x+y = x+z$ implies $(x+y, y) = (x+z, y)$, and consequently $(y, y) = (y, z)$. Also, $(x+y, z) = (x+z, z)$ and so $(y, z) = (z, z)$. Hence $y = z$ by Q_3.

THEOREM 58.3. *Distances* θx, $x \in \Sigma$, *have the following properties*:
(a) $\theta x = (x, x)^{\frac{1}{2}} \geq 0$,
(b) $\theta x = 0$ *if and only if* $x = \theta$,
(c) $\mathrm{dist}[\theta, \lambda.x] = |\lambda|\theta x$,
(d) $\mathrm{dist}[\theta, x+y] \leq \theta x + \theta y$.

Proof. The relation $\theta x = (x, x)^{\frac{1}{2}}$ is given in the second paragraph of page 141, and (b) follows from Q_2 and Theorem 57.2.

Since
$$\mathrm{dist}[\theta, \lambda.x] = [(\theta, \theta) + (\lambda.x, \lambda.x) - 2(\theta, \lambda.x)]^{\frac{1}{2}}$$
$$= (\lambda.x, \lambda.x)^{\frac{1}{2}} = |\lambda|(x, x)^{\frac{1}{2}} = |\lambda|\theta x,$$
property (c) is proved.

To establish (d) we have from $G(x, y) \geq 0$,
$$[(x, x)^{\frac{1}{2}} + (y, y)^{\frac{1}{2}}]^2 \geq (x, x) + (y, y) + 2(x, y),$$
or
$$\theta x + \theta y \geq (x+y, x+y)^{\frac{1}{2}} = \mathrm{dist}[\theta, x+y]. \qquad (*)$$

We are justified by Theorem 58.3 in defining the *norm* $||x||$ of an element x of Σ as the distance θx. Noting that
$$(x-y, x-y)^{\frac{1}{2}} = [(x+(-1)y, x+(-1)y]^{\frac{1}{2}}$$
$$= [(x, x) + (y, y) - 2(x, y)]^{\frac{1}{2}}$$
$$= xy,$$
or
$$xy = ||y-x|| = \mathrm{dist}[\theta, y-x],$$
it follows that for each three elements x, y, z of Σ
$$\mathrm{dist}[y-x, z-x] = ||(z-x) - (y-x)|| = ||z-y|| = yz.$$

Remark. The space Σ is metric.

THEOREM 58.4. *The space* $\{\Sigma: Q_1, Q_2, Q_3, S_1, S_2, S_3, E_1, E_2, E_3\}$ *is a generalized euclidean space*.

Proof. Postulates Q_1, Q_2 and Theorems 57.2, 57.5, 57.14 prove that the quasi inner product (x, y) has the properties (1)–(5) of an inner product (§ 56), and Theorems 57.6, 57.7, 58.2 establish that Σ is a linear space with respect to the addition $x+y$ and the scalar multiplication $\lambda.x$.

By Theorem 58.3 the space Σ is normed by taking $||x|| = \theta x$, and since $\theta x = (x, x)^{\frac{1}{2}}$, Σ satisfies all the requirements of Definition 56.1 for a generalized euclidean space.

We develop now some further properties of Σ.

THEOREM 58.5. *Metric linearity and algebraic linearity are equivalent in Σ.*

Proof. By the Remark following Definition 57.3, algebraic linearity implies metric linearity. To prove the converse it suffices to show that if $x, y, z \in \Sigma$, then $B(x, y, z) = 0$ implies $G(x, y, z) = 0$. This is obvious in case the elements are not pairwise distinct, or if one of the elements is θ. Suppose neither alternative holds.

It is easily seen that $x, y, z \approx \theta, y-x, z-x$, and so the metric linearity of x, y, z implies that of $\theta, y-x, z-x$. Consequently
$$G(y-x, z-x) = -\tfrac{1}{4}D(\theta, y-x, z-x) = 0,$$
and application of Theorem 57.8 gives
$$z-x = \lambda.(y-x)$$
$$z = (1-\lambda).x + \lambda.y,$$
from which $G(x, y, z) = 0$ follows by computation.

THEOREM 58.6. *The space Σ has the weak euclidean four-point property.*

Proof. Let x, y, z, w be any four elements of Σ (pairwise distinct) with x, y, z metrically linear. Since
$$x, y, z, w \approx x-w, y-w, z-w, \theta,$$
it suffices to prove the latter quadruple imbeddable congruently in the plane. The space Σ being metric, this follows upon showing
$$D(\theta, x-w, y-w, z-w) = 0.$$

Now the metric linearity of x, y, z implies that of $x-w, y-w, z-w$ and hence (Theorem 58.5) this triple is algebraically linear; that is
$$G(x-w, y-w, z-w) = 0.$$
But $\qquad D(\theta, x-w, y-w, z-w) = 8G(x-w, y-w, z-w) = 0,$
and the proof is complete.

DEFINITION 58.1. *An element x of Σ is a limit of an infinite sequence $\{x_n\}$ of elements of Σ provided $\lim(x, x_n) = \lim(x_n, x_n) = (x, x)$.*

Remark. Writing $(x, x) + (x_n, x_n) - 2(x, x_n)$ as in (†) of §57, it is seen that $\lim x_n = x$ if and only if $\lim xx_n = 0$.

Cauchy sequences and separability are now easily defined in terms of inner product, and we have the following theorems.

THEOREM 58.7. *The space Σ is congruent with Hilbert space provided it is separable, complete, and for each positive integer k a subset $x_1, x_2, \ldots, x_k$ of Σ exists with $G(x_1, x_2, \ldots, x_k) \neq 0$.*

THEOREM 58.8. *The space Σ is congruent with the n-dimensional euclidean space E_n provided it is complete, and n is the smallest positive integer such that each $n+1$ elements of Σ have a vanishing determinant G.*

Since Σ is metrically convex, externally convex, *and has the weak euclidean four-point property*, these theorems are immediate consequences of Theorem 55.1 and Theorem 50.4, respectively, when we observe (as noted previously) that for each positive integer k,

$$D(\theta, x_1, x_2, ..., x_k) = (-1)^{k+1} 2^k G(x_1, x_2, ..., x_k).$$

We note, in conclusion, that a congruence of Σ with E_n or with Hilbert space *preserves inner products if and only if the element θ' corresponding to θ in the congruence is the origin (or null vector) of the space*. Since this can always be brought about by a translation, Theorems 58.7, 58.8 can be restated in terms of a *vectorial application* of the 'vector' space Σ onto the euclidean or Hilbert vector space.†

EXERCISE

Show that the points $x^{(k)} = (x_1^{(k)}, x_2^{(k)}, ..., x_n^{(k)}, ...)$ $x_i^{(k)} = 1$ $(k = i)$, $x_i^{(k)} = 0$ $(k \neq i)$ $(i, k = 1, 2, ...)$ form a metric basis for Hilbert space. Denoting by $v^{(k)}$ the vector obtained by joining the origin to $x^{(k)}$ $(k = 1, 2, ...)$, the vectors $v^{(1)}, v^{(2)}, ..., v^{(n)}$, ... form a vector basis for Hilbert vector space, but cease to do so if any of the vectors are omitted.

REFERENCES

§ 48. Theorem 48.1 appears in Menger [1] with a different proof.
§§ 49–50. Compare with Wilson [3].
§ 51. Blumenthal [9].
§ 52. Blumenthal [11], Ville [1].
§ 53. Theorem 53.1 is in Menger [5]; Theorem 53.2 in Schoenberg [3].
§ 54. Schoenberg [3].
§§ 56–58. Blumenthal [12].

SUPPLEMENTARY PAPERS

Von Neumann [1] and Schoenberg, Schoenberg [4, 5, 6] deal with the congruent imbedding in Hilbert space of metric transforms of euclidean space.
Busemann [1, 2] concern the space problem for euclidean, hyperbolic, and spherical spaces.
Birkhoff [2], Gillam [1]—metric postulates for geometry.

† We thus obtain a new set of postulates for Hilbert space in terms of just *one* primitive operation.

CHAPTER VI

CONGRUENCE INDICES OF SOME EUCLIDEAN SUBSETS

59. Introductory remarks

We have seen how important is the notion of congruence indices in solving the imbedding and characterization problems of euclidean space. The systematic metric study of a given space necessitates the solutions of similar problems concerning certain subsets of the space, in which the set of comparison spaces may be the class $\{S\}$ of semimetric spaces or the class $\{\Sigma\}$ of subsets of the given space. Thus, for example, if one subjected the subset of the plane consisting of two mutually perpendicular lines to metric study, problems of finding congruence indices of this figure with respect to each of the classes $\{S\}$ and $\{\Sigma\}$ would arise.

In this chapter such problems are solved for certain linear and planar subsets of particular interest. The solutions contribute not only to the metric theory of those subsets, but they also serve, perhaps, to make the nature of the concept of congruence indices better understood.

We shall be mostly concerned with congruence indices for which the second index is zero (that is, with *congruence order*), and unless stated otherwise, the class of comparison spaces is the class $\{S\}$ of semimetric spaces. It is convenient to list here some obvious properties of congruence order.

1. If $\{S^*\}$ is a sub-class of the class $\{S\}$ and R is a space with indices $(n, 0)$ with respect to $\{S\}$, then R has indices $(m, 0)$ with respect to $\{S^*\}$, with $m \leqslant n$.
2. A space R of exactly n points has congruence order $n+1$.
3. If R has indices $(n, 0)$ with respect to $\{S\}$ and Σ is a subset of R with indices $(m, 0)$ with respect to the class $\{\Sigma_R\}$ of subsets of R, then Σ has indices $(k, 0)$ with respect to $\{S\}$, where $k = \max(m, n)$.
4. A space R with finite congruence order with respect to $\{S\}$ may contain a subset without this property.

    The E_1, for example, has congruence order 4 with respect to $\{S\}$, but the ray has no finite congruence order with respect to $\{S\}$ since for each positive integer n, each n points of E_1 are imbeddable in a given ray while the E_1 is not.
5. A space with minimum congruence order n may contain a subset with a larger minimum congruence order. (For example, the space

of the six points of E_1 with coordinates 0, 1, 2, 3, 5, 6 has minimum congruence order exceeding 4.)

EXERCISES

1. Find the minimum congruence order of the space of six points of E_1 mentioned above.
2. What linear subsets have congruence order 1 with respect to the class $\{\Sigma\}$ of all linear subsets?
3. Show that (a) the closed interval $a \leqslant x \leqslant b$ and (b) the interval $a \leqslant x < b$ have minimum congruence order 2 with respect to subsets of E_1. What are the best congruence indices of a line segment with respect to the class $\{S\}$ of semimetric spaces?
4. Show that the set R of rational points of E_1 has minimum congruence order 2 with respect to subsets of E_1. What is the minimum congruence order of R with respect to $\{S\}$?
5. Find the minimum congruence orders of (a) the subset $0, \pm 1, \pm 2, ..., \pm n, ...,$ and (b) the subset $1, 2, 3, ..., n$ of E_1 with respect to the class of all subsets of E_1.
6. What is a necessary and sufficient condition that a triple of E_1 have congruence order 2 with respect to linear sets?

60. Hyperfinite and transfinite congruence orders of linear sets. A general theorem

Linear subsets with arbitrarily large finite *minimum* congruence orders exist. If, for example, k is any integer greater than 1, the linear subset $(1, 2, ..., k, k+2, ..., 2k)$ has the minimum congruence order $m \leqslant 2k$ by property 2 (§ 59). On the other hand, $m > k$, since each k points of the linear subset $(1, 2, ..., k, k+1)$ are imbeddable in the first set while the set itself is not.

DEFINITION 60.1. *If a semimetric space R is such that (1) for each integer k a semimetric space S exists which is not congruently imbeddable in R though each k-tuple of its points is and (2) each semimetric space S is imbeddable in R whenever every finite subset of S is, then R has hyperfinite congruence order with respect to $\{S\}$.*

THEOREM 60.1. *Each bounded and closed subset of E_1 has either finite or hyperfinite congruence order with respect to $\{S\}$.*

Proof. Let Σ be a bounded and closed subset of E_1 with no finite congruence order with respect to $\{S\}$ and let S be a semimetric space with each finite subset imbeddable in Σ. Then clearly $S \approx X \subset E_1$.

Denoting the diameter of Σ by $\delta(\Sigma)$, translate Σ so that it is contained in the segment $[\delta(\Sigma), 2\delta(\Sigma)]$, and obtain by reflection in the origin the set Σ^*, symmetric to Σ. It is apparent that each finite subset of X can be *translated* into a subset of Σ or of Σ^*.

Let $x_1, x_2, ..., x_n$ be n points of X. A translation $x' = x+t$ that carries a point x_i into a point s of $\Sigma + \Sigma^*$ is given by $t = s - x_i$, and the set

$\{t(x_i)\}$ of numbers t obtained by letting s vary over $\Sigma+\Sigma^*$ is evidently congruent with the latter set. Hence $\{t(x_i)\}$ is bounded and closed ($i = 1, 2,..., n$), and since $(x_1, x_2,..., x_n) \subseteqq \Sigma+\Sigma^*$, by translation, it follows that the product
$$\prod_{i=1}^{n} \{t(x_i)\}$$
is non-empty. We may conclude that there is a point common to all the sets $\{t(x)\}$, $x \in X$, for F. Riesz has shown that any family of bounded and closed subsets of E_n has a common point whenever each finite subfamily has.

Hence a translation parameter t' exists such that the translation $x' = x+t'$ carries X into $\Sigma+\Sigma^*$. But since $\delta(X)$ is clearly at most equal to $\delta(\Sigma)$, then X is carried into Σ or Σ^* and so $X \subseteqq \Sigma$.

An example of a subset of E_1 with hyperfinite congruence order with respect to $\{S\}$ is obtained by deleting from the segment $[0, 1]$ the *open* intervals $(i/2^k, i/2^k+1/2 . 4^k)$ ($k = 1, 2,...; i = 1, 2,..., 2^k-1$). The resulting set H is bounded and closed, and has no finite congruence order with respect to $\{S\}$. If any one of the ten points $i/4^2$ ($i = 0, 1,..., 9$) other than an end-point be deleted, a gap of width $\frac{1}{8}$ is present in the remaining set and there are at most four points on one of the two sides of this gap. Translating and/or reflecting this set so that this gap falls on the gap $(\frac{1}{2}, \frac{1}{2}+\frac{1}{8})$ in H, with the side of the set that contains at most four points falling to the right, it is seen that all nine points fall on points of H because none of the 16th points on either side of the gap are taken out by subsequent deletions. It follows that each nine of the ten points are imbeddable in H but the ten points obviously are not.

Similar considerations show that each $2^{k+2}+1$ of the points $i/4^{k+1}$ ($i = 0, 1,..., 2^{k+2}+1$) are imbeddable in H, for every positive integer k, but the $2^{k+2}+2$ points are not and hence H has no finite congruence order. It follows from the preceding theorem that H has hyperfinite congruence order.

THEOREM 60.2. *Every compact metric space M has finite or hyperfinite congruence order with respect to the class of all separable semimetric spaces.*

Proof. Suppose M does not have a finite congruence order with respect to the given class of spaces. Let S be any *separable* semimetric space, $P = (p_1, p_2,..., p_n,...)$ a subset of S (denumerable at most) with $\overline{P} = S$, and suppose that each finite subset of S is imbeddable in M. We wish to show that $S \subseteqq M$.

This is obvious if S has only a finite number of points. In the contrary case, there corresponds to each positive integer n an n-tuple $q_1^{(n)}, q_2^{(n)}, ..., q_n^{(n)}$ of M such that

$$p_1, p_2, ..., p_n \approx q_1^{(n)}, q_2^{(n)}, ..., q_n^{(n)} \quad (n = 1, 2, ...).$$

Since M is compact it contains points $q_1, q_2, ...$ such that

$$q_i = \lim q_i^{(k_1, k_2, ..., k_i)} \quad (i = 1, 2, ...),$$

where $\{q_1^{(k_1)}\}$ is a properly selected subsequence of the sequence $\{q_1^{(n)}\}$ and $\{q_i^{(k_1, k_2, ..., k_i)}\}$ is a subsequence of $\{q_i^{(k_1, k_2, ..., k_{i-1})}\}$ ($i = 2, 3, ...$).

Clearly $\quad p_1, p_2, ..., p_n, ... \approx q_1, q_2, ..., q_n, ... \quad$ (*)

since
$$p_i p_j = q_i^{(k_1, k_2, ..., k_j)} q_j^{(k_1, k_2, ..., k_j)}$$
$$= \lim q_i^{(k_1, k_2, ..., k_j)} q_j^{(k_1, k_2, ..., k_j)}$$
$$= q_i q_j,$$

($i, j = 1, 2, ...; i < j$), from the continuity of the metric.

Extending congruence (*) to the closures of the sets

$$P \text{ and } Q = (q_1, q_2, ..., q_n, ...),$$

we obtain $\overline{P} = S \approx \overline{Q} \subset M$, and the theorem is proved.

This theorem is, of course, more widely applicable than Theorem 60.1. The class of comparison spaces it utilizes, though less general than that considered in the former theorem, is sufficiently extensive for most investigations.

If a space does not have finite or hyperfinite congruence order, but every semimetric space is imbeddable in it whenever each of its denumerable subsets has that property, the space has congruence order $\aleph_0$ with respect to $\{S\}$. The open interval $I = (a, b)$ is a space with congruence order $\aleph_0$, for since each finite subset of the interval $[a, b]$, $a \leqslant x < b$, is congruently contained in I while the semi-closed interval is not, then I does not have finite or hyperfinite congruence order with respect to $\{S\}$. On the other hand, if each denumerable subset of a semimetric space S is imbeddable in I then S is congruent to a subset L of E_1 with diameter at most $b-a$. If $\delta(L) = b-a$ then L does not contain either its greatest lower bound or its least upper bound, for in the contrary case a sequence of points of L can be selected that is not imbeddable in I. It follows that L is congruently imbeddable in I.

THEOREM 60.3. *The complement of a finite or denumerable subset of E_1 is a space with minimum congruence order $\mathfrak{c}$ with respect to $\{S\}$.*

Proof. Let D denote a finite or denumerable subset of E_1 and let $X = E_1 - D$. Clearly each semimetric space S is congruently contained in X whenever each $\mathfrak{c}$ of its points are, for the power of such a space does not exceed $\mathfrak{c}$ since it is imbeddable in E_1. It remains to show that a semimetric space exists which is not imbeddable in X though each of its subsets with power $\aleph_0$ has that property.

Such a space is E_1 itself, for consider any denumerable subset

$$Q = (q_1, q_2, \ldots, q_n, \ldots)$$

of E_1. If Q is translated along the line, only translations with parameters $(p_i - q_j)$, $p_i \in D$, can carry a point of Q into a point of D. This parameter set is obviously denumerable, so a real number t exists, $t \neq (p_i - q_j)$, such that the translation $x' = x + t$ carries Q into a subset of X. Hence each denumerable subset of E_1 is imbeddable in X, while X contains no subset congruent to E_1. Thus, X has minimum congruence order $\mathfrak{c}$ with respect to $\{S\}$.

EXERCISES

1. Show that the rational points of any interval form a space that has neither finite nor hyperfinite congruence order with respect to $\{S\}$.
2. What is the minimum congruence order with respect to $\{S\}$ of the punctured interval $0 \leqslant x \leqslant 1$, $x \neq \frac{1}{2}$?
3. Show that every countable subset of E_1 has finite, hyperfinite, or congruence order $\aleph_0$ with respect to $\{S\}$.

61. One-dimensional subsets of E_2

If A, $B \subset E_2$ and $A \subseteqq B$, a subset B^* of E_2 exists such that $B \approx B^*$ and $A \subset B^*$. We refer to B^* as a *congruent image* of B and say that A is *coverable* by a congruent image of B.

LEMMA 61.1. *If a k-tuple of a planar set P is coverable by at most m pairwise distinct congruent images of a planar set Q, and each $k+m$ points of P are congruently contained in Q, then P is imbeddable in Q.*

Proof. Let $p_1, p_2, \ldots, p_k$ be a k-tuple of P that is coverable by each of the pairwise distinct congruent images $Q_1, Q_2, \ldots, Q_j$ of Q ($j \leqslant m$), and by no others. Then for at least one index i ($1 \leqslant i \leqslant j$), $P \subset Q_i$. In the contrary case, points x_i of P exist such that $x_i \bar{\in} Q_i$ ($i = 1, 2, \ldots, j$). But then the $k+j$ points $p_1, p_2, \ldots, p_k, x_1, x_2, \ldots, x_j$ of P are not coverable by any congruent image of Q, and since $k+j \leqslant k+m$ this contradicts the imbeddability in Q of each $k+m$ points of P.

If θ ($0 \leqslant \theta \leqslant \frac{1}{2}\pi$) is the smallest angle made by two distinct lines, the figure formed by the lines is called a θ-*cross*. In seeking a congruence order for a two-dimensional space it is sometimes useful to determine the minimum congruence order of such a figure.

THEOREM 61.1. *The minimum congruence order with respect to* $\{S\}$ *of any θ-cross in E_2 is five.*

Proof. Any semimetric space S with each five points imbeddable in a θ-cross X_θ is congruent with a subset P of E_2 (corollary, Theorem 38.1).

If each quintuple of P is linear, then P is linear and hence imbeddable in X_θ. If $p_1, p_2,..., p_5 \in P$, not linear, then (since these points are coverable by a congruent image of the θ-cross) some three, say p_1, p_2, p_3, are linear and a fourth point p_4 is not on their line. If only one congruent image of X_θ covers p_1, p_2, p_3, p_4 then P is imbeddable in the θ-cross by Lemma 61.1. In no case are there more than two congruent images $X_\theta^{(1)}$, $X_\theta^{(2)}$ of X_θ covering p_1, p_2, p_3, p_4, and P is contained in one of them. For if the contrary be assumed then points q_1, q_2 of P exist such that $q_1 \in P \cdot X_\theta^{(1)}$, $q_1 \bar{\in} X_\theta^{(2)}$, and $q_2 \in P \cdot X_\theta^{(2)}$, $q_2 \bar{\in} X_\theta^{(1)}$.

The angle between the lines $L(p_1, p_3)$, $L(q_1, q_2)$ is then different from θ and so the five points p_1, p_2, p_3, q_1, q_2 are not imbeddable in X_θ contrary to hypothesis.

The reader may easily show by examples that X_θ does not have congruence order 4 with respect to planar subsets. Hence 5 is the minimum congruence order of X_θ with respect to $\{S\}$, as well as with respect to subsets of the plane.

THEOREM 61.2. *A regular polygon P_n of E_2, $n \geqslant 3$, has congruence order $2n+1$ with respect to* $\{S\}$.

Proof. Since $n \geqslant 3$ it suffices to show that each subset Q of E_2 is imbeddable in P_n whenever all of its $(2n+1)$-tuples are. If $q_1, q_2,..., q_{2n+1}$ are pairwise distinct points of Q then clearly three of these points, say q_1, q_2, q_3 are linear. If Q is a subset of the line $L(q_1, q_3)$ then Q is congruently imbeddable in a side of P_n. Suppose, on the other hand, that Q contains points not on $L(q_1, q_3)$.

Case 1. *n is odd*. If $q_4 \in Q$, $q_4 \bar{\in} L(q_1, q_3)$, then at most two congruent images of P_n can cover q_1, q_2, q_3, q_4, so that (Lemma 61.1) Q is coverable by a congruent image of P_n if each 6 of its points are. Since $n \geqslant 3$, $2n+1 > 6$ and so the theorem is proved in this case.

Case 2. *n is even*. If Q contains a point q with dist$\{q, L(q_1, q_3)\}$ not 0 or d, where d is the distance between two opposite sides of P_n, the argument in Case 1 applies. Suppose, on the other hand, that each element of Q has distance from $L(q_1, q_3)$ either zero or d. Since the case in which $Q \subset L(q_1, q_3)$ has already been disposed of, we may assume that

$$Q \subset L(q_1, q_3) + L^*, \qquad Q \cdot L^* \neq 0,$$

where L^* is *one* of the lines of E_2 parallel to $L(q_1, q_2)$ and distant d from it.

It is now readily seen that Q is coverable by two opposite sides of P_n, and the theorem is proved.

THEOREM 61.3. *A conic C has congruence order 6 with respect to $\{S\}$.*

Proof. If C is degenerate it has congruence order 4 or 5 according as it is a pair of coincident or distinct lines (Theorem 61.1). Suppose C non-degenerate and let S have each six of its points imbeddable in C. Then $S \approx Q \subset E_2$, and if $q_1, q_2, \ldots, q_5$ are any pairwise distinct points of Q there is exactly one congruent image of C that covers them. It follows by Lemma 61.1 that $Q \lessapprox C$.

The reader is asked to consider the problem of determining whether or not 6 is the *minimum* congruence order of C with respect to $\{S\}$. Consider, in particular, the possibility of indices $(5, 2)$.

LEMMA 61.2. *If each three of four distinct points of E_2 are coverable by the (chord) circle $C_{1,r}$, but the four points are not, then they form an orthocentric quadruple (that is, each is the orthocentre of the triangle determined by the other three).*

Proof. Let p_1, p_2, p_3, p_4 denote the four points, and q_i the circumcentre of the triple obtained by omitting the i-th point $(i = 1, 2, 3, 4)$. It follows from the hypotheses that q_1, q_2, q_3 are reflections of q_4 in the lines $L(p_2, p_3)$, $L(p_1, p_3)$, $L(p_1, p_2)$ containing the sides of the triangle (p_1, p_2, p_3), and $p_4 q_1 = p_4 q_2 = p_4 q_3 = r$. Thus p_4 is the isogonal conjugate of the circumcentre q_4 of triangle (p_1, p_2, p_3) and hence is the orthocentre of the triangle.

If four points form an orthocentric quadruple, the four triples have circumcircles of the same radius. But the four points are not concyclic, for since p_4, for example, is the isogonal conjugate of the circumcentre q_4 of triangle (p_1, p_2, p_3), it follows that q_4 is the isogonal conjugate of p_4 (with respect to the same triangle). Then p_4 cannot lie on the circumcircle of that triangle (Exercise 3, §45). We have shown, therefore, that *a necessary and sufficient condition that four pairwise distinct points of E_2 be not coverable by $C_{1,r}$, though each triple of the points is coverable, is that the four points form an orthocentric quadruple.*

THEOREM 61.4. *The $C_{1,r}$ has best congruence indices $(3, 1)$ with respect to the class of subsets of E_2.*

Proof. To show $C_{1,r}$ possesses indices $(3, 1)$ it evidently suffices to prove that any planar quintuple is coverable by $C_{1,r}$ whenever each three of its points are. If, now, $p_1, p_2, \ldots, p_5$ is such a quintuple, an assumption that it is not coverable by $C_{1,r}$ implies that at least two of its quadruples are not coverable. According to the foregoing, each of these quadruples forms an orthocentric set, and since they have a triple in common, the fourth points in each quadruple coincide since each is the orthocentre of the common triple.

It is obvious that the indices (3, 1) are best, since it is trivial to observe that the first index cannot be reduced, while the example of an orthocentric quadruple shows that the second index cannot be replaced by zero.

LEMMA 61.3. *If four non-coplanar points have each triple coverable by $C_{1,r}$ they are the vertices of an isosceles tetrahedron (that is, a tetrahedron whose opposite edges are equal).*

Proof. Let o denote the centre of the sphere circumscribing the four points and R its radius. The feet of the perpendiculars from o to the faces of the tetrahedron are the circumcentres of these faces and so the distances of o from each face is $(R^2 - r^2)^{\frac{1}{2}}$. Thus o is also the centre of the inscribed sphere and the assertion of the lemma follows.

LEMMA 61.4. *The $C_{1,r}$ has best congruence indices (3, 1) with respect to the class of subsets of E_3.*

Proof. Let P be any subset of E_3 containing more than four points, each three of which are coverable by $C_{1,r}$, and let $p_1, p_2,..., p_5$ be any five points of P. If these five points are not coverable by $C_{1,r}$ then, by Theorem 61.4, the quintuple is not planar and hence contains at most one planar quadruple. (Why?) Applying the preceding lemma to the (at least) four non-planar quadruples, one sees that all of the ten distances determined by the five points are equal. But E_3 contains no equilateral five-point, and the lemma follows, for since the indices (3, 1) cannot be bettered with respect to subsets of E_2 the same is, *a fortiori*, true for subsets of E_3.

THEOREM 61.5. *The $C_{1,r}$ has best congruence indices $(3, n-2)$ with respect to subsets of E_n $(n > 2)$.*

Proof. This has been proved for $n = 3$. Make the inductive assumption of the theorem's validity in E_n, $n > 2$, and let $p_1, p_2,..., p_{n+3}$ be any $(n+3)$-tuple of a subset P of E_{n+1} containing more than $n+2$ points, with each triple coverable by $C_{1,r}$. Assuming that this $(n+3)$-tuple is not coverable by $C_{1,r}$ then (by the inductive hypothesis) the $n+3$ points are not contained in E_n. If no $n+2$ of the points $p_1, p_2,..., p_{n+3}$ are in E_n then no four of them are in E_2 and application of Lemma 61.3 shows that the $(n+3)$-tuple is equilateral, which is impossible. If, on the other hand, at least one $(n+2)$-tuple of $p_1, p_2,..., p_{n+3}$ lies in E_n, it is coverable by $C_{1,r}$ and hence lies in E_2. But then the $n+3$ points lie in

E_3, which (since $n > 2$) contradicts the conclusion reached above that those points are not contained in E_n.

Hence each $n+3$ points of P are coverable by $C_{1,r}$, the set P, consequently, is planar and coverable by $C_{1,r}$ (by Theorem 61.4). Thus, for each $n > 2$, $C_{1,r}$ has congruence indices $(3, n-2)$ with respect to subsets of E_n. The presence of equilateral $(n+1)$-tuples in E_n, with each triple coverable by $C_{1,r}$, shows that those indices are best.

Remark. Equilateral semimetric spaces of arbitrarily high power show that for no number k, finite or transfinite, does $C_{1,r}$ have congruence indices $(3, k)$ with respect to the class $\{S\}$.

Remark. It would be interesting to determine if $C_{1,r}$ is *characterized* among all simple closed curves (homeomorphs of the circle) of E_2 by having indices $(3, 1)$ with respect to planar subsets. Such a characterization of $C_{1,r}$ could be easily proved if it were shown that *corresponding to each plane simple closed curve Γ there exists a circle with each three of its points coverable by Γ*.

EXERCISES

1. Show that each three points of the inscribed circle of a triangle are congruently contained in the triangle. What circle has this property for a rectangle?
2. Show that the chord n-sphere $C_{n,r}$ has congruence indices $(n+2, 1)$ with respect to subsets of E_{n+1}.
3. DEFINITION. *A subset P_n of E_2 formed by a sum of n segments*
$$S_{p_1,p_2}+S_{p_2,p_3}+\ldots+S_{p_{n-1},p_n} \quad (p_n \text{ may coincide with } p_1)$$
that have pairwise at most end-points in common is a simple broken line. The segments are called 'sides'. A simple broken line P_n is convex provided each straight line of E_2 not containing a side of P_n has at most two points in common with P_n.

 Prove that if no two sides of a convex simple broken line P_n are parallel, it has congruence order $4(n+1)$ with respect to $\{S\}$. (The restriction covering parallel sides may be dispensed with, but the proof of the unrestricted assertion is somewhat more complicated.)
4. Construct examples to show that X_θ does not have congruence indices $(4, 0)$ with respect to planar subsets. Investigate the possibility of indices $(4, k)$, k a positive integer.
5. Is $2n+1$ the minimum congruence order of a regular polygon P_n with respect to $\{S\}$?
6. If each three points of a planar subset of more than four points is coverable by *translating* a given ellipse, show that the whole set is so coverable.

62. Two-dimensional subsets of E_2. Monomorphic sets. Characterization of the circular disk

When we examine two-dimensional subsets of E_2 for congruence order with respect to $\{S\}$ or even with respect to subsets of E_2 the results are mostly negative. Thus neither of the two portions of E_2 bounded by a hyperbola or a non-degenerate parabola have a finite congruent order with respect to subsets of E_2 since each such portion contains congruently every finite subset of E_2 without containing congruently E_2 itself. The same is obviously true for the points of E_2 exterior to or on an ellipse.

A degenerate parabola of rank 2 (pair of parallel lines) bounds a strip of E_2 and two half-planes. It is evident that the two half-planes have no finite congruence order. To see that this is also the case for the strip, let m be any positive integer and n an odd integer greater than m. It is easy to see that a regular n-gon exists with each $(n-1)$ vertices imbeddable in the (closed) strip, but not the n points. Hence for no integer m does the strip have congruence order m with respect to planar subsets. In a similar manner it is seen that the portion of E_2 that lies interior to or on an ellipse has no finite congruence order.

DEFINITION 62.1. *A subset R of a metric space is monomorphic provided it is not congruent with a proper subset of itself; that is, $P \approx R$ and $P \subset R$ imply $P = R$.*

The euclidean ray $x \geqslant 0$ is congruent with its proper subset $x \geqslant 1$ and hence is not monomorphic. The *bounded* subset of E_2 with polar coordinates $(1, n\alpha)$, where α is the radian measure of an angle incommensurable with π and $(n = 0, 1, 2, ...)$, is congruent with the proper subset $(1, n\alpha)$ $(n = 1, 2, ...)$, and so is not monomorphic. An interesting non-monomorphic planar set R is given by the set of all complex numbers representable by a polynomial $P(e^i)$ in e^i, with positive integral (or zero) coefficients.† If the 'constant term' of $P(e^i)$ is zero, the number represented belongs to the subset Q of R; in the contrary case, to the subset T. One easily shows that $Q \cdot T = 0$ and

$$Q \approx R = Q + T \approx T,$$

for a rotation of the plane about the origin through one radian makes R coincide with Q, while a translation of one unit in the positive direction of the axis of reals makes R coincide with T.

Lindenbaum has shown that a subset of a compact metric space is monomorphic provided it is both an F_σ and a G_δ. It follows that *every bounded open subset O of E_n is monomorphic*, for O is a subset of the compact metric space $\mathscr{D} = \bar{O}$; it is a G_δ since it is an open subset of $\mathscr{D}$, and an F_σ since every open subset of a metric space is an F_σ.

THEOREM 62.1. *A bounded open subset of E_n has no finite congruence order with respect to subsets of E_n.*

† Here i denotes $\sqrt{(-1)}$.

Proof. Let O be a bounded open subset of E_n, $p \in B(O)$, the boundary of O, and $O^* = O + (p)$. Since O is monomorphic, O^* is not congruently contained in O.

If $p_1, p_2, ..., p_m$ is any finite subset of O^*, with $p_i \neq p$ $(i = 1, 2, ..., m)$, then obviously the m-tuple is congruently contained in O. Suppose on the other hand, that one of the points, say p_m, is p. Now for each $i = 1, 2, ..., m-1$, positive numbers ρ_i exist such that $U(p_i; \rho_i) \subset O$. If $\rho = \min(\rho_1, \rho_2, ..., \rho_{m-1})$, then (since p is an accumulation point of O) a point q of O exists with $0 < pq < \rho$. The vector $\overrightarrow{pq}$ defines a translation of $p_1, p_2, ..., p_{m-1}, p$ into a subset $p'_1, p'_2, ..., p'_{m-1}, p'$ of O with which, of course, it is congruent.

Hence for every positive integer m, each m-tuple of O^* is congruently imbeddable in O. But O^* is not, and the theorem is proved.

By a *domain* $\mathscr{D}$ of E_2 we shall mean the closure $\bar{O}$ of a bounded open subset O of E_2.

THEOREM 62.2. *A domain of E_2 whose boundary is a simple broken line has no finite congruence order with respect to subsets of E_2.*

Proof. Let $\mathscr{D}$ be a domain of E_2 whose boundary $B(\mathscr{D})$ consists of n segments.

ASSERTION. *There is a largest (non-degenerate) circle C contained in $\mathscr{D}$.*

Since $\mathscr{D}$ is bounded the least upper bound $\bar{r}$ of the radii of all circles contained in $\mathscr{D}$ is finite, and since $\mathscr{D}$ contains the open set O, $\bar{r} > 0$. Let $\{C_i\}$ be a sequence of circles, $C_i \subset \mathscr{D}$, with centre p_i, radius r_i $(i = 1, 2, ...)$, and $\lim r_i = \bar{r}$. A subsequence $\{C_{i_n}\}$ exists such that $\lim p_{i_n} = p$, a point of $\mathscr{D}$, and $\lim r_{i_n} = \bar{r}$. The circle $C(p; \bar{r})$, with centre p and radius $\bar{r}$, is the desired circle; for since its points are clearly accumulation points of the closed set $\mathscr{D}$ then $C(p; \bar{r}) \subset \mathscr{D}$, while from the definition of $\bar{r}$ no circle of larger radius is part of $\mathscr{D}$.

Introduce polar coordinates in E_2 with p as pole and consider a circle $C(p; r)$ of radius $r \geq \bar{r}$. Denoting by $l_C(r)$ the length of that part of $C(p; r)$ not contained in $\mathscr{D}$, we wish to show that $l_C(r)$ is continuous at $r = \bar{r}$.

Since $l_C(\bar{r}) = 0$, let us assume that $l_C(r)$ is bounded away from zero as $r \to \bar{r}$, that is, a positive constant ρ exists such that $l_C(r) \geq \rho$ as $r \to \bar{r}$. Select r so that $[r^2 - \bar{r}^2]^{\frac{1}{2}} < \rho/4n^2$. Now $l_C(r)$ is the sum of the lengths of at most n arcs of $C(p; r)$ that lie outside $\mathscr{D}$ (with end-points on $B(\mathscr{D})$) and hence at least one of these arcs has length greater than or equal to ρ/n. Denote by a, b the end-points of such an arc. These points belong to $B(\mathscr{D})$ and so are connected by a broken line of the boundary which can contain at most n segments and which lies between the circles $C(p;\bar{r})$ and $C(p;r)$.

If $r-\bar{r}$ is sufficiently small the chord $[a,b]$ of $C(p;r)$ has length ab that exceeds half its arc, and so $ab > \rho/2n$. Hence at least one segment I of the broken line joining a, b has length exceeding $\rho/2n^2$. But since I lies between the two circles $C(p;\bar{r})$, $C(p;r)$ its length is at most

$$2[r^2-\bar{r}^2]^{\frac{1}{2}} < \rho/2n^2.$$

This contradiction establishes the continuity of $l_C(r)$ at $r = \bar{r}$.

If, now, m is any positive integer, a circle $C(p;r^*)$ exists ($r^* > \bar{r}$) such that $0 < l_C(r^*) < 2\pi r^*/m$. Denote by $a_i^* < \theta < b_i^*$ ($i = 1,2,...,k$) those arcs of $C(p;r^*)$ outside $\mathscr{D}$ (with end-points on $B(\mathscr{D})$), and select an arbitrary set of m points $p_1, p_2,..., p_m$ of $C(p,r^*)$ with vectorial angles $\theta_1, \theta_2,..., \theta_m$, respectively. The m points $p_i(\alpha) = (r^*, \theta_i+\alpha)$ ($i = 1,2,...,m$), of $C(p;r^*)$ are obtained by rotating $p_1, p_2,..., p_m$ about the pole through an angle α, and so

$$p_1(\alpha), p_2(\alpha),..., p_m(\alpha) \approx p_1, p_2,..., p_m.$$

The point p_i is rotated into a point outside $\mathscr{D}$ by any value of α for which $a_j^* < \theta_i+\alpha < b_j^*$ or $a_j^*-\theta_i < \alpha < b_j^*-\theta_i$, for some $j = 1, 2,..., k$. But for a fixed index i the sum of the lengths of the intervals

$$a_j^*-\theta_i < \alpha < b_j^*-\theta_i$$

equals $l_C(r^*)/r^* < 2\pi/m$. Hence the set of values of α ($0 \leqslant \alpha \leqslant 2\pi$) such that at least one of the points $p_1(\alpha), p_2(\alpha),..., p_m(\alpha)$ is outside $\mathscr{D}$ is a *proper* subset of the interval $[0, 2\pi]$.

Hence there corresponds to each m-tuple $p_1, p_2,..., p_m$ of $C(p;r^*)$ an angle α such that the points $p_1(\alpha), p_2(\alpha),..., p_m(\alpha)$ belong to $\mathscr{D}$. Since

$$p_1, p_2,..., p_m \approx p_1(\alpha), p_2(\alpha),..., p_m(\alpha),$$

then each m-tuple of $C(p;r^*)$ is congruently contained in $\mathscr{D}$. But since $r^* > \bar{r}$ clearly $C(p;r^*)$ is not contained in $\mathscr{D}$, and the theorem is proved.

By *a circular disk with centre p and radius r* we mean the set of all points x of E_2 with $px \leqslant r$. In view of the negative results concerning finite congruence order of two-dimensional subsets of E_2 that we have established in this section, the behaviour of the circular disk in this respect is very striking.

THEOREM 62.3. *The circular disk D_r of radius r has congruence order three with respect to subsets of E_2.*

Proof. Let P be any subset of E_2 with each triple coverable by D_r, and consider the family of such disks with centres at the points of P. If $D_r(p), D_r(q), D_r(s)$ are any three members of this family, with centres

p, q, s, respectively, then since p, q, s are coverable by D_r a point x of E_2 exists with px, qx, sx at most r. Hence $x \in D_r(p)D_r(q)D_r(s)$ and consequently each three members of the family of disks have a point in common. It follows from the well-known theorem of Helly† on convex bodies that a point z of E_2 exists that belongs to every member of the family. Thus the centre of each disk of the family has distance at most r from z, and so $P \subset D_r(z)$.

A subset C of a topological space is called *connected* provided C is not the sum of two non-null, mutually exclusive sets, neither of which contains an accumulation element of the other.

Consider, now, a connected domain $\mathscr{D}$ of E_2 with the property that no simple closed curve in $\mathscr{D}$ contains in its interior a point of the boundary of $\mathscr{D}$. Such a domain is said to be *simply connected*.

THEOREM 62.4. *A connected, simply connected domain $\mathscr{D}$ of E_2 is a circular disk if and only if $\mathscr{D}$ has congruence order three with respect to subsets of E_2.*

Proof. The necessity is proved in the preceding theorem.

To prove the sufficiency, we remark first that $\mathscr{D}$ contains a largest circle $C(o; r)$ and since $\mathscr{D}$ is simply connected the disk D_r bounded by this circle is a subset of $\mathscr{D}$. *Suppose, now, that $\mathscr{D}$ contains a point p not belonging to D_r.* Since a continuous real function defined in a connected subset of a metric space takes every value between any two values it assumes, a point p' of $\mathscr{D}$ exists such that $r < op' \leqslant \min(op, r\sqrt{3})$.

Introduce rectangular cartesian coordinates in E_2, with origin o and ray op' as positive half of the x-axis. Then D_r is given by the inequality
$$x^2 + y^2 \leqslant r^2, \tag{*}$$
and the circle K through (o, r), $(o, -r)$, and p' has equation
$$\left[x - \left(\frac{v^2 - r^2}{2v}\right)\right]^2 + y^2 = \left(\frac{v^2 + r^2}{2v}\right)^2, \tag{**}$$
where (v, o) are the coordinates of p' ($v > r$). It should be remarked that from $op' \leqslant r\sqrt{3}$ it follows that $\theta = \angle op'b \geqslant 30°$, where b is the label of the point with coordinates (o, r).

We observe that K has the following properties:

(α) *The radius of K exceeds r*; for since (o, r), $(o, -r)$ are diametral points of $C(o; r)$, this circle is obviously the smallest one containing these points.

---

† An interesting new proof of the Helly theorem is given in Rademacher and Schoenberg, 'Convex domains and Tchebycheff's approximation problem', *Canadian Journal of Mathematics*, vol. II, No. 2 (1950), pp. 245–56.

Ch. VI, § 62   SOME EUCLIDEAN SUBSETS   161

(β) *The arc of K to the left of the Y-axis is at least* $120°$; for the arc is $2(2\theta) = 4\theta \geq 120°$ by the above remark.

(γ) *The arc of K to the left of the Y-axis is contained in D_r*; for from (**)

$$x^2 + y^2 = r^2 + x\left(\frac{v^2 - r^2}{v}\right), \qquad (\dagger)$$

and so (*) is satisfied for $x \leq 0$.

(δ) *The maximum distance of a point of K from the origin is v*; for by ($\dagger$) $x^2 + y^2$ is a maximum for a point of K with largest abscissa, and clearly $(v, 0)$ is that point.

Now let p_1, p_2, p_3 be any three points of K. Two of these points, say p_1, p_2, lie on an arc of at most $120°$, and K may be rotated about its centre so that p_1, p_2 fall to the left of the Y-axis and hence lie in D_r by (γ). From (δ), $op_3 \leq v$ and so there exists a point p'_3 of the connected domain $\mathscr{D}$ such that $op'_3 = op_3$. Rotating K about the origin o so that p_3 coincides with p'_3, sends p_1, p_2 into points p'_1, p'_2 with the same, respective, distances from o as p_1, p_2, and hence $p'_1, p'_2 \in D_r$. Thus each three points p_1, p_2, p_3 of K are congruent with points p'_1, p'_2, p'_3 of $\mathscr{D}$, and since $\mathscr{D}$ is assumed to have congruence order three with respect to subsets of E_2, it follows that $K \subseteq \mathscr{D}$. This is impossible by (γ), and so the assumption that $\mathscr{D}$ contains a point not belonging to D_r leads to a contradiction, and the proof is complete.

REFERENCES

§§ 60–62. Blumenthal [9], Lindebaum [1] (monomorphic sets), Robinson [3, 4].

PART III
THE NON-EUCLIDEAN SPACES

CHAPTER VII

IMBEDDING AND CHARACTERIZATION THEOREMS FOR SPHERICAL SPACE CHARACTERIZATION OF PSEUDO-SPHERICAL SETS

63. Solution of the imbedding problem for $S_{n,r}$

THE space $S_{n,r}$ whose metric theory is to be investigated in this part of our study is the convex n-sphere defined in Example 2, § 9, and shown there to be metric. The problem of determining necessary and sufficient conditions, expressed wholly and explicitly in terms of the metric, in order that an arbitrary semimetric space may be congruently imbedded in $S_{n,r}$ is quickly solved by utilizing the results established in Part II.

We have already seen that $S_{n,r}$ has congruence order $n+3$ with respect to $\{S\}$; that is, any semimetric space is congruently contained in $S_{n,r}$ whenever each $n+3$ of its points are (Theorem 39.2), and hence it suffices to find necessary and sufficient conditions for imbeddability of semimetric $(n+3)$-tuples in $S_{n,r}$ in order to solve the imbedding problem. We shall, however, proceed more directly by proving the following theorem.

THEOREM 63.1. *A semimetric space S is congruently imbeddable in $S_{n,r}$ if and only if no pair of points of S has distance exceeding πr and*

(i) *S contains an $(m+1)$-tuple $p_1, p_2, \ldots, p_{m+1}$ ($m \leqslant n$) such that*

$$\Delta_k(p_1, p_2, \ldots, p_k) = |\cos p_i p_j / r| \quad (i,j = 1, 2, \ldots, k)$$

is positive ($k = 1, 2, \ldots, m+1$),

(ii) *for every pair x, y of elements of S,*

$$\Delta_{m+2}(p_1, p_2, \ldots, p_{m+1}, x) = \Delta_{m+2}(p_1, p_2, \ldots, p_{m+1}, y)$$
$$= \Delta_{m+3}(p_1, p_2, \ldots, p_{m+1}, x, y) = 0.$$

Then S is irreducibly imbeddable in $S_{m,r} \subset S_{n,r}$.

Proof. Let $\bar{S}$ denote the metric transform of S by $\phi(x) = 2r . \sin(x/2r)$; that is, $\bar{S}$ arises from S by defining the distance $\overline{pq}$ of elements p, q of S as $\overline{pq} = 2r . \sin(pq/2r)$, where pq is the distance of p, q in S. Let S^*

denote the space obtained by adjoining a point p_0 to the pointset of $\bar{S}$ and defining $p_0 p = r$ for each element p of $\bar{S}$. Since $0 \leqslant pq < 2\pi r$ ($p, q \in S$), the space S^* is semimetric.

Now it is clear upon reflection that since $0 \leqslant pq \leqslant \pi r$ ($p, q \in S$), $S \congeq S_{n,r}$ if and only if $\bar{S} \congeq C_{n,r}$ (the 'chord' n-sphere), and surely $\bar{S} \congeq C_{n,r}$ if and only if $S^* \congeq E_{n+1}$. By the corollary to Theorem 42.2, $S^* \congeq E_{n+1}$ if and only if S^* contains $m+1$ points $p_1, p_2, \ldots, p_m, p_{m+1}$ ($m \leqslant n$), such that (i) $\operatorname{sgn} D(p_0, p_1, p_2, \ldots, p_k) = (-1)^{k+1}$ ($k = 1, 2, \ldots, m+1$), while (ii) for every pair x, y of points of S^*,

$$D(p_0, p_1, \ldots, p_{m+1}, x) = D(p_0, p_1, \ldots, p_{m+1}, y)$$
$$= D(p_0, p_1, \ldots, p_{m+1}, x, y) = 0.$$

Then S^* is irreducibly imbeddable in E_{m+1}.

By formula (†) of §43

$$D(p_0, p_1, \ldots, p_k) = (-1)^{k+1}(2r^2)^k \Delta_k(p_1, p_2, \ldots, p_k),$$

since $\cos p_0 \colon p_i, p_j = \cos p_i p_j / r$, where the distances $p_i p_j$ appearing in Δ_k are distances of p_i, p_j, in S, while those occurring in D are the distances $\overline{p_i p_j}$ of p_i, p_j in $\bar{S}$ ($i, j = 1, 2, \ldots, k$), and $p_0 p_j = r$ ($j = 1, 2, \ldots, k$).

Hence $S^* \congeq E_{n+1}$ if and only if S contains $m+1$ points $p_1, p_2, \ldots, p_{m+1}$ ($m \leqslant n$), such that $\Delta_k(p_1, p_2, \ldots, p_k) > 0$ ($k = 1, 2, \ldots, m+1$), while for every pair of elements x, y of S

$$\Delta_{m+2}(p_1, p_2, \ldots, p_{m+1}, x) = \Delta_{m+2}(p_1, p_2, \ldots, p_{m+1}, y)$$
$$= \Delta_{m+3}(p_1, p_2, \ldots, p_{m+1}, x, y) = 0,$$

and the theorem is proved.

In the distance geometry of $S_{n,r}$ the determinant Δ functions in a role analogous to that played by the determinant D in the development of the metric theory of euclidean space.

COROLLARY. *A finite semimetric space of k points $p_1, p_2, \ldots, p_k$ is congruently imbeddable in $S_{n,r}$ if and only if $p_i p_j \leqslant \pi r$ ($i, j = 1, 2, \ldots, k$), and $\Delta_k(p_1, p_2, \ldots, p_k)$ has rank at most $n+1$, with all non-vanishing principal minors positive.*

The proof is left to the reader.

EXERCISE

How may criteria for imbedding finite semimetric spaces in $S_{n,r}$ be described in terms of quadratic form theory?

64. The Σ_r space. Some lemmas

If a complete convex semimetric space S satisfies the hypotheses of Theorem 63.1 with $m = n$, and if, moreover, S is *diameterized* (that is, if S contains for each of its points p at least one point p^* such that $pp^* = \pi r$), then S is congruent with $S_{n,r}$. (We shall refer to points p, p^* with $pp^* = \pi r$ as *diametral*.) For by that theorem, $S \approx S' \subset S_{n,r}$ and points $p'_1, p'_2, \ldots, p'_{n+1}$ of S' exist which are not contained in any $S_{n-1,r}$. It follows easily that (i) $p'_1 p'_2 \neq \pi r$, (ii) S' contains the $S_{1,r}(p'_1, p'_2)$ determined by p'_1, p'_2, (iii) S' contains the $S_{2,r}(p'_1, p'_2, p'_3)$ determined by p'_3 and $S_{1,r}(p'_1, p'_2)$, and (by induction) (iv) S' contains the $S_{n,r}$ determined by p'_{n+1} and $S_{n-1,r}(p'_1, p'_2, \ldots, p'_n)$.†

It turns out, however, that if S be assumed complete, convex, and diameterized, the hypotheses of the imbedding theorem may be materially reduced. We shall establish this by a procedure very similar to that developed in §§ 49, 50. There does not, however, seem to be any direct method by which the desired characterization of $S_{n,r}$ can be obtained from the *result* of § 50, since the $S_{n,r}$ is *not* a subset of the E_{n+1}. If the $S_{n,r}$ be metrically transformed by replacing the distance pq by $\overline{pq} = 2r \cdot \sin(pq/2r)$ the resulting space—the chord sphere $C_{n,r}$—is, indeed, a subset of E_{n+1}, but it is *not* (metrically) convex.

DEFINITION 64.1. *A semimetric space S has the weak spherical four-point property provided any four pairwise distinct points of S are congruently contained in $S_{2,r}$ whenever three of the points are imbeddable in $S_{1,r}$.*

Remark. Let S be a semimetric space of diameter at most πr. If S has the weak spherical four-point property and if $p_1, p_2, p_3 \in S$ with $\Delta(p_1, p_2, p_3) = 0$, then for every point p_4 of S, $\Delta(p_1, p_2, p_3, p_4) = 0$.

Let, now, Σ_r denote a semimetric space with the following properties:

   I. If $p, q \in \Sigma_r$, then $pq \leqslant \pi r$.
  II. *The space Σ_r has the weak spherical four-point property.*
 III. *The space Σ_r is complete, convex, and diameterized.*

LEMMA 64.1. *Each three points of Σ_r are imbeddable in $S_{2,r}$.*

Proof. The proof follows easily from I, II, and III.

LEMMA 64.2. *If $p, p^* \in \Sigma_r$ and $pp^* = \pi r$, then $p, q, p^* \subseteqq S_{1,r}$ for each $q \in \Sigma_r$, and pqp^* holds if $p \neq q \neq p^*$.*

Proof. This is an immediate consequence of the preceding lemma.

LEMMA 64.3. *Each two distinct non-diametral points of Σ_r are endpoints of a unique segment.*

Proof. It follows from Lemma 64.1 that the space Σ_r is metric and hence (by III) if $p, q \in \Sigma_r$ ($p \neq q$) a segment $S_{p,q} \subset \Sigma_r$ exists. The

† The $S_{2,r}(p'_1, p'_2, p'_3)$, for example, is the set of all points of $S_{n,r}$ on a great circle with p'_3 and p, for each point p of $S_{1,r}(p'_1, p'_2)$.

uniqueness of $S_{p,q}$ is easily seen, for suppose two points x, y of Σ_r exist with $px+xq = pq = py+yq$, and $px = py = c \cdot pq$, with c a constant, $0 < c < 1$. Then $p, x, y, q \approx p', x', y', q'$, of $S_{2,r}$, since p, x, q are evidently imbeddable in $S_{1,r}$. But clearly $p', x', y', q' \in S_{1,r}$ and from

$$p'x' = p'y' = c \cdot p'q' \quad (p'q' = pq \neq \pi r),$$

follows $x' = y'$. Then $x = y$ and the lemma is proved.

LEMMA 64.4. *To each point p of Σ_r there corresponds a unique point p^* of Σ_r with $pp^* = \pi r$.*

Proof. By III at least one such point p^* exists. If p^{**} is another, then

$$\Delta(p, p^*, p^{**}) = -[1-\cos(p^*p^{**}/r)]^2.$$

Whence $\cos(p^*p^{**}/r) = 1$ and $p^* = p^{**}$.

LEMMA 64.5. *If p, q are two distinct non-diametral points of Σ_r, and p^*, q^* their respective diametral points, the segments $S_{p,q}, S_{q,p^*}, S_{p^*,q^*}, S_{q^*,p}$ have pairwise at most end-points in common.*

We leave the proof (which is similar to that of the assertion in Theorem 17.1) to the reader.

DEFINITION 64.2. *If p, q are two distinct, non-diametral points of Σ_r and p^*, q^* their respective diametral points, the subset*

$$C_r(p,q) = S_{p,q} + S_{q,p^*} + S_{p^*,q^*} + S_{q^*,p}$$

is called the great circle of Σ_r with base points p, q.

By Lemmas 64.3, 64.4, $C_r(p,q)$ is uniquely determined by p, q.

THEOREM 64.1. *If p, q are two distinct, non-diametral points of Σ_r and p', q' elements of $S_{1,r}$ with $p, q \approx p', q'$, then $C_r(p,q) \approx S_{1,r}(p',q')$.*

Proof. If $\bar{p}, \bar{q}$ are points of $S_{1,r}(p',q')$ diametral, respectively, to p', q', then it follows readily from Property II of Σ_r and Lemmas 64.2, 64.4 that $p, q, p^*, q^* \approx p', q', \bar{p}, \bar{q}$, where p^*, q^* are the unique points of Σ_r diametral, respectively, to p, q. By Lemmas 64.3, 64.5, the congruences

$$S_{p,q} \approx S_{p',q'}, \quad S_{q,p^*} \approx S_{q',\bar{p}}, \quad S_{p^*,q^*} \approx S_{\bar{p},\bar{q}}, \quad S_{q^*,p} \approx S_{\bar{q},p}$$

map $C_r(p,q)$ onto $S_{1,r}(p',q')$. We show the mapping is a congruence.

If $x, y \in C_r(p, q)$ and x', y' are their respective corresponding elements in $S_{1,r}(p',q')$, $xy = x'y'$ follows immediately from

$$S_{p,p^*} = S_{p,q} + S_{q,p^*} \approx S_{p',q'} + S_{q',\bar{p}} = S_{p',\bar{p}}, \text{ etc.,}$$

except in the one case (apart from labelling) in which $x \in S_{p,q}$, $p \neq x \neq q$, and $y \in S_{p^*,q^*}$, $p^* \neq y \neq q^*$. Then $x' \in S_{p',q'}$, $p' \neq x' \neq q'$ and $y' \in S_{\bar{p},\bar{q}}$, $\bar{p} \neq y' \neq \bar{q}$.

From pq^*p^* and p^*yq^* (Lemma 64.2) we have pq^*y, and so

$$p, q, q^*, y \subseteq S_{1,r}.$$

Then $p, q, y \subseteq S_{1,r}$ and since pxq holds, $p, q, x, y \approx p'', q'', x'', y''$ of $S_{1,r}(p', q')$. Any motion of this circle that carries p'', q'' into p', q', respectively, sends x'' into x' and y'' into a uniquely determined point y^*. Thus $p, q, x, y \approx p', q', x', y^*$.

Examining the quintuple p, q, p^*, q^*, y we see that each of its quadruples is congruently contained in $S_{1,r}$ and so (Theorem 39.2)

$$p, q, p^*, q^*, y \approx p'', q'', \bar{p}'', \bar{q}'', y''$$

of $S_{1,r}$ (where the 'seconds' are not necessarily the same as those encountered above). By a motion we have

$$p, q, p^*, q^*, y \approx p', q', \bar{p}, \bar{q}, \bar{y}$$

and since $p'\bar{y} = py = p'y^*$, $q'\bar{y} = qy = q'y^*$, with $p'q' \neq \pi r$, we conclude that $\bar{y} = y^*$. But then $\bar{p}y^* = p^*y = \bar{p}y'$ and $\bar{q}y^* = q^*y = \bar{q}y'$ (since $p^*, q^*, y \approx \bar{p}, \bar{q}, y'$) and so (since $\bar{p}\bar{q} = p'q' \neq \pi r$) $y^* = y'$.

Hence $p, q, x, y \approx p', q', x', y'$ and $xy = x'y'$.

COROLLARY. *Each great circle C_r of Σ_r is congruent with $S_{1,r}$.*

Remark. There is a unique congruence of C_r with $S_{1,r}$ which contains the congruence $p, q \approx p', q'$, where p, q are distinct, non-diametral points of C_r and $p', q' \in S_{1,r}$ (for two such congruences imply the existence of distinct points of $S_{1,r}$ with the same distances from two distinct, non-diametral points of $S_{1,r}$, which is impossible).

THEOREM 64.2. *Any two distinct, non-diametral points of a great circle C_r of Σ_r are base points of the circle.*

Proof. Suppose $u, v \in C_r(p, q)$, $0 < uv < \pi r$, and $u', v' \in S_{1,r}(p', q')$ corresponding, respectively, to u, v in the congruence

$$C_r(p, q) \approx S_{1,r}(p', q').$$

The points of $C_r(p, q)$ corresponding by this congruence to the points of $S_{1,r}(p', q')$ that are, respectively, diametral to u', v' are clearly u^*, v^* (diametral, respectively, to u, v). Thus $u^*, v^* \in C_r(p, q)$ which, being compact and convex, contains the unique segments $S_{u,v}, S_{v,u^*}, S_{u^*,v^*}, S_{v^*,u}$. Consequently, $C_r(p, q) \supset C_r(u, v)$, and since each is congruent to $S_{1,r}$ it follows that $C_r(p, q) = C_r(u, v)$.

Combining the two preceding theorems we have the important result:

THEOREM 64.3. *Each two distinct, non-diametral points of Σ_r are on one and only one great circle of Σ_r.*

EXERCISES

1. Prove the remark following Definition 64.1.
2. Establish Lemma 64.5.

65. Additional lemmas. Subspaces of Σ_r

The proofs of the following lemmas differ so little from those of the corresponding lemmas in § 49 that it is unnecessary to exhibit them.

LEMMA 65.1. *If p_0, p_1, $p_2 \in \Sigma_r$ $(0 \neq p_1 p_2 \neq \pi r)$, and p'_0, p'_1, $p'_2 \in S_{2,r}$ with p_0, p_1, $p_2 \approx p'_0$, p'_1, p'_2, this congruence has a unique extension to*

$$(p_0)+C_r(p_1,p_2) \approx (p'_0)+S_{1,r}(p'_1,p'_2).$$

LEMMA 65.2. *Let p_0, p_1, p_2 be pairwise distinct points of Σ_r with $p_0 p_1$ and $p_1 p_2$ different from πr. If p'_0, p'_1, $p'_2 \in S_{2,r}$ with p_0, p_1, $p_2 \approx p'_0$, p'_1, p'_2, this congruence has a unique extension to*

$$C_r(p_0,p_1)+C_r(p_1,p_2) \approx S_{1,r}(p'_0,p'_1)+S_{1,r}(p'_1,p'_2).$$

LEMMA 65.3. *If p_0 is any point and $C_r(p_1,p_2)$ any great circle of Σ_r, there exists a foot $f(p_0)$ of p_0 on $C_r(p_1,p_2)$ which is unique unless $p_0 x = \frac{1}{2}\pi r$ for every point x of $C_r(p_1,p_2)$.*

Proof. Since points p'_0, p'_1, p'_2 of $S_{2,r}$ exist with p_0, p_1, $p_2 \approx p'_0$, p'_1, p'_2, the proof follows immediately from the congruence asserted in Lemma 65.1.

LEMMA 65.4. *Let p be a point and C_r a great circle of Σ_r. The congruence*

$$(p)+C_r \approx (p')+S_{1,r} \quad (p' \in S_{2,r} \supset S_{1,r})$$

is valid if and only if $\operatorname{dist}(p, C_r) = \operatorname{dist}(p', S_{1,r})$.

Proof. The validity of Theorem 64.3 permits application of the argument used to establish Lemma 49.5, with the modification that the point q of that lemma is now selected on C_r so that $0 \neq qf(p) \neq \pi r$.

We define n-dimensional subspaces $\Sigma_{n,r}$ of Σ_r inductively. Define a zero-dimensional subspace to be any subset of Σ_r consisting of two diametral points of Σ_r. If, now, $\Sigma_{k,r}$ is a k-dimensional subspace of Σ_r and $p \in \Sigma_r$, $p \bar{\in} \Sigma_{k,r}$, the set of all points of Σ_r belonging to the great circles $C_r(p,x)$, for each point x of $\Sigma_{k,r}$ defines a $(k+1)$-dimensional subspace $\Sigma_{k+1,r}$. We say $\Sigma_{k+1,r}$ is *generated* by p and $\Sigma_{k,r}$ and write

$$\Sigma_{k+1,r} = \{p; \Sigma_{k,r}\}.$$

It is clearly uniquely determined by p and $\Sigma_{k,r}$.

Remark. The class of one-dimensional subspaces $\Sigma_{1,r}$ of Σ_r is identical with the class of great circles C_r of Σ_r.

We have shown in the foregoing that for $k = 1$, (i) if $p_0, p_1,..., p_k \in \Sigma_r$, not belonging to any $\Sigma_{k-1,r}$, there is one and only one $\Sigma_{k,r}$ of Σ_r containing them, (ii) $\Sigma_{k,r}$ contains for each two of its distinct, non-diametral points the unique $\Sigma_{1,r}$ to which they belong, and (iii) $\Sigma_{k,r} \approx S_{k,r}$.

66. Metric characterization of $S_{n,r}$

We make now the inductive assumption that the properties (i), (ii), (iii) of $\Sigma_{k,r}$, stated in the concluding paragraph of the preceding section (and which have been proved for $k = 1$) hold for every $k = 1, 2,..., n$.

LEMMA 66.1. *For any $k = 1, 2,..., n$ let $\Sigma_{k,r}$ be a k-dimensional subspace of Σ_r and p an element of Σ_r. If p' is a point and $S_{k,r}$ a k-dimensional subspace of $S_{k+1,r}$ then*
$$(p) + \Sigma_{k,r} \approx (p') + S_{k,r}$$
if and only if $\text{dist}(p, \Sigma_{k,r}) = \text{dist}(p', S_{k,r})$.

Proof. The necessity is clear, while if
$$\text{dist}(p, \Sigma_{k,r}) = \text{dist}(p', S_{k,r}) = 0,$$
the sufficiency follows at once from $\Sigma_{k,r} \approx S_{k,r}$ (inductive hypothesis), for then $p \in \Sigma_{k,r}$, $p' \in S_{k,r}$.

To complete the proof we suppose $\text{dist}(p, \Sigma_{k,r}) = \text{dist}(p', S_{k,r}) > 0$, and let $f(p)$, $f'(p)$ denote feet of p, p', respectively, on $\Sigma_{k,r}$ and $S_{k,r}$. (The existence of $f(p)$, $f'(p)$ follows from Lemma 16.2, for since $\Sigma_{k,r} \approx S_{k,r}$, then $\Sigma_{k,r}$ is closed and compact.) Then $pf(p) = p'f(p') > 0$. Let
$$\Sigma_{k,r} \approx S_{k,r} \tag{*}$$
be any congruence between the two k-dimensional subspaces in which $f(p)$ and $f(p')$ are corresponding points. Then (*) together with the correspondence $p \leftrightarrow p'$ maps $(p) + \Sigma_{k,r}$ onto $(p') + S_{k,r}$. Suppose $q \in \Sigma_{k,r}$, $q' \in S_{k,r}$ and q, q' correspond by the above mapping. We wish to show that $pq = p'q'$.

If $q = f(p)$, then $q' = f(p')$ and $pq = p'q'$ by the above. If
$$qf(p) = q'f(p') = \pi r,$$
then from the congruent imbedding of $p, f(p), q$ in $S_{2,r}$ we have
$$pq = qf(p) - pf(p) = q'f(p') - p'f(p') = p'q'.$$
If, finally, $0 \neq qf(p) \neq \pi r$, then
$$\Sigma_{k,r} \supset \Sigma_{1,r}(q, f(p)) \approx S_{1,r}(q', f(p')) \subset S_{k,r},$$
and clearly $f(p), f(p')$ are feet of p, p' on $\Sigma_{1,r}(q, f(p)), S_{1,r}(q', f(p'))$, respectively. Hence
$$(p) + \Sigma_{1,r}(q, f(p)) \approx (p') + S_{1,r}(q', f(p'))$$
by Lemma 65.4, and so $pq = p'q'$.

THEOREM 66.1. *For every non-negative integer k, each k-dimensional subspace $\Sigma_{k,r}$ of Σ_r is congruent with $S_{k,r}$.*

Proof. It suffices to prove the assertion for $k = n+1$ ($n > 0$).

If $\Sigma_{n+1,r} = \{p; \Sigma_{n,r}\}$ is an $(n+1)$-dimensional subspace of Σ_r, let $f(p)$ denote a foot of p on $\Sigma_{n,r}$. Then $pf(p) \leqslant \tfrac{1}{2}\pi r$, for in the contrary case, the points p, $f(p)$, $f^*(p)$ ($f^*(p)$ is diametral to $f(p)$ and hence belongs to $\Sigma_{n,r}$) have distances $pf(p) > \tfrac{1}{2}\pi r$, $pf^*(p) > \tfrac{1}{2}\pi r$, $f(p)f^*(p) = \pi r$, and hence are not congruently contained in $S_{2,r}$, contrary to Lemma 64.1.

Since $pf(p) \leqslant \tfrac{1}{2}\pi r$, a point p' and a subspace $S_{n,r}$ of $S_{n+1,r}$ exist such that $\operatorname{dist}(p', S_{n,r}) = pf(p) = \operatorname{dist}(p, \Sigma_{n,r})$. It follows (Lemma 66.1) that

$$(p) + \Sigma_{n,r} \approx (p') + S_{n,r}. \tag{*}$$

To obtain a mapping of $\Sigma_{n+1,r}$ onto $S_{n+1,r}$, let x be a point of $\Sigma_{n+1,r}$. If x is p or p^*, then x' is taken as p' or its diametral point of $S_{n+1,r}$, respectively. If $0 \neq px \neq \pi r$, let u be a point of $\Sigma_{n,r}$ on $\Sigma_{1,r}(p, x)$ and let u' be the point of $S_{n,r}$ corresponding to u by (*). The congruence

$$p, u \approx p', u'$$

is extended uniquely to

$$\Sigma_{1,r}(p, u) \approx S_{1,r}(p', u') \quad \text{(clearly } 0 \neq pu \neq \pi r\text{)}, \tag{\dag}$$

and we let x' be the (unique) point of $S_{1,r}(p', u')$ corresponding to x in (†). The mapping is easily seen to be one-to-one.

To show that this mapping is a congruence, let x, y, elements of $\Sigma_{n+1,r}$, correspond, respectively, in this mapping to x', y', elements of $S_{n+1,r}$, with u', v' of $S_{n,r}$ corresponding by (*) to u, v (where $u \in \Sigma_{1,r}(p, x) \cdot \Sigma_{n,r}$, $v \in \Sigma_{1,r}(p, y) \cdot \Sigma_{n,r}$).† By (*) $p, u, v \approx p', u', v'$, and since $0 \neq pu \neq \pi r$, $0 \neq pv \neq \pi r$, the congruence of triples is uniquely extended to

$$\Sigma_{1,r}(p, u) + \Sigma_{1,r}(p, v) \approx S_{1,r}(p', u') + S_{1,r}(p', v').$$

In this congruence x and x', as well as y and y', are corresponding points. Consequently $xy = x'y'$, and the proof is complete.

COROLLARY. *For every positive integer k, each $\Sigma_{k,r}$ contains the $\Sigma_{1,r}$ determined by any two of its distinct, non-diametral points.*

THEOREM 66.2. *If k is any positive integer and $p_0, p_1, \ldots, p_k$ are $k+1$ points of Σ_r that are not contained in any $(k-1)$-dimensional subspace, these points belong to a unique k-dimensional subspace of Σ_r.*

Proof. In view of Theorem 64.3, the remark following Lemma 65.4, and the inductive hypothesis, it suffices to prove the theorem for $k = n+1$. The argument used to establish Theorem 50.2 applies here,

---

† The cases in which px or py equals zero or πr present no difficulty.

with the corollary to Theorem 66.1 replacing the linearity of subspaces $\bar{L}_k$ employed in the proof of the earlier theorem.

THEOREM 66.3. *For every non-negative integer k, each $(k+1)$-tuple of Σ_r is congruently imbeddable in $S_{k,r}$; that is, Σ_r has the spherical $(k+1)$-point property, for $k = 0, 1, 2,\ldots$.*

Proof. The theorem is an immediate consequence of the two preceding theorems.

THEOREM 66.4. *A space Σ_r is congruent with the $S_{n,r}$ if and only if (i) Σ_r contains an $(n+1)$-tuple $p_0, p_1,\ldots, p_n$ with non-vanishing determinant Δ_{n+1}, and (ii) the determinant Δ_{n+2} of each $(n+2)$-tuple of Σ_r containing $p_0, p_1,\ldots, p_n$ vanishes.*

Proof. If $\Sigma_r \approx S_{n,r}$, then Σ_r is irreducibly imbeddable in $S_{n,r}$ and properties (i), (ii) of Σ_r follow from Theorem 63.1.

Assuming, now, properties (i), (ii) of Σ_r, we obtain from (i) that the $(n+1)$-tuple $p_0, p_1,\ldots, p_n$ does not belong to any $(n-1)$-dimensional subspace of Σ_r (since $\Sigma_{n-1,r} \approx S_{n-1,r}$ and the determinant Δ_{n+1} formed for any $n+1$ points of $S_{n-1,r}$ vanishes) and hence it is contained in a unique n-dimensional subspace $\Sigma_{n,r}(p_0, p_1,\ldots, p_n)$ of Σ_r.

If Σ_r contains a point p, $p \bar{\in} \Sigma_{n,r}(p_0, p_1,\ldots, p_n)$, then

$$\Sigma_r \supset \Sigma_{n+1,r} = \{p, \Sigma_{n,r}(p_0, p_1,\ldots, p_n)\} \approx S_{n+1,r},$$

and the points $p'_0, p'_1,\ldots, p'_n, p'$ of $S_{n+1,r}$ corresponding to $p_0, p_1,\ldots, p_n, p$, respectively, in this congruence have

$$\Delta_{n+2}(p'_0, p'_1,\ldots, p'_n, p') = \Delta_{n+2}(p_0, p_1,\ldots, p_n, p) \neq 0,$$

since p' does not belong to the $S_{n,r}$ determined by $p'_0, p'_1,\ldots, p'_n$.

Hence $\Sigma_r \equiv \Sigma_{n,r} \approx S_{n,r}$, and the theorem is proved.

COROLLARY. *A space Σ_r is congruent with $S_{n,r}$ if and only if n is the smallest non-negative integer such that every $(n+2)$-tuple of Σ_r has vanishing determinant Δ_{n+2}.*

We have, then, obtained the following metric characterization of the convex n-sphere $S_{n,r}$, of metric diameter πr, with respect to the class of all semimetric spaces.

THEOREM 66.5. *A semimetric space S is congruent with $S_{n,r}$ if and only if*

(i) *S is complete, convex, and diameterized,*

(ii) *no two points of S have distance exceeding πr,*

(iii) *S has the weak spherical four-point property,*

(iv) *S contains an $(n+1)$-tuple $p_0, p_1,\ldots, p_n$ with $\Delta_{n+1}(p_0, p_1,\ldots, p_n) \neq 0$,*

(v) *if $p \in S$ then $\Delta_{n+2}(p_0, p_1,\ldots, p_n, p) = 0$.*

If (v) is suppressed and (iv) is amended to state that for every positive integer n, S contains an $(n+1)$-tuple with non-vanishing determinant Δ_{n+1}, we obtain the metric characterization of the convexly metrized sphere in Hilbert space provided we assume, further, that S is separable.

67. Basic properties of $S_{n,r}$

A semimetric space S forms a pseudo-$S_{n,r}$ set provided (1) S is not congruently contained in $S_{n,r}$, but (2) each $(n+2)$-tuple of S is imbeddable in $S_{n,r}$. The reader will recall that pseudo-E_n sets were similarly defined (Definition 44.1). It was established (§ 46) that pseudo-E_n sets consist of exactly $n+3$ pairwise distinct points, and the metric structure of such $(n+3)$-tuples was determined.

Though, as we shall see, the structure of pseudo-$S_{n,r}$ $(n+3)$-tuples is so closely related to that of pseudo-E_n $(n+3)$-tuples as to present nothing essentially new, the study of pseudo-$S_{n,r}$ sets of *more than $n+3$ points* which (in contrast to the euclidean case) do exist, has no analogue in our discussion of pseudo-euclidean sets.

To obtain the characterization of pseudo-$S_{n,r}$ sets of arbitrary power (exceeding $n+2$) we shall need properties of the $S_{n,r}$ with which the reader may not be acquainted and which (in any event) are certainly less 'intuitive' than their analogues in the geometry of euclidean n-space. Nor is there any textbook devoted to the geometry of the $S_{n,r}$ to which the reader might be referred for their proofs.

It seems desirable, therefore, to list all those properties of $S_{n,r}$ which we shall use in the investigations of this chapter, and to exhibit *five basic properties of $S_{n,r}$ from which all the others that we are interested in here can be derived*. The reader may then make the derivations for himself.

Since it is clear that the only pseudo-$S_{0,r}$ sets are equilateral, with common (non-zero) distance πr, we shall suppose in the following that $n \geqslant 1$.

Basic Properties of $S_{n,r}$

The $S_{n,r}$ is a semimetric space of diameter $d = \pi r$, with the following properties:

I. *The determinant*

$$\Delta_{n+2}(s_1, s_2, ..., s_{n+2}) = |\cos(s_i s_j / r)| \quad (i,j = 1, 2, ..., n+2)$$

vanishes for each set of $n+2$ points $s_1, s_2, ..., s_{n+2}$ of $S_{n,r}$.

II. *There exists at least one set of $n+1$ points of $S_{n,r}$ with non-vanishing determinant Δ_{n+1}.*

III. *Each finite subset of $S_{n,r}$ has its Δ determinant non-negative.*

These three properties are, of course, immediate consequences of Theorem 63.1, corollary. Defining a k-dimensional subspace $S_{k,r}$ of $S_{n,r}$ ($k \leq n$), generated by $p_1, p_2,..., p_{k+1}$ ($\Delta_{k+1}(p_1, p_2,..., p_{k+1}) \neq 0$), as the set of all points p of $S_{n,r}$ such that $\Delta_{k+2}(p_1, p_2,..., p_{k+1}, p) = 0$, Properties I, II, III permit us to prove that (i) Δ_{k+2} vanishes for *each* $(k+2)$-tuple of an $S_{k,r}$, and (ii) any $S_{k,r}$ is uniquely determined by any of its $(k+1)$-tuples with non-vanishing determinant Δ_{k+1}.

A subset of $m+1$ points of $S_{n,r}$ ($m \geq 1$), is called *dependent* or *independent* according as its determinant Δ_{m+1} vanishes or not; that is (by the above remarks), according as the subset belongs or does not belong to an $(m-1)$-dimensional subspace $S_{m-1,r}$ of $S_{n,r}$. A semimetric $(m+1)$-tuple is $S_{n,r}$-dependent or $S_{n,r}$-independent according as it is congruent to a dependent or independent $(m+1)$-tuple of $S_{n,r}$.

Properties I, II, III do *not* imply the congruence of two k-dimensional subspaces of $S_{n,r}$, nor do they suffice to prove the congruence order $k+3$ of each $S_{k,r}$ ($k = 0, 1,..., n$), with respect to $\{S\}$. Both of these may be established, however, when we adjoin the following fourth property of $S_{n,r}$.

IV. If $s_1, s_2,..., s_{k+1} \approx t_1, t_2,..., t_{k+1}$ are two congruent sets of $k+1$ points (not necessarily pairwise distinct) of two (coincident or distinct) k-dimensional subspaces of $S_{n,r}$, $k \leq n$, then to each point s of the subspace containing $s_1, s_2,..., s_{k+1}$ there corresponds at least one point t of the other subspace such that
$$s_1, s_2,..., s_{k+1}, s \approx t_1, t_2,..., t_{k+1}, t.$$

Finally, in order to establish determinant conditions for the imbedding in $S_{k,r}$ ($k = 1, 2,..., n$), of semimetric $(k+2)$-tuples and $(k+3)$-tuples we use a fifth 'reflection' property.

V. If $s_1, s_2,..., s_k$ are k independent points of $S_{k,r}$ ($k = 1, 2,..., n$), there corresponds to each point s of $S_{k,r}$, with $\Delta_{k+1}(s_1, s_2,..., s_k, s) \neq 0$, at least one point s' of $S_{k,r}$ such that $s' \neq s$ and
$$s_1, s_2,..., s_k, s \approx s_1, s_2,..., s_k, s'.$$

68. Derived properties of $S_{n,r}$

We list now twelve properties of $S_{n,r}$ which are consequences of basic properties I–V and to which reference will be made in our study of pseudo-$S_{n,r}$ sets.

(1) A semimetric set of $k+2$ points is imbeddable in $S_{k,r}$, $k \leq n$, if and only if each $k+1$ of the points are congruently contained in $S_{k,r}$ and the determinant Δ_{k+2} of the $k+2$ points vanishes.

(2) A semimetric $(k+3)$-tuple is imbeddable in $S_{k,r}$, $k \leqslant n$, if and only if each $k+2$ of the points are congruently contained in $S_{k,r}$ and the determinant Δ_{k+3} of the $k+3$ points vanishes.

Properties (1), (2) follow also, of course, from Theorem 63.1, corollary.

(3) If $s_1, s_2, \ldots, s_{k+1} \approx t_1, t_2, \ldots, t_{k+1}$ are two congruent sets of $k+1$ independent points of two (coincident or distinct) k-dimensional subspaces of $S_{n,r}$, $k \leqslant n$, and if s, s' are points of the subspace containing $s_1, s_2, \ldots, s_{k+1}$, while t, t' are points of the other subspace such that

$$s_1, s_2, \ldots, s_{k+1}, s \approx t_1, t_2, \ldots, t_{k+1}, t,$$

$$s_1, s_2, \ldots, s_{k+1}, s' \approx t_1, t_2, \ldots, t_{k+1}, t',$$

then $ss' = tt'$.

Remark. It follows from (3) that a point t corresponding to a point s by Property IV is *unique* whenever $s_1, s_2, \ldots, s_{k+1}$ are independent, and there is at most one point of $S_{k,r}$ with prescribed distances from $k+1$ independent points of $S_{k,r}$, $k \leqslant n$.

(4) Any subset (of at least two points) of an independent k-tuple of $S_{n,r}$ is an independent set.

(5) If $s_1, s_2, \ldots, s_k$ are k independent points of $S_{k,r}$ ($k = 1, 2, \ldots, n$), then corresponding to each point s of $S_{k,r}$ with $s_1, s_2, \ldots, s_k$, s independent, there is exactly one point s' of $S_{k,r}$ such that $s' \neq s$ and

$$s_1, s_2, \ldots, s_k, s \approx s_1, s_2, \ldots, s_k, s'.$$

(6) Let $s_1, s_2, \ldots, s_k$ be k independent points of $S_{k,r}$ ($k = 1, 2, \ldots, n$), and let s, s' and t, t' be two pairs of points of $S_{k,r}$ such that either $s_1, s_2, \ldots, s_k$, t are dependent or $t \neq s$, and either $s_1, s_2, \ldots, s_k$, t' are dependent or $t' \neq s'$, while

$$s_1, s_2, \ldots, s_k, s \approx s_1, s_2, \ldots, s_k, t,$$

$$s_1, s_2, \ldots, s_k, s' \approx s_1, s_2, \ldots, s_k, t'.$$

Then $ss' = tt'$.

(7) Let $s_1, s_2, \ldots, s_{k+1}$ be $k+1$ independent points of $S_{k,r}$, $k \leqslant n$, and suppose s is a point of $S_{k,r}$ not common to any two of the $k+1$ subspaces $S_{k-1,r}$ determined by the k points

$$s_1, s_2, \ldots, s_{i-1}, s_{i+1}, \ldots, s_{k+1} \quad (i = 1, 2, \ldots, k+1).$$

If $s^{(i)}$ is a point of $S_{k,r}$ such that

$$s_1, s_2, \ldots, s_{i-1}, s^{(i)}, s_{i+1}, \ldots, s_{k+1} \approx s_1, s_2, \ldots, s_{i-1}, s, s_{i+1}, \ldots, s_{k+1}$$

$$(i = 1, 2, \ldots, k+1),$$

where $s^{(i)} \neq s$ if $s_1, s_2, \ldots, s_{i-1}, s, s_{i+1}, \ldots, s_{k+1}$ are an independent set, then there are not two independent points s', t' of $S_{k,r}$ such that

$$s's^{(1)} = s's^{(2)} = \ldots = s's^{(k+1)}$$

and
$$t's^{(1)} = t's^{(2)} = \ldots = t's^{(k+1)}.$$

Remark. There are at most two points satisfying the conditions of (7).

(8) The $S_{k,r}$ has best congruence indices $(k+3, 0)$ $(k \leqslant n)$. This will be shown in the next section.

(9) If $s_1, s_2, \ldots, s_{k+1}$ are $k+1$ independent points of $S_{k,r}$, $k \leqslant n$, and s is a point of $S_{k,r}$ such that for each integer i $(i = 1, 2, \ldots, k)$, the points $s_1, s_2, \ldots, s_{i-1}, s, s_{i+1}, \ldots, s_{k+1}$ are dependent, then the points s, s_{k+1} are dependent.

(10) If s, t, u are any three distinct points of $S_{k,r}$, $k \leqslant n$, such that $st = tu = su$, and if $s_1, s_2, \ldots, s_k$ are k points of $S_{k,r}$ with

$$s_i s = s_i t = s_i u \quad (i = 1, 2, \ldots, k),$$

then the points $s_1, s_2, \ldots, s_k$ are dependent.

(11) Let s, t be any two distinct points of $S_{k,r}$ $(k = 1, 2, \ldots, n)$, and let $s_1, s_2, \ldots, s_k$ be k independent points of $S_{k,r}$ such that

$$s_i s = s_i t \quad (i = 1, 2, \ldots, k).$$

The $(k-1)$-dimensional subspace $S_{k-1,r}$ determined by $s_1, s_2, \ldots, s_k$ is the locus of points of $S_{k,r}$ equidistant from s and t.

(12) Two distinct k-dimensional subspaces $S_{k,r}$, $k \leqslant n$, have at most k independent points in common.

Remark. Let S be any semimetric space with positive (finite) diameter δ, and let $\phi(pq/\rho)$ be a real, monotonically decreasing, function defined for all the values pq $(p, q \in S)$, where ρ denotes a fixed positive parameter. Suppose, moreover, that $\phi(0) = 1$, and $\phi(\delta/\rho) = -1$. (The function ϕ is an 'abstract cosine' over S.)

Replacing $\cos(x/r)$ by $\phi(x/\rho)$ in I–V, we may assume the resulting five statements as postulates, and denote the resulting space by $\mathfrak{S}_{n,\rho}$. Then $\mathfrak{S}_{n,\rho}$ has all the properties (1)–(12), and the characterization we shall obtain for pseudo-$S_{n,r}$ sets will be valid for pseudo-$\mathfrak{S}_{n,\rho}$ sets also, for any space $\mathfrak{S}_{n,\rho}$, provided we avoid using in our work any special properties of the cosine function which are not valid for the function ϕ.

EXERCISES

1. Derive property (3).
2. Derive properties (4), (5).
3. Show that the chord n-sphere $C_{n,r}$ is a $\mathfrak{S}_{n,r}$ space.

69. Pseudo-$S_{k,r}$ $(k+3)$-tuples, $k \leqslant n$

We have already seen that $S_{k,r}$ has congruence indices $(k+3, 0)$ with respect to the class $\{S\}$; that is, each semimetric space is congruently contained in $S_{k,r}$ whenever every $k+3$ of its points has that property. To show that those indices are best, we prove first a lemma.

LEMMA 69.1. *The $S_{k,r}$, $k \leqslant n$, contains an equilateral $(k+2)$-tuple with 'edge' equal to $r\cos^{-1}[-1/(k+1)]$. Each two equilateral $(k+2)$-tuples of $S_{k,r}$ are congruent; that is, equilateral $(k+2)$-tuples of $S_{k,r}$ are metrically unique.*

Proof. Consider the determinant

$$\Delta_m(a) = |a_{ij}|, \qquad a_{ii} = 1, \qquad a_{ij} = a \quad (i \neq j),$$

$(i, j = 1, 2, \ldots, m; m > 1)$. Adding the 2nd, 3rd,…, mth columns of this determinant to the first column, and subtracting the first row of the determinant so obtained from each of the other rows, we get

$$\Delta_m(a) = (1-a)^{m-1}[1+(m-1)a]. \tag{*}$$

Hence $\Delta_m(a) = 0$ for $a = -1/(m-1)$, and

$$\Delta_i(-1/(m-1)) > 0 \quad (i = 1, 2, \ldots, m-1),$$
$$\Delta_i(-1/(m-1)) < 0 \quad (i > m).$$

It follows that the semimetric $(k+2)$-tuple $p_1, p_2, \ldots, p_{k+2}$ with

$$p_i p_j = r\cos^{-1}[-1/(k+1)]$$

is congruently imbeddable in $S_{k,r}$ (corollary, Theorem 63.1). Since, moreover, the only other root of $\Delta_m(a) = 0$ is $a = 1$, the only other $(k+2)$-tuple of $S_{k,r}$ with all distances equal consists of $k+2$ coincident points. We do not call such sets equilateral (cf. definition of equilateral triple, §16) and so the lemma is proved.

THEOREM 69.1. *The $S_{k,r}$, $k \leqslant n$, has best congruence indices $(k+3, 0)$ with respect to the class $\{S\}$ of semimetric spaces.*

Proof. It suffices to consider a semimetric space S with power exceeding $k+m+2$, with $m \geqslant 1$, and $pq = r\cos^{-1}[-1/(k+1)]$, $p, q \in S, p \neq q$. By the preceding lemma each $(k+2)$-tuple of S is congruently contained in $S_{k,r}$, but S is *not*, for since $\Delta_{k+3}(-1/(k+1)) < 0$, no $(k+3)$-tuple of S is imbeddable in $S_{k,r}$. Hence $S_{k,r}$ does not have indices $(k+2, m)$, m finite or transfinite, and the theorem follows.

Thus pseudo-$S_{k,r}$ sets of arbitrarily high power (exceeding $k+2$) exist. We are concerned in this section with determining the metric structure of pseudo-$S_{k,r}$ $(k+3)$-tuples.

THEOREM 69.2. *A pseudo-$S_{k,r}$ $(k+3)$-tuple $p_1, p_2,..., p_{k+3}$ contains at least one $S_{k,r}$-independent set of $k+1$ points.*

Proof. By Property (1) the determinant Δ_{k+2} of each $(k+2)$-tuple of the $k+3$ points vanishes. If Δ_{k+1} vanishes for each $(k+1)$-tuple, then the rank of $\Delta_{k+3}(p_1, p_2,..., p_{k+3})$ is at most k and (Property (2)) the $k+3$ points are imbeddable in $S_{k,r}$, contrary to hypothesis. Since each $(k+1)$-tuple of the $k+3$ points is imbeddable in $S_{k,r}$, the theorem follows.

THEOREM 69.3. *The determinant $\Delta_{k+3}(p_1,p_2,...,p_{k+3})$ of a pseudo-$S_{k,r}$ $(k+3)$-tuple is negative.*

Proof. By Property (2), $\Delta_{k+3}(p_1,p_2,...,p_{k+3}) \neq 0$. If $p_1, p_2,..., p_{k+1}$ is an independent $(k+1)$-tuple, then since $\Delta_{k+2}(p_1,p_2,...,p_{k+2}) = 0$, we have

$$\Delta_{k+3}(p_1,p_2,...,p_{k+3}) = -[k+2,k+3]^2/\Delta_{k+1}(p_1,p_2,...,p_{k+1}),$$

where $[k+2, k+3]$ denotes the co-factor of the element in the $(k+2)$th row and $(k+3)$th column of $\Delta_{k+3}(p_1,p_2,...,p_{k+3})$. Since $p_1, p_2,..., p_{k+1}$ are $S_{k,r}$-independent, Property III gives $\Delta_{k+1}(p_1,p_2,...,p_{k+1}) > 0$, and the theorem is proved.

THEOREM 69.4. *If $p_1, p_2,..., p_{k+3}$ form a pseudo-$S_{k,r}$ set, with the points $p_1, p_2,..., p_{k+1}$ $S_{k,r}$-independent, then the $S_{k,r}$ contains $k+3$ points $s_1, s_2,..., s_{k+3}$ such that*

$$p_1, p_2,..., p_{k+1}, p_{k+2} \approx s_1, s_2,..., s_{k+1}, s_{k+2},$$
$$p_1, p_2,..., p_{k+1}, p_{k+3} \approx s_1, s_2,..., s_{k+1}, s_{k+3},$$

and $p_{k+2}p_{k+3} \neq s_{k+2}s_{k+3}$.

Proof. Since $p_1, p_2,..., p_{k+3}$ form a pseudo-$S_{k,r}$ set, there exist two sets $s_1, s_2,..., s_{k+1}, s_{k+2}$ and $t_1, t_2,..., t_{k+1}, t_{k+3}$ of $k+2$ points of $S_{k,r}$ such that

$$p_1, p_2,..., p_{k+1}, p_{k+2} \approx s_1, s_2,..., s_{k+1}, s_{k+2},$$
$$p_1, p_2,..., p_{k+1}, p_{k+3} \approx t_1, t_2,..., t_{k+1}, t_{k+3}.$$

The points $p_1, p_2,..., p_{k+1}$ being $S_{k,r}$-independent, then

$$\{s_i\} \quad \text{and} \quad \{t_i\} \quad (i = 1, 2,..., k+1),$$

are two congruent sets of $k+1$ independent points of $S_{k,r}$, and hence (remark following Property (3)) the $S_{k,r}$ contains exactly one point s_{k+3} such that

$$t_1, t_2,..., t_{k+1}, t_{k+3} \approx s_1, s_2,..., s_{k+1}, s_{k+3}.$$

Then we have

$$p_1, p_2,..., p_{k+1}, p_{k+3} \approx s_1, s_2,..., s_{k+1}, s_{k+3},$$

and since the $k+3$ points $p_1, p_2,..., p_{k+3}$ form a pseudo-$S_{k,r}$ set,

$$p_{k+2}p_{k+3} \neq s_{k+2}s_{k+3}.$$

Ch. VII, § 69   THEOREMS FOR SPHERICAL SPACE   177

The $k+3$ points $s_1, s_2,\ldots, s_{k+3}$ of $S_{k,r}$ are said to be *almost* congruent to the pseudo-$S_{k,r}$ set $p_1, p_2,\ldots, p_{k+3}$, and we write

$$p_1, p_2,\ldots, p_{k+3} \sim s_1, s_2,\ldots, s_{k+3}.$$

THEOREM 69.5. *Let $p_1, p_2,\ldots, p_{k+3}$ form a pseudo-$S_{k,r}$ $(k+3)$-tuple with the $S_{k,r}$-independent $(k+1)$-tuple $p_1, p_2,\ldots, p_{k+1}$. Then the $k+1$ points $p_2, p_3,\ldots, p_{k+2}$ are $S_{k,r}$-independent.*

Proof. Let $s_1, s_2,\ldots, s_{k+3}$ be $k+3$ points of $S_{k,r}$ almost congruent to $p_1, p_2,\ldots, p_{k+3}$; that is,

$$p_1, p_2,\ldots, p_{k+1}, p_{k+2} \approx s_1, s_2,\ldots, s_{k+1}, s_{k+2},$$

$$p_1, p_2,\ldots, p_{k+1}, p_{k+3} \approx s_1, s_2,\ldots, s_{k+1}, s_{k+3},$$

and $p_{k+2} p_{k+3} \neq s_{k+2} s_{k+3}$. Since each $k+2$ points of the set $p_1, p_2,\ldots, p_{k+3}$ are congruent with $k+2$ points of $S_{k,r}$, we have

$$p_2, p_3,\ldots, p_{k+2}, p_{k+3} \approx t_2, t_3,\ldots, t_{k+2}, t_{k+3},$$

and hence $\quad t_2, t_3,\ldots, t_{k+1}, t_{k+2} \approx s_2, s_3,\ldots, s_{k+1}, s_{k+2},$

with the points $s_2, s_3,\ldots, s_{k+1}$ independent since they belong to the independent $(k+1)$-tuple $s_1, s_2,\ldots, s_{k+1}$ (Property (4)).

Property IV applied to the congruence

$$t_2, t_3,\ldots, t_{k+1} \approx s_2, s_3,\ldots, s_{k+1}$$

entails the existence of two points s, s' of $S_{k,r}$ such that

$$t_2, t_3,\ldots, t_{k+1}, t_{k+2}, t_{k+3} \approx s_2, s_3,\ldots, s_{k+1}, s, s'.$$

Then

$$s_2, s_3,\ldots, s_{k+1}, s \approx t_2, t_3,\ldots, t_{k+1}, t_{k+2} \approx s_2, s_3,\ldots, s_{k+1}, s_{k+2}.$$

Suppose, now, that the points $p_2, p_3,\ldots, p_{k+1}, p_{k+2}$ are $S_{k,r}$-dependent. Then $s_2, s_3,\ldots, s_{k+1}, s_{k+2}$ are dependent and applying the remark following Property (3) to the $S_{k-1,r}$ determined by the k independent points $s_2, s_3,\ldots, s_{k+1}$, the above congruence gives $s = s_{k+2}$, and hence

$$t_2, t_3,\ldots, t_{k+1}, t_{k+2}, t_{k+3} \approx s_2, s_3,\ldots, s_{k+1}, s_{k+2}, s'.$$

Now

$$s_2, s_3,\ldots, s_{k+1}, s_{k+3} \approx p_2, p_3,\ldots, p_{k+1}, p_{k+3} \approx t_2, t_3,\ldots, t_{k+1}, t_{k+3},$$

which gives $s_2, s_3,\ldots, s_{k+1}, s_{k+3} \approx s_2, s_3,\ldots, s_{k+1}, s'$. If s_{k+3} is not dependent on $s_2, s_3,\ldots, s_{k+1}$, then by Property (5) the point s' may be distinct from s_{k+3}. But then the last congruence together with the congruence

$$s_2, s_3,\ldots, s_{k+1}, s_{k+2} \approx s_2, s_3,\ldots, s_{k+1}, s_{k+2}$$

shows that $s_{k+2} s_{k+3} = s_{k+2} s'$, according to Property (6). If, on the other hand, s_{k+3} is dependent on $s_2, s_3,\ldots, s_{k+1}$, then the congruence

$$s_2, s_3,\ldots, s_{k+1}, s_{k+3} \approx s_2, s_3,\ldots, s_{k+1}, s'$$

evidently implies $s' = s_{k+3}$, and $s_{k+2}s_{k+3} = s_{k+2}s'$ as before. Hence in any case, $s_{k+2}s_{k+3} = s_{k+2}s' = t_{k+2}t_{k+3} = p_{k+2}p_{k+3}$, which gives the desired contradiction and establishes the theorem.

Remark. It is clear that the method of the preceding theorem may be used to show that each of the $(k+1)$-tuples

$$p_1, p_2, ..., p_{i-1}, p_{i+1}, ..., p_{k+1}, p_{k+2},$$

$$p_1, p_2, ..., p_{i-1}, p_{i+1}, ..., p_{k+1}, p_{k+3}$$

$(i = 1, 2, ..., k+1)$ is $S_{k,r}$-independent.

THEOREM 69.6. *Let* $p_1, p_2, ..., p_{k+2}, p_{k+3}$ *form a pseudo-*$S_{k,r}$ *set with the* $S_{k,r}$-*independent* $(k+1)$-*tuples* $p_1, p_2, ..., p_k, p_{k+1}$;

$$p_1, p_2, ..., p_{i-1}, p_{i+1}, ..., p_{k+1}, p_{k+2}; \qquad p_1, p_2, ..., p_{i-1}, p_{i+1}, ..., p_{k+1}, p_{k+3}$$

$(i = 1, 2, ..., k+1)$. *Then the* $(k+1)$-*tuple* $p_3, p_4, ..., p_{k+2}, p_{k+3}$ *is* $S_{k,r}$-*independent.*

Proof. The proof follows the lines of the proof of the preceding theorem, with the $S_{k,r}$-independent $(k+1)$-tuple $p_1, p_3, p_4, ..., p_{k+1}, p_{k+2}$ in the role of the $(k+1)$-tuple $p_1, p_2, ..., p_{k+1}$. Thus, the $S_{k,r}$ contains $k+3$ points $s_1, s_2, ..., s_{k+2}, s_{k+3}$ such that

$$p_1, p_3, p_4, ..., p_{k+1}, p_{k+2}, p_{k+3} \approx s_1, s_3, ..., s_{k+1}, s_{k+2}, s_{k+3},$$

$$p_1, p_3, p_4, ..., p_{k+1}, p_{k+2}, p_2 \approx s_1, s_3, ..., s_{k+1}, s_{k+2}, s_2,$$

with $p_2 p_{k+3} \neq s_2 s_{k+3}$, and the same procedure as in Theorem 69.5 leads to the desired result.

Remark. In a similar manner it is seen that the $(k+1)$-tuples

$$p_1, p_2, ..., p_{i-1}, p_{i+1}, ..., p_{j-1}, p_{j+1}, ..., p_{k+1}, p_{k+2}, p_{k+3}$$

$(i, j = 1, 2, ..., k+1; i \neq j)$ are all $S_{k,r}$-independent $(k+1)$-tuples.

We have thus obtained the following result:

THEOREM 69.7. *If* $p_1, p_2, ..., p_{k+2}, p_{k+3}$ *form a pseudo-*$S_{k,r}$ $(k+3)$-*tuple, then each of the* $\frac{1}{2}(k+2)(k+3)$ *sets of* $k+1$ *points contained in this* $(k+3)$-*tuple is an* $S_{k,r}$-*independent set.*

If $p_1, p_2, ..., p_{k+3}$ form a pseudo-$S_{k,r}$ $(k+3)$-tuple, and $s_1, s_2, ..., s_{k+3}$ are $k+3$ points of $S_{k,r}$ almost congruent to them, consider the points

$$p_1, p_2, ..., p_{i-1}, p_{i+1}, ..., p_{k+3} \quad (i = 1, 2, ..., k+1).$$

The $S_{k,r}$ contains $k+2$ points $t_1, t_2, ..., t_{i-1}, t_{i+1}, ..., t_{k+3}$ such that for each $i = 1, 2, ..., k+1$,

$$p_1, p_2, ..., p_{i-1}, p_{i+1}, ..., p_{k+3} \approx t_1, t_2, ..., t_{i-1}, t_{i+1}, ..., t_{k+3}.$$

Then
$$s_1, s_2,..., s_{i-1}, s_{i+1},..., s_{k+1}, s_{k+2} \approx t_1, t_2,..., t_{i-1}, t_{i+1},..., t_{k+1}, t_{k+2},$$
with each of the sets of $k+1$ points $s_1, s_2,..., s_{i-1}, s_{i+1},..., s_{k+1}, s_{k+2}$ ($i = 1, 2,..., k+1$) being independent (Theorem 69.7). It follows that the $S_{k,r}$ contains exactly one point $s_{k+3}^{(i)}$ such that
$$s_1, s_2,..., s_{i-1}, s_{i+1},..., s_{k+1}, s_{k+2}, s_{k+3}^{(i)}$$
$$\approx t_1, t_2,..., t_{i-1}, t_{i+1},..., t_{k+1}, t_{k+2}, t_{k+3} \quad (i = 1, 2,..., k+1),$$
and hence
$$p_1, p_2,..., p_{i-1}, p_{i+1},..., p_{k+2}, p_{k+3} \approx s_1, s_2,..., s_{i-1}, s_{i+1},..., s_{k+2}, s_{k+3}^{(i)}$$
$$(i = 1, 2,..., k+1).$$
Since
$$s_1, s_2,..., s_{i-1}, s_{i+1},..., s_{k+1}, s_{k+3}^{(i)}$$
$$\approx p_1, p_2,..., p_{i-1}, p_{i+1},..., p_{k+1}, p_{k+3}$$
$$\approx s_1, s_2,..., s_{i-1}, s_{i+1},..., s_{k+1}, s_{k+3},$$
the point $s_{k+3}^{(i)}$ has the same distances from the points
$$s_1, s_2,..., s_{i-1}, s_{i+1},..., s_{k+1}$$
as the point s_{k+3}, but $s_{k+3}^{(i)} \neq s_{k+3}$ for $s_{k+2} s_{k+3}^{(i)} = p_{k+2} p_{k+3} \neq s_{k+2} s_{k+3}$. We have
$$s_{k+2} s_{k+3}^{(1)} = s_{k+2} s_{k+3}^{(2)} = ... = s_{k+2} s_{k+3}^{(k+1)} = p_{k+2} p_{k+3}.$$

Now s_{k+2} is not uniquely determined by the above equalities, but by Property (7) there is at most one other point s_{k+2}^* equidistant from the $k+1$ points $s_{k+3}^{(1)}, s_{k+3}^{(2)},..., s_{k+3}^{(3)}$, and $s_{k+2} s_{k+2}^* = d = \pi r$.

In a similar manner it is seen that $S_{k,r}$ contains $k+1$ points
$$s_{k+2}^{(1)}, s_{k+2}^{(2)},..., s_{k+2}^{(k+1)}$$
such that
$$p_1, p_2,..., p_{i-1}, p_{i+1},..., p_{k+1}, p_{k+2} \approx s_1, s_2,..., s_{i-1}, s_{i+1},..., s_{k+2}, s_{k+2}^{(i)}$$
$$(i = 1, 2,..., k+1),$$
with $\quad s_{k+2}^{(1)} s_{k+3} = s_{k+2}^{(2)} s_{k+3} = ... = s_{k+2}^{(k+1)} s_{k+3} = p_{k+2} p_{k+3}.$

Thus s_{k+3} is equidistant from the $k+1$ points $s_{k+2}^{(1)}, s_{k+2}^{(2)},..., s_{k+2}^{(k+1)}$, and there are at most two (diametral) points with that property.

Calling the points s_{k+2}, s_{k+3} *spherical isogonal conjugates with respect to the spherical simplex whose vertices are* $s_1, s_2,..., s_{k+1}$, we have the following characterization of pseudo-$S_{k,r}$ $(k+3)$-tuples:

THEOREM 69.8. *A semimetric set of $k+3$ points $p_1, p_2, ..., p_{k+3}$ is a pseudo-$S_{k,r}$ $(k+3)$-tuple if and only if $k+3$ points $s_1, s_2, ..., s_{k+3}$ of $S_{k,r}$ exist with*

$$p_1, p_2, ..., p_{k+2}, p_{k+3} \sim s_1, s_2, ..., s_{k+2}, s_{k+3},$$

where s_{k+2}, s_{k+3} are spherical isogonal conjugates with respect to the spherical simplex with vertices $s_1, s_2, ..., s_{k+1}$, and $p_{k+2}p_{k+3}$ equals the common distance of s_{k+3} from the $k+1$ reflections of s_{k+2} in the 'faces' of the simplex.

The necessity is established in the preceding argument, and the reader may readily prove the sufficiency.

Remark. For each $\epsilon > 0$ there exists a pseudo-$S_{k,r}$ $(k+3)$-tuple with diameter less than ϵ.

THEOREM 69.9. *Let $P \equiv (p_1, p_2, ..., p_{k+3})$ and $Q \equiv (q_1, q_2, ..., q_{k+3})$ be two pseudo-$S_{k,r}$ $(k+3)$-tuples such that*

$$p_1, p_2, ..., p_{k+2} \approx q_1, q_2, ..., q_{k+2}.$$

Then either $p_i p_{k+3} = q_i q_{k+3}$ $(i = 1, 2, ..., k+2)$, and the two $(k+3)$-tuples are congruent or

$$\cos(p_i p_{k+3}/r) + \cos(q_i q_{k+3}/r) = 0 \quad (i = 1, 2, ..., k+2).\dagger$$

Proof. From the hypothesis the following congruences are valid:

$$p_1, p_2, ..., p_{k+1}, p_{k+2} \approx s_1, s_2, ..., s_{k+1}, s_{k+2},$$
$$p_1, p_2, ..., p_{k+1}, p_{k+3} \approx s_1, s_2, ..., s_{k+1}, s_{k+3},$$
$$q_1, q_2, ..., q_{k+1}, q_{k+2} \approx s_1, s_2, ..., s_{k+1}, s_{k+2},$$
$$q_1, q_2, ..., q_{k+1}, q_{k+3} \approx s_1, s_2, ..., s_{k+1}, t_{k+3},$$

where the points on the right-hand sides of these congruences are in $S_{k,r}$ and

$$p_{k+2}p_{k+3} \neq s_{k+2}s_{k+3}, \quad q_{k+2}q_{k+3} \neq s_{k+2}t_{k+3}.$$

From the first two congruences follows, as has been seen, the existence of points $s_{k+2}^{(i)}$ $(i = 1, 2, ..., k+1)$, of $S_{k,r}$ such that

$$p_{k+2}p_{k+3} = s_{k+2}^{(1)}s_{k+3} = s_{k+2}^{(2)}s_{k+3} = ... = s_{k+2}^{(k+1)}s_{k+3}.$$

Similarly, from the last two congruences we may write

$$q_{k+2}q_{k+3} = s_{k+2}^{(1)}t_{k+3} = s_{k+2}^{(2)}t_{k+3} = ... = s_{k+2}^{(k+1)}t_{k+3}.$$

It follows (Property (7)) that either $s_{k+3} = t_{k+3}$ or $s_{k+3}t_{k+3} = d$. In the first case, clearly, $p_i p_{k+3} = q_i q_{k+3}$ $(i = 1, 2, ..., k+2)$, and $P \approx Q$. In the second case, $s_{k+3}t_{k+3} = d$ implies (Property III) $\Delta_3(s_i, s_{k+3}, t_{k+3}) = 0$, and hence (upon expanding)

$$\cos(s_i s_{k+3}/r) + \cos(s_i t_{k+3}/r) = 0 \quad (i = 1, 2, ..., k+1).$$

† With a view towards the abstraction presented in the remark concluding § 68, we shall not use any properties of the cosine function that are not shared by the abstract cosine function ϕ. Hence we do not write $p_i p_{k+3} + q_i q_{k+3} = \pi r$.

Then, by the above congruences,
$$\cos(p_i p_{k+3}/r) + \cos(q_i q_{k+3}/r) = 0 \quad (i = 1, 2, \ldots, k+1).$$
Finally, $s_{k+3} t_{k+3} = d$ implies $\Delta_3(s^{(1)}_{k+2}, s_{k+3}, t_{k+3})$ vanishes; that is,
$$\cos(s^{(1)}_{k+2} s_{k+3}/r) + \cos(s^{(1)}_{k+2} t_{k+3}/r) = 0.$$
Then (Theorem 69.8)
$$\cos(p_{k+2} p_{k+3}/r) + \cos(q_{k+2} q_{k+3}/r) = 0,$$
and the theorem is proved.

70. Some properties of pseudo-$S_{k,r}$ ($k+4$)-tuples without diametral point-pairs

Since every ($k+1$) tuple of a pseudo-$S_{k,r}$ set of $k+3$ points is $S_{k,r}$-independent (Theorem 69.7), it follows that no two points of such a set are diametral. A pseudo-$S_{k,r}$ ($k+4$)-tuple may, however, contain a diametral point-pair. An example is the semimetric quintuple $p_1, p_2, \ldots, p_5$ with $p_1 p_2 = p_3 p_4 = \frac{1}{3}d$, $p_1 p_3 = p_3 p_5 = p_2 p_4 = \frac{1}{2}d$, $p_1 p_4 = p_2 p_3 = \frac{1}{6}d$, $p_1 p_5 = d$, $p_2 p_5 = \frac{2}{3}d$, $p_4 p_5 = \frac{5}{6}d$. The reader may easily verify that this quintuple is pseudo-$S_{1,r}$, where $d = \pi r$.

We shall be concerned in this and the following section with the class of pseudo-$S_{k,r}$ sets *without diametral point-pairs* and containing more than $k+3$ pairwise distinct points.

LEMMA 70.1. *Any pseudo-$S_{k,r}$ set P of $k+4$ pairwise distinct points $p_1, p_2, \ldots, p_{k+4}$, no two of which are diametral, contains at least three pseudo-$S_{k,r}$ ($k+3$)-tuples.*

Proof. Since, by Property (8), the $S_{k,r}$ has congruence order $k+3$, the set P contains at least one pseudo-$S_{k,r}$ ($k+3$)-tuple. The labelling may be assumed so that $p_1, p_2, \ldots, p_{k+3}$ is such a set. In case P does not contain at least two ($k+3$)-tuples congruent with $k+3$ points of $S_{k,r}$, the lemma is surely valid. In the contrary case, let $p_1, p_2, \ldots, p_{k+1}, p_{k+2}, p_{k+4}$ and $p_1, p_2, \ldots, p_{k+1}, p_{k+3}, p_{k+4}$ be congruent with two ($k+3$)-tuples of $S_{k,r}$. Since $p_1, p_2, \ldots, p_{k+3}$ is a pseudo-$S_{k,r}$ set, we have
$$p_1, p_2, \ldots, p_{k+1}, p_{k+2} \approx s_1, s_2, \ldots, s_{k+1}, s_{k+2},$$
$$p_1, p_2, \ldots, p_{k+1}, p_{k+3} \approx s_1, s_2, \ldots, s_{k+1}, s_{k+3},$$
with $p_{k+2} p_{k+3} \neq s_{k+2} s_{k+3}$, and it follows that
$$p_1, p_2, \ldots, p_{k+1}, p_{k+2}, p_{k+4} \approx s_1, s_2, \ldots, s_{k+1}, s_{k+2}, s_{k+4},$$
$$p_1, p_2, \ldots, p_{k+1}, p_{k+3}, p_{k+4} \approx s_1, s_2, \ldots, s_{k+1}, s_{k+3}, s_{k+4}, \tag{I}$$
with the point s_{k+4} of $S_{k,r}$ uniquely determined since its distances from the $k+1$ independent points $s_1, s_2, \ldots, s_{k+1}$ of $S_{k,r}$ are fixed. Now, by

hypothesis, each pair of points of P is independent and hence the points s_{k+4} and s_i ($i = 1, 2, ..., k+3$) are independent. It follows that at least two of the $k+1$ sets

$$s_1, s_2, ..., s_{i-1}, s_{i+1}, ..., s_{k+1}, s_{k+4} \quad (i = 1, 2, ..., k+1)$$

are independent $(k+1)$-tuples, for in the contrary case, the k dependent $(k+1)$-tuples have, in addition to the point s_{k+4}, one of the points $s_1, s_2, ..., s_{k+1}$ in common. This point is then, by Property (9), diametral to (or coincident with) the point s_{k+4}, in contradiction with the preceding remark.

Let, then, $s_2, s_3, ..., s_{k+1}, s_{k+4}$ and $s_1, s_3, ..., s_{k+1}, s_{k+4}$ be independent $(k+1)$-tuples. We show that the two $(k+3)$-tuples

$$p_2, p_3, ..., p_{k+1}, p_{k+2}, p_{k+3}, p_{k+4}$$

and
$$p_1, p_3, ..., p_{k+1}, p_{k+2}, p_{k+3}, p_{k+4}$$

are pseudo-$S_{k,r}$ sets. (A similar procedure gives the desired result in case two other $(k+1)$-tuples are independent.)

We make the assumption that

$$p_2, p_3, ..., p_{k+1}, p_{k+2}, p_{k+3}, p_{k+4} \approx t_2, t_3, ..., t_{k+1}, t_{k+2}, t_{k+3}, t_{k+4},$$

points of $S_{k,r}$. Then (using congruences (I)) it follows that

$$s_2, s_3, ..., s_{k+1}, s_{k+2}, s_{k+4} \approx t_2, t_3, ..., t_{k+1}, t_{k+2}, t_{k+4},$$

$$s_2, s_3, ..., s_{k+1}, s_{k+3}, s_{k+4} \approx t_2, t_3, ..., t_{k+1}, t_{k+3}, t_{k+4},$$

with $s_2, s_3, ..., s_{k+1}, s_{k+4}$ an independent $(k+1)$-tuple. It follows that $s_{k+2}s_{k+3} = t_{k+2}t_{k+3} = p_{k+2}p_{k+3}$ (the first equality resulting from Property (3)) which gives the desired contradiction. In a similar way, the set $p_1, p_3, p_4, ..., p_{k+1}, p_{k+2}, p_{k+3}, p_{k+4}$ is shown to be a pseudo-$S_{k,r}$ $(k+3)$-tuple, and the proof of the lemma is complete.

THEOREM 70.1. *If a pseudo-$S_{k,r}$ set P of $k+4$ pairwise distinct points has no diametral point-pairs, then each $(k+3)$-tuple contained in P is a pseudo-$S_{k,r}$ set.*

Proof. From the lemma, P contains at least three pseudo-$S_{k,r}$ $(k+3)$-tuples, say

$$p_1, p_2, ..., p_{k+1}, p_{k+2}, p_{k+3}; \quad p_1, p_2, ..., p_{k+1}, p_{k+2}, p_{k+4};$$

$$p_1, p_2, ..., p_{k+1}, p_{k+3}, p_{k+4}.$$

The first and second, and the first and third of these pseudo-$S_{k,r}$ $(k+3)$-tuples contain congruent $(k+2)$-tuples. An application of Theorem 69.9

Ch. VII, § 70   THEOREMS FOR SPHERICAL SPACE   183

shows at once that the distances determined by the points of P satisfy one of the following four sets of relations:

Case I
$$p_i p_{k+2} = p_i p_{k+3} = p_i p_{k+4} \ (i = 1, 2, ..., k+1),$$
$$p_{k+2} p_{k+3} = p_{k+3} p_{k+4} = p_{k+2} p_{k+4}.$$

Case II
$$\cos(p_i p_{k+2}/r) = \cos(p_i p_{k+3}/r) = -\cos(p_i p_{k+4}/r) \ (i = 1, 2, ..., k+1),$$
$$\cos(p_{k+2} p_{k+3}/r) = -\cos(p_{k+3} p_{k+4}/r) = -\cos(p_{k+2} p_{k+4}/r).$$

Case III
$$\cos(p_i p_{k+2}/r) = -\cos(p_i p_{k+3}/r) = -\cos(p_i p_{k+4}/r) \ (i = 1, 2, ..., k+1),$$
$$\cos(p_{k+2} p_{k+3}/r) = -\cos(p_{k+3} p_{k+4}/r) = \cos(p_{k+2} p_{k+4}/r).$$

Case IV
$$\cos(p_i p_{k+2}/r) = -\cos(p_i p_{k+3}/r) = \cos(p_i p_{k+4}/r) \ (i = 1, 2, ..., k+1),$$
$$\cos(p_{k+2} p_{k+3}/r) = \cos(p_{k+3} p_{k+4}/r) = -\cos(p_{k+2} p_{k+4}/r).$$

To show, for example, that $p_2, p_3, ..., p_{k+1}, p_{k+2}, p_{k+3}, p_{k+4}$ is a pseudo-$S_{k,r}$ $(k+3)$-tuple, assume the contrary and let $s_2, s_3, ..., s_{k+1}, s_{k+2}, s_{k+3}, s_{k+4}$ be $k+3$ points of $S_{k,r}$ congruent to them. Then the distances determined by these $k+3$ points of $S_{k,r}$ satisfy one of the above four sets of relations (with the index i taking on the values $2, 3, ..., k+1$).

Case I. Applying Property (10) to $s_2, s_3, ..., s_{k+4}$, it is seen that $s_2, s_3, ..., s_{k+1}$ are dependent. Then the k points $p_2, p_3, ..., p_{k+1}$ congruent to them are $S_{k-1,r}$-dependent, which is not possible since these k points are contained in the $k+1$ points $p_1, p_2, ..., p_{k+1}$ which are $S_{k,r}$-independent because they belong to the pseudo-$S_{k,r}$ $(k+3)$-tuple $p_1, p_2, ..., p_{k+3}$.

Case II. The points $s_2, s_3, ..., s_{k+1}$ are each equidistant from the points s_{k+2}, s_{k+3} and since these k points are independent it follows from Property (11) that the $(k-1)$-dimensional subspace $S^*_{k-1,r}$ determined by them is the locus of points of $S_{k,r}$ equidistant from s_{k+2} and s_{k+3}. Now the points $s_2, s_3, ..., s_{k+1}$ are also contained in the locus of points s of $S_{k,r}$ such that $\cos(ss_{k+2}/r) + \cos(ss_{k+4}/r) = 0$. Since $s_{k+2} s_{k+4} = p_{k+2} p_{k+4}$, the points s_{k+2}, s_{k+4} are distinct and not diametral. We prove now the following assertion.

ASSERTION. *The locus of points t of $S_{k,r}$ such that*
$$\cos(ts_{k+2}/r) + \cos(ts_{k+4}/r) = 0$$
is the $(k-1)$-dimensional subspace of $S_{k,r}$ determined by $s_2, s_3, ..., s_{k+1}$.

To prove this assertion, it is shown first that if $t_1, t_2, \ldots, t_{k+1}$ are $k+1$ points of $S_{k,r}$ such that $\cos(t_i s_{k+2}/r)+\cos(t_i s_{k+4}/r) = 0$, then the determinant $\Delta_{k+1}(t_1, t_2, \ldots, t_{k+1})$ of these points vanishes, and hence the points are in an $S_{k-1,r}$. For suppose this determinant does not vanish, and consider $\Delta_{k+3}(t_1, t_2, \ldots, t_{k+1}, s_{k+2}, s_{k+4})$, which clearly equals zero. Adding the elements of the last row (column) to the corresponding elements of the preceding row (column) we obtain, after some obvious reductions,

$$[1+\cos(s_{k+2} s_{k+4}/r)] \cdot \Delta_{k+1}(t_1, t_2, \ldots, t_{k+1}) = 0.$$

Since $s_{k+2} s_{k+4} \neq d$, it follows that $\Delta_{k+1}(t_1, t_2, \ldots, t_{k+1})$ vanishes, contrary to our supposition. Hence each set of $k+1$ points of the locus is a dependent set, and the locus is at most $(k-1)$-dimensional. But the locus contains the k independent points $s_2, s_3, \ldots, s_{k+1}$ of $S_{k,r}$, and so the locus is exactly $(k-1)$-dimensional.

Finally, let $s \neq s_i$ ($i = 2, 3, \ldots, k+1$) be any element of the $(k-1)$-dimensional subspace determined by $s_2, s_3, \ldots, s_{k+1}$. Now the determinant $\Delta_{k+3}(s_2, s_3, \ldots, s_{k+1}, s_{k+2}, s_{k+4}, s)$ is zero, and every principal minor of order $k+2$ vanishes. It follows that *every* $(k+2)$th-order minor of the determinant is zero. Adding the $(k+2)$th row (column) to the preceding row (column)—a transformation of the determinant which leaves the rank unaltered—and expanding, as above, the vanishing $(k+2)$th-order minor $[k+2, k+2]$ of the resulting determinant, we obtain,

$$2[1+\cos(s_{k+2} s_{k+4}/r)] \cdot \Delta_{k+1}(s_2, s_3, \ldots, s_{k+1}, s) -$$
$$- [\cos(ss_{k+2}/r)+\cos(ss_{k+4}/r)]^2 \cdot \Delta_k(s_2, s_3, \ldots, s_{k+1}) = 0.$$

Since $\Delta_{k+1}(s_2, s_3, \ldots, s_{k+1}, s) = 0$ and $\Delta_k(s_2, s_3, \ldots, s_{k+1})$ does not vanish, we have $\cos(ss_{k+2}/r)+\cos(ss_{k+4}/r) = 0$, and s is a point of the locus. Hence the assertion is proved, and the locus in question *identified with the $(k-1)$-dimensional subspace $S^*_{k-1,r}$ determined by the k independent points* $s_2, s_3, \ldots, s_{k+1}$ (Property (12)).

But this is impossible, for since

$$\cos(s_{k+2} s_{k+3}/r)+\cos(s_{k+3} s_{k+4}/r) = 0,$$

the point s_{k+3} belongs to the above locus, though it surely does *not* belong to $S^*_{k-1,r}$, for s_{k+3} is not equidistant from the points s_{k+2}, s_{k+3}.

This contradiction shows that the distances determined by the points $s_2, s_3, \ldots, s_{k+1}, s_{k+2}, s_{k+3}, s_{k+4}$ do not satisfy the relations of Case II.

Interchanging s_{k+2} with s_{k+4} in Case III, and s_{k+3} with s_{k+4} in Case IV reduces those cases to Case II, and hence the assumption that

$$p_2, p_3, \ldots, p_{k+1}, p_{k+2}, p_{k+3}, p_{k+4}$$

is not a pseudo-$S_{k,r}$ set leads to the distances determined by those points not satisfying the relations of any one of the above four cases. This contradiction proves that the $k+3$ points form a pseudo-$S_{k,r}$ set. A similar procedure is used to show that the k remaining $(k+3)$-tuples of the set P are each pseudo-$S_{k,r}$ sets, and the theorem is established.

COROLLARY. *If P is a pseudo-$S_{k,r}$ set containing more than $k+3$ pairwise distinct points, no pair of which is diametral, then every set of $m \geqslant k+3$ points of P is a pseudo-$S_{k,r}$ set.*

Proof. Since P is a pseudo-$S_{k,r}$ set there is at least one pseudo-$S_{k,r}$ $(k+3)$-tuple $q_1, q_2, \ldots, q_{k+3}$ contained in P (Property (8)). Clearly, any subset of P that contains these $k+3$ points is a pseudo-$S_{k,r}$ set. Suppose, now, that $p_1, p_2, \ldots, p_{k+3}, \ldots, p_m$ is a set of $m \geqslant k+3$ points of P not containing any of the points $q_1, q_2, \ldots, q_{k+3}$. Then $q_1, q_2, \ldots, q_{k+3}, p_1$ is a pseudo-$S_{k,r}$ $(k+4)$-tuple without diametral points and hence, by the preceding theorem, each set of $k+3$ of these points (in particular, the set $q_2, q_3, \ldots, q_{k+3}, p_1$) is a pseudo-$S_{k,r}$ set. Then $q_2, q_3, \ldots, q_{k+3}, p_1, p_2$ is a pseudo-$S_{k,r}$ $(k+4)$-tuple without diametral points. It is clear that this process may be continued until the points $q_1, q_2, \ldots, q_{k+3}$ are replaced by the points $p_1, p_2, \ldots, p_{k+3}$ forming a pseudo-$S_{k,r}$ $(k+3)$-tuple. Then the points $p_1, p_2, \ldots, p_{k+3}, \ldots, p_m$ surely form a pseudo-$S_{k,r}$ set.

Finally, if $i \leqslant k+2$ of the points $q_1, q_2, \ldots, q_{k+3}$ occur among the m-tuple of points $p_1, p_2, \ldots, p_{k+3}, \ldots, p_m$, we have, with convenient labelling, $p_j = q_j$ $(j = 1, 2, \ldots, i)$, and the above process, starting with the pseudo-$S_{k,r}$ $(k+3)$-tuple $q_{i+1}, q_{i+2}, \ldots, q_{k+3}, p_1, p_2, \ldots, p_i$, is applied as before to complete the proof of the corollary.

LEMMA 70.2. *Let $P = (p_1, p_2, \ldots, p_{k+3}, p_{k+4})$ be a pseudo-$S_{k,r}$ $(k+4)$-tuple without diametral points. Then $p_i p_m = p_j p_m$ or*

$$\cos(p_i p_m/r) + \cos(p_j p_m/r) = 0 \quad (i,j,m = 1,2,\ldots,k+4;\ i \neq m \neq j).$$

Proof. If $i = j$, the lemma is trivial. Suppose, then, $i \neq j$, and consider the two $(k+3)$-tuples

$$p_1, p_2, \ldots, p_{j-1}, p_{j+1}, \ldots, p_{k+4}; \qquad p_1, p_2, \ldots, p_{i-1}, p_{i+1}, \ldots, p_{k+4}$$

obtained from P by omitting, in turn, the points p_j and p_i, respectively. According to Theorem 70.1 these two $(k+3)$-tuples are pseudo-$S_{k,r}$ sets, and since the $(k+2)$-tuple $p_1, p_2, \ldots, p_{i-1}, p_{i+1}, \ldots, p_{j-1}, p_{j+1}, \ldots, p_{k+4}$ is common to both sets, an application of Theorem 69.9 gives at once the desired result.

THEOREM 70.2. *Let* $P \equiv (p_1, p_2, ..., p_{k+4})$ *be a pseudo-$S_{k,r}$ $(k+4)$-tuple without diametral points. Then* $p_i p_m = p_j p_n$ *or*

$$\cos(p_i p_m/r) + \cos(p_j p_n/r) = 0 \quad (i \neq m; j \neq n),$$

for each pair $p_i p_m$, $p_j p_n$ *of the* $\frac{1}{2}(k+3)(k+4)$ *distances determined by the points of* P.

Proof. If any one of the indices i, m equals one of the indices j, n the theorem reduces to the preceding lemma. Suppose this is not the case. According to the lemma, $p_i p_m = p_j p_m$ or $\cos(p_i p_m/r) + \cos(p_j p_m/r) = 0$, and $p_j p_m = p_j p_n$ or $\cos(p_j p_m/r) + \cos(p_j p_n/r) = 0$. A consideration of the four possibilities thus presented leads at once to the theorem.

71. Characterization of pseudo-$S_{n,r}$ sets without diametral point-pairs

Consider, now, the symmetric determinant

$$\Delta_{n+4}(p_1, p_2, ..., p_{n+4}) = |\cos(p_i p_j/r)| \quad (i, j = 1, 2, ..., n+4),$$

of order $n+4$, formed for the $n+4$ pairwise distinct points $p_1, p_2, ..., p_{n+4}$ constituting a pseudo-$S_{n,r}$ set without diametral point-pairs. By Theorem 70.2 any two elements of this determinant outside the principal diagonal *differ at most in sign*. Since each $n+2$ of the $n+4$ points are imbeddable in $S_{n,r}$, all $(n+2)$th-order principal minors of Δ_{n+4} vanish, and hence no element of the determinant is zero (see (*) of Lemma 69.1). Also, since $p_i \neq p_j$ $(i \neq j)$, $\cos(p_i p_j/r) \neq 1$ $(i \neq j)$.

Remark. Apart from the elements in the first and the j-th rows $(j = 2, 3, ..., n+4)$, either each element in the first column of Δ_{n+4} is equal to the corresponding element in the j-th column, or each element in the first column is the negative of the corresponding element in the j-th column.

Consider, for example, the first and second columns. According to Theorem 70.1 the two $(n+3)$-tuples

$$p_1, p_3, ..., p_{n+3}, p_{n+4}; \quad p_2, p_3, ..., p_{n+3}, p_{n+4}$$

are pseudo-$S_{n,r}$ sets. Since they obviously contain congruent $(n+2)$-tuples, it follows from Theorem 68.9 that either

$$p_i p_1 = p_i p_2 \quad (i = 3, 4, ..., n+4),$$

or $$\cos(p_i p_1/r) = -\cos(p_i p_2/r) \quad (i = 3, 4, ..., n+4).$$

The same consideration applied to the first and the jth columns verifies the remark in general.

A closer examination of the determinant Δ_{n+4}, in the light of the foregoing observations, enables one to evaluate each of its elements (to within sign), and to describe the possible distribution of the signs. For suppose that the first column of the determinant contains, in addition to the element 1, exactly p positive elements, $0 \leqslant p \leqslant n+3$. If $p = 0$, then the elements $\cos(p_1 p_j/r)$ are negative ($j = 2, 3,..., n+4$), and from the above remark either the elements $\cos(p_2 p_j/r)$ are *all* negative or they are *all* positive ($j = 3, 4,..., n+4$). In the first case it follows from the symmetry of the determinant (and the remark) that *every* element outside the principal diagonal is negative. Then the vanishing of the $(n+2)$th-order principal minor $\Delta_{n+2}(p_1, p_2,..., p_{n+2})$ yields at once that $\cos(p_i p_j/r) = -1/(n+1)$ ($i, j = 1, 2,..., n+4; i \neq j$). In the second case the same procedure shows that all the elements $\cos(p_i p_j/r)$ are positive ($i, j = 2, 3,..., n+4; i \neq j$). But this is impossible, for the principal minor $\Delta_{n+2}(p_3, p_4,..., p_{n+4})$ of order $n+2$ then fails to vanish. Hence the first case alone is possible, and it may be concluded that if $p = 0$, then each element of Δ_{n+4} outside the principal diagonal has the value $-1/(n+1)$.

There is no difficulty in applying the same argument to the general case. The following two cases, represented schematically, are obtained:

Case A

$$\overbrace{\begin{vmatrix} 1 & + & + & . & + & - & - & . & - \\ + & 1 & + & . & + & - & - & . & - \\ + & + & 1 & . & + & - & - & . & - \\ . & . & . & . & . & . & . & . & . \\ + & + & + & . & 1 & - & - & . & - \\ - & - & - & . & - & 1 & + & . & + \\ - & - & - & . & - & + & 1 & . & + \\ . & . & . & . & . & . & . & . & . \\ - & - & - & . & - & + & + & . & 1 \end{vmatrix}}^{p+1}$$

Case B

$$\overbrace{\begin{vmatrix} 1 & + & + & . & + & - & - & . & - \\ + & 1 & - & . & - & + & + & . & + \\ + & - & 1 & . & - & + & + & . & + \\ . & . & . & . & . & . & . & . & . \\ + & - & - & . & 1 & + & + & . & + \\ - & + & + & . & + & 1 & - & . & - \\ - & + & + & . & + & - & 1 & . & - \\ . & . & . & . & . & . & . & . & . \\ - & + & + & . & + & - & - & . & 1 \end{vmatrix}}^{p+1}$$

In Case B, multiplication of the 2nd, 3rd,..., $(p+1)$th rows and columns by -1 makes every element outside the principal diagonal negative, and we find, as before, that each such element has the value $-1/(n+1)$. Case A is seen to be impossible, for multiplication of the last $n-p+3$ rows and columns by -1 makes all the elements of the determinant positive, and the same contradiction as before is encountered.

THEOREM 71.1. *A semimetric set* $p_1, p_2,..., p_{n+4}$ *of* $n+4$ *pairwise distinct points, no two of which have distance* πr, *is a pseudo-*$S_{n,r}$ ($n+4$)-*tuple if and only if* (i) *for every pair of distinct points* p_i, p_j,

$$\cos(p_i p_j/r) = \pm 1/(n+1),$$

and (ii) *the plus and minus signs are so distributed that upon multiplication by* -1 *of appropriate rows and the same numbered columns, the determinant*

$$\Delta_{n+4}(p_1, p_2, ..., p_{n+4}) = |\cos(p_i p_j/r)| \quad (i, j = 1, 2, ..., n+4),$$

is transformed into the determinant $\Delta_{n+4}(-1/(n+1))$, *which has each element outside the principal diagonal equal to* $-1/(n+1)$.

Proof. The necessity follows from applying the preceding remarks to the results obtained in Theorem 70.2, while the sufficiency is immediate from Lemma 69.1 and the fact (shown in the proof of Theorem 69.1) that $S_{n,r}$ does not contain any equilateral sets of $n+3$ points.

We now extend the results obtained for pseudo-$S_{n,r}$ sets of $n+4$ pairwise distinct points, without diametral point-pairs, to *any* such pseudo-$S_{n,r}$ set containing more than $n+3$ pairwise distinct points. *In the following section we shall drop the restriction that no two points have a distance d.*

THEOREM 71.2. *Let P be a pseudo-$S_{n,r}$ set of more than $n+3$ points, no two of which have distance πr. If p, q are any two distinct points of P, then* $\cos(pq/r) = \pm 1/(n+1)$.

Proof. If P consists of exactly $n+4$ pairwise distinct points, the conclusion follows from Theorem 71.1. Suppose P contains at least $n+5$ points and select any $n+4$ pairwise distinct points containing p and q. By the corollary to Theorem 70.1, these $n+4$ points form a pseudo-$S_{n,r}$ $(n+4)$-tuple, and Theorem 71.1 again gives the desired conclusion.

LEMMA 71.1. *If $p_1, p_2, ..., p_j$ and $q_1, q_2, ..., q_j$ are any two pseudo-$S_{n,r}$ sets without diametral point-pairs and*

$$p_1, p_2, ..., p_{j-1} \approx q_1, q_2, ..., q_{j-1},$$

then either $\qquad p_i p_j = q_i q_j \quad (i = 1, 2, ..., j-1)$

or $\qquad \cos(p_i p_j/r) + \cos(q_i q_j/r) = 0 \quad (i = 1, 2, ..., j-1).$

Proof. Since the two sets are pseudo-$S_{n,r}$ sets, then $j \geqslant n+3$. The two $(n+3)$-tuples $p_{j-n-2}, p_{j-n-1}, ..., p_{j-1}, p_j$; $q_{j-n-2}, q_{j-n-1}, ..., q_{j-1}, q_j$ are (corollary, Theorem 70.1), pseudo-$S_{n,r}$ sets. It follows from the hypothesis that

$$p_{j-n-2}, p_{j-n-1}, ..., p_{j-1} \approx q_{j-n-2}, q_{j-n-1}, ..., q_{j-1},$$

and hence (Theorem 69.9)

$$p_i p_j = q_i q_j \quad (i = j-n-2, j-n-1, ..., j-1), \tag{A_1}$$

or $\quad \cos(p_i p_j/r) + \cos(q_i q_j/r) = 0 \quad (i = j-n-2, j-n-1, ..., j-1). \tag{B_1}$

Applying the same reasoning to the two $(n+3)$-tuples

$$p_{j-n-3}, p_{j-n-1}, \ldots, p_{j-1}, p_j \quad \text{and} \quad q_{j-n-3}, q_{j-n-1}, \ldots, q_{j-1}, q_j,$$

we obtain

$$p_j p_{j-n-3} = q_j q_{j-n-3}, \quad p_j p_{j-n-1} = q_j q_{j-n-1}, \quad \ldots, \quad p_j p_{j-1} = q_j q_{j-1}, \tag{A$_2$}$$

or

$$\cos(p_j p_{j-n-3}/r) + \cos(q_j q_{j-n-3}/r) = 0,$$
$$\cos(p_j p_{j-n-1}/r) + \cos(q_j q_{j-n-1}) = 0, \ldots, \tag{B$_2$}$$
$$\cos(p_j p_{j-1}/r) + \cos(q_j q_{j-1}/r) = 0.$$

It is now easily seen that if the alternative (A$_1$) subsists, then the alternative (A$_2$) holds, while the validity of (B$_1$) implies that of (B$_2$). Thus the alternatives (A$_1$), (B$_1$) have been extended from

$$i = j-n-2, j-n-1, \ldots, j-1 \quad \text{to} \quad i = j-n-3, j-n-2, \ldots, j-1.$$

Continuing in this manner, the index i is made to recede to 1, and the lemma is established.

THEOREM 71.3. *A semimetric set P of arbitrary power exceeding $n+3$ and containing no diametral point-pairs is a pseudo-$S_{n,r}$ set if and only if for every integer $i > 1$ the determinant Δ_i formed for each set of i points (pairwise distinct) of P has, upon multiplication of appropriate rows and the same numbered columns by -1, all elements outside the principal diagonal equal to $-1/(n+1)$.*

Proof. The sufficiency is clear. To prove the necessity let $p_1, p_2, \ldots, p_i$ be any set of $i > 1$ pairwise distinct points of P. If $i = n+4$, the conclusion follows from Theorem 71.1 and the corollary to Theorem 70.1.

Case 1. If $i < n+4$, then the i points $p_1, p_2, \ldots, p_i$ form part of a set of $n+4$ points which is a pseudo-$S_{n,r}$ set. Hence, the determinant Δ_{n+4} of these $n+4$ points has, upon multiplication of appropriate rows and the same numbered columns by -1, all elements outside the principal diagonal equal to $-1/(n+1)$. The determinant $\Delta_i(p_1, p_2, \ldots, p_i)$ being a principal minor of this determinant is then transformed by these elementary operations to the form specified in the theorem.

Case 2. If $i > n+4$, then by using the preceding lemma in the same manner that Theorem 69.9 was applied to the proof of Theorem 71.1, the method utilized to prove the latter theorem may be adopted without change to establish the present theorem in the case under consideration.

72. Characterization of general pseudo-$S_{n,r}$ sets

We are now in a position to characterize pseudo-$S_{n,r}$ sets without the additional condition that they be free from diametral point-pairs.

It might well be conjectured that dropping this restriction will greatly increase the variety of these sets, so that no simple characterization analogous to that given in Theorem 71.2 or Theorem 71.3 will be valid for them. Also, in view of the difficulties attendant upon the derivation of those two theorems it might reasonably appear that the characterization of general pseudo-$S_{n,r}$ sets would entail even more involved considerations.

In view of these remarks it is surprising that (1) the structure of general pseudo-$S_{n,r}$ sets is, in fact, no more complicated than that of the restricted type, and (2) the derivation of their characterization theorem is obtained from the theorems of § 71 by *very simple means*. We establish first three lemmas.

LEMMA 72.1. *If p, p^* are diametral points of a pseudo-$S_{n,r}$ set P ($n > 0$), then for each point x of P, $px+xp^* = d$ ($d = \pi r$).*

Proof. Since $n > 0$, each three points of P are imbeddable in $S_{n,r}$. The lemma results from applying this remark to p, p^*, x.

LEMMA 72.2. *If p, p^* are diametral points of a pseudo-$S_{n,r}$ set P ($n > 0$), the set $Q = P-(p^*)$ is a pseudo-$S_{n,r}$ set.*

Proof. Assume the lemma false. Since each $(n+2)$-tuple of P is imbeddable in $S_{n,r}$, the same is true for each $(n+2)$-tuple of Q and it follows from our assumption that Q is imbeddable in $S_{n,r}$.

Let T be a subset of $S_{n,r}$ such that
$$Q \approx T, \qquad (*)$$
and let t be the point of T corresponding by $(*)$ to the point p of Q. By Lemma 72.1 the extension of congruence $(*)$ to the congruence
$$P = Q+(p^*) \approx T+(t^*),$$
where t^* is the point of $S_{n,r}$ diametral to t, is clearly valid, and P is imbeddable in $S_{n,r}$ contrary to hypothesis.

Let R be a subset of a pseudo-$S_{n,r}$ set P. If $r \in R$, adjoin to P an element r^*, defining $r^*p = d-rp$ for each element p of P. Denoting the set of adjoined elements by R^*, the semimetric space $S = P+R^*$ is said to be obtained from P by the *d-process*. Thus S is a diametrization (partial or complete) of P. (Define $x^*y^* = xy$ for $x^*, y^* \in R^*$.)

LEMMA 72.3. *Every pseudo-$S_{n,r}$ set S is obtainable from a pseudo-$S_{n,r}$ set P without diametral point-pairs by the d-process.*

Proof. Deleting from S exactly one point of each of the diametral point-pairs contained in S yields a pseudo-$S_{n,r}$ set P (Lemma 72.2) without diametral point-pairs from which S may evidently be obtained by the d-process.

It has been noted (§ 70) that a pseudo-$S_{n,r}$ $(n+3)$-tuple does not contain any diametral point-pairs. From this it readily follows that a pseudo-$S_{n,r}$ set of $n+3+k$ pairwise distinct points can have *at most* k diametral point-pairs. If a pseudo-$S_{n,r}$ set S of $n+3+k$ points $(k > 1)$ contains at most $k-1$ diametral point-pairs, then the pseudo-$S_{n,r}$ set P obtained from S by deleting exactly one of the points of each diametral point-pair has at least $n+3+k-(k-1) = n+4$ pairwise distinct points and so every non-zero distance in P equals $r.\cos^{-1}[\pm 1/(n+1)]$ (Theorem 71.2). Hence *every non-zero distance in S is d or $r.\cos^{-1}[\pm 1/(n+1)]$*.

Whenever $k > n+3$ the deleted set P always contains more than $n+3$ pairwise distinct points, and the above conclusion holds. We are thus led to the following characterization theorem of general pseudo-$S_{n,r}$ sets.

THEOREM 72.1. *A pseudo-$S_{n,r}$ set S of more than $2(n+3)$ pairwise distinct points has every non-zero distance equal to d or $r.\cos^{-1}[\pm 1/(n+1)]$. The same conclusion is valid for a pseudo-$S_{n,r}$ set S of $n+3+k$ pairwise distinct points $(k > 0)$, provided S contains at most $k-1$ diametral point-pairs.*

EXERCISES

1. Prove that, apart from labelling, there is a unique pseudo-$S_{n,r}$ set of $2n+2$ pairwise distinct points $(n > 1)$, free from diametral point-pairs, which does not contain an equilateral set of $n+2$ points.
2. Show that every pseudo-$S_{n,r}$ set of $2n+2$ pairwise distinct points $(n > 1)$, without diametral point-pairs, contains an equilateral $(n+1)$-tuple with edge $r.\cos^{-1}[-1/(n+1)]$.
3. Prove that every pseudo-$S_{n,r}$ set of more than $2n+2$ pairwise distinct points, no two of which have distance πr, contains an equilateral $(n+2)$-tuple with edge $r.\cos^{-1}[-1/(n+1)]$.
4. Using Theorem 69.8 show that a pseudo-$S_{1,r}$ quadruple Q has either all triples linear, no triples linear, or exactly three triples linear. Prove that (1) in the first case Q is pseudo-linear with no distance equal to πr; (2) in the second case, Q has 'opposite' distances equal; (3) in the third case, the points of Q can be labelled p, q, r, s so that psq, qsr, psr subsist, with p, q, r non-linear.
5. Formulate two analogues of Theorem 71.3 for general pseudo-$S_{n,r}$ sets.

REFERENCES

§ 63. Blumenthal [13] and Garrett.
§§ 64–66. Blumenthal [14].
§§ 67–68. Blumenthal [15].
§§ 69–71. Blumenthal [16] and Thurman.
§ 72. Blumenthal [17] and Stamey.

CHAPTER VIII

INTERSECTION THEOREMS FOR CONVEX SUBSETS OF $S_{n,r}$. CONGRUENCE INDICES OF SPHERICAL CAPS

73. Preliminary definitions and remarks

THE problems dealt with in this chapter are intimately connected with questions concerning the intersection of convex subsets of $S_{n,r}$. We would like to know, for example, for what integer k (if any) the intersection of each k members of any family of such subsets implies the existence of a point that belongs to every member of the family. Similar questions for subsets of E_n were answered by F. Riesz and E. Helly, the former of whom showed that any family $\mathscr{F}$ of bounded and closed subsets of E_n has a non-null product whenever each finite sub-family of $\mathscr{F}$ has, while the latter proved that if the members of $\mathscr{F}$ are also convex the conclusion follows when each $n+1$ of them have a point in common. What is the analogue of Helly's theorem in $S_{n,r}$?

Using Theorem 14.1, a *closed* subset C of $S_{n,r}$ is metrically convex if and only if each two points of C are end-points of a great circle arc of length at most πr contained in C. Hence $S_{n,r}$ is metrically convex, along with each of its subspaces $S_{k,r}$ ($k = 1, 2,..., n$), but $S_{0,r}$ (which consists of a diametral point-pair) is not a metrically convex subset of $S_{n,r}$ ($n > 0$). Now it is easy to show that, *apart from the $S_{n,r}$*, each closed metrically convex subset of $S_{n,r}$ is contained in a hemisphere; *while, apart from the $S_{0,r}$*, each subset of $S_{n,r}$ which is a product of hemispheres is metrically convex (Exercises 1, 2).† Since, moreover, it is immaterial whether or not we consider $S_{n,r}$ to be convex in all that concerns products of its subsets, while our results are more conveniently obtained and described in terms of a notion of convexity with respect to which the $S_{0,r}$ is convex, we make the following definitions.

DEFINITION 73.1. *A spherical cap $K_{n,\rho}$ of $S_{n,r}$ ($0 < \rho \leqslant \tfrac{1}{2}\pi r$) is the locus of points of $S_{n,r}$ with distance at most ρ from a fixed point called its centre. If $\rho < \tfrac{1}{2}\pi r$, $K_{n,\rho}$ is a small cap, while if $\rho = \tfrac{1}{2}\pi r$, $K_{n,\rho}$ is a hemisphere, denoted by $H_{n,r}$.*

Those points of a cap $K_{n,\rho}$ with distance ρ from the centre form its *rim*. The rim is evidently an $(n-1)$-sphere, which is an $(n-1)$-dimensional

---
† A definition of 'hemisphere' is given below.

subspace $S_{n-1,r}$ of $S_{n,r}$ if and only if $\rho = \tfrac{1}{2}\pi r$. The E_n containing the rim of $K_{n,\rho}$ is called its *base plane*. The base plane of a cap passes through the origin o of the cartesian coordinate system in E_{n+1} with respect to which the points $(x_1, x_2, ..., x_{n+1})$ of $S_{n,r}$ are given by $\sum_{i=1}^{n+1} x_i^2 = r^2$, if and only if the cap is a hemisphere.

DEFINITION 73.2. *A subset C of $S_{n,r}$ is convex provided C is the product of a family of hemispheres of $S_{n,r}$.*

As immediate consequences of this definition of convexity, we have (1) *each convex subset of $S_{n,r}$ is closed*, (2) *the product of any two convex subsets of $S_{n,r}$ is a convex subset of $S_{n,r}$*, (3) *$S_{0,r}$ is a convex subset of $S_{n,r}$* ($n > 0$), and (4) *the $S_{n,r}$ is not convex*.

DEFINITION 73.3. *A set of $m+1$ hemispheres ($m \geq 1$) of $S_{n,r}$ is called dependent or independent according as the centres of the hemispheres form a dependent or independent $(m+1)$-tuple of $S_{n,r}$, respectively.*

If $a^{(i)}: (a_1^{(i)}, a_2^{(i)}, ..., a_{n+1}^{(i)})$ ($i = 1, 2, ..., m+1$) are the centres of $m+1$ hemispheres of $S_{n,r}$ and $v^{(i)} = \overrightarrow{oa^{(i)}}$ is the vector with initial point o and terminal point $a^{(i)}$ ($i = 1, 2, ..., m+1$), the $m+1$ hemispheres are dependent if and only if these $m+1$ vectors belong to an E_m (containing o) and hence are linearly dependent. This is the case if and only if the matrix

$$||a_j^{(i)}|| \quad (i = 1, ..., m+1; j = 1, 2, ..., n+1)$$

has rank less than $m+1$. Since the $m+1$ base planes of the hemispheres have equations

$$\sum_{j=1}^{n+1} a_j^{(i)} x_j = 0 \quad (i = 1, 2, ..., m+1),$$

it follows that *$m+1$ hemispheres are dependent if and only if the rank of the coefficient matrix of their base planes is less than $m+1$*.

A bounded, closed, and convex subset of E_n is called a *convex body*. Of fundamental importance in the theory of convex bodies is the theorem that through each boundary point p of such a set there passes an E_{n-1}, one side of which contains no points of the body. Such an E_{n-1} is called a *supporting hyperplane* of the body at p.

EXERCISES

1. Prove that each closed metrically convex proper subset of $S_{n,r}$ is contained in a hemisphere of $S_{n,r}$.
2. Prove that each product of hemispheres of $S_{n,r}$, different from $S_{0,r}$, is metrically convex.
3. Show that an independent set of $n+1$ hemispheres of $S_{n,r}$ do not have a pair of diametral points of $S_{n,r}$ in common.

74. Some lemmas

Let C be a convex subset of $S_{n,r}$ that does not contain a diametral point-pair, and let $\{H_{n,r}\}$ be a family of hemispheres whose product is C. Then $\{H_{n,r}\}$ contains an independent set of $n+1$ hemispheres; for in the contrary case, the subset of $S_{n,r}$ consisting of the centres of the members of $\{H_{n,r}\}$ has each $n+1$ of its points in an $S_{n-1,r}$ and consequently the subset is contained in an $S_{n-1,r}$. But then $C = \prod H_{n,r}$ has a pair of diametral points (namely those points in which the line through o perpendicular to the E_n through o which contains that $S_{n-1,r}$ intersects $S_{n,r}$).

If
$$\sum_{i=1}^{n+1} a_i^{(j)} x_i = 0 \quad (j = 1, 2, ..., n+1), \tag{*}$$

are the equations of the base planes of $n+1$ independent hemispheres of $\{H_{n,r}\}$, then the system of equations (∗) *has only the trivial solution* $(0, 0, ..., 0)$ (Exercise 3, § 73), and we may suppose the coefficients in (∗) so chosen that

$$\sum_{i=1}^{n+1} a_i^{(j)} x_i \geqslant 0 \quad (j = 1, 2, ..., n+1)$$

for each point $(x_1, x_2, ..., x_{n+1})$ of C. It follows that the linear function

$$\sum_{i=1}^{n+1} \Big(\sum_{j=1}^{n+1} a_i^{(j)}\Big) x_i = \sum_{i=1}^{n+1} a_i^{(1)} x_i + \sum_{i=1}^{n+1} a_i^{(2)} x_i + ... + \sum_{i=1}^{n+1} a_i^{(n+1)} x_i$$

is *positive* for $(x_1, x_2, ..., x_{n+1})$ a point of C, and hence assumes on C (closed and compact) a positive minimum m. Since the linear function

$$\sum_{i=1}^{n+1} \Big(\sum_{j=1}^{n+1} a_i^{(j)}\Big) x_i - m$$

is non-negative for points of C and has the value $-m$ at the origin, the hyperplane

$$\sum_{i=1}^{n+1} \Big(\sum_{j=1}^{n+1} a_i^{(j)}\Big) x_i - m = 0$$

is the base of a small cap $K_{n,\rho}$ containing C. We have proved, therefore, the following lemma.

LEMMA 74.1. *A convex subset of $S_{n,r}$ that does not possess a diametral point-pair is contained in a small cap.*

If the hyperplane $\sum_{i=1}^{n+1} a_i x_i - c = 0 \ (c > 0)$, is the base of a small cap $K_{n,\rho}$ of $S_{n,r}$, that cap is contained in the half-E_{n+1} formed by the points $(x_1, x_2, ..., x_{n+1})$ of E_{n+1} for which

$$\sum_{i=1}^{n+1} a_i x_i - c \geqslant 0.$$

Ch. VIII, § 74   CONVEX SUBSETS OF $S_{n,r}$ 195

It is convenient to denote the left-hand member of this inequality by $P(x)$, and to refer to the half-E_{n+1} $P(x) \geq 0$ as the half-space of the small cap. Similarly, the half-E_{n+1} containing a hemisphere is called the half-space of the hemisphere.

LEMMA 74.2. *If each $n+1$ members of a family of hemispheres of $S_{n,r}$ have a point in common with the product of $n+1$ fixed independent hemispheres of $S_{n,r}$, then all the hemispheres have a point in common.*

Proof. Let F be the product of the half-spaces of the $n+1$ fixed independent hemispheres and let $C = F.S_{n,r}$. Since C is the product of the $n+1$ fixed independent hemispheres, it has no diametral point-pairs and is consequently contained in a small cap with base hyperplane π and half-space $Q(x) \geq 0$ (Lemma 74.1). Let $H^\alpha_{n,r}$ be the hemispheres of the given family, with half-spaces H^α, and consider the subsets $C^\alpha = \pi.F.H^\alpha$ of the euclidean n-space π.

It is clear that the sets C^α are closed, convex, and bounded (each point of C^α is interior to or on $S_{n,r}$), and since each $n+1$ of the hemispheres $H^\alpha_{n,r}$ have a point in common with C, it follows that each $n+1$ of the sets C^α have a non-null product. For

$$C . \prod_{i=1}^{n+1} H^{\alpha_i}_{n,r} = F.S_{n,r} \prod_{i=1}^{n+1} H^{\alpha_i},$$

and a point p of the (non-null) left-hand member belongs to $F . \prod_{i=1}^{n+1} H^{\alpha_i}$. Since the origin o and p ($p \in C$) are not on the same side of π, the ray op intersects π in a point p' which belongs to $\pi.F. \prod_{i=1}^{n+1} H^{\alpha_i}$.

We may conclude from the Helly theorem that a point q' of π exists that belongs to each of the sets C^α, and the intersection of $S_{n,r}$ with the ray oq' is a point q belonging to all the hemispheres of the theorem.

This lemma is generalized in the one that follows.

LEMMA 74.3. *Let k be any one of the integers $0, 1, 2,..., n+1$. If each $n+1+k$ members of a family of hemispheres of $S_{n,r}$ have a point in common with the product of a fixed set of $n+1-k$ independent hemispheres of $S_{n,r}$, then all the hemispheres have a point in common.*

Proof. If $k = 0$, the lemma reduces to the preceding one. Suppose $k > 0$ and assume the lemma true for each non-negative integer less than k.

Let $H^{(1)}_{n,r}, H^{(2)}_{n,r},..., H^{(n+1-k)}_{n,r}$ denote the fixed set of $n+1-k$ independent hemispheres of $S_{n,r}$, and denote the given family of hemispheres by $\{\mathscr{H}_{n,r}\}$.†

Case 1. For each element $\mathscr{H}_{n,r}$ of $\{\mathscr{H}_{n,r}\}$ the hemispheres

$$H_{n,r}^{(1)}, H_{n,r}^{(2)},\ldots, H_{n,r}^{(n+1-k)}, \mathscr{H}_{n,r}$$

are a dependent set.

Then each $\mathscr{H}_{n,r}$ contains the $S_{k-1,r}$ common to the hemispheres $H_{n,r}^{(i)}$ ($i = 1, 2,\ldots, n+1-k$), and hence all the hemispheres have this $S_{k-1,r}$ in common. Since $k > 0$, the lemma is proved in this case.

Case 2. A greatest integer r^* exists ($0 \leqslant r^* < k$) such that the $n+1-r^*$ hemispheres

$$H_{n,r}^{(1)}, H_{n,r}^{(2)},\ldots, H_{n,r}^{(n+1-k)}, \mathscr{H}_{n,r}^{(1)},\ldots, \mathscr{H}_{n,r}^{(k-r^*)} \tag{*}$$

form an independent set, with $\mathscr{H}_{n,r}^{(i)} \in \{\mathscr{H}_{n,r}\}$ ($i = 1, 2,\ldots, k-r^*$).

Now it is clear that any $n+1+r^*$ hemispheres selected from the set of hemispheres obtained from the family $\{\mathscr{H}_{n,r}\}$ by deleting

$$\mathscr{H}_{n,r}^{(i)} \quad (i = 1, 2,\ldots, k-r^*)$$

has a point in common with the product of the $n+1-r^*$ hemispheres in (*) because, by hypothesis, any $n+1+k = (k-r^*)+(n+1+r^*)$ members of $\{\mathscr{H}_{n,r}\}$ have a point in common with the product of the $n+1-k$ hemispheres $H_{n,r}^{(i)}$ ($i = 1, 2,\ldots, n+1-k$). Since $0 \leqslant r^* < k$, it follows by the inductive hypothesis that all the hemispheres have a point in common, and the lemma is proved.

If a finite subset of E_n contains an independent $(n+1)$-tuple, the convex extension of the subset is called an n-dimensional *convex polyhedron* of E_n. A boundary point p of an n-dimensional convex polyhedron P ($n > 1$) is a *vertex* of the polyhedron provided the set of all rays with initial point p, each of which is normal to a supporting hyperplane of P at p, is not contained in a hyperplane. A one-dimensional convex polyhedron has exactly two vertices, the end-points of the line segment that is the convex extension of the given finite subset of the E_1.

The convex extension of a subset of the vertices of an n-dimensional convex polyhedron P is a *sub-polyhedron* of P. A sub-polyhedron of P is a *face* provided it is contained in the boundary of P. One-dimensional faces of P are called *edges*.

† This notation for a family of hemispheres occurs only in the proof of this theorem, and the previously adopted use of this symbolism (Example 4, § 9) is hardly likely to lead to any confusion.

LEMMA 74.4. *If q is an interior point of an n-dimensional convex polyhedron P with vertices $p_1, p_2,..., p_m$ ($m \geq 2n-1$), then either* (1) *q is interior to an n-dimensional sub-polyhedron of P with $2n-1$ or fewer vertices, or* (2) *P has exactly $2n$ vertices collinear in pairs with q.*

Proof. Since each one-dimensional convex polyhedron has exactly two vertices, and they are collinear with each interior point of the polyhedron, alternative (2) holds and the lemma is valid for $n = 1$. Assume the lemma valid for k-dimensional convex polyhedra, where $k = 1, 2,..., n-1$, and suppose $m \geq 2n$ (if $m = 2n-1$, the lemma is trivial). Denote by p'_i the point in which the ray $p_i q$ intersects the boundary of P ($p'_i \neq p_i$) ($i = 1, 2,..., m$).

Case 1 *One of the points p'_i is interior to a k-dimensional face P_k of P*, $1 \leq k < n$. Then by the inductive hypothesis, p'_i is *interior* to a k-dimensional sub-polyhedron P'_k of P_k which has $2k$ or fewer vertices, and it follows that q is interior to the $(k+1)$-dimensional sub-polyhedron of P which is the convex extension $C(p_i, P'_k)$ of p_i together with those vertices of P whose convex extension is P'_k. Then q is interior to a $(k+1)$-dimensional sub-polyhedron of P with $2k+1 = 2(k+1)-1$ or fewer vertices. If $k+1 = n$, then alternative (1) holds and the lemma is proved.

Suppose $k+1 < n$. Now $C(p_i, P'_k)$ is not part of the boundary of P since $q \in C(p_i, P'_k)$ and q is an interior point of P. Two $(k+1)$-dimensional faces of P have the k-dimensional face P_k in common. Selecting vertices p_s, p_t of these $(k+1)$-dimensional faces, respectively, not belonging to P_k, the at least $(k+2)$-dimensional convex polyhedron $C(p_s, p_t, p_i, P'_k)$ of $2k+3 = 2(k+2)-1$ or fewer vertices clearly contains q in its interior. If $k+2 = n$, the lemma is proved; if $k+2 < n$, the process is repeated, and after at most $n-k$ steps an n-dimensional convex sub-polyhedron of P with $2n-1$ or fewer vertices is obtained containing q in its interior. Hence the lemma is proved in Case 1.

Case 2. *The situation described in Case 1 does not occur.* Then all the points p'_i are vertices of P, and m is even. If $m = 2n$, alternative (2) holds and the lemma is proved. Since by hypothesis $m \geq 2n-1$, then $m \geq 2n+2$ in case $m \neq 2n$. In that event select $2n$ vertices

$$p_1, p'_1, p_2, p'_2,..., p_n, p'_n$$

not lying in any E_{n-1}, with p_i, q, p'_i collinear ($i = 1, 2,..., n$). (Let the reader verify that such a choice is possible.) The n hyperplanes determined by the n sets of $n-1$ rays qp_i ($i = 1, 2,..., n$) have only q in common and consequently no $(2n+1)$th vertex p_{n+1} of P lies in all of these hyperplanes. Suppose $p_{n+1} \bar{\in} E_{n-1}(q, p_1, p_2, ..., p_{n-1})$.

Now p_{n+1} and p_n (or p_{n+1} and p'_n) are on opposite sides of the hyperplane $E_{n-1}(q, p_1, p_2, ..., p_{n-1})$ and so q is an interior point of the n-dimensional polyhedron that is the convex extension of the vertices

$$p_1, p'_1, p_2, p'_2, ..., p_{n-1}, p'_{n-1}, p_n, p_{n+1}$$

(or the vertices $p_1, p'_1, p_2, p'_2, ..., p_{n-1}, p'_{n-1}, p'_n, p_{n+1}$) of P and the ray $p_{n+1}q$ does not meet the boundary of that polyhedron in a vertex. Hence for $m \geqslant 2n+2$ the argument of Case 1 may be applied to show that alternative (1) holds, and the proof is complete.

75. Intersection theorems

A family of six hemispheres of $S_{2,r}$ may have a null intersection even though each five members of the family have a point in common. This is instanced by the six hemispheres of $S_{2,r}$ with centres $(\pm r, 0, 0)$, $(0, \pm r, 0)$, $(0, 0, \pm r)$, and this example has an obvious extension to a family of $2n+2$ hemispheres of $S_{n,r}$ with a null product, though each $2n+1$ members have a non-vacuous intersection. It is noteworthy, however, that if a family of hemispheres contains *more than* $2n+2$ members, there is a point common to all whenever there are points common to each $2n+1$ members of the family. This is proved in the following theorem.

THEOREM 75.1. *A family of more than $2n+2$ hemispheres of $S_{n,r}$ has a common point whenever each $2n+1$ members of the family have that property.*

Proof. Since the hemispheres of any family have a point in common whenever each $2n+2$ of them have a non-empty product (Lemma 74.3 with $k = n+1$), the theorem is proved when it is shown that each $2n+2$ members of the family have a common point.

Let $H_{n,r}^{(1)}, H_{n,r}^{(2)}, ..., H_{n,r}^{(2n+2)}$ be any $2n+2$ hemispheres of the given family, and adjoin a $(2n+3)$th hemisphere $H_{n,r}^{(2n+3)}$ of the family to the set. Let p_i be the centre of $H_{n,r}^{(i)}$ ($i = 1, 2, ..., 2n+3$) and consider the following two cases:

Case 1. The points $p_1, p_2, ..., p_{2n+3}$ lie on a hemisphere of $S_{n,r}$. If p is the centre of such a hemisphere, then $p_i p \leqslant \frac{1}{2}\pi r$ and hence p is a point of each of the hemispheres $H_{n,r}^{(i)}$ ($i = 1, 2, ..., 2n+3$).

Case 2. The points $p_1, p_2, ..., p_{2n+3}$ do not lie on a hemisphere of $S_{n,r}$. The $(n+1)$-dimensional convex polyhedron P obtained as the convex extension of the points $p_1, p_2, ..., p_{2n+3}$ has each of these points as a vertex and contains o in its interior. Applying Lemma 74.4 (for $n+1$) then clearly alternative (2) does not hold and so o is interior to an $(n+1)$-dimensional sub-polyhedron of P with $2n+1$ or fewer vertices. But the vertices of any such sub-polyhedron evidently do *not* lie on a hemisphere of $S_{n,r}$, which they must do since each $2n+1$ of the hemispheres intersect, and hence the set of centres has each $(2n+1)$-tuple on a hemisphere.

THEOREM 75.2. *Let k be any one of the integers $0, 1, 2,..., n$. If each $n+k+2$ members of a family of convex subsets of the $S_{n,r}$ have a common point, and if at least one member of the family does not contain an $S_{k,r}$, then there is a point common to all the sets of the family.*

Proof. Let $\{C\}$ denote the given family of convex subsets of $S_{n,r}$, and suppose $C_0 \in \{C\}$ that contains no $S_{k,r}$. Denote by $\{H_{n,r}\}$ the set of all hemispheres of $S_{n,r}$ each of which contains a member of $\{C\}$.

There is an independent set of $n-k+1$ members of $\{H_{n,r}\}$, say $H_{n,r}^{(i)}$ ($i = 1, 2,..., n-k+1$), whose product contains C_0, for in the contrary case the centres of all those hemispheres of $H_{n,r}$ whose product is C_0 lie in an S_{n-k-1} and hence their product C_0 contains an $S_{k,r}$ contrary to hypothesis.

Consider a set of hemispheres

$$H_{n,r}^{(1)}, H_{n,r}^{(2)},..., H_{n,r}^{(n-k+1)}, H_{n,r}^{(n-k+2)},..., H_{n,r}^{(2n+2)},$$

where the $n+k+1$ hemispheres adjoined to the first $n-k+1$ are selected arbitrarily from the set $\{H_{n,r}\}$, and contain, respectively, sets

$$C_{n-k+2}, C_{n-k+3},..., C_{2n+2}$$

of the family $\{C\}$. By hypothesis the product

$$C_0 \cdot C_{n-k+2} \cdot C_{n-k+3} \cdots C_{2n+2} \neq 0,$$

and so
$$\prod_{i=1}^{2n+2} H_{n,r}^{(i)} \neq 0.$$

It follows that the set $\{H_{n,r}\}$ of hemispheres satisfies the hypotheses of Lemma 74.3 and hence a point exists that belongs to each member of the set. Such a point evidently belongs to each set of the family $\{C\}$, and the theorem is proved.

THEOREM 75.3. *A family of more than $2n+2$ convex subsets of $S_{n,r}$ has a common point provided each $2n+1$ of the subsets has a non-null product.*

Proof. If each member of the given family $\{C\}$ is a hemisphere, the theorem follows from Theorem 75.1. If, on the other hand, a member C_0 of the family is not a hemisphere, then either C_0 does not contain any $S_{n-1,r}$ or C_0 is an $S_{n-1,r}$. In the former event, the theorem follows from Theorem 75.2 with $k = n-1$; in the latter case we make the inductive hypothesis of the theorem's validity for $S_{n-1,r}$ (with the induction anchored by the obvious truth of the theorem for the $S_{0,r}$).

Intersecting each member of $\{C\}$ with $C_0 = S^*_{n-1,r}$, each member of the family $\{C \cdot S^*_{n-1,r}\}$, $C \in \{C\}$, is a convex subset of $S^*_{n-1,r}$ *except for those members C of $\{C\}$ such that* $C \cdot S^*_{n-1,r} = S^*_{n-1,r}$ (for, according to Definition 73.2, an $S_{k,r}$ is *not* convex when it constitutes the whole space considered). Clearly $C \cdot S^*_{n-1,r} = S^*_{n-1,r}$ can occur for at most three (distinct) members of $\{C\}$; namely $S^*_{n-1,r}$ itself and the two (opposite) hemispheres of $S_{n,r}$ whose rim is $S^*_{n-1,r}$. Denote by $\{C\}'$ the family of sets obtained by deleting from $\{C\}$ each of those three sets that may be present.

The family $\{C \cdot S^*_{n-1,r}\}$, $C \in \{C\}'$, of convex subsets of $S^*_{n-1,r}$ is such that each $2n$ of them have a point in common, since the product of $S^*_{n-1,r}$ with each $2n$ of the members of $\{C\}$ is non null by hypothesis. In case $\{C \cdot S^*_{n-1,r}\}$ has at most $2n$ members then evidently there is a point common to all of its members, and any such point belongs to each member of the family $\{C\}$. If $\{C \cdot S^*_{n-1,r}\}$ has more than $2n$ members, a common point is ensured by the inductive hypothesis, and the theorem is proved.

THEOREM 75.4. *If a family of convex subsets $\{C\}$ of $S_{n,r}$ has a null product while each $2n+1$ members of the family have a common point, then the family consists of $n+1$ pairs of opposite hemispheres. If*

$$p_1, p_1^*, p_2, p_2^*, \ldots, p_{n+1}, p_{n+1}^*$$

are the centres of these hemispheres $(p_i p_i^* = \pi r \ (i = 1, 2, \ldots, n+1))$, the points $p_1, p_2, \ldots, p_{n+1}$ form an independent $(n+1)$-tuple.

Proof. It follows at once from Theorem 75.3 and the hypothesis that $\{C\}$ contains exactly $2n+2$ members. If any member C_0 of $\{C\}$ is not a hemisphere, an argument similar to that used in the preceding theorem (but simplified here since $\{C\}$ consists of exactly $2n+2$ sets) may be employed to show that C_0 cannot be an $S_{n-1,r}$. Hence any member of $\{C\}$ different from a hemisphere does not contain an $S_{n-1,r}$ and so the family has a non-null product by Theorem 75.2 (with $k = n-1$) in case such a member of $\{C\}$ exists.

Thus each of the $2n+2$ members of $\{C\}$ is a hemisphere, and since their product is null the $2n+2$ centres do not lie on a hemisphere. The convex extension of these $2n+2$ points of $S_{n,r}$ has o as an interior point. Since each $2n+1$ of the hemispheres intersect, each $(2n+1)$-tuple of the centres lies on a hemisphere and so Alternative (1) of Lemma 74.4 cannot hold. Hence Alternative (2) does, and the $2n+2$ points are formed by $n+1$ diametral point-pairs, $p_1, p_1^*, p_2, p_2^*, \ldots, p_{n+1}, p_{n+1}^*$.

Ch. VIII, § 75   CONVEX SUBSETS OF $S_{n,r}$ 201

If, finally, $p_1, p_2,..., p_{n+1}$ were dependent (and hence contained in an $S_{n-1,r}$), it is clear that the $2n+2$ hemispheres with centres

$$p_1, p_1^*, p_2, p_2^*,..., p_{n+1}, p_{n+1}^*$$

would have in common the $S_{0,r}$ formed by the diametral point-pair of $S_{n,r}$ consisting of the two poles of the $S_{n-1,r}$ containing $p_1, p_2,..., p_{n+1}$.

THEOREM 75.5. *Let $\{C\}$ denote a family of convex subsets of $S_{n,r}$, the diameter of each member of which is less than $\tfrac{2}{3}\pi r$. If every $n+1$ of the sets have a common point, then there is a point that belongs to every member of the family.*

Proof. Consider, first, a family $\{C\}$ of arcs of $S_{1,r}$ with $\delta(C) < \tfrac{2}{3}\pi r$, $C \in \{C\}$. If C_0 is a member of $\{C\}$, with middle point c, let c^* denote the point of $S_{1,r}$ diametral to c, and T the tangent to $S_{1,r}$ at c. Clearly c^* does not belong to any arc of $\{C\}$ and by projecting $\{C\}$ on T from c^* we obtain a family of segments on T, each two of which intersect. By Helly's theorem a point p of T exists that belongs to all of those segments, and the line joining p to c^* meets $S_{1,r}$ in a point that lies in every member of the family $\{C\}$. Hence the theorem is valid for $n = 1$, which anchors an inductive hypothesis of its validity for $S_{n-1,r}$.

Let, now, $\{C\}$ be a family of convex subsets of $S_{n,r}$ satisfying the hypotheses of the theorem. The desired conclusion follows from Theorem 75.2 (with $k = 0$) when it is shown that each $n+2$ members of $\{C\}$ intersect, since obviously no member of $\{C\}$ contains a diametral point-pair.

If $C_0, C_1,..., C_{n+1}$ are any $n+2$ sets of $\{C\}$, suppose $C_0 . D = 0$, where $D = C_1 . C_2 ... C_{n+1} \neq 0$. Then D is a convex subset of $S_{n,r}$ with no points in common with C_0 and it follows that a subspace $S_{n-1,r}^*$ of $S_{n,r}$ exists *separating* C_0 and D (that is, $C_0 . S_{n-1,r}^* = D . S_{n-1,r}^* = 0$, and C_0, D are contained respectively, in $H_{n,r}, H_{n,r}^*$, the two opposite hemispheres of $S_{n,r}$ with rim $S_{n-1,r}^*$).

Each of the $n+1$ sets

$$D_1 = C_2 . C_3 ... C_{n+1}, \quad D_2 = C_1 . C_3 ... C_{n+1} \quad ..., \quad D_{n+1} = C_1 . C_2 ... C_n$$

contains the non-null set D and a point of C_0 (since each $n+1$ members of $\{C\}$ have a non-null product). Consequently each of these sets must intersect $S_{n-1,r}^*$ and so $D_i' = D_i . S_{n-1,r}^* \neq 0$ ($i = 1, 2,..., n+1$). It follows that each n of the $n+1$ convex subsets $C_1 . S_{n-1,r}^*, C_2 . S_{n-1,r}^*,..., C_{n+1} . S_{n-1,r}^*$ of $S_{n-1,r}^*$ (each with diameter less than $\tfrac{2}{3}\pi r$) has a non-null product, and so (by the inductive hypothesis) the product

$$\prod_{i=1}^{n+1} C_i \cdot S_{n-1,r}^* = S_{n-1,r}^* \prod_{i=1}^{n+1} C_i = S_{n-1,r}^* \cdot D \neq 0.$$

This contradiction establishes the theorem.

Theorem 75.5 shows that *locally* the $S_{n,r}$ has the same 'Helly number' as the E_n. For $n = 1$ the diameter bound $\tfrac{2}{3}\pi r$ cannot be increased. It may be shown that for $n = 2$ the largest diameter bound for which the Helly number $n+1$ is valid is $r \cdot \cos^{-1}(-3^{-\frac{1}{2}})$.

EXERCISE

If C_1, C_2 are mutually exclusive, convex subsets of $S_{n,r}$, neither of which contains a diametral point-pair, then an $(n-1)$-dimensional subspace $S_{n-1,r}$ of $S_{n,r}$ exists such that $C_1 \cdot S_{n-1,r} = C_2 \cdot S_{n-1,r} = 0$, and C_1, C_2 are contained, respectively, in $H_{n,r}$, $H_{n,r}^*$, the two opposite hemispheres of $S_{n,r}$ with rim $S_{n-1,r}$.

76. Congruence indices of hemispheres and small caps

The intersection theorems of the preceding section have immediate application to the determination of congruence indices for certain subsets of $S_{n,r}$ ($n > 0$).

THEOREM 76.1. *A hemisphere $H_{n,r}$ of $S_{n,r}$ has best congruence indices $(2n+1, 1)$ with respect to the class $\{S\}$ of semimetric spaces.*

Proof. Since $H_{1,r}$ is a metric segment, it has best indices $(3, 1)$ with respect to $\{S\}$ (Exercise 3, § 59), so the theorem is valid for $n = 1$. Suppose $n > 1$ and let S be any semimetric space of more than $2n+2$ points, with each $(2n+1)$-tuple imbeddable in $H_{n,r}$. Since $n > 1$, $2n+1 \geqslant n+3$ and hence $S \Subset S_{n,r}$ (Theorem 39.2). We write

$$S \approx S' \subset S_{n,r}.$$

Let $\{H_{n,r}\}$ be the family of hemispheres with the points of S' as centres. Since each $(2n+1)$-tuple of S' lies in a hemisphere, each $2n+1$ of the hemispheres of the family $\{H_{n,r}\}$ have a non-null product. Applying Theorem 75.1, at least one point p exists that belongs to every hemisphere of the family, and the hemisphere of $S_{n,r}$ with p as centre clearly contains every point of S'. Hence S is congruently imbeddable in a hemisphere of $S_{n,r}$ and so $H_{n,r}$ has indices $(2n+1, 1)$ with respect to $\{S\}$.

It is clear that $H_{n,r}$ does not have congruence indices $(2n+1, 0)$ with respect to $\{S\}$ (see Remark 1 that follows). Let the reader complete the proof of the theorem by showing that $H_{n,r}$ does not have indices $(2n, k)$ with respect to $\{S\}$.

Remark 1. It follows from the preceding theorem that if S' is any subset of $S_{n,r}$ containing more than $2n+2$ points, and each $(2n+1)$-tuple of S' can be *covered* by some hemisphere of $S_{n,r}$, then the whole set S' can be covered by a hemisphere of $S_{n,r}$. Whence any subset of $S_{n,r}$ can be covered by a hemisphere whenever each of its $(2n+2)$-tuples has that property. The $2n+2$ points $(\pm r, 0,..., 0)$, $(0, \pm r,..., 0),..., (0, 0,..., \pm r)$ of $S_{n,r}$ have each $(2n+1)$-tuple coverable by an $H_{n,r}$ without itself having that property, so the requirement that S' contain *more than* $2n+2$ points is essential; that is, $H_{n,r}$ does not have congruence order $2n+1$ (indices $(2n+1, 0)$) with respect to subsets of $S_{n,r}$.

Remark 2. Any set of $2n+2$ points of $S_{n,r}$ that is not contained in a hemisphere though each of its $(2n+1)$-tuples is, consists of $n+1$ diametral point-pairs. This is an immediate consequence of Lemma 74.4 (applied to the origin o of the E_{n+1}), for since each $(2n+1)$-tuple of the $2n+2$ points is contained in some $H_{n,r}$, Alternative (1) of that lemma cannot hold, and so Alternative (2) does.

COROLLARY. *The $H_{n,r}$ has congruence order $2n+2$ with respect to $\{S\}$.*

LEMMA 76.1. *A small cap $K_{n,\rho}$ of $S_{n,r}$ ($\rho < \frac{1}{2}\pi r$) has congruence order $n+2$ with respect to the class of subsets of $S_{n,r}$.*

Proof. If A is a subset of $S_{n,r}$ with each of its $(n+2)$-tuples imbeddable in $K_{n,\rho}$, then each $n+2$ members of the set of all caps $K_{n,\rho}(p)$ with centres at the points p of A intersect. Denoting by $C(p)$ the convex extension in the E_{n+1} of the cap $K_{n,\rho}(p)$ with centre p, the family $\{C(p)\}$, $p \in A$, is a family of convex bodies of E_{n+1} with each $n+2$ members intersecting. It follows by the Helly theorem that there is a point c that belongs to each member of the family.

The point c is distinct from the origin o (the centre of the $S_{n,r}$), since o does not belong to any of the sets $C(p)$, and the ray $\overrightarrow{oc}$ intersects $S_{n,r}$ in a point c' which is clearly common to all the caps $K_{n,\rho}(p)$, $p \in A$. Hence the distances $c'p$, $p \in A$, are at most ρ, and so A is contained in the cap $K_{n,\rho}(c')$.

We have shown that if each $n+2$ points of any subset A of $S_{n,r}$ is coverable by a spherical cap $K_{n,\rho}$ of fixed radius ρ less than $\frac{1}{2}\pi r$, then the whole set A may be covered by such a cap. Since each $n+1$ points of an equilateral $(n+2)$-tuple of $S_{n,r}$ is coverable by a small cap, but the whole set is not, it follows that $K_{n,\rho}$ does not have congruence order $n+1$ with respect to subsets of $S_{n,r}$.

THEOREM 76.2. *A small cap $K_{n,\rho}$ of $S_{n,r}$ ($\rho < \frac{1}{2}\pi r$) has congruence indices $(n+2, n)$ with respect to the class $\{S\}$ of semimetric spaces.*

Proof. The theorem is an immediate consequence of the preceding lemma when it is shown that each semimetric space S containing more than $2n+2$ points and having each $(n+2)$-tuple imbeddable in $K_{n,\rho}$, is imbeddable in $S_{n,r}$. Now if such a semimetric space S is not imbeddable in $S_{n,r}$, it is a pseudo-$S_{n,r}$ set without diametral point-pairs, and hence the set S contains an equilateral $(n+2)$-tuple with edge $r\cos^{-1}[-1/(n+1)]$ (Exercise 3, § 72). But such an $(n+2)$-tuple is surely not imbeddable in $K_{n,\rho}$. To establish this, suppose that p'_0 is the centre of a small cap $K_{n,\rho}$ containing an equilateral $(n+2)$-tuple $p'_1, p'_2, \ldots, p'_{n+2}$ with edge

$$r\cos^{-1}[-1/(n+1)].$$

In the vanishing determinant

$$\Delta_{n+3}(p'_0, p'_1, \ldots, p'_{n+2}) = |\cos p'_i p'_j/r| \quad (i, j = 0, 1, \ldots, n+2)$$

the elements $\cos p'_0 p'_k/r = \cos p'_k p'_0/r$ $(k = 1, 2, \ldots, n+2)$ are all *positive*, and $\cos p'_i p'_j/r = \cos p'_j p'_i/r = -1/(n+1)$ $(i, j = 1, 2, \ldots, n+2)$. Adding the 2nd, 3rd,..., $(n+2)$th rows (columns) to the last row (column) yields

$$\Delta_{n+3}(p'_0, p'_1, \ldots, p'_{n+2}) = -\Big(\sum_{k=1}^{n+2} \cos p'_0 p'_k/r\Big)^2 \Delta_{n+1}(-1/(n+1)).$$

But since

$$\Delta_{n+1}(-1/(n+1)) = (n+2)^n/(n+1)^{n+1} \neq 0,$$

and

$$\cos p'_0 p'_k/r > 0 \quad (k = 1, 2, \ldots, n+2),$$

the determinant $\Delta_{n+3}(p'_0, p'_1, \ldots, p'_{n+2})$ cannot vanish. This contradiction completes the proof of the theorem.

THEOREM 76.3. *The cap $K_{n,\rho}$ ($\rho < r\cos^{-1}[1/(n+1)]$, $n > 1$) has congruence indices $(n+2, n-1)$ with respect to the class $\{S\}$ of semimetric spaces.*

Proof. It suffices to show that if S is any semimetric space of more than $2n+1$ pairwise distinct points, and each $(n+2)$-tuple of S is imbeddable in $K_{n,\rho}$ ($\rho < r\cos^{-1}[1/(n+1)]$), then S is imbeddable in $S_{n,r}$.

If the contrary be assumed, then S is pseudo-$S_{n,r}$ with no two points diametral and, by Exercise 2, § 72, contains an equilateral $(n+1)$-tuple with edge $r\cos^{-1}[-1/(n+1)]$. But the smallest cap containing such an equilateral $(n+1)$-tuple has radius $\rho = r\cos^{-1}[1/(n+1)]$. (This is the spherical radius of the cap of $S_{n,r}$ whose rim is the small $(n-1)$-dimensional hypersphere in which $S_{n,r}$ is cut by the E_n determined by these $n+1$ points.) Hence S is imbeddable in $S_{n,r}$ and the theorem follows from Lemma 76.1.

Applying the preceding theorem for $n = 2$, we see that a cap $K_{2,\rho}$ ($\rho < r\cos^{-1}\frac{1}{3}$) of $S_{2,r}$ has indices $(4, 1)$ with respect to $\{S\}$ and hence behaves in that regard like the euclidean plane. A little consideration shows that the indices $(4, 1)$ are the best possible.

It is not the case, however, that the indices given by $(n+2, n-1)$ are the best for every value of n. Thus when $n = 3$ the indices $(5, 2)$ given for the cap $K_{3,\rho}$ ($\rho < r\cos^{-1}\frac{1}{4}$) can indeed be bettered, since it may be shown that the symbol $(5, 1)$ is valid for such a cap. If $n = 4$, the indices $(6, 3)$ given by Theorem 76.3 can be reduced to $(6, 2)$ and perhaps even to $(6, 1)$.

It would be of interest to find the best congruence indices for caps $K_{n,\rho}$, $\rho < r\cos^{-1}[1/(n+1)]$, and to ascertain for each n the radius of the largest cap of $S_{n,r}$ which, like euclidean n-space, has $(n+2, 1)$ as its best congruence symbol with respect to the class of all semimetric spaces.

EXERCISES

1. Classify those quadruples of $S_{2,r}$ for which every distance is $r\cos^{-1}(\pm\frac{1}{3})$.
2. Show that the congruence symbol $(4, 2)$ given by Theorem 76.2 for caps $K_{2,\rho}$ of $S_{2,r}$ with $\rho < \frac{1}{2}\pi r$ is the best symbol when $\cos^{-1}\frac{1}{3} \leqslant \rho < \frac{1}{2}\pi r$. Such caps have also the symbol $(5, 0)$, not implied by $(4, 2)$, and the two symbols form a complete set (§ 37).
3. Prove that $K_{3,\rho}$, $\rho < r\cos^{-1}\frac{1}{4}$, has the congruence symbol $(5, 1)$.

REFERENCE

§ 73–76. Robinson [5].

CHAPTER IX

IMBEDDING AND CHARACTERIZATION THEOREMS FOR ELLIPTIC SPACE

77. The notion of δ-supplementation

IF Σ denotes the pointset of a semimetric space S with finite diameter d, let Σ^2 be the set of all unordered pairs of points of Σ and

$$\Sigma^2 = \Sigma_1 + \Sigma_2, \qquad \Sigma_1 \cdot \Sigma_2 = 0,$$

a decomposition of Σ^2 into two mutually exclusive subsets. Corresponding to this decomposition, a real, non-negative function $F(x)$ is defined:

$$\begin{aligned} F(x) &= x, & x &= pq; & (p,q) &\in \Sigma_1, \\ F(x) &= \delta - x, & x &= pq; & (p,q) &\in \Sigma_2, \end{aligned} \qquad (*)$$

where δ is a fixed real number ($\delta \geq d$), and pq denotes the distance of p, q in S.

Consider, now, the metric transform $S^* = F(S)$ of S by the function $F(x)$ (§ 52). It is clear that points of Σ with positive distance in S may have zero distance in S^*, while two coincident points of Σ might have a positive distance in S^*. If $p, q \in \Sigma$ with $pq^* = F(pq) = 0$, identification of p and q is permissible *provided for each point y of Σ we have $py^* = qy^*$*. This is assured in case $\delta > d$, for then $pq^* = 0$ implies $pq = 0$ and so $p = q$ in Σ. Consequently for each y of Σ the point-pairs $(p,y), (q,y)$ are identical and $py^* = qy^*$. We denote by Σ^* the pointset of S^* after the identification of those (and only those) points p, q for which

$$pq^* = F(pq) = 0.$$

In general, $\Sigma^* \neq \Sigma$.

The identification of points p, q with $pq^* = 0$ may be valid even when $\delta = d$. An important example is furnished by $S = S_{n,r}$ and $S^* = F(S)$, where $(p,q) \in \Sigma_1$ or $(p,q) \in \Sigma_2$ according as $pq \leq \tfrac{1}{2}\pi r$ or $pq > \tfrac{1}{2}\pi r$, respectively, and

$$\begin{aligned} F(x) &= x, & x &= pq; & (p,q) &\in \Sigma_1, \\ F(x) &= \pi r - x, & x &= pq; & (p,q) &\in \Sigma_2. \end{aligned} \qquad (**)$$

If $pq^* = F(pq) = 0$ and $p \neq q$ in $S_{n,r}$, then $pq = \pi r$, and so for each point y of $S_{n,r}$ $py + yq = \pi r$. If $py \leq \tfrac{1}{2}\pi r$, then $qy \geq \tfrac{1}{2}\pi r$; whence

$$py^* = py = \pi r - qy = qy^*.$$

In case $qy \leq \tfrac{1}{2}\pi r$ the result is obtained in a similar manner. Identifying those points in S^* with $pq^* = 0$ gives the n-dimensional elliptic space

$\mathscr{E}_{n,r}$ (Example 3, §9). We shall denote the special function $F(x)$ defined in (**) by $\mathfrak{F}(x)$.

DEFINITION 77.1. *A space $S^* = F(S)$ with a pointset Σ^* in which points p, q are identified if and only if $pq^* = F(pq) = 0$, and $F(x)$ is defined as in (*), is called a δ-supplement of S (more precisely, the δ_F-supplement of S).*

We shall write $S^* = \sup \delta_F(S)$. It follows from the above remarks that $\mathscr{E}_{n,r} = \sup \pi r_{\mathfrak{F}}(S_{n,r})$.

EXERCISES

1. Prove that if $\delta > d$, each δ-supplement of a semimetric space of finite diameter d is semimetric.
2. Show that $S_{n,r}$ is a πr supplement of $\mathscr{E}_{n,r}$.

78. Imbedding theorems for $\mathscr{E}_{n,r}$

The observation made in the preceding section that the elliptic space $\mathscr{E}_{n,r}$ is the $\pi r_{\mathfrak{F}}$-supplement of the convex n-sphere $S_{n,r}$ is the basis for the imbedding theorems developed in this section.

THEOREM 78.1. *A semimetric space S is congruently contained in $\mathscr{E}_{n,r}$ if and only if (i) no distance in S exceeds $\tfrac{1}{2}\pi r$, and (ii) there exists at least one function F such that the δ_F-supplement of S is congruently imbeddable in $S_{n,r}$, where $\delta = \pi r$.*

Proof. If $S \subseteqq \mathscr{E}_{n,r}$, let E be a subset of $\mathscr{E}_{n,r}$ such that $S \approx E$. Since

$$\mathrm{diam}(S) = \mathrm{diam}(E) \leqslant \mathrm{diam}(\mathscr{E}_{n,r}) = \tfrac{1}{2}\pi r,$$

condition (i) is satisfied. Further, from $S_{n,r} = \sup \pi r(\mathscr{E}_{n,r})$ (Exercise 2, §77), it follows that a subset S' of $S_{n,r}$ exists such that

$$S' = \sup \pi r(E) \approx \sup \pi r(S),$$

which is condition (ii).

On the other hand, suppose that a semimetric space S satisfies (i), (ii) and let S' be a subset of $S_{n,r}$ such that $S' \approx \sup \pi r(S)$. Now a subset E of $\mathscr{E}_{n,r}$ exists with

$$E = \sup \pi r_{\mathfrak{F}}(S') \approx \sup \pi r_{\mathfrak{F}}[\sup \pi r(S)].$$

The proof is completed by showing that

$$S = \sup \pi r_{\mathfrak{F}}[\sup \pi r(S)].$$

To this end, denote distance of p, q in $\sup \pi r(S)$ by $d(p,q)$ and in $\sup \pi r_{\mathfrak{F}}[\sup \pi r(S)]$ by $d^*(p,q)$. The same argument shows that the two pointsets and the two distance-sets are identical. For if $p = q$ in S, then $pq = 0$ and $d(p,q) = 0$ or $d(p,q) = \pi r$. In either case $d^*(p,q) = 0$, and so $p = q$ in S implies $p = q$ in $\sup \pi r_{\mathfrak{F}} [\sup \pi r(S)]$.

Finally, if p, q are distinct points of S, then $0 < pq \leqslant \tfrac{1}{2}\pi r$. If $d(p,q) = pq$, then since $pq \leqslant \tfrac{1}{2}\pi r$, $d^*(p,q) = pq$; while if

$$d(p,q) = \pi r - pq \geqslant \tfrac{1}{2}\pi r,$$

then $d^*(p,q) = \pi r - (\pi r - pq) = pq > 0$, and the theorem is proved.

Thus imbedding in $\mathscr{E}_{n,r}$ is referred to imbedding πr-supplements of semimetric spaces in $S_{n,r}$.

THEOREM 78.2. *A semimetric m-tuple* $p_1, p_2, \ldots, p_m$ *is congruently contained in* $\mathscr{E}_{n,r}$ *if and only if* (i) $p_i p_j \leqslant \tfrac{1}{2}\pi r$ $(i,j = 1, 2, \ldots, m)$, *and* (ii) *there exists a symmetric square matrix* $\epsilon = (\epsilon_{ij})$, $\epsilon_{ij} = \epsilon_{ji} = \pm 1$, $\epsilon_{ii} = 1$ $(i,j = 1, 2, \ldots, m)$ *such that the determinant* $|\epsilon_{ij} \cos p_i p_j / r|$ $(i,j = 1, 2, \ldots, m)$ *has rank not exceeding* $n+1$, *with all non-vanishing principal minors positive*.

Proof. Suppose conditions (i), (ii) satisfied. Since $\epsilon_{ij} \cos p_i p_j / r$ equals either $\cos p_i p_j / r$ or $\cos(\pi - p_i p_j / r)$, it follows that a πr-supplement of the m-tuple $p_1, p_2, \ldots, p_m$ is congruently imbeddable in $S_{n,r}$ (corollary, Theorem 63.1). Hence the m-tuple is congruently imbeddable in $\mathscr{E}_{n,r}$ (Theorem 78.1).

If, on the other hand, $p_1, p_2, \ldots, p_m \Subset \mathscr{E}_{n,r}$, then (i) is obviously satisfied and a subset $p'_1, p'_2, \ldots, p'_m$ of $S_{n,r}$ exists which is a πr-supplement of $p_1, p_2, \ldots, p_m$. Then $p'_i p'_j / r$ equals either $p_i p_j / r$ or $\pi - p_i p_j / r$ and

$$\cos p'_i p'_j / r = \epsilon_{ij} \cos p_i p_j / r, \qquad \epsilon_{ij} = \epsilon_{ji} = \pm 1, \qquad \epsilon_{ii} = 1$$

$(i,j = 1, 2, \ldots, m)$. Applying the corollary, Theorem 63.1, to the subset $p'_1, p'_2, \ldots, p'_m$ of $S_{n,r}$ shows that

$$|\epsilon_{ij} \cos p_i p_j / r| = |\cos p'_i p'_j / r| \quad (i,j = 1, 2, \ldots, m)$$

has the properties demanded in (ii).

We shall refer to (ϵ_{ij}) as an *epsilon matrix*.

Since $\mathscr{E}_{n,r}$ is compact and metric, each separable semimetric space is imbeddable in $\mathscr{E}_{n,r}$ whenever each of its finite subsets has that property (Theorem 60.2). From this fact and Theorem 78.2, we have the first solution of the imbedding problem for elliptic space (with respect to the class of *separable* semimetric spaces).†

---

† Theorem 78.1, which expresses the imbedding of *any* semimetric space in $\mathscr{E}_{n,r}$ in terms of the existence of a function F, does *not* characterize metrically subsets of $\mathscr{E}_{n,r}$ with respect to the class $\{S\}$ of semimetric spaces in the sense of Definition 36.1.

EXERCISES

1. If p_1, p_2, p_3 are a semimetric triple with $p_i p_j \leqslant \frac{1}{2}\pi r$ ($i, j = 1, 2, 3; r > 0$), let $\Delta^*(p_1, p_2, p_3)$ denote the determinant $|\epsilon_{ij} \cos p_i p_j/r|$ ($i, j = 1, 2, 3$) with every $\epsilon_{ij} = 1$ except $\epsilon_{23} = \epsilon_{32} = -1$. Show that $p_1, p_2, p_3 \in \mathscr{E}_{1,r}$ if and only if

$$\Delta(p_1, p_2, p_3) = |\cos p_i p_j/r| \quad (i, j = 1, 2, 3)$$

vanishes or $\Delta^* = 0$. Prove that each three points of $\mathscr{E}_{1,r}$ are imbeddable in $S_{2,r}$ and in $S_{1, \frac{1}{2}r}$.

2. Prove that $\mathscr{E}_{n,r} \approx S_{n,r'}$ if and only if $n = 1$ and $r' = \frac{1}{2}r$.

79. Some metric peculiarities of elliptic space

Though elliptic space is, as a πr-supplement of $S_{n,r}$ by the function $\mathfrak{F}$, closely related to the $S_{n,r}$ (the two spaces being, in fact, indistinguishable in the small) they differ in many important respects (both topological and metric) when considered globally. We note in this section some of the peculiarities of elliptic space that greatly complicate its metric study.

1. *Distinction between congruence and superposability.* In previous sections concerned with the euclidean and spherical spaces we made much use of the property enjoyed by those spaces that *any congruence between any two subsets can be extended to a motion* (e.g. Property IV, § 38), and so the notions of 'congruence' and 'superposability' are logically identical in them. In another chapter we shall make an intensive comparative study of these two notions in elliptic space, but we note here that not only do congruences of subsets of $\mathscr{E}_{2,r}$ exist which cannot be extended to a motion, but two congruent subsets of $\mathscr{E}_{2,r}$ may not even be superposable. If, for example, $p, q, s \in S_{2,r}$ with $pq = 6\pi r/18$, $qs = 7\pi r/18$, $ps = 8\pi r/18$, and $p', q', s' \in S_{2,r}$ with $p'q' = pq$, $q's' = qs$, $p's' = 10\pi r/18$, the process of πr-supplementation with respect to $\mathfrak{F}$ yields two congruent but not superposable triples of $\mathscr{E}_{2,r}$.†

A remarkable example of congruence without superposability is furnished by two triples p, q, s and p', q', s' of $S_{2,r}$ with *spherical distances* $pq = qs = ps = p'q' = q's' = \frac{1}{3}\pi r$, $p's' = \frac{2}{3}\pi r$. As points of $\mathscr{E}_{2,r}$ these triples are both equilateral with edge $\frac{1}{3}\pi r$ and hence are congruent to each other. But p', q', s' lie in an elliptic line $\mathscr{E}_{1,r}$ while p, q, s do not. The triples are clearly not superposable.

2. *Distinction between 'contained in' and 'congruently contained in' for subsets of $\mathscr{E}_{n,r}$.* If a subset E of a euclidean space E_n is congruently contained in a subspace E_k ($k \leqslant n$) (that is, E is congruent with a subset of an E_k), then there is a subspace E_k of E_n that contains E. Hence for the E_n (and also for the $S_{n,r}$) 'contained in' and 'congruently

---

† We shall refer to $\mathscr{E}_{1,r}$ and $\mathscr{E}_{2,r}$ as an elliptic line and plane, respectively.

contained in' are logically equivalent when applied to *subsets* of the space.

This is not the case in elliptic space. The second of the two examples in (1) exhibits three points p, q, s that are congruently contained in an elliptic line but do not lie in any elliptic line. So far as the elliptic line is concerned, the difficulty brought about by this circumstance is not serious, for we shall show later that if *four* points of $\mathscr{E}_{n,r}$ are congruently contained in an elliptic line there is an $\mathscr{E}_{1,r}$ containing them. But for $n > 1$ the distinction between the two notions in $\mathscr{E}_{n,r}$ cannot be eliminated by considering subsets containing 'enough' points, for we shall see in a later chapter that *for every integer $k > 1$ and every cardinal number m ($k+1 < m \leqslant \mathfrak{c}$) each k-dimensional elliptic space $\mathscr{E}_{k,r}$ contains a subset of power m which is congruent with a subset of $\mathscr{E}_{n,r}$ (n, finite) that does not lie in any $\mathscr{E}_{k,r}$.*

3. *Dependence in the usual sense not a congruence invariant.* In spaces E_n and $S_{n,r}$ a set of m points is dependent provided it is contained in a subspace of dimension $m-2$. From (2) it is clear that dependence in this sense is not a congruence invariant in $\mathscr{E}_{n,r}$. Six points, for example, that lie in an $\mathscr{E}_{5,r}$ and are not contained in any lower dimensional subspace may, nevertheless, be congruent with six points of $\mathscr{E}_{2,r}$. Thus there is no possibility of ascertaining the dimension of the lowest-dimensional subspace containing an elliptic set from a complete knowledge of the metric structure of the set alone. This suggests the desirability of defining in $\mathscr{E}_{n,r}$ a metrically invariant notion of dependence which will lead to metric definitions of the subspaces.

4. *The non-linearity of the equidistant locus.* The locus of points of $\mathscr{E}_{2,r}$ equidistant from two distinct points consists of two mutually perpendicular lines. This circumstance affects the character of a metric basis for the plane, since clearly no set contained in two such lines can form such a basis. In particular, no three points of $\mathscr{E}_{2,r}$ form a metric basis for the plane.

An important fifth peculiarity of elliptic space will be discussed in the next section.

EXERCISES

1. State some metric and topological differences of the $\mathscr{E}_{n,r}$ and $S_{n,r}$, $n > 1$.
2. Can any congruence between any two subsets of Hilbert space be extended to a motion?
3. Verify the assertion made in (1) concerning the non-superposability of the two pairs of congruent triples defined there.
4. Show that each quadruple of $\mathscr{E}_{1,r}$ contains at least two linear triples.

5. Prove that the locus of points of $\mathscr{E}_{2,r}$ equidistant from two distinct points consists of two mutually perpendicular lines. What is this locus for two distinct points of $\mathscr{E}_{n,r}$?

6. Show that congruence and superposability are not equivalent in the space $M_n^{(p)}$ of Exercise 6, § 9 for $p = 1$.

80. Equilateral subsets

A complete analysis of equilateral subsets of $\mathscr{E}_{n,r}$ has not yet been obtained for $n > 3$; indeed, the interesting novelties presented by such subsets in $\mathscr{E}_{2,r}$ and $\mathscr{E}_{3,r}$ have only recently come to light.

Since $\mathscr{E}_{1,r}$ is congruent with the convex circle $S_{1,\frac{1}{2}r}$, the equilateral subsets of the former are merely those of the latter; that is, all equilateral triples of $\mathscr{E}_{1,r}$ have edge $\frac{1}{3}\pi r$, and each two are superposable.

Let us call two equilateral subsets of $\mathscr{E}_{2,r}$ of the *same type* provided they are superposable when they are congruent. Then all equilateral triples of $\mathscr{E}_{2,r}$ fall into exactly two types. These two types are characterized by the matrices

$$(\epsilon_{ij}), \quad \epsilon_{ij} = \epsilon_{ji} = 1 \quad (i,j = 1, 2, 3)$$

and (ϵ_{ij}), $\epsilon_{12} = \epsilon_{13} = \epsilon_{21} = \epsilon_{31} = 1$, $\epsilon_{23} = \epsilon_{32} = -1$,

in a manner to be described later. For each number c, $0 < c \leq \frac{1}{2}\pi r$, the $\mathscr{E}_{2,r}$ contains equilateral triples of edge c.

Turning to equilateral quadruples, if $p_1, p_2, p_3, p_4 \in \mathscr{E}_{2,r}$ with

$$\cos p_i p_j / r = a, \quad 0 \leq a < 1 \quad (i,j = 1, 2, 3, 4; i \neq j),$$

then by Theorem 78.2 an epsilon matrix (ϵ_{ij}) $(i,j = 1, 2, 3, 4)$ exists such that the determinant

$$\begin{vmatrix} 1 & a & a & a \\ a & 1 & \epsilon_{23}a & \epsilon_{24}a \\ a & \epsilon_{23}a & 1 & \epsilon_{34}a \\ a & \epsilon_{24}a & \epsilon_{34}a & 1 \end{vmatrix}$$

has rank 3, with all non-vanishing principal minors positive, since obviously $\epsilon_{12}, \epsilon_{13}, \epsilon_{14}$, may be taken as 1.

Examining the eight possibilities presented by choosing

$$\epsilon_{ij} = \pm 1 \quad (i,j = 2, 3, 4; i \neq j)$$

it is found that the determinant has one of the values

$$(1-a)^3(1+3a),$$

$$(1+a)^3(1-3a),$$

$$(1-a^2)(1-5a^2),$$

and since $0 \leqslant a < 1$, the determinant vanishes if and only if $a = \frac{1}{3}$ or $a = 1/\sqrt{5}$. For each of these values all non-vanishing principal minors are positive, and so the $\mathscr{E}_{2,r}$ contains equilateral quadruples with edge $r\cos^{-1}\frac{1}{3}$ and also equilateral quadruples with edge $r\cos^{-1}1/\sqrt{5}$, but no others. Two equilateral quadruples with edge $r\cos^{-1}1/\sqrt{5}$ are superposable, though not every congruence between two such quadruples can be extended to a motion. On the other hand, every congruence between two equilateral quadruples with edge $r\cos^{-1}\frac{1}{3}$ can be extended to a motion.

THEOREM 80.1. *The $\mathscr{E}_{2,r}$ contains a metrically unique equilateral quintuple. It has edge $r\cos^{-1}1/\sqrt{5}$.*

Proof. Since each quadruple of an equilateral quintuple is equilateral, it follows from the above remarks that any equilateral quintuple of $\mathscr{E}_{2,r}$ has $r\cos^{-1}\frac{1}{3}$ or $r\cos^{-1}1/\sqrt{5}$ for edge.

If $\epsilon_{i,j}$ ($i, j = 2, 3, 4, 5$) can be chosen so that for $a = \frac{1}{3}$ the determinant

$$\begin{vmatrix} 1 & a & a & a & a \\ a & 1 & \epsilon_{23}a & \epsilon_{24}a & \epsilon_{25}a \\ a & \epsilon_{23}a & 1 & \epsilon_{34}a & \epsilon_{35}a \\ a & \epsilon_{24}a & \epsilon_{34}a & 1 & \epsilon_{45}a \\ a & \epsilon_{25}a & \epsilon_{35}a & \epsilon_{45}a & 1 \end{vmatrix} \qquad (*)$$

has rank 3, then the vanishing of the cofactor of the element in the last row and column gives $\epsilon_{23} = \epsilon_{24} = \epsilon_{34} = -1$. Examining the cofactor $[4, 4]$ yields $\epsilon_{25} = \epsilon_{35} = -1$, and from the vanishing of $[3, 3]$ we have $\epsilon_{45} = -1$. But then the determinant has the value $-(\frac{1}{3})(\frac{4}{3})^4$ and fails to vanish.

Examining the determinant in $(*)$ for $a = 1/\sqrt{5}$ it is seen that for every choice of the epsilons that results in each fourth-order principal minor vanishing, the determinant may (by interchange of rows (columns) and/or the multiplication of appropriate rows and the same numbered columns by -1) be put in the form

$$\begin{vmatrix} 1 & a & a & a & a \\ a & 1 & -a & a & -a \\ a & -a & 1 & -a & a \\ a & a & -a & 1 & a \\ a & -a & a & a & 1 \end{vmatrix}, \qquad a = 1/\sqrt{5}.$$

It is easily seen that this determinant has rank 3, with all non-vanishing principal minors positive, and the theorem is proved.

Remark. Any two equilateral quintuples of $\mathscr{E}_{2,r}$ are superposable.

THEOREM 80.2. *The $\mathscr{E}_{2,r}$ contains a metrically unique equilateral sextuple. It has edge $r\cos^{-1} 1/\sqrt{5}$.*

Proof. If any equilateral sextuple exists in $\mathscr{E}_{2,r}$, it has edge $r\cos^{-1} 1/\sqrt{5}$ by Theorem 80.1. That such a sextuple does exist is seen by noting that the determinant

$$\begin{vmatrix} 1 & a & a & a & a & a \\ a & 1 & a & -a & -a & a \\ a & a & 1 & a & -a & -a \\ a & -a & a & 1 & a & -a \\ a & -a & -a & a & 1 & a \\ a & a & -a & -a & a & 1 \end{vmatrix}, \quad a = 1/\sqrt{5}, \qquad (\dagger)$$

has rank 3 with all third-order principal minors positive.

Remark. The vertices of a regular icosahedron inscribed in $S_{2,r}$ may be labelled $p_1, p_2, ..., p_6, p_1^*, p_2^*, ..., p_6^*$, $p_i p_i^* = \pi r$ ($i = 1, 2, ..., 6$) with $|\cos(p_i, p_j)/r|$ identical with the determinant (†), where (p_i, p_j) denotes the distance of p_i, p_j in $S_{2,r}$.

THEOREM 80.3. *The $\mathscr{E}_{2,r}$ does not contain an equilateral septuple.*

Proof. If $p_1, p_2, ..., p_7$ form an equilateral septuple of $\mathscr{E}_{2,r}$ the labelling may be assumed so that p_7 is the centre (pole) of a small circle C of $S_{2,r}$, with radius $r\cos^{-1} 1/\sqrt{5}$, containing $p_1, p_2, ..., p_6$. Then at least *three* of the spherical distances of p_1 from $p_2, p_3, ..., p_6$ are $r\cos^{-1} 1/\sqrt{5}$ or $r\cos^{-1}(-1/\sqrt{5})$. But the circle C contains at most two points with the same spherical distance from p_1, and hence the points $p_2, p_3, ..., p_6$ cannot be pairwise distinct.

We have seen that the $S_{2,r}$ contains six points whose distance set consists of exactly two numbers (different from zero) and if the sum of these numbers is πr, then the sextuple is metrically unique (Theorem 80.2). The problem arises of ascertaining all subsets of $S_{2,r}$ with the distance of each two of its distinct points one of two positive numbers. It has been shown that the $S_{2,r}$ contains exactly three distinct types of two-distance sextuples. They are given by the vertices of (a) a regular inscribed octahedron, (b) an inscribed triangular prism with square faces, and (c) a truncated regular icosahedron.†

If, now, $p_1, p_2, ..., p_7$ form an equilateral septuple of $\mathscr{E}_{3,r}$, then the labelling may be assumed so that p_7 has equal spherical distances from the remaining points. This distance is not $\tfrac{1}{2}\pi r$, for if it were the septuple

† The twelve vertices of a regular inscribed icosahedron are diametral in pairs. Six of the vertices, no two diametral, are vertices of a *truncated* icosahedron.

would have elliptic edge $\frac{1}{2}\pi r$ and obviously no epsilon matrix exists such that $|\epsilon_{ij} \cos p_i p_j / r|$ $(i,j = 1, 2,..., 7)$ has rank 4.

Hence the six points $p_1,..., p_6$ lie on a small sphere $S_{2,r'}$ of $S_{3,r}$ $(r > r')$. Since distances in $S_{2,r'}$ *exceed* the corresponding distances in $S_{3,r}$, and $p_1, p_2,..., p_6$ form a two-distance set in $S_{3,r}$ with the sum of the two different distances equal to πr, the first property of the sextuple holds in $S_{2,r'}$ but *not* the second. That is, $p_1, p_2,..., p_6$ form a two-distance set of $S_{2,r'}$, but the points are not the vertices of a truncated regular icosahedron of $S_{2,r'}$.

Now it may be shown that neither are $p_1, p_2,..., p_6$ vertices of a regular octahedron or a triangular prism with square faces inscribed in $S_{2,r'}$, and so we have

THEOREM 80.4. *The $\mathscr{E}_{3,r}$ does not contain an equilateral septuple.*

This result is surprising, since it is natural to expect that an increase in dimension will yield equilateral sets of higher power than those already existing in the elliptic plane.

EXERCISES

1. Verify the remark made in the paragraph immediately preceding the statement of Theorem 80.4.
2. Investigate two-distance sets in the euclidean plane and show that no two-distance sextuples exist in E_2.
3. Show that $\mathscr{E}_{n,r}$ contains an equilateral $2n$-tuple.

81. The space problem for the $\mathscr{E}_{n,r}$. A lemma

The objective of the next few sections is to develop metric characterizations of elliptic spaces of finite dimensions. The methods are similar to those used first in §§ 49 and 50 for the metric characterization of euclidean space and applied later (§§ 64–66), with suitable modifications, to the characterization of the $S_{n,r}$. Though we shall assume that the reader is sufficiently well acquainted with those methods by now to permit us to proceed more rapidly, the metric peculiarities of elliptic space discussed in the two preceding sections may well be expected to introduce additional considerations.

One of these considerations is necessitated by the possibility that two congruent subsets of $\mathscr{E}_{n,r}$ may not be superposable. Though, as previously remarked, we shall treat this anomaly systematically in a later chapter, it is convenient for us to have at hand the following *sufficient* condition for two congruent *triples* of $\mathscr{E}_{2,r}$ to be superposable.

LEMMA 81.1. *Two congruent triples p_1, p_2, p_3 and p'_1, p'_2, p'_3 of the elliptic plane $\mathscr{E}_{2,r}$ are superposable if* (i) *one of the distances*

$$p_i p_j = p'_i p'_j = \tfrac{1}{2}\pi r \quad (i,j = 1,2,3),$$

or (ii) *the determinant*

$$\Delta^*(p_1, p_2, p_3) = |\epsilon_{ij} \cos p_i p_j / r| \quad (i,j = 1,2,3)$$

is negative, where every $\epsilon_{ij} = \epsilon_{ji} = 1$, *except* $\epsilon_{23} = \epsilon_{32} = -1$.

Proof. If alternative (ii) holds, then, since $p_1, p_2, p_3 \in \mathscr{E}_{2,r}$,

$$\Delta(p_1, p_2, p_3) = |\cos p_i p_j / r| \quad (i,j = 1,2,3)$$

is non-negative (Theorem 78.2), and hence the elliptic distances are also spherical distances on $S_{2,r}$. The two congruent elliptic triples are then congruent spherical triples, and they are superposable by a motion of $S_{2,r}$. Since such a motion is also evidently a motion of the $\mathscr{E}_{2,r}$, the lemma is established in this case.

If (i) holds, suppose $p_1 p_2 = p'_1 p'_2 = \tfrac{1}{2}\pi r$. Clearly a motion of $S_{2,r}$ (and hence of $\mathscr{E}_{2,r}$) exists carrying p'_1, p'_2 into p_1, p_2, respectively. Let such a motion send p'_3 into $\bar{p}_3$. If $\bar{p}_3 \neq p_3$, then either both p_1 and p_2 lie on one branch of the equidistant locus of $p_3, \bar{p}_3$ or they lie on different branches (§ 70, Exercise 5). In the first case a rotation of $S_{2,r}$ 180° about a pole of the great circle containing p_1, p_2 keeps them fixed and sends $\bar{p}_3$ into p_3. In the second case the condition $p_1 p_2 = \tfrac{1}{2}\pi r$ implies that at least one of the points p_1, p_2 lies on the great circle containing $p_3, \bar{p}_3$. Rotating the sphere 180° about such a point accomplishes the objective.

The condition $\Delta^* < 0$ is used in our work to give a precise meaning to such phrases as 'triples whose diameters are smaller than a fixed number, depending on the constant of the space', 'points sufficiently close together', etc., which are frequently encountered in work on elliptic space. The requirement is equivalent, upon expansion of the determinant, to the inequality $p_1 p_2 + p_2 p_3 + p_1 p_3 < \pi r$.

EXERCISES

1. Prove that if $p_1, p_2, p_3 \in \mathscr{E}_{n,r}$, $\Delta^*(p_1, p_2, p_3) < 0$ if and only if

$$p_1 p_2 + p_2 p_3 + p_1 p_3 < \pi r.$$

2. Let p_1, p_2, p_3 be a linear triple imbeddable in $\mathscr{E}_{n,r}$ and (ϵ_{ij}) an epsilon matrix with $|\epsilon_{ij} \cos p_i p_j / r| \geqslant 0$ $(i,j = 1,2,3)$. Show that $|\epsilon_{ij} \cos p_i p_j / r| = 0$ $(i,j = 1,2,3)$.

82. Definition and elementary properties of an $\mathscr{E}_r$ space

Let r be a given positive number. A complete convex semimetric space with diameter at most $\tfrac{1}{2}\pi r$ (and containing at least two distinct points) is called an $\mathscr{E}_r$ space provided it has the following two properties:

PROPERTY I. *Let $p_0, p_1, ..., p_4$ be any five pairwise distinct points with* (i) *two triples linear, and* (ii) *the determinant Δ^* formed for three of the points (one of which is common to the two linear triples) negative. Then an epsilon matrix (ϵ_{ij}) exists such that all principal minors of*

$$|\epsilon_{ij} \cos p_i p_j / r| \quad (i, j = 0, 1, ..., 4)$$

are non-negative.

PROPERTY II. *If p, q are any two distinct points with $pq \neq \frac{1}{2}\pi r$, points $d(p), d(q)$ exist such that q is between p and $d(p)$, p is between q and $d(q)$, and $pd(p) = qd(q) = \frac{1}{2}\pi r$.*

It is observed that Property I has both a local and a global character, the local character being derived from the assumption that Δ^* is negative for three of the five points.

We shall show that $\mathscr{E}_r$ *is congruent with an elliptic space of finite or infinite dimension*, and shall, moreover, find necessary and sufficient metric conditions in order that $\mathscr{E}_r$ be congruent with an elliptic space $\mathscr{E}_{n,r}$ of given dimension n.

THEOREM 82.1. *Five points of $\mathscr{E}_r$ that satisfy conditions* (i), (ii) *of Property* I *are congruently imbeddable in $\mathscr{E}_{2,r}$.*

Proof. Let $p_0, p_1, ..., p_4$ be five such points and (ϵ_{ij}) an epsilon matrix such that $|\epsilon_{ij} \cos p_i p_j / r|$ $(i, j = 0, 1, ..., 4)$ has all principal minors non-negative (Property I). If the two linear triples have exactly one point in common, we may assume the labelling so that p_0, p_1, p_3 and p_0, p_2, p_4 are linear triples and $\Delta^*(p_0, p_1, p_2) < 0$.

Now the two third-order principal minors formed for the two linear triples vanish (§ 81, Exercise 2). Observing that each principal minor containing a vanishing principal minor is *negative or zero*, it follows that each such principal minor is zero. If, now, $|\epsilon_{ij} \cos p_i p_j / r|$ $(i, j = 0, 1, 2)$ is not zero, then clearly the rank of the fifth-order determinant is 3 and the five points are imbeddable in $\mathscr{E}_{2,r}$. If $|\epsilon_{ij} \cos p_i p_j / r|$ $(i, j = 0, 1, 2)$ vanishes, then $|\epsilon_{ij} \cos p_i p_j / r|$ $(i, j = 0, 1, 2, 3)$ has rank 2. The vanishing of all of its third-order principal minors implies the vanishing of all fourth-order principal minors of $|\epsilon_{ij} \cos p_i p_j / r|$ $(i, j = 0, 1, ..., 4)$. Hence that determinant has rank *at most* 3, and the five points are imbeddable in $\mathscr{E}_{2,r}$ as before.

A similar argument is used in case the two linear triples have two points in common.

THEOREM 82.2. *Each four pairwise distinct points of $\mathscr{E}_r$ containing a linear triple is congruent with four points of $\mathscr{E}_{2,r}$.*

Proof. Let q_0, q_1, q_2, q_3 be such a quadruple with the labelling selected so that $q_0 q_1 q_2$ holds. Then each of the distances $q_0 q_1, q_1 q_2$ is less than $\tfrac{1}{2}\pi r$.

Case 1. The point q_3 is not between q_0 and q_1. Since $\mathscr{E}_r$ is convex it contains a point q_4 such that $q_0 q_4 q_1$ holds. Then q_4 is distinct from q_i $(i = 0, 1, 2, 3)$ ($q_4 \neq q_2$ since $q_0 q_4 < q_0 q_1 < q_0 q_2$ and $q_4 \neq q_3$ since q_4 is between q_0, q_1 while q_3 is not) and the five pairwise distinct points $q_0, q_1, \ldots, q_4$ satisfy conditions (i), (ii) of Property I. For q_0, q_1, q_2 and q_0, q_1, q_4 are linear and since $q_0 q_4 + q_4 q_1 = q_0 q_1 < \tfrac{1}{2}\pi r$, then

$$q_0 q_4 + q_4 q_1 + q_0 q_1 < \pi r$$

and so $\Delta^*(q_0, q_1, q_4) < 0$. Hence $q_0, q_1, \ldots, q_4 \Subset \mathscr{E}_{2,r}$ (Theorem 82.1) and consequently $q_0, q_1, q_2, q_3 \Subset \mathscr{E}_{2,r}$.

Case 2. The point q_3 is between q_0 and q_1. Let q_4 be a point of $\mathscr{E}_r$ between q_0 and q_3. Then $q_4 \neq q_0, q_1, q_2, q_3$ ($q_4 \neq q_1$ since $q_0 q_4 < q_0 q_3 < q_0 q_1$, and $q_4 \neq q_2$ since $q_0 q_4 < q_0 q_1 < q_0 q_2$). Also,

$$q_0 q_3 + q_3 q_4 + q_0 q_4 = 2 q_0 q_3 < 2 q_0 q_1 < \pi r,$$

so $\Delta^*(q_0, q_3, q_4) < 0$. Applying Theorem 82.1 we conclude that q_0, q_1, q_2, q_3 are imbeddable in $\mathscr{E}_{2,r}$.

THEOREM 82.3. *Each triple of points of $\mathscr{E}_r$ is congruently contained in $\mathscr{E}_{2,r}$.*

Proof. If $q_0, q_1, q_2 \in \mathscr{E}_r$ and q_0, q_1, q_2 are not imbeddable in $\mathscr{E}_{2,r}$, then the triple is surely not linear. A point q_3 of $\mathscr{E}_r$ between q_0 and q_1 forms with q_0, q_1, q_2 four pairwise distinct points containing a linear triple and so

$$q_0, q_1, q_2, q_3 \Subset \mathscr{E}_{2,r}.$$

COROLLARY. *The space $\mathscr{E}_r$ is metric.*

THEOREM 82.4. *Two distinct non-diametral points of $\mathscr{E}_r$ are joined by exactly one segment.*

Proof. Since $\mathscr{E}_r$ is complete, convex, and metric, each pair of its (distinct) points is joined by at least one segment. If a pair of distinct, non-diametral, points p, q are joined by two segments, then a constant c $(0 < c < 1)$ exists such that the relations $px + xq = pq, px = c \cdot pq$ are satisfied by two distinct points a, b of $\mathscr{E}_r$. The four pairwise distinct points p, q, a, b are congruent with points p', q', a', b' of $\mathscr{E}_{2,r}$ (Theorem 82.2) with a' and b' on the unique elliptic line determined by p', q'. But an elliptic line does not contain two distinct points with the same distances from two of its non-diametral points, and so $a' = b'$. Then $ab = a'b' = 0$ and hence $a = b$.

COROLLARY. *The congruence* $p, q \approx p', q'$ $(p', q' \in \mathscr{E}_{n,r})$, $0 < pq < \frac{1}{2}\pi r$, *uniquely determines the congruence* $\mathrm{seg}(p,q) \approx \mathrm{seg}(p',q')$ *(that is, there is one and only one congruence between the unique segments determined by p, q and p', q', respectively, that maps p onto p' and q onto q').*

If $pqd(p)$ holds $(pd(p) = \frac{1}{2}\pi r)$, then $pq < \frac{1}{2}\pi r$, $qd(p) < \frac{1}{2}\pi r$ and unique segments $\mathrm{seg}(p,q)$, $\mathrm{seg}(q, d(p))$ are determined which have only q in common (for $pqd(p)$ and pxq imply (since $\mathscr{E}_r$ is metric) $xqd(p)$, and so $qxd(p)$ does not hold). Hence $\mathrm{seg}(p,q) + \mathrm{seg}(q,d(p))$ is a segment (Lemma 15.1) and so diametral point-pairs p, $d(p)$ are joined by segments (of length $\frac{1}{2}\pi r$). Moreover, if $q \in \mathscr{E}_r$ with $pqd(p)$ holding, there is only one $\mathrm{seg}(p,q,d(p))$.

THEOREM 82.5. *If $p, q \in \mathscr{E}_r$ ($0 < pq < \frac{1}{2}\pi r$), there is exactly one point $d(p)$ of $\mathscr{E}_r$ with $pqd(p)$.*

Proof. By Property II at least one point with this property exists. If $d_1(p), d_2(p)$ are two distinct points with $pqd_1(p)$, $pqd_2(p)$, then the four pairwise distinct points p, q, $d_1(p)$, $d_2(p)$ are congruent with p', q', $d_1(p')$, $d_2(p')$ of $\mathscr{E}_{2,r}$, and q' belongs to both elliptic lines $\mathscr{E}_{1,r}(p', d_1(p'))$, $\mathscr{E}_{1,r}(p', d_2(p'))$ which are distinct since $d_1(p')d_2(p') = d_1(p)d_2(p) \neq 0$. Hence $p' = q'$, which contradicts $p'q' = pq > 0$.

If, now, $p, q \in \mathscr{E}_r$ ($0 < pq < \frac{1}{2}\pi r$), consider the four pairwise distinct points p, q, $d(p)$, $d(q)$, where $d(p)$, $d(q)$ are the unique (and evidently distinct) points of $\mathscr{E}_r$ such that $pqd(p)$ and $qpd(q)$ hold (Theorem 82.5). Since
$$p, q, d(p), d(q) \approx p', q', d(p'), d(q')$$
of $\mathscr{E}_{1,r}$, we have $p'd(q')d(p')$ and $q'd(p')d(q')$. Consequently, $pd(q)d(p)$ and $qd(p)d(q)$ hold, and none of the distances pq, $qd(p)$, $d(p)d(q)$, $d(q)p$ equals $\frac{1}{2}\pi r$. The unique segments $\mathrm{seg}(p,q)$, $\mathrm{seg}(q,d(p))$, $\mathrm{seg}(d(p), d(q))$, $\mathrm{seg}(d(q), p)$ have pairwise at most end-points in common. It follows that the two segments
$$\mathrm{seg}(p, q, d(p)) = \mathrm{seg}(p,q) + \mathrm{seg}(q, d(p)),$$
$$\mathrm{seg}(p, d(q), d(p)) = \mathrm{seg}(p, d(q)) + \mathrm{seg}(d(q), d(p))$$
have only the points p, $d(p)$ in common.

DEFINITION 82.1. *If $p, q \in \mathscr{E}_r$ ($0 < pq < \frac{1}{2}\pi r$), then the sum*
$$\mathrm{seg}(p, q, d(p)) + \mathrm{seg}(p, d(q), d(p)),$$
where $d(p)$, $d(q)$ are the unique points diametral to p, q, respectively, such that $pqd(p)$, $qpd(q)$ hold, is called a one-dimensional subspace $\mathscr{E}_r^1$ of $\mathscr{E}_r$ with base points p, q.

We have shown that $\mathscr{E}_r^1$ is uniquely determined by base points p, q. It should be observed that $\mathscr{E}_r^1$ *is a simple closed curve*.

EXERCISE

Show that the $S_{2,\frac{1}{2}r}$ is *not* an $\mathscr{E}_r$ space.

83. One-dimensional subspaces (lines) of $\mathscr{E}_r$

In this section and the next we obtain properties of one-dimensional subspaces of $\mathscr{E}_r$ that serve to anchor an inductive assumption to be made later.

THEOREM 83.1. *A one-dimensional subspace $\mathscr{E}_r^1$ of $\mathscr{E}_r$ is congruent with the elliptic line $\mathscr{E}_{1,r}$.*

Proof. Denoting base points of $\mathscr{E}_r^1$ by p, q,
$$\mathscr{E}_r^1(p,q) = \operatorname{seg}(p,q,d(p)) + \operatorname{seg}(p,d(q),d(p)).$$
Points p', q', $d(p')$, $d(q')$ of $\mathscr{E}_{2,r}$ with
$$p, q, d(p), d(q) \approx p', q', d(p'), d(q')$$
lie, as already observed, on a line $\mathscr{E}_{1,r}(p',q')$, and the two congruences
$$\operatorname{seg}(p,q,d(p)) \approx \operatorname{seg}(p',q',d(p')), \tag{1}$$
$$\operatorname{seg}(p,d(q),d(p)) \approx \operatorname{seg}(p',d(q'),d(p')) \tag{2}$$
establish a mapping of $\mathscr{E}_r^1(p,q)$ onto $\mathscr{E}_{1,r}(p',q')$.

To show the mapping a congruent one, suppose $x, y \in \mathscr{E}_r^1(p,q)$ mapped on x', y', respectively, of $\mathscr{E}_{1,r}(p',q')$. If x, y are both in the same component segment, then $xy = x'y'$ by one of the above congruences. In the contrary case, suppose $x \in \operatorname{seg}(p,q,d(p))$, $y \in \operatorname{seg}(p,d(q),d(p))$, and let p_1, p_2 be points of the first and second of these segments, respectively, distinct from q, $d(q)$, x, y, such that $pp_1 q$, $pp_2 d(q)$ hold and $\Delta^*(p,p_1,p_2) < 0$, with p_1', p_2' the points corresponding to p_1, p_2 by means of congruences (1), (2), respectively.† The pairwise distinct points $p, p_1, p_2, q, d(q)$ satisfy the conditions of Property I and consequently are congruent with points $p'', p_1'', p_2'', q'', d(q'')$ of $\mathscr{E}_{2,r}$. Since these points necessarily lie on an elliptic segment with end-points q'', $d(q'')$, the points p'', p_1'', p_2'' are linear. It follows that $p_1 p p_2$ holds and so
$$p_1 p_2 = p_1'' p_2'' = p_1'' p'' + p'' p_2'' = p_1 p + p p_2 = p_1' p' + p' p_2' = p_1' p_2'.$$
Hence $p, p_1, p_2 \approx p', p_1', p_2'$.

Applying Theorem 82.1 to p, p_1, p_2, x, y (the triples p, p_1, x and p, p_2, y are linear)
$$p, p_1, p_2, x, y \approx p'', p_1'', p_2'', x'', y'' \text{ of } \mathscr{E}_{2,r}$$

† Since $\Delta^*(p, p_1, p_2) < 0$ whenever $pp_1 + p_1 p_2 + pp_2 < \pi r$, it is clear that points p_1, p_2 exist satisfying all these conditions.

(the 'double-primed' points not necessarily the same as those above). Since $p'', p_1'', p_2'' \approx p, p_1, p_2 \approx p', p_1', p_2'$ and $\Delta^*(p, p_1, p_2) < 0$, a congruent transformation of $\mathscr{E}_{2,r}$ onto itself exists (Lemma 81.1) mapping p'', p_1'', p_2'' onto p', p_1', p_2', respectively. This transformation maps x'' on a point x^* of the line $\mathscr{E}_{1,r}(p', p_1') = \mathscr{E}_{1,r}(p', q')$ such that

$$p'x^* = p''x'' = px = p'x' \quad \text{and} \quad p_1'x^* = p_1''x'' = p_1x = p_1'x';$$

that is, x^* has the same distances as x' from the non-diametral points p', p_1' of $\mathscr{E}_{1,r}(p', q')$ and hence $x^* = x'$. Similarly, y'' is mapped on y', $p, p_1, p_2, x, y \approx p', p_1', p_2', x', y'$ and so $xy = x'y'$.

COROLLARY. *Each one-dimensional subspace of $\mathscr{E}_r$ has length πr.*

LEMMA 83.1. *If $s, t \in \mathscr{E}_r^1$ ($0 < st < \tfrac{1}{2}\pi r$), then $\mathrm{seg}(s, t) \subset \mathscr{E}_r^1$.*

Proof. By Theorem 83.1, $\mathscr{E}_r^1 \approx \mathscr{E}_{1,r}$ and the points s', t' of $\mathscr{E}_{1,r}$ congruent to s, t determine the elliptic segment $\mathrm{seg}(s', t')$ which is a subset of $\mathscr{E}_{1,r}$. Since a congruence transforms segments into segments, the subset of $\mathscr{E}_r^1$ congruent to $\mathrm{seg}(s', t')$ by the above congruence is a segment with endpoints s, t and hence (Theorem 82.4) is $\mathrm{seg}(s, t)$.

LEMMA 83.2. *A diametral point-pair of $\mathscr{E}_r$ is contained in one and only one subspace $\mathscr{E}_r^1$.*

Proof. If $p, d(p)$ is such a pair and q any point of $\mathscr{E}_r$ between p and $d(p)$, the unique subspace $\mathscr{E}_r^1(p, q)$ contains p and $d(p)$. Let $\mathscr{E}_r^{1*}$ be any one-dimensional subspace containing $p, d(p)$. If $x \in \mathscr{E}_r^{1*}$, then the congruence of $\mathscr{E}_r^{1*}$ with an elliptic line shows that $pxd(p)$ holds. Select points p_1, p_2 of $\mathscr{E}_r^1(p, q)$ (distinct from p, x) such that $p_1 \in \mathrm{seg}(p, q, d(p))$, $p_2 \in \mathrm{seg}(p, d(q), d(p))$, and $0 < pp_1 = pp_2 < \tfrac{1}{6}\pi r$.

By Theorem 82.1, $p, p_1, p_2, d(p), x \approx p', p_1', p_2', d(p'), x'$, points of $\mathscr{E}_{2,r}$, and it is clear (since $p'x'd(p')$ holds) that the 'primed' points all lie on a line $\mathscr{E}_{1,r}$. Hence x' belongs to one of the segments $\mathrm{seg}(p', p_1')$, $\mathrm{seg}(p_1', d(p'))$, $\mathrm{seg}(d(p'), p_2')$, $\mathrm{seg}(p', p_2')$ and so x is a point of one of the segments $\mathrm{seg}(p, p_1)$, $\mathrm{seg}(p_1, d(p))$, $\mathrm{seg}(d(p), p_2)$, $\mathrm{seg}(p, p_2)$. From the preceding lemma, $x \in \mathscr{E}_r^1(p, q)$ and $\mathscr{E}_r^{1*} \subset \mathscr{E}_r^1(p, q)$. Since each subspace is congruent with $\mathscr{E}_{1,r}$, it follows that $\mathscr{E}_r^{1*} = \mathscr{E}_r^1(p, q)$ and the lemma is proved.†

LEMMA 83.3. *A non-diametral point-pair is contained in one and only one subspace $\mathscr{E}_r^1$.*

Proof. Each subspace $\mathscr{E}_r^1$ containing the non-diametral points p, q evidently contains the unique point $d(p)$ such that $pqd(p)$ holds, and the lemma follows from the preceding one.

† $\mathscr{E}_r^1$ is compact and hence monomorphic (§ 62); that is, $\mathscr{E}_r^{1*} \approx \mathscr{E}_r^1$ and $\mathscr{E}_r^{1*} \subset \mathscr{E}_r^1$ imply $\mathscr{E}_r^{1*} = \mathscr{E}_r^1$.

Combining these lemmas, we have:

THEOREM 83.2. *There is one and only one subspace $\mathscr{E}_r^1$ containing any given pair of distinct points of $\mathscr{E}_r$.*

Thus a one-dimensional subspace contains (is identical with) the one-dimensional subspace determined by any two of its points. We refer to $\mathscr{E}_r^1$ as a *line* of $\mathscr{E}_r$.

COROLLARY 1. *Two distinct lines of $\mathscr{E}_r$ have at most one point in common.*

COROLLARY 2. *Any two non-diametral points of a one-dimensional subspace may be taken as base points of the subspace.*

84. Imbedding line-sums of $\mathscr{E}_r$ in the elliptic plane

We consider next certain subsets of $\mathscr{E}_r$ consisting of intersecting lines and show them to be imbeddable in $\mathscr{E}_{2,r}$.

THEOREM 84.1. *Let p_0, p_1, p_2 be three pairwise distinct points of $\mathscr{E}_r$ with $\Delta^*(p_0, p_1, p_2) < 0$. If p_0', p_1', p_2' are any points of the elliptic plane, with $p_0, p_1, p_2 \approx p_0', p_1', p_2'$, then the congruences*

(a) $\quad \mathscr{E}_r^1(p_0, p_1) \approx \mathscr{E}_{1,r}(p_0', p_1'),\quad$ (b) $\quad \mathscr{E}_r^1(p_0, p_2) \approx \mathscr{E}_{1,r}(p_0', p_2')$

determine uniquely the congruence

$$\mathscr{E}_r^1(p_0, p_1) + \mathscr{E}_r^1(p_0, p_2) \approx \mathscr{E}_{1,r}(p_0', p_1') + \mathscr{E}_{1,r}(p_0', p_2').$$

Proof. Since $\Delta^*(p_0, p_1, p_2) < 0$, then the determinant

$$\Delta(p_0, p_1, p_2) = |\cos(p_i p_j / r)| \quad (i, j = 0, 1, 2)$$

is non-negative, and it follows that no one of the distances

$$p_i p_j \quad (i, j = 0, 1, 2)$$

is $\tfrac{1}{2}\pi r$. Hence p_0, p_1 and p_0, p_2 may be taken as base points of lines $\mathscr{E}_r^1(p_0, p_1), \mathscr{E}_r^1(p_0, p_2)$, respectively, of $\mathscr{E}_r$. The notation of the congruences (a), (b) indicates that p_i and p_i' ($i = 0, 1, 2$) are corresponding points, and since $p_0 p_i \neq \tfrac{1}{2}\pi r$ ($i = 1, 2$), these congruences are unique. If $\Delta(p_0, p_1, p_2) = 0$, the points p_0, p_1, p_2 are linear, $p_2 \in \mathscr{E}_r^1(p_0, p_1)$, and so $\mathscr{E}_r^1(p_0, p_1) = \mathscr{E}_r^1(p_0, p_2)$. Similarly $\mathscr{E}_{1,r}(p_0', p_1') = \mathscr{E}_{1,r}(p_0', p_2')$, and the theorem follows from Theorem 83.1.

If, on the other hand $\Delta(p_0, p_1, p_2) = \Delta(p_0', p_1', p_2')$ is positive, then since $\Delta^*(p_0', p_1', p_2') < 0$, the points p_0', p_1', p_2' neither lie on an elliptic line nor are they congruent with points of such a line, and consequently p_0, p_1, p_2 do not lie on a line of $\mathscr{E}_r$. The congruences (a), (b) ensured by Theorem 83.1 give a mapping of the set $\mathscr{E}_r^1(p_0, p_1) + \mathscr{E}_r^1(p_0, p_2)$ onto the set $\mathscr{E}_{1,r}(p_0', p_1') + \mathscr{E}_{1,r}(p_0', p_2')$. To prove the mapping a congruent one it evidently suffices to examine the case in which $x \in \mathscr{E}_r^1(p_0, p_1)$,

$y \in \mathscr{E}_r^1(p_0, p_2)$ and x', y' are the points of the elliptic lines corresponding by congruences (a), (b).

Suppose, first, that p_0, p_1, x and p_0, p_2, y are both linear triples, i.e. x and y are not interior points of the segments

$$\operatorname{seg}(d_1(p_0), d_1(p_1)), \qquad \operatorname{seg}(d_2(p_0), d_2(p_2)),$$

respectively, where $d_1(p_0)$, $d_1(p_1)$ are points of $\mathscr{E}_r^1(p_0, p_1)$ which are diametral to p_0 and p_1, respectively, and $d_2(p_0)$, $d_2(p_2)$ are similarly defined points of $\mathscr{E}_r^1(p_0, p_2)$, with $x \neq p_0, p_1$ and $y \neq p_0, p_2$.

By Theorem 82.1

$$p_0, p_1, p_2, x, y \approx p_0'', p_1'', p_2'', x'', y'',$$

points of $\mathscr{E}_{2,r}$. Since

$$p_0'', p_1'', p_2'' \approx p_0, p_1, p_2 \approx p_0', p_1', p_2' \quad \text{and} \quad \Delta^*(p_0, p_1, p_2) < 0,$$

a congruent transformation of $\mathscr{E}_{2,r}$ on itself exists mapping p_i'' onto p_i' ($i = 0, 1, 2$). Points p_0, p_1, x being linear, so are p_0'', p_1'', x'' and consequently they lie on an elliptic line. As the transformation sends lines into lines, it maps x'' into the unique point x^* of $\mathscr{E}_r^1(p_0', p_1')$ with given distances $(p_0 x, p_1 x)$ from the non-diametral points p_0', p_1'. Then

$$p_0' x^* = p_0'' x'' = p_0 x = p_0' x' \quad \text{and} \quad p_1' x^* = p_1'' x'' = p_1 x = p_1' x'.$$

Since both x^* and x' are points of $\mathscr{E}_{1,r}(p_0', p_1')$, it follows that $x^* = x'$. In the same manner it is seen that y'' maps into y' and hence

$$p_0, p_1, p_2, x, y \approx p_0', p_1', p_2', x', y',$$

from which $xy = x'y'$.

Thus the set

$$\{\mathscr{E}_r^1(p_0, p_1) - \operatorname{int} \operatorname{seg}(d_1(p_0), d_1(p_1))\} +$$
$$+ \{\mathscr{E}_r^1(p_0, p_2) - \operatorname{int} \operatorname{seg}(d_2(p_0), d_2(p_2))\}$$

is congruently contained in $\mathscr{E}_{1,r}(p_0', p_1') + \mathscr{E}_{1,r}(p_0', p_2')$.†

Let now q_1, q_2 be points of $\mathscr{E}_r^1(p_0, p_1)$, $\mathscr{E}_r^1(p_0, p_2)$, respectively, with $0 < p_0 q_1 < \epsilon$, $0 < p_0 q_2 < \epsilon$ ($\epsilon > 0$, arbitrarily small) and let q_1', q_2' be the corresponding points on $\mathscr{E}_{1,r}(p_0', p_1')$, $\mathscr{E}_{1,r}(p_0', p_2')$, respectively, by means of (a), (b). Then exactly as above

$$\{\mathscr{E}_r^1(p_0, q_1) - \operatorname{int} \operatorname{seg}(d_1(p_0), d_1(q_1))\} +$$
$$+ \{\mathscr{E}_r^1(p_0, q_2) - \operatorname{int} \operatorname{seg}(d_2(p_0), d_2(q_2))\}$$

is congruently contained in $\mathscr{E}_{1,r}(p_0', q_1') + \mathscr{E}_{1,r}(p_0', q_2')$.

---

† The cases $x = p_1$, $y = p_2$ are taken care of by the continuity of the metric. They may also be proved separately by use of Theorem 82.2.

Ch. IX, § 84   THEOREMS FOR ELLIPTIC SPACE   223

Since $\mathscr{E}_r^1(p_0, q_i) = \mathscr{E}_r^1(p_0, p_i)$ and $\mathscr{E}_{1,r}(p'_0, q'_i) = \mathscr{E}_{1,r}(p'_0, p'_i)$ ($i = 1, 2$) we see that with the *exception of the interiors of two segments, each of arbitrarily small length* ϵ, the sum $\mathscr{E}_r^1(p_0, p_1) + \mathscr{E}_r^1(p_0, p_2)$ is congruently contained in the sum of the two corresponding elliptic lines. It follows by continuity that $\mathscr{E}_r^1(p_0, p_1) + \mathscr{E}_r^1(p_0, p_2)$ is congruently contained in

$$\mathscr{E}_{1,r}(p'_0, p'_1) + \mathscr{E}_{1,r}(p'_0, p'_2)$$

and the mapping of the one set onto the given set by congruences (a), (b) is a congruent one. The uniqueness of the congruence follows from that of the two defining congruences.

COROLLARY 1. *The sum of any two intersecting lines of $\mathscr{E}_r$ is congruently imbeddable in $\mathscr{E}_{2,r}$.*

COROLLARY 2. *If p_0, p_1, p_2 are pairwise distinct points of $\mathscr{E}_r$ with $\Delta^*(p_0, p_1, p_2) < 0$, and $p_0, p_1, p_2 \approx p'_0, p'_1, p'_2$, points of $\mathscr{E}_{2,r}$, then the three congruences*

$$\mathscr{E}_r^1(p_i, p_{i+1}) \approx \mathscr{E}_{1,r}(p'_i, p'_{i+1}) \quad (i = 0, 1, 2; \ p_3 = p_0)$$

determine uniquely the congruence

$$\mathscr{E}_r^1(p_0, p_1) + \mathscr{E}_r^1(p_1, p_2) + \mathscr{E}_r^1(p_0, p_2)$$
$$\approx \mathscr{E}_{1,r}(p'_0, p'_1) + \mathscr{E}_{1,r}(p'_1, p'_2) + \mathscr{E}_{1,r}(p'_0, p'_2).$$

Proof. The three congruences of the pairs of lines determine a mapping of the sum of the three lines of $\mathscr{E}_r$ onto the sum of the corresponding three elliptic lines. By the preceding theorem the mapping defines uniquely a congruence between the sum of any two of the three lines of $\mathscr{E}_r$ and the sum of the corresponding two lines of $\mathscr{E}_{2,r}$. Since points x, y of the sum of all three lines necessarily are contained in the sum of two of the lines, the corollary follows at once.

COROLLARY 3. *Each set of five points of $\mathscr{E}_r$ containing two linear triples is congruently imbeddable in $\mathscr{E}_{2,r}$.*

Proof. Each such set is evidently contained in two intersecting lines of $\mathscr{E}_r$.

THEOREM 84.2. *Let p_0, $p_{0,1}$, p_1, p_2 be pairwise distinct points of $\mathscr{E}_r$ such that $p_{0,1} \in \mathscr{E}_r^1(p_0, p_1)$, $p_2 \bar{\in} \mathscr{E}_r^1(p_0, p_1)$, $p_0 p_2 \neq \frac{1}{2}\pi r$. If $p'_0, p'_{0,1}, p'_1, p'_2$ are any points of $\mathscr{E}_{2,r}$ such that $p'_{0,1} \in \mathscr{E}_{1,r}(p'_0, p'_1)$ and*

$$p_0, p_{0,1}, p_1, p_2 \approx p'_0, p'_{0,1}, p'_1, p'_2,$$

then the two congruences

(a) $\mathscr{E}_r^1(p_0, p_2) \approx \mathscr{E}_{1,r}(p'_0, p'_2)$,   (b) $\mathscr{E}_r^1(p_0, p_{0,1}, p_1) \approx \mathscr{E}_{1,r}(p'_0, p'_{0,1}, p'_1)$

determine the unique congruence

$$\mathscr{E}_r^1(p_0, p_{0,1}, p_1) + \mathscr{E}_r^1(p_0, p_2) \approx \mathscr{E}_{1,r}(p'_0, p'_{0,1}, p'_1) + \mathscr{E}_{1,r}(p'_0, p'_2).$$

Proof. The points p'_0, $p'_{0,1}$, p'_1, p'_2 of $\mathscr{E}_{2,r}$ are not on a line, for if so they would contain two linear triples, and hence a linear triple with p'_2 as an element. But then p_2 forms with two of the points p_0, $p_{0,1}$, p_1 (which lie on a line) a linear triple and so $p_2 \in \mathscr{E}^1_r(p_0,p_1)$, contrary to hypothesis. Thus the lines $\mathscr{E}_{1,r}(p'_0,p'_{0,1},p'_1)$, $\mathscr{E}_{1,r}(p'_0,p'_2)$ are distinct.

Let x, y be elements of $\mathscr{E}^1_r(p_0,p_{0,1},p_1) + \mathscr{E}^1_r(p_0,p_2)$ and x', y' the elements of $\mathscr{E}_{1,r}(p'_0,p'_{0,1},p'_1) + \mathscr{E}_{1,r}(p'_0,p'_2)$ corresponding to them in the mapping of the one sum onto the other given by congruences (a), (b). We may suppose x a point of the first and y a point of the second summand.

By Corollary 1 of the preceding theorem,

$$p_0,\ p_{0,1},\ p_1,\ p_2,\ x,\ y \approx p''_0,\ p''_{0,1},\ p''_1,\ p''_2,\ x'',\ y'',$$

with p''_0, $p''_{0,1}$, p''_1, x'' on one line of $\mathscr{E}_{2,r}$ and p''_0, p''_2, y'' on another. Since p''_0, $p''_{0,1}$, p''_1, $p''_2 \approx p_0, p_{0,1}, p_1, p_2 \approx p'_0, p'_{0,1}, p'_1, p'_2$ and the two congruent triples $p'_0, p'_{0,1}, p'_1$ and $p''_0, p''_{0,1}, p''_1$ both lie on lines, there is a congruent transformation of $\mathscr{E}_{2,r}$ on itself mapping $p''_0, p''_{0,1}, p''_1, p''_2$ on $p'_0, p'_{0,1}, p'_1, p'_2$, respectively. This transformation maps x'' on that unique point x^* of $\mathscr{E}_{1,r}(p'_0,p'_{0,1},p'_1)$ with distances from $p'_0, p'_{0,1}, p'_1$ the same as those of x', and so $x^* = x'$. Similarly, y'' is sent into that point y^* of $\mathscr{E}_{1,r}(p'_0,p'_2)$ whose distances from the two non-diametral points p'_0, p'_2 are the same as those of y'. Thus $y^* = y'$, $xy = x''y'' = x'y'$, and the two line-sums are congruent. That this congruence between the two sets is the only one in which $p_0, p_{0,1}, p_1, p_2$ correspond to $p'_0, p'_{0,1}, p'_1, p'_2$, respectively, follows at once, since this correspondence determines congruences (a) and (b).

DEFINITION. *If p is any point and $\mathscr{E}^1_r$ any line of $\mathscr{E}_r$, $f(p)$ a foot of p on $\mathscr{E}^1_r$, then the point $d(f(p))$, diametral to $f(p)$ on $\mathscr{E}^1_r(p,f(p))$, is called a pole of $\mathscr{E}^1_r$, for $p \neq f(p)$.*

THEOREM 84.3. *A pole of a line $\mathscr{E}^1_r$ has distance $\tfrac{1}{2}\pi r$ from each point of the line, and hence distance $\tfrac{1}{2}\pi r$ from the line.*

Proof. Let $f(p)$ be a foot of p on $\mathscr{E}^1_r$, $d(f(p))$ a pole of $\mathscr{E}^1_r$, and q any point of $\mathscr{E}^1_r$. By Corollary 1, Theorem 84.1, the sum of the two intersecting lines $\mathscr{E}^1_r$, $\mathscr{E}^1_r(p,f(p), d(f(p)))$ is imbeddable in $\mathscr{E}_{2,r}$ and hence

$$\mathscr{E}^1_r(q,f(p)) + \mathscr{E}^1_r(p,f(p), d(f(p))) \approx \mathscr{E}_{1,r}(q',f(p')) + \mathscr{E}_{1,r}(p',f(p'), d(f(p'))),$$

with $f(p')$ a foot of p' on $\mathscr{E}_{1,r}(q',f(p'))$ and $d(f(p'))$ the point of

$$\mathscr{E}_{1,r}(p',f(p'), d(f(p')))$$

diametral to $f(p')$. By an elementary property of the elliptic plane, it follows that $q'd(f(p')) = qd(f(p)) = \tfrac{1}{2}\pi r$.

THEOREM 84.4. *The sums $\mathscr{E}_r^1+(p)$ and $\mathscr{E}_{1,r}+(p')$ are congruent if and only if* dist$(p, \mathscr{E}_r^1) =$ dist$(p', \mathscr{E}_{1,r})$.

Proof. The theorem follows from Theorem 83.1 in case dist$(p, \mathscr{E}_r^1) = 0$, and is obvious if the distance is $\frac{1}{2}\pi r$. Suppose, now,

$$0 < \text{dist}(p, \mathscr{E}_r^1) = \text{dist}(p', \mathscr{E}_{1,r}) < \tfrac{1}{2}\pi r,$$

let $f(p), f(p')$ be feet of p, p' on $\mathscr{E}_r^1, \mathscr{E}_{1,r}$, respectively, and $d(f(p)), d(f(p'))$ poles on $\mathscr{E}_r^1(p, f(p)), \mathscr{E}_{1,r}(p', f(p'))$, respectively. If $q \in \mathscr{E}_r^1$ $(q \neq f(p))$, let q' of $\mathscr{E}_{1,r}$ correspond to q in any congruence between $\mathscr{E}_r^1$ and $\mathscr{E}_{1,r}$ which maps $f(p)$ on $f(p')$.

By the preceding theorem, $qd(f(p)) = \tfrac{1}{2}\pi r = q'd(f(p'))$ and the two quadruples $p, q, f(p), d(f(p))$ and $p', q', f(p'), d(f(p'))$ are seen to have all corresponding distances equal except, perhaps, pq and $p'q'$. But

$$p, q, f(p), d(f(p)) \approx p'', q'', f(p''), d(f(p''))$$

of $\mathscr{E}_{2,r}$ (Theorem 82.1), and hence this elliptic quadruple has five of its six distances equal to the corresponding five distances of the elliptic quadruple $p', q', f(p'), d(f(p'))$. It follows that the sixth pair of corresponding distances are also equal and the two quadruples are congruent.†
Hence $p'q' = p''q'' = pq$ and $\mathscr{E}_r^1+(p) \approx \mathscr{E}_{1,r}+(p')$. Since the necessity is obvious, the theorem is established.

85. Linear subspaces of $\mathscr{E}_r$

Defining a zero-dimensional subspace $\mathscr{E}_r^0$ of $\mathscr{E}_r$ as consisting of a single point, it follows from the preceding section that for $k = 1$, (i) a k-dimensional subspace $\mathscr{E}_r^k$ is the locus of all points of $\mathscr{E}_r$ on a line with a point of $\mathscr{E}_r^{k-1}$ and a point not belonging to it, (ii) if $p_0, p_1, \ldots, p_k$ are not elements of a $(k-1)$-dimensional subspace $\mathscr{E}_r^{k-1}$, there is one and only one subspace $\mathscr{E}_r^k$ of k dimensions containing them, (iii) $\mathscr{E}_r^k$ is a linear space, (iv) $\mathscr{E}_r^k$ is congruent with $\mathscr{E}_{k,r}$, and (v) two distinct $(k-1)$-dimensional subspaces contained in an $\mathscr{E}_r^k$ have an $\mathscr{E}_r^{k-2}$ in common.‡

If, for a given integer k, $\mathscr{E}_r$ contains a subspace $\mathscr{E}_r^k$ and a point p not belonging to it, a subspace $\mathscr{E}_r^{k+1}$, sometimes denoted by $\{p; \mathscr{E}_r^k\}$ is defined as the locus of all points of $\mathscr{E}_r$ on a line with p and a point of $\mathscr{E}_r^k$. As in §§ 49 and 66, we make the inductive hypothesis that all properties of $\mathscr{E}_r^1$ proved in the foregoing (in particular, those properties listed above) are valid for every $\mathscr{E}_r^k$ $(k = 1, 2, \ldots, n)$.

The procedure is now so similar to that employed in §§ 50 and 66 that it is perhaps unnecessary to give all the details. It is noted, however,

---

† The two quadruples are even superposable.
‡ For $k = 1$, $\mathscr{E}_r^{k-2}$ is *null*.

that the argument here is not exactly the same as that used in those earlier sections since (1) it is based on an application of Theorem 84.2, and (2) it must take account of the complicating possibility

$$\text{dist}(p_0, \mathscr{E}_r^n) = \tfrac{1}{2}\pi r.$$

THEOREM 85.1. *The sums $\mathscr{E}_r^n + (p)$ and $\mathscr{E}_{n,r} + (p')$ are congruent if and only if $\text{dist}(p, \mathscr{E}_r^n) = \text{dist}(p', \mathscr{E}_{n,r})$.*

Proof. This is easily deduced as a consequence of Theorem 84.4.

THEOREM 85.2. *Each $(n+1)$-dimensional subspace $\mathscr{E}_r^{n+1}$ of $\mathscr{E}_r$ is congruent with the elliptic space $\mathscr{E}_{n+1,r}$.*

Proof. Let $\mathscr{E}_r^{n+1} = \{p_0; \mathscr{E}_r^n\}$ and consider first the case in which $0 < \text{dist}(p_0, \mathscr{E}_r^n) < \tfrac{1}{2}\pi r$. Select in $\mathscr{E}_{n+1,r}$ a subspace $\mathscr{E}_{n,r}$ and a point p_0' such that $\text{dist}(p_0', \mathscr{E}_{n,r}) = \text{dist}(p_0, \mathscr{E}_r^n)$, and let $f(p_0), f(p_0')$ be feet of p_0, p_0' on $\mathscr{E}_r^n, \mathscr{E}_{n,r}$, respectively.

Consider the congruence $\mathscr{E}_r^n + (p_0) \approx \mathscr{E}_{n,r} + (p_0')$ (Theorem 85.1) which associates $f(p_0)$ with $f(p_0')$. If, now, $x \in \mathscr{E}_r^{n+1}$, let p be the point of $\mathscr{E}_r^n$ such that p_0, p, x are on a line, and p' the point of $\mathscr{E}_{n,r}$ corresponding to p in the above congruence. Since $p_0 p = p_0' p'$, $\mathscr{E}_r^1(p_0, p) \approx \mathscr{E}_{1,r}(p_0', p')$. Assuming for the present that $p_0 p \neq \tfrac{1}{2}\pi r$, let x' be the (unique) point of $\mathscr{E}_{1,r}(p_0', p')$ corresponding to x in this congruence. We wish to show this mapping (defined for all points x of $\mathscr{E}_r^{n+1}$ such that $p_0 p \neq \tfrac{1}{2}\pi r$) a congruent one.

Let x, x' and y, y' be corresponding points in this mapping, with $p_0, p, x \approx p_0', p', x'$ and $p_0, q, y \approx p_0', q', y'$ (all triples lying on lines). If $p = q$, then $x, y \in \mathscr{E}_r^1(p_0, p)$ and $xy = x'y'$ by an above congruence. If $p \neq q$, let s be a point of $\mathscr{E}_r^1(p, q)$, $p \neq s \neq q$, and s' the corresponding point of $\mathscr{E}_{1,r}(p', q')$.† Noting that the two congruent quadruples p_0, p, q, s and p_0', p', q', s' satisfy the conditions of Theorem 84.2 we have

$$\mathscr{E}_r^1(p_0, p) + \mathscr{E}_r^1(p, q, s) \approx \mathscr{E}_{1,r}(p_0', p') + \mathscr{E}_{1,r}(p', q', s')$$

and hence $qx = q'x'$. Applying the same theorem to the congruent quadruples p_0, p, x, q and p_0', p', x', q' gives

$$\mathscr{E}_r^1(p_0, q) + \mathscr{E}_r^1(p_0, p, x) \approx \mathscr{E}_{1,r}(p_0', q') + \mathscr{E}_{1,r}(p_0', p', x')$$

and hence $xy = x'y'$.‡

It remains to define the mapping (and prove it a congruence) for points x of $\mathscr{E}_r^{n+1}$ such that $p_0 p = \tfrac{1}{2}\pi r$. Now there exists an infinite sequence $\{x_i\}$ of distinct points of $\mathscr{E}_r$ with $\lim_{i \to \infty} x_i = x$ and $p_0 p_i \neq \tfrac{1}{2}\pi r$,

---

† Since $\mathscr{E}_r^n$ is a linear space (inductive hypothesis), $s \in \mathscr{E}_r^1(p, q)$, $p, q \in \mathscr{E}_r^n$ imply $s \in \mathscr{E}_r^n$.
‡ The cases $p = x$ and/or $q = y$ offer no difficulty.

where $p_i \in \mathscr{E}_r^n$ with p_0, p_i, x_i on a line ($i = 1, 2,...$). For if no such sequence exists, then $\mathscr{E}_r^n$ contains a neighbourhood N of p, each point of which is diametral to p_0. By Theorem 85.1 each point of N', the neighbourhood of p', which corresponds to N in the congruence $\mathscr{E}_r^n{+}(p_0) \approx \mathscr{E}_{n,r}{+}(p_0')$, has distance $\tfrac{1}{2}\pi r$ from p_0' and hence p_0' is the pole of $\mathscr{E}_{n,r}$; i.e. $p_0' x' = \tfrac{1}{2}\pi r$ for every point x' of $\mathscr{E}_{n,r}$. This is impossible since

$$p_0' f(p_0') = p_0 f(p_0) = \mathrm{dist}(p_0, \mathscr{E}_r^n) < \tfrac{1}{2}\pi r.$$

Let x_i' be the unique point of $\mathscr{E}_{n+1,r}$ corresponding to x_i by means of the mapping previously defined. Since $\{x_i\}$ has limit x, it is a Cauchy sequence and since (by the first part of the proof) $x_i x_j = x_i' x_j'$ ($i,j = 1, 2,...$), sequence $\{x_i'\}$ is also a Cauchy sequence. The elliptic space $\mathscr{E}_{n,r}$ being complete contains a unique point $x' = \lim_{i\to\infty} x_i'$. If we let x and x' correspond, the continuity of the metric ensures $xy = x'y'$. Thus the existence of a congruent mapping of $\mathscr{E}_r^{n+1}$ onto $\mathscr{E}_{n+1,r}$ is established, and the theorem is proved in this case.

It remains to consider the case in which

$$\mathscr{E}_r^{n+1} = \{p_0; \mathscr{E}_r^n\}, \qquad \mathrm{dist}(p_0, \mathscr{E}_r^n) = \tfrac{1}{2}\pi r.$$

Let q_0 be a point of $\mathscr{E}_r^1(p_0, f(p_0))$, $q_0 \neq p_0$, $f(p_0)$, and $\mathscr{E}_r^{*n+1} = \{q_0; \mathscr{E}_r^n\}$. Then $f(p_0)$ is a foot of q_0 on $\mathscr{E}_r^n$ (for if $p \in \mathscr{E}_r^n$, then

$$q_0 p \geqslant p_0 p - p_0 q_0 \geqslant p_0 f(p_0) - p_0 q_0 = q_0 f(p_0))$$

and since $0 < q_0 f(p_0) \neq \tfrac{1}{2}\pi r$, then $0 < \mathrm{dist}(q_0, \mathscr{E}_r^n) < \tfrac{1}{2}\pi r$. It follows from the above that $\mathscr{E}_r^{*n+1} \approx \mathscr{E}_{n+1,r}$. If, now, $x \in \mathscr{E}_r^{n+1} = \{p_0; \mathscr{E}_r^n\}$ and $p \in \mathscr{E}_r^n$ with p_0, p, x on a line, the above congruence maps these three points onto p_0', p', x' (since it maps $\mathscr{E}_r^1(p_0, p)$ onto $\mathscr{E}_{1,r}(p_0', p')$). Since $\mathscr{E}_{1,r}(q_0', x')$ intersects $\mathscr{E}_{n,r}$, then $\mathscr{E}_r^1(q_0, x)$ meets $\mathscr{E}_r^n$ and so $x \in \{q_0; \mathscr{E}_r^n\}$. In this way it is seen that $\mathscr{E}_r^{n+1} = \mathscr{E}_r^{*n+1} \approx \mathscr{E}_{n+1,r}$ and the proof of the theorem is complete.

COROLLARY. *The $(n{+}1)$-dimensional subspace $\mathscr{E}_r^{n+1}$ is a linear space.*

THEOREM 85.3. *Any $n{+}2$ points $p_0, p_1,..., p_{n+1}$ of $\mathscr{E}_r$ which are not contained in any n-dimensional subspace lie in one and only one $(n{+}1)$-dimensional subspace.*

The reader is asked to supply the proof.

86. Characterization theorems

We have seen that for every positive integer n, each set of $n{+}1$ points of $\mathscr{E}_r$ either lies in an $(n{-}1)$-dimensional linear subspace or determines uniquely an n-dimensional linear subspace, and all linear subspaces of $\mathscr{E}_r$ are congruent with elliptic spaces of corresponding dimensions and space constant r. It follows that the space $\mathscr{E}_r$ is congruent with an elliptic space of finite or infinite dimension and space constant r.

An additional property is needed for the congruence of $\mathscr{E}_r$ with an elliptic space of *given* finite dimension. We introduce two properties (one global and the other a localization of it), the adjunction of either of which to the properties defining $\mathscr{E}_r$ yields a space congruent with the elliptic space $\mathscr{E}_{n,r}$.

PROPERTY III (GLOBAL). *There is a positive integer k such that each set of $k+2$ points $p_0, p_1,..., p_{k+1}$ of $\mathscr{E}_r$ has the property that if an ϵ-matrix (ϵ_{ij}) exists for which no principal minor of*

$$|\epsilon_{ij}\cos(p_i p_j/r)| \quad (i,j = 0, 1,..., k+1)$$

is negative, then there is at least one ϵ-matrix for which the determinant vanishes and has no principal minor negative.

Denoting by Property III_n (Global) the above property in case n is the smallest member of the class $\{k\}$ of integers described in it, we prove:

THEOREM 86.1. *A necessary and sufficient condition that $\mathscr{E}_r$ be congruent with the n-dimensional elliptic space $\mathscr{E}_{n,r}$ is that it have Property III_n (Global).*

Proof. The necessity follows at once from Theorem 78.2. To prove the sufficiency, note first that $\mathscr{E}_r$ contains $n+1$ points which are not imbeddable in $\mathscr{E}_{n-1,r}$; for in the contrary case it follows (Theorem 78.2) that Property III_{n-1} (Global) holds in $\mathscr{E}_r$, contrary to hypothesis. Since $\mathscr{E}_r^{n-1} \approx \mathscr{E}_{n-1,r}$, a set of $n+1$ points not imbeddable in $\mathscr{E}_{n-1,r}$ is not congruently contained (and hence not contained) in any $\mathscr{E}_r^{n-1}$ and therefore determines uniquely an n-dimensional subspace $\mathscr{E}_r^n$ of $\mathscr{E}_r$ (Theorem 85.3) which is congruent with $\mathscr{E}_{n,r}$.

To complete the proof we show that each point of $\mathscr{E}_r$ belongs to $\mathscr{E}_r^n$. In the contrary case, suppose $p \in \mathscr{E}_r$ and $p \bar{\in} \mathscr{E}_r^n$. Then p and $\mathscr{E}_r^n$ generate a unique $(n+1)$-dimensional subspace $\mathscr{E}_r^{n+1}$ of $\mathscr{E}_r$ which is congruent to $\mathscr{E}_{n+1,r}$, and therefore contains $n+2$ points $p_0, p_1,..., p_{n+1}$ with

$$p_i p_j = \tfrac{1}{2}\pi r \quad (i,j = 0, 1,..., n+1;\ i \neq j).$$

Now the ϵ-matrix (ϵ_{ij}), $\epsilon_{ij} = \epsilon_{ji} = 1$ $(i,j = 0, 1,..., n+1)$ is such that $|\epsilon_{ij}\cos(p_i p_j/r)|$ $(i,j = 0, 1,..., n+1)$ has no principal minor negative, but obviously no ϵ-matrix exists for which the determinant vanishes. Thus $\mathscr{E}_r = \mathscr{E}_r^n \approx \mathscr{E}_{n,r}$ and the theorem is proved.

Localizing Property III_n (Global), suppose there is a point q_0 of $\mathscr{E}_r$

and a spherical neighbourhood $U(q_0; \frac{1}{6}\pi r)$ in which it holds. Then (as in the proof of the preceding theorem) $U(q_0; \frac{1}{6}\pi r)$ contains $n+1$ points $p_0, p_1, \ldots, p_n$ which are not congruently imbeddable in $\mathscr{E}_{n-1,r}$ and so are not contained (congruently or actually) in any $\mathscr{E}_r^{n-1}$. These points determine a unique n-dimensional subspace $\mathscr{E}_r^n(p_0, p_1, \ldots, p_n)$ which we show contains $U(q_0; \frac{1}{6}\pi r)$.

Let x be any element of $U(q_0; \frac{1}{6}\pi r)$. If $p_0, p_1, \ldots, p_n, x$ are in any $\mathscr{E}_r^n$, they are in $\mathscr{E}_r^n(p_0, p_1, \ldots, p_n)$. If they do not belong to any n-dimensional subspace, they determine a unique $\mathscr{E}_r^{n+1}(p_0, p_1, \ldots, p_n, x)$. Denoting by $p_0', p_1', \ldots, p_n', x'$ the points of $\mathscr{E}_{n+1,r}$ corresponding to $p_0, p_1, \ldots, p_n, x$ in a congruence $\mathscr{E}_r^{n+1} \approx \mathscr{E}_{n+1,r}$, it is clear that the 'primed' points are not in any $\mathscr{E}_{n,r}$. On the other hand, $p_0, p_1, \ldots, p_n, x \in U(q_0; \frac{1}{6}\pi r)$ implies (using Property III$_n$ (Global) (assumed to hold in this neighbourhood), the congruence of all linear subspaces with elliptic spaces, and Theorem 78.2) the existence of points $p_0'', p_1'', \ldots, p_n'', x''$ of an $\mathscr{E}_{n,r}$ congruent to them. Thus $p_0', p_1', \ldots, p_n', x' \approx p_0'', p_1'', \ldots, p_n'', x''$ with the 'primed' points in $\mathscr{E}_{n+1,r}$ and no lower-dimensional elliptic space, and the 'double primed' points in $\mathscr{E}_{n,r}$.†

It follows that ϵ-matrices (ϵ_{ij}), (ϵ_{ij}') exist such that

(i) $|\epsilon_{ij} \cos(p_i p_j / r)|$ $(i, j = 0, 1, \ldots, n+1; p_{n+1} = x)$ vanishes,

(ii) $|\epsilon_{ij}' \cos(p_i p_j / r)|$ $(i, j = 0, 1, \ldots, n+1; p_{n+1} = x)$ does not vanish,

and in each determinant all non-vanishing principal minors are positive. Making the (obviously permissible) selection

$$\epsilon_{0j} = \epsilon_{j0} = \epsilon_{0j}' = \epsilon_{j0}' = 1 \quad (j = 0, 1, \ldots, n+1)$$

and noting that, since $p_0, p_1, \ldots, p_n, x$ are points of $U(q_0; \frac{1}{6}\pi r)$, $\Delta^* < 0$ for each three of these points, the reader is asked to show that every ϵ_{ij} and ϵ_{ij}' equals 1 $(i, j = 0, 1, \ldots, n+1)$, and the properties (i) and (ii) are contradictory. Thus $p_0, p_1, \ldots, p_n, x$ are elements of an n-dimensional subspace and so $U(q_0; \frac{1}{6}\pi r)$ is a subset of $\mathscr{E}_r^n(p_0, p_1, \ldots, p_n)$.

If, now, $x \in \mathscr{E}_r$, $x \neq q_0$, then $x \in \mathscr{E}_r^1(q_0, x)$. Since this line clearly has two distinct points in common with $U(q_0; \frac{1}{6}\pi r)$, it has two distinct points in common with the linear subspace $\mathscr{E}_r^n(p_0, p_1, \ldots, p_n)$ and therefore is contained in this subspace. Thus $\mathscr{E}_r^n(p_0, p_1, \ldots, p_n)$ contains every point of $\mathscr{E}_r$, which gives $\mathscr{E}_r = \mathscr{E}_r^n \approx \mathscr{E}_{n,r}$. These considerations lead to the following property and theorem:

PROPERTY III (LOCAL). *There exists an integer k, a point q_0 of $\mathscr{E}_r$, and a spherical neighbourhood $U(q_0; \frac{1}{6}\pi r)$ such that each $k+2$ points $p_0, p_1, \ldots, p_{k+1}$ of $U(q_0; \frac{1}{6}\pi r)$ have the property that if an ϵ-matrix (ϵ_{ij}) exists for which no*

† This is not *a priori* impossible (see § 79).

principal minor of the determinant $|\epsilon_{ij} \cos(p_i p_j/r)|$ $(i,j = 0, 1,..., k+1)$ is negative, then there is at least one ϵ-matrix for which no principal minor of the corresponding determinant is negative and the determinant vanishes.

THEOREM 86.2. *A necessary and sufficient condition that $\mathscr{E}_r$ be congruent with the n-dimensional elliptic space $\mathscr{E}_{n,r}$ is that it have Property* III_n (*Local*).

EXERCISES

1. Show that the elliptic n-space $\mathscr{E}_{n,r}$ has best congruence order $n+2$ with respect to the class of spaces $\mathscr{E}_r$.
2. Investigate the class of spaces $\mathscr{E}_r^*$ obtained by replacing Property I by the weak elliptic four-point property proved in Theorem 82.2. Is Theorem 85.2 valid for $\mathscr{E}_r^*$ spaces?

REFERENCES

§§ 77–78. Blumenthal [10].
§§ 79–80. Blumenthal [18], Blumenthal [19], and Kelly. Equilateral subsets of $\mathscr{E}_{2,r}$ and $\mathscr{E}_{3,r}$ (§ 80) were analysed independently in Haantjes [4].
§§ 81–86. Blumenthal [18].

CHAPTER X

CONGRUENCE AND SUPERPOSABILITY IN ELLIPTIC SPACE

87. Preliminary remarks and definitions

It was remarked in § 79 that elliptic space differs from euclidean and spherical spaces in the important respect that congruence of two subsets of $\mathscr{E}_{n,r}$ does not imply their superposability; that is, a congruence f between two subsets of $\mathscr{E}_{n,r}$ is not necessarily extendible to the whole space. Even if two subsets A, B of $\mathscr{E}_{n,r}$ are superposable, we cannot conclude that every congruence between A and B is extendible to a *motion* (a congruence of $\mathscr{E}_{n,r}$ with itself).†

These circumstances give rise to the two principal problems with which this chapter is concerned: (1) *to find necessary and sufficient conditions for the superposability of two congruent subsets of $\mathscr{E}_{n,r}$, and* (2) *to determine when a given congruence between two subsets of $\mathscr{E}_{n,r}$ may be extended to the whole space.*

We denote by $A \sim_f B$, $A \approx_f B$, and $A \cong_f B$, respectively, the one-to-one correspondence, congruence, and superposability of sets A, B established by the function f. By

$$(p_1, p_2, ..., p_m) \cong (q_1, q_2, ..., q_m)$$

we shall mean that the two m-tuples are superposable without implying that p_i and q_i are corresponding points ($i = 1, 2, ..., m$). The latter implication is carried by the symbol obtained from the above by omitting the parentheses.

Let P, Q be two subsets of $\mathscr{E}_{n,r}$ with $P \sim_f Q$. Subsets A_P, B_Q of P, Q, respectively, are *f-superposable* ($A_P \cong_f B_Q$) provided there exists a motion Γ of $\mathscr{E}_{n,r}$ such that $A_P \sim_f B_Q$ is identical with $A_P \sim_\Gamma B_Q$. If $A_P \approx_f B_Q$ and a motion Γ exists such that $A_P \sim_\Gamma B_Q$ is identical with $A_P \sim_f B_Q$, we say f is extendible to the motion Γ.

If an m-tuple of $\mathscr{E}_{n,r}$ is ordered so that no one of the points has distance $\tfrac{1}{2}\pi r$ from *each* preceding (succeeding) point, the m-tuple is said to be in left-hand (right-hand) *apolar* order.

We shall refer to the convex n-sphere $S_{n,r}$, from which we obtain the $\mathscr{E}_{n,r}$ by supplementation (§ 77), as the *associated* n-sphere.

† See § 38.

88. Congruent subsets, one of which is contained in $\mathscr{E}_{1,r}$

Since the elliptic line $\mathscr{E}_{1,r}$ is congruent with the convex circle $S_{1,\frac{1}{2}r}$ (Exercise 2, § 78), it is clear that any two congruent subsets of $\mathscr{E}_{1,r}$ are superposable. We have seen, however, that the elliptic line contains three points p_1, p_2, p_3 with $p_i p_j = \frac{1}{3}\pi r$ ($i,j = 1, 2, 3$) which are congruent with three points of $\mathscr{E}_{2,r}$ that do not lie on any $\mathscr{E}_{1,r}$. These two congruent triples are, of course, not superposable, for any motion in $\mathscr{E}_{2,r}$ is a rotation about one of its points and consequently points on a line are sent into collinear points.† We show in this section that this anomalous behaviour is restricted to triples; that is, a subset of $\mathscr{E}_{1,r}$ which is congruent with an elliptic subset not contained in an elliptic line must consist of exactly three points.

THEOREM 88.1. *Let P be a subset of $\mathscr{E}_{n,r}$ containing more than three points. If $P' \subset \mathscr{E}_{1,r}$ and $P' \approx P$, then P lies on an elliptic line.*

Proof. Let p, q be any two distinct elements of P, and consider the quadruple p, q, x, y of pairwise distinct elements of P. This quadruple contains at least two linear triples since it is congruent with a quadruple of P' which lies on an $\mathscr{E}_{1,r}$ (Exercise 4, § 79). But each *linear* triple of $\mathscr{E}_{n,r}$ necessarily lies on an $\mathscr{E}_{1,r}$ and two distinct lines of $\mathscr{E}_{n,r}$ have at most one point in common.

It follows that every two points of P (distinct from each other and from p, q) lie on the elliptic line determined by p, q and hence $P \subset \mathscr{E}_{1,r}(p, q)$.

THEOREM 88.2. *If $p_1, p_2, p_3 \in \mathscr{E}_{1,r}$, then $p_1, p_2, p_3 \approx p'_1, p'_2, p'_3$ with p'_1, p'_2, p'_3 irreducibly contained in $\mathscr{E}_{2,r}$ if and only if*

$$p_i p_j \neq \tfrac{1}{2}\pi r \quad (i,j = 1, 2, 3)$$

and
$$p_1 p_2 + p_2 p_3 + p_1 p_3 = \pi r.$$

Proof. If $p_1, p_2, p_3 \approx p'_1, p'_2, p'_3$ ($p'_1, p'_2, p'_3 \in \mathscr{E}_{1,r}$), then clearly p_1, p_2, p_3 are not linear. It follows that

$$p_i p_j \neq \tfrac{1}{2}\pi r \quad (i,j = 1, 2, 3)$$

and $\Delta(p_1, p_2, p_3) \neq 0$. Then, since $p_1, p_2, p_3 \in \mathscr{E}_{1,r}$, $\Delta^*(p_1, p_2, p_3)$ must vanish and we readily obtain $p_1 p_2 + p_2 p_3 + p_1 p_3 = \pi r$.

If, on the other hand, $p_1, p_2, p_3 \in \mathscr{E}_{1,r}$ with $p_i p_j \neq \tfrac{1}{2}\pi r$ ($i,j = 1, 2, 3$) and $p_1 p_2 + p_2 p_3 + p_1 p_3 = \pi r$, then $0 < p_i p_j < \tfrac{1}{2}\pi r$ ($i,j = 1, 2, 3$) and $p_1 p_2/r$, $p_2 p_3/r$, $p_1 p_3/r$ satisfy the strict triangle *in*equality. It follows that a *proper* triangle of the associated $S_{2,r}$ exists whose vertices are

---

† Coxeter [1, p. 109]. It might also be observed that if the two triples were superposable, the pole of the line bearing p_1, p_2, p_3 would be sent into a point with distance $\tfrac{1}{2}\pi r$ from each point of the other triple. But since that triple is not contained in a line, no such point exists.

congruent to p_1, p_2, p_3. Since each of the three spherical distances determined by these vertices is less than $\tfrac{1}{2}\pi r$, these distances are also the distances of the points in the elliptic plane $\mathscr{E}_{2,r}$ and the theorem is proved.

EXERCISE

Show that $\Delta^*(p_1,p_2,p_3) = -4\cos A \cdot \cos B \cdot \cos C \cdot \cos D$,
where
$$A = (p_1p_2+p_2p_3+p_1p_3)/2r, \quad B = (p_1p_2+p_2p_3-p_1p_3)/2r,$$
$$C = (p_1p_2-p_2p_3+p_1p_3)/2r, \quad D = (-p_1p_2+p_2p_3+p_1p_3)/2r.$$

89. Non-superposable congruent subsets of $\mathscr{E}_{n,r}$

Since the determinant $|r_{ij}|$, $r_{ij} = r_{ji} = \tfrac{1}{3}$ ($i,j = 1,2,3,4$; $i \neq j$), $r_{ii} = 1$ ($i = 1,2,3,4$), and all of its principal minors are positive, the $S_{3,r}$ contains an equilateral quadruple with edge $r\arccos\tfrac{1}{3}$ which does not lie in any $S_{2,r}$. Hence the elliptic three-space $\mathscr{E}_{3,r}$ contains, irreducibly, such an equilateral quadruple. On the other hand, we have seen in § 80 that the elliptic plane $\mathscr{E}_{2,r}$ contains an equilateral quadruple with edge $r\arccos\tfrac{1}{3}$. These two congruent subsets of $\mathscr{E}_{3,r}$ are evidently not superposable.

But in view of Theorem 88.1 it might be conjectured that subsets of $\mathscr{E}_{2,r}$ which are congruent with subsets of $\mathscr{E}_{n,r}$ not contained in any elliptic plane consist of exactly four points or, failing this, that the power of such sets cannot exceed a finite integer k. The following theorem shows, however, that such conjectures are not valid.

THEOREM 89.1. *For every cardinal number m ($3 < m \leq \mathfrak{c}$) the elliptic plane contains a subset of power m that is congruent with a subset of $\mathscr{E}_{n,r}$ not contained in any $\mathscr{E}_{2,r}$.*

Proof. We have already given an example of such a planar subset for $m = 4$. The elliptic plane contains five points p_1, p_2,..., p_5 with p_1, p_2, p_3 on a line, $p_i p_j = \tfrac{1}{3}\pi r$ ($i,j = 1,2,3$), p_4 and p_5 reflections of each other in this line, with
$$p_1p_4 = p_2p_4 = p_1p_5 = p_2p_5 = \tfrac{1}{3}\pi r,$$
and
$$p_3p_4 = p_3p_5 = \tfrac{1}{2}\pi r, \quad p_4p_5 = r\arccos\tfrac{1}{3}.$$
The reader may verify this by noting that the determinant $|\epsilon_{ij}\cos p_i p_j/r|$ ($i,j = 1,2,...,5$) has rank 3 with all non-vanishing principal minors positive, where
$$\epsilon_{13} = \epsilon_{31} = \epsilon_{45} = \epsilon_{54} = -1$$
and the remaining $\epsilon_{ij} = 1$.

But the determinant $|\epsilon'_{ij}\cos p_i p_j/r|$ ($i,j = 1,2,...,5$), with $\epsilon'_{ij} = 1$ ($i,j \neq 1,3$), $\epsilon_{13} = \epsilon_{31} = -1$, is seen to have rank 4 (with all non-vanishing principal minors positive) and so p_1, p_2,..., p_5 are congruent with five points irreducibly contained in $\mathscr{E}_{3,r}$.

It is clear now that if P is any subset of the line containing p_1, p_2, p_3, the planar subset $(p_1, p_2, ..., p_5) + P$ is congruent with a subset of $\mathscr{E}_{3,r}$ not contained in any plane, and the theorem is proved.

The procedure given above may be directly extended. If, for example, four points p_1, p_2, p_3, p_4 of $\mathscr{E}_{2,r}$ be selected, not lying on a line, a fifth point p_5 not in their plane determines with them a unique $\mathscr{E}_{3,r}$. Let p_6 be the reflection of p_5 in $\mathscr{E}_{2,r}(p_1, p_2, p_3, p_4)$, and suppose p_5 has been so chosen that the spherical distance of p_5, p_6 (in the associated $S_{3,r}$) exceeds $\tfrac{1}{2}\pi r$. Then $p_1, p_2, ..., p_6$ are congruent with six points irreducibly contained in $\mathscr{E}_{4,r}$, and adjoining any subset P of $\mathscr{E}_{2,r}(p_1, p_2, p_3, p_4)$ to the sextuple shows that Theorem 89.1 is valid for the $\mathscr{E}_{3,r}$.† In this manner, the following result is obtained:

THEOREM 89.2. *For every integer $k > 1$ and every cardinal number m ($k+1 < m \leqslant \mathfrak{c}$) each k-dimensional elliptic space contains a subset of power m which is congruent with a subset of $\mathscr{E}_{n,r}$ that is not contained in any k-dimensional elliptic space $\mathscr{E}_{k,r}$.*

Thus certain subsets of a given elliptic space cannot be *metrically* distinguished from certain subsets contained irreducibly in an elliptic space of different dimension (and same space constant r)! The property of being contained in a subspace of given dimension is not a congruence invariant in elliptic geometry. An m-tuple that is 'dependent' in the classical sense may be congruent with an 'independent' m-tuple.

Congruent subsets of $\mathscr{E}_{n,r}$ which are contained irreducibly in subspaces of different dimensions are obviously not superposable, since a motion of $\mathscr{E}_{n,r}$ maps each k-dimensional subspace onto a k-dimensional subspace, for every value of k. But the phenomenon of congruent non-superposable sets is not restricted to such cases.

The elliptic plane $\mathscr{E}_{2,r}$, for example, contains points p_1, p_2, p_3, p_4, p_4' such that

$$p_1 p_2 = \tfrac{1}{3}\pi r, \qquad p_1 p_4 = p_2 p_3 = \tfrac{1}{6}\pi r, \qquad p_1 p_3 = r \arccos(3^{\frac{1}{2}}/6),$$

$$p_2 p_4 = r \arccos(3^{\frac{1}{2}}/10), \qquad p_3 p_4 = r \arccos(2^{\frac{3}{2}}/15),$$

and
$$p_1 p_4' = p_1 p_4, \qquad p_2 p_4' = p_2 p_4,$$

with
$$\text{spher dist}(p_3, p_4') = \sup \text{spher dist}(p_3, p_4)$$
$$= r \arccos(-2^{\frac{3}{2}}/15).$$

---

† Rotating the associated $S_{3,r}$ about the $S_{2,r}$ associated with $\mathscr{E}_{2,r}(p_1, p_2, p_3, p_4)$, a point p_6' of $S_{4,r}$ may be found whose spherical distance from p_5 *is the supplement of the spherical distance of p_5, p_6.* Then $p_1, p_2, p_3, p_4, p_5, p_6 \approx p_1, p_2, p_3, p_4, p_5, p_6'$, with the latter sextuple not in any $\mathscr{E}_{3,r}$ (since p_6' is *not* in the $\mathscr{E}_{3,r}$ determined by $p_1, p_2, ..., p_5$).

The reader may verify this by showing that each of the determinants

$$|\epsilon_{ij} \cos p_i p_j/r|, \qquad |\epsilon'_{ij} \cos p_i p_j/r|,$$

$\epsilon_{ij} = \epsilon_{ji} = 1$ $(i,j = 1,2,3,4)$, $\epsilon'_{ij} = \epsilon'_{ji} = 1$ $(i,j = 1,2,3,4; i,j \neq 3,4)$, $\epsilon'_{23} = \epsilon'_{32} = -1$, vanishes, with all non-vanishing principal minors positive.

The points p_1, p_2, p_3 are on the locus of points of $\mathscr{E}_{2,r}$ equidistant from p_4 and p'_4 but they are not on any $\mathscr{E}_{1,r}$. The quadruples p_1, p_2, p_3, p_4 and p_1, p_2, p_3, p'_4 are congruent but not superposable, for any motion of $\mathscr{E}_{2,r}$ that sent one quadruple into the other would leave each of the three lines

$$\mathscr{E}_{1,r}(p_i, p_j) \quad (i,j = 1,2,3; i \neq j)$$

pointwise invariant (since $p_i p_j \neq \tfrac{1}{2}\pi r$ $(i,j = 1,2,3)$). Since the lines are pairwise distinct, this motion would leave the $\mathscr{E}_{2,r}$ itself pointwise invariant.

This example may be extended to sets of arbitrary power (not exceeding c) by adjoining points of $\mathscr{E}_{1,r}(p_1, p_2)$. Similar examples are found in each $\mathscr{E}_{k,r}$ $(k > 1)$.

EXERCISES

1. Let Q denote the quadratic form

$$\sum_{i=1}^{m} x_i^2 + 2 \sum_{i<j=1}^{m} \epsilon_{ij} a_{ij} x_i x_j, \quad \epsilon_{ij} = \pm 1; \quad 0 \leqslant a_{ij} < 1$$

$$(i,j = 1,2,...,m; i < j, m > 3).$$

If each sub-form in four variables obtained from Q by setting equal to zero all but four of the m variables can be given rank 2 by choosing properly the epsilons appearing in it, prove that

(a) a selection of ϵ_{ij} $(i,j = 1,2,...,m)$ is possible for which Q has rank 2, and

(b) every positive semi-definite form Q' obtained from Q by a choice of the epsilons has rank 2.

2. Is the quadratic form theorem of the preceding exercise valid for $m = 3$? Is the extension obtained upon replacing 'four' by '$k+2$' and 'rank 2' by 'rank k', $k > 2$, valid?

3. Show that the only choices of the epsilons which might result in the quadratic form Q' being positive semi-definite are those that are equivalent to replacing certain of the m variables by their negatives (that is, to reversing the directions of certain of the m axes). (This result makes (b) obvious.)

90. Spherical sets and matrices associated with an elliptic m-tuple

Since $\mathscr{E}_{n,r}$ is the metric transform of its associated convex n-sphere $S_{n,r}$ by the πr-supplementation function F (defined in (**) of §77), there corresponds to each ordered m-tuple $p_1, p_2,..., p_m$ of $\mathscr{E}_{n,r}$ the set of ordered m-tuples of $S_{n,r}$, each one of which is transformed into the ordered elliptic m-tuple by F. Each ordered m-tuple of the set is called an *ordered associated spherical m-tuple of* $p_1, p_2,..., p_m$. It is seen that if $p_1', p_2',..., p_m'$ is any one of these associated spherical m-tuples ($p_i' \sim p_i$ ($i = 1, 2,..., m$)), then the set consists of all those m-tuples, *free from diametral point-pairs*, contained in the *associated spherical* $2m$-tuple $p_1', p_2',..., p_m', p_1^*, p_2^*,..., p_m^*$ formed by adjoining those points p_i^* of $S_{n,r}$ diametral to p_i' ($i = 1, 2,..., m$).

A *strictly elementary* transformation of a matrix is one which is effected by (a) the interchange of two rows and the same numbered columns, and/or (b) the multiplication of the elements of a row and those of the same numbered column by -1. Two matrices are called *strictly* equivalent provided one may be transformed into the other by strictly elementary transformations.

THEOREM 90.1. *If* $p_1', p_2',..., p_m'$ *and* $p_1'', p_2'',..., p_m''$ *are ordered associated spherical m-tuples of the elliptic subset* $p_1, p_2,..., p_m$, *then the two matrices* $(\cos p_i' p_j'/r)$ *and* $(\cos p_i'' p_j''/r)$ ($i,j = 1, 2,..., m$) *are strictly equivalent*.

Proof. It follows from the above that each point p_i'' of the second m-tuple is either identical with the corresponding point p_i' of the first m-tuple or is diametral to it. Hence the two sets may be labelled, respectively,

$$p_{i_1}', p_{i_2}',..., p_{i_k}', p_{i_{k+1}}',..., p_{i_m}' \quad \text{and} \quad p_{i_1}'', p_{i_2}'',..., p_{i_k}'', p_{i_{k+1}}'',..., p_{i_m}'',$$

with $\quad p_{i_1}'' = p_{i_1}', \quad p_{i_2}'' = p_{i_2}', \quad ..., \quad p_{i_k}'' = p_{i_k}'$

and $\quad p_{i_{k+1}}'' = p_{i_{k+1}}^*, \quad ..., \quad p_{i_m}'' = p_{i_m}^* \quad (0 \leqslant k \leqslant m)$,

where $p_{i_j}^*$ is the point of $S_{n,r}$ diametral to p_{i_j}' ($j = k+1, k+2,..., m$).

Forming the two symmetric matrices $(\cos p_{i_s}' p_{i_t}'/r)$, $(\cos p_{i_s}'' p_{i_t}''/r)$ ($s,t = 1, 2,..., m$) it is observed that since $\cos(p_{i_s}' p_{i_t}'/r) = \cos(p_{i_s}'' p_{i_t}''/r)$ for $(s,t = 1, 2,..., k)$ as well as for $(s,t = k+1, k+2,..., m)$, while the sum of the cosines is zero for $(s = 1, 2,..., k; t = k+1, k+2,..., m)$ and $(s = k+1, k+2,..., m; t = 1, 2,..., k)$, multiplication by -1 of the elements of the last $(m-k)$ rows and columns of one matrix transforms it into the other. Hence the matrices

$$(\cos p_{i_s}' p_{i_t}'/r), \quad (\cos p_{i_s}'' p_{i_t}''/r) \quad (s,t = 1, 2,..., m)$$

are strictly equivalent, and since these two matrices are strictly equivalent to $(\cos p_i' p_j'/r)$ and $(\cos p_i'' p_j''/r)$ ($i,j = 1, 2,..., m$), respectively, the theorem is proved.

Attached to an ordered elliptic m-tuple $p_1, p_2,..., p_m$ is the matrix

$(\cos p_i'p_j'/r)$ $(i,j = 1, 2,..., m)$ of an ordered associated spherical m-tuple $p_1', p_2',..., p_m'$. Clearly

$$(\cos p_i'p_j'/r) = (\epsilon_{ij} \cos p_i p_j/r) \quad (i,j = 1, 2,..., m),$$

where (ϵ_{ij}) is an epsilon matrix of order m. We agree to put $\epsilon_{ij} = 1$ in case $p_i p_j = \frac{1}{2}\pi r$.

To each m-tuple of $\mathscr{E}_{n,r}$ there corresponds (1) the set of all ordered spherical m-tuples associated with the orderings of the elliptic m-tuple, and (2) the set of all the corresponding attached matrices. This set of matrices forms the set of *associated matrices* of the elliptic m-tuple.

THEOREM 90.2. *Let $i_1, i_2,..., i_m$ be a permutation of the integers $1, 2,..., m$. If a matrix $(\epsilon_{s,t} \cos p_{i_s} p_{i_t}/r)$ $(s,t = 1, 2,..., m)$ is an associated matrix of an m-tuple $p_1, p_2,..., p_m$ of $\mathscr{E}_{n,r}$, then its rank is at most $n+1$ and all non-vanishing principal minors are positive.*

The proof is left to the reader.

EXERCISE
Is the converse of Theorem 90.2 valid?

91. First superposability theorems

The notion of associated matrix of an elliptic m-tuple is the basis of our criteria for superposability.

THEOREM 91.1. *Two m-tuples of $\mathscr{E}_{n,r}$ are superposable if and only if an associated matrix of one m-tuple is strictly equivalent to an associated matrix of the other.*

Proof. To prove the necessity, let $(p_1, p_2,..., p_m)$ and $(q_1, q_2,..., q_m)$ be superposable m-tuples of $\mathscr{E}_{n,r}$ with the labelling selected so that

$$p_1, p_2,..., p_m \cong q_1, q_2,..., q_m.$$

If Γ is a motion of $\mathscr{E}_{n,r}$ with $p_i = \Gamma(q_i)$ $(i = 1, 2,..., m)$, then the associated spherical $2m$-tuple $(q_1', q_2',..., q_m', q_1^*, q_2^*,..., q_m^*)$ of $q_1, q_2,..., q_m$ is carried into the associated spherical $2m$-tuple $(p_1', p_2',..., p_m', p_1^*, p_2^*,..., p_m^*)$ of $p_1, p_2,..., p_m$. Then, clearly, ordered associated spherical m-tuples of $p_1, p_2,..., p_m$ and $q_1, q_2,..., q_m$, respectively, exist with identical matrices, and the necessity is proved.

If, on the other hand, the elliptic subsets $(p_1, p_2,..., p_m)$ and $(q_1, q_2,..., q_m)$ have an associated matrix of one strictly equivalent to an associated matrix of the other, then there is an associated spherical m-tuple of one of these sets which is *congruent* with an associated spherical m-tuple of the other. A motion of $S_{n,r}$ that brings these two associated m-tuples into coincidence evidently induces a motion in $\mathscr{E}_{n,r}$ which superposes the two elliptic subsets, and the theorem is established.

Remark. We observe that Lemma 81.1 is an immediate consequence of the preceding theorem. For if p_1, p_2, p_3 and q_1, q_2, q_3 are congruent elliptic subsets and a pair of points of one triple has distance $\tfrac{1}{2}\pi r$, any two epsilon matrices $(\epsilon_{ij} \cos p_i p_j / r)$, $(\epsilon'_{ij} \cos q_i q_j / r)$ $(i,j = 1, 2, 3)$ are evidently strictly equivalent. If the determinant

$$\Delta^*(p_1, p_2, p_3) = |\epsilon_{ij} \cos p_i p_j / r| \quad (i,j = 1, 2, 3)$$

is negative, where every $\epsilon_{ij} = \epsilon_{ji} = 1$, *except* $\epsilon_{23} = \epsilon_{32} = -1$, then no associated matrix of either triple is strictly equivalent to the matrix of Δ^* (since, according to Theorem 90.2, the determinant of each associated matrix of any elliptic m-tuple is non-negative). It follows that each associated matrix of each triple is strictly equivalent to the matrix $(\cos p_i p_j / r)$ $(i,j = 1, 2, 3)$, which completes the proof of the remark.

THEOREM 91.2. *Let* $p_1, p_2, \ldots, p_m \approx q_1, q_2, \ldots, q_m$ *be two f-congruent m-tuples of* $\mathscr{E}_{n,r}$. *The two ordered m-tuples are f-superposable if and only if there exists an associated matrix of one m-tuple which equals an associated matrix of the other.*

Proof. If the two m-tuples are f-superposable, then ordered associated spherical m-tuples $p'_1, p'_2, \ldots, p'_m$ and $q'_1, q'_2, \ldots, q'_m$ exist with

$$p'_1, p'_2, \ldots, p'_m \approx q'_1, q'_2, \ldots, q'_m,$$

and the two matrices $(\cos p'_i p'_j / r)$, $(\cos q'_i q'_j / r)$ $(i,j = 1, 2, \ldots, m)$ are equal.

On the other hand, if an associated matrix of one of the ordered m-tuples equals an associated matrix of the other, then again associated spherical m-tuples exist with $p'_1, p'_2, \ldots, p'_m \approx q'_1, q'_2, \ldots, q'_m$. Since this is a congruence between two subsets of the spherical space $S_{n,r}$, it may be extended to the whole space; that is, a congruent mapping of $S_{n,r}$ onto itself exists which maps p'_i onto q'_i ($i = 1, 2, \ldots, m$). This mapping induces a congruence of $\mathscr{E}_{n,r}$ with itself which maps p_i onto q_i ($i = 1, 2, \ldots, m$), and the two m-tuples are f-superposable.

THEOREM 91.3. *If two triples of* $\mathscr{E}_{n,r}$ *are superposable, any congruence between them can be extended to the whole space.*

Proof. Let $(p_1, p_2, p_3) \cong (q_1, q_2, q_3)$ and suppose $p_1, p_2, p_3 \approx_f q_1, q_2, q_3$. Selecting associated matrices $(\epsilon_{ij} \cos p_i p_j / r)$, $(\epsilon'_{ij} \cos q_i q_j / r)$ $(i,j = 1, 2, 3)$, with $\epsilon_{1j} = \epsilon_{j1} = \epsilon'_{1j} = \epsilon'_{j1} = 1$ ($j = 1, 2, 3$), it follows from the superposability of the two triples that these matrices are strictly equivalent.

Then their determinants are equal, and expanding yields

$$\epsilon_{23} \cos(p_1 p_2/r)\cos(p_2 p_3/r)\cos(p_1 p_3/r)$$
$$= \epsilon'_{23} \cos(q_1 q_2/r)\cos(q_2 q_3/r)\cos(q_1 q_3/r).$$

If neither $p_1 p_2$ nor $p_1 p_3$ equals $\tfrac{1}{2}\pi r$, then $\epsilon_{23} = \epsilon'_{23}$ ($\epsilon_{23} = \epsilon'_{23} = 1$ in case $p_2 p_3 = \tfrac{1}{2}\pi r$) and the two associated matrices are *equal*; while if $p_1 p_2$ or $p_1 p_3$ equals $\tfrac{1}{2}\pi r$, then the existence of equal associated matrices of the ordered triples p_1, p_2, p_3 and q_1, q_2, q_3 is obvious. Hence the two triples are f-superposable.

Remark. We shall see later that Theorem 91.3 cannot be extended to m-tuples, $m > 3$, and so the superposability of two m-tuples of $\mathscr{E}_{n,r}$ ($m > 3$, $n > 1$) does not imply that *every* congruence between the m-tuples can be extended to a motion.

EXERCISES

1. In the example of two congruent, non-superposable quadruples given at the end of § 89, what triples of p_1, p_2, p_3, p_4 are superposable with triples of p_1, p_2, p_3, p'_4?
2. Give an example of two congruent, non-superposable quadruples of $\mathscr{E}_{2,r}$ with three pairs of superposable triples.

92. Congruent m-tuples with corresponding triples superposable

We are concerned in this and the next few sections with determining the smallest integer k such that each two congruent elliptic subsets are superposable whenever corresponding k-tuples of the subsets have that property. It will be seen that, in general, the desired integer k varies with the dimension of the elliptic space containing the two sets, but for a rather large class of sets the number is independent of the dimension of the containing space and is *three*.

THEOREM 92.1. *Let* $p_1, p_2, ..., p_m \approx_f q_1, q_2, ..., q_m$ *be two congruent subsets of* $\mathscr{E}_{n,r}$ *such that* (i) *there exists a point* p_k *of the first m-tuple with* $p_i p_k \neq \tfrac{1}{2}\pi r$ ($i = 1, 2, ..., m$), *and* (ii) *each triple of* $(p_1, p_2, ..., p_m)$ *containing* p_k *is superposable* (*not necessarily f-superposable*) *with the corresponding triple of* $(q_1, q_2, ..., q_m)$. *Then the two m-tuples are f-superposable.*

Proof. Writing the above congruence

$$p_k, p_1, p_2, ..., p_{k-1}, p_{k+1}, ..., p_m \approx q_k, q_1, q_2, ..., q_{k-1}, q_{k+1}, ..., q_m,$$

and making the cyclic substitution $(1, 2, ..., k)$ on the first k indices of *both* sets yields
$$p_1, p_2, ..., p_m \approx_f q_1, q_2, ..., q_m,$$
$$p_1 p_i \neq \tfrac{1}{2}\pi r \quad (i = 1, 2, ..., m). \tag{*}$$

Let $(\epsilon_{ij}\cos p_i p_j/r)$, $(\epsilon'_{ij}\cos q_i q_j/r)$ $(i,j = 1, 2,..., m)$ be two associated matrices of the respective ordered m-tuples, with

$$\epsilon_{1j} = \epsilon'_{1j} = 1 \quad (j = 1, 2,..., m).$$

Since $(p_1, p_s, p_t) \cong (q_1, q_s, q_t)$ $(s, t = 2, 3,..., m)$, the respective third-order principal sub-matrices of the above matrices which involve these triples are strictly equivalent, and hence the corresponding third-order principal minors are equal. Expanding and taking into account (∗), together with $\epsilon_{ij} = 1$ if $p_i p_j = \frac{1}{2}\pi r$, gives $\epsilon_{st} = \epsilon'_{st}$ $(s, t = 2, 3,..., m)$. Then $\epsilon_{ij} = \epsilon'_{ij}$ $(i, j = 1, 2,..., m)$ and the two associated matrices are equal. Hence the ordered m-tuples are f-superposable.

The reader will observe that the restriction imposed by hypothesis (i) of the theorem is essential to our argument, for if $p_1 p_s = \frac{1}{2}\pi r$ the superposability of triples p_1, p_s, p_t and q_1, q_s, q_t does not imply that $\epsilon_{st} = \epsilon'_{st}$. It is, moreover, essential for the validity of the theorem, for an example of two congruent non-superposable quadruples of $\mathscr{E}_{2,r}$ with three pairs of superposable triples is easily given (§ 91, Exercise 2).

A weaker property than that proved in the preceding theorem is of interest. We state it as a corollary.

COROLLARY. *If P, Q are m-tuples of $\mathscr{E}_{n,r}$ and no two points of P have distance $\frac{1}{2}\pi r$, then any congruence between P and Q can be extended to a motion provided corresponding triples are superposable.*

It will be seen later that we may dispense with the restriction of P and Q to *finite* subsets. Thus superposability is reduced to a problem of *triples* for a large class of subsets of $\mathscr{E}_{n,r}$.

THEOREM 92.2. *Two f-congruent m-tuples*

$$p_1, p_2,..., p_m \approx_f q_1, q_2,..., q_m$$

of $\mathscr{E}_{n,r}$ are f-superposable if and only if, for a pair p_k, q_k of corresponding points, arbitrarily small spherical neighbourhoods $U(p_k; \rho)$, $U(q_k; \rho)$ of radius ρ exist with the property that corresponding to each point p of $U(p_k; \rho)$ with $p_j p \neq \frac{1}{2}\pi r$ $(j = 1, 2,..., k-1, k+1,..., m)$ there is a point q of $U(q_k; \rho)$ such that the two congruent m-tuples

$$p_1, p_2,..., p_{k-1}, p, p_{k+1},..., p_m \approx q_1, q_2,..., q_{k-1}, q, q_{k+1},..., q_m$$

have those corresponding triples containing p and q superposable.

Proof. If $p_1, p_2,..., p_m \cong_f q_1, q_2,..., q_m$, then a motion Γ of the space, with $\Gamma(p_i) = q_i$ $(i = 1, 2,..., m)$, carries arbitrarily small neighbourhoods $U(p_k; \rho)$ into congruent arbitrarily small neighbourhoods $U(q_k; \rho)$, and hence takes each p of $U(p_k;\rho)$ into a corresponding point q of $U(q_k;\rho)$. Then

$$p_1, p_2, \ldots, p_{k-1}, p, p_{k+1}, \ldots, p_m \cong q_1, q_2, \ldots, q_{k-1}, q, q_{k+1}, \ldots, q_m,$$

and the conclusion of the theorem is valid *a fortiori*.

Conversely, let $\{U_i(p_k;\rho_i)\}$, $\{U_i(q_k;\rho_i)\}$ be sequences of neighbourhoods of p_k, q_k, respectively, with $\rho_i < 1/i$ ($i = 1, 2, \ldots$), which have the properties stated in the theorem. Now each $U_i(p_k;\rho_i)$ contains a point p'_i such that $p'_i p_j \neq \frac{1}{2}\pi r$ ($j = 1, 2, \ldots, k-1, k+1, \ldots, m$), and hence a corresponding point q'_i of $U_i(q_k;\rho_i)$ exists such that

$$p_1, p_2, \ldots, p_{k-1}, p'_i, p_{k+1}, \ldots, p_m \approx q_1, q_2, \ldots, q_{k-1}, q'_i, q_{k+1}, \ldots, q_m$$

with corresponding triples containing p'_i and q'_i, respectively, superposable.

It follows from Theorem 92.1 that

$$p_1, p_2, \ldots, p_{k-1}, p'_i, p_{k+1}, \ldots, p_m \cong q_1, q_2, \ldots, q_{k-1}, q'_i, q_{k+1}, \ldots, q_m;$$

and so an infinite sequence $\{\Gamma_i\}$ of motions exists with

$$\Gamma_i(p_j) = q_j \quad (j = 1, 2, \ldots, k-1, k+1, \ldots, m),$$
and
$$\Gamma_i(p'_i) = q'_i \ (i = 1, 2, \ldots).$$

From a theorem of van Dantzig [1] and van der Waerden, the infinite sequence $\{\Gamma_i\}$ contains a convergent subsequence $\{\Gamma_{j_i}\}$ with a motion Γ for limit; that is,

$$\Gamma_{j_i}(p_1, p_2, \ldots, p_{k-1}, p'_{j_i}, p_{k+1}, \ldots, p_m) = (q_1, q_2, \ldots, q_{k-1}, q'_{j_i}, q_{k+1}, \ldots, q_m)$$

($i = 1, 2, \ldots$). Since $\lim\limits_{i \to \infty} q'_{j_i} = q_k$, $\lim\limits_{i \to \infty} \Gamma_{j_i}(p'_{j_i}) = \Gamma(p_k)$, then

$$\Gamma(p_1, p_2, \ldots, p_{k-1}, p_k, p_{k+1}, \ldots, p_m) = (q_1, q_2, \ldots, q_{k-1}, q_k, q_{k+1}, \ldots, q_m),$$

with $\Gamma(p_i) = q_i$ ($i = 1, 2, \ldots, m$), and the theorem is proved.

93. Superposability of infinite subsets of $\mathscr{E}_{n,r}$

We show in this section that any two elliptic sets are superposable whenever there exists a one-to-one correspondence between their points such that, for every positive integer m, each two corresponding m-tuples are superposable.

THEOREM 93.1. *Let P, Q be two subsets of $\mathscr{E}_{n,r}$ whose points are in a one-to-one correspondence such that each finite subset of P is superposable with its corresponding subset of Q. Then this property can be extended to the closures $\bar{P}$, $\bar{Q}$ of those sets.*

Proof. Since corresponding pairs of points of P and Q are superposable, the one-to-one correspondence is a congruence which can be extended in the usual manner to obtain $\bar{P} \approx \bar{Q}$.

Let $p_1, p_2, \ldots, p_m$ and $q_1, q_2, \ldots, q_m$ be corresponding m-tuples in that congruence. Then, for each p_i and q_i, infinite sequences $p_{i1}, p_{i2}, \ldots$, and $q_{i1}, q_{i2}, \ldots$ of corresponding points of P and Q, respectively, exist with $\lim_{k \to \infty} p_{ik} = p_i$, $\lim_{k \to \infty} q_{ik} = q_i$ ($i = 1, 2, \ldots, m$). If Γ_k denotes the congruent mapping of $\mathscr{E}_{n,r}$ onto itself which carries the points $p_{1k}, p_{2k}, \ldots, p_{mk}$ into the points $q_{1k}, q_{2k}, \ldots, q_{mk}$, respectively ($k = 1, 2, \ldots$), it follows from the van Dantzig–van der Waerden theorem cited above that the infinite sequence $\{\Gamma_k\}$ of motions contains a subsequence $\{\Gamma_{j_k}\}$ with limit Γ, a motion.

Selecting appropriate subsequences, $\lim_{k \to \infty} p_{ij_k} = p_i$, $\lim_{k \to \infty} q_{ij_k} = q_i$, and

$$\lim_{k \to \infty} \Gamma_{j_k}(p_{1j_k}, p_{2j_k}, \ldots, p_{mj_k}) = \lim_{k \to \infty} (q_{1j_k}, q_{2j_k}, \ldots, q_{mj_k})$$
$$= (q_1, q_2, \ldots, q_m).$$

Now, by the triangle inequality,

$$0 \leqslant \Gamma_{j_k}(p_{ij_k})\Gamma(p_i) \leqslant \Gamma_{j_k}(p_{ij_k})\Gamma_{j_k}(p_i) + \Gamma_{j_k}(p_i)\Gamma(p_i)$$
$$\leqslant p_{ij_k}p_i + \Gamma_{j_k}(p_i)\Gamma(p_i)$$

($i = 1, 2, \ldots, m$), and hence $\lim_{k \to \infty} \Gamma_{j_k}(p_{ij_k})\Gamma(p_i) = 0$, since $\lim_{k \to \infty} \Gamma_{j_k}(p) = \Gamma(p)$ for every element p of $\mathscr{E}_{n,r}$.

Thus, $\lim_{k \to \infty} \Gamma_{j_k}(p_{ij_k}) = \Gamma(p_i)$ ($i = 1, 2, \ldots, m$); that is, $\Gamma(p_i) = q_i$ ($i = 1, 2, \ldots, m$), and the theorem is proved.

THEOREM 93.2. *If P, Q are two subsets of $\mathscr{E}_{n,r}$ whose points are in a one-to-one correspondence f such that each finite subset of P is f-superposable with the corresponding subset of Q, then P and Q are f-superposable.*

Proof. The theorem being trivial for P and Q finite subsets, suppose them infinite. As in the preceding theorem, the correspondence is seen to be a congruence $P \approx_f Q$ which is extended to $\bar{P} \approx_{\bar{f}} \bar{Q}$, where $\bar{f}$ denotes the extension of the congruence f.

The closed subset $\bar{P}$ of the compact metric space $\mathscr{E}_{n,r}$ is the closure of one of its denumerable subsets $p_1, p_2, \ldots, p_k, \ldots$ (corollary, Theorem 10.6). If we denote by $q_1, q_2, \ldots, q_k, \ldots$ the subset of $\bar{Q}$ corresponding to $p_1, p_2, \ldots, p_k, \ldots$ by the above congruence of $\bar{P}$ and $\bar{Q}$ ($q_i = \bar{f}(p_i)$ ($i = 1, 2, \ldots$)), then by the preceding theorem there exists, for every positive integer m, a motion Γ_m such that

$$\Gamma_m(p_i) = q_i = \bar{f}(p_i) \quad (i = 1, 2, \ldots, m).$$

Letting $\{\Gamma_{i_m}\} \to \Gamma$, it follows that $\Gamma(p_i) = \lim_{m \to \infty} \Gamma_{i_m}(p_i) = q_i$ ($i = 1, 2, \ldots$), and so

$$p_1, p_2, \ldots, p_k, \ldots \cong q_1, q_2, \ldots, q_k, \ldots .$$

The motion Γ which superposes these two denumerable sets evidently superposes their respective closures $\bar{P}$ and $\bar{Q}$. From $\bar{P} \cong_f \bar{Q}$ follows $P \cong_f Q$, and the theorem is proved.

94. First reduction theorem

We have shown in the preceding section that any two f-congruent subsets of $\mathscr{E}_{n,r}$ are f-superposable provided each finite subset of the one is f-superposable with the corresponding subset of the other. This section establishes a reduction theorem that permits replacing the assumption of superposability of each two (corresponding) finite subsets by the much milder restriction that corresponding $(n+2)$-tuples be f-superposable.

THEOREM 94.1. *If P, Q are any two subsets of $\mathscr{E}_{n,r}$ with $P \approx_f Q$, then $P \cong_f Q$ if and only if each two corresponding $(n+2)$-tuples of P and Q are f-superposable.*

Proof. The necessity is obvious. To prove the sufficiency it suffices to show that each two f-corresponding finite subsets of P and Q (containing more than $n+2$ pairwise distinct points) are f-superposable.

Let P^*, Q^* be f-corresponding subsets of P, Q, respectively, consisting of $m+1$ points, $m > n+1$. We wish to prove the assertion $P^* \cong_f Q^*$. From the congruence of $\mathscr{E}_{1,r}$ with $S_{1,\frac{1}{2}r}$ and the f-superposability of any two f-congruent subsets of $S_{1,\frac{1}{2}r}$, the assertion is true for $n = 1$. This anchors an inductive hypothesis (on the dimension n) of its validity for each positive integer $k \leqslant n-1$, by virtue of which the assertion is established when both P^* and Q^* lie in $(n-1)$-dimensional subspaces of $\mathscr{E}_{n,r}$.†

Let us suppose the labelling so that P^* does not lie in any $\mathscr{E}_{n-1,r}$. Now make the inductive assumption that f-corresponding m-tuples of P^* and Q^* are f-superposable—an assumption that is valid by hypothesis for $m = n+2$. Two cases are distinguished:

Case 1. There is a labelling $p_1, p_2,..., p_i, p_{i+1},..., p_{m+1}$ of the points of P^* ($1 \leqslant i \leqslant m$) such that each of the points $p_{i+1}, p_{i+2},..., p_{m+1}$ has distance $\frac{1}{2}\pi r$ from each of the points $p_1, p_2,..., p_i$.

We label the points of Q^* so that $q_j = f(p_j)$ ($j = 1, 2,..., m+1$), and so

$$p_1, p_2,..., p_i, p_{i+1},..., p_{m+1} \approx_f q_1, q_2,..., q_i, q_{i+1},..., q_{m+1}.$$

† If they lie in different $(n-1)$-dimensional subspaces of $\mathscr{E}_{n,r}$, a motion of $\mathscr{E}_{n,r}$ will carry them into the same $\mathscr{E}_{n-1,r}$.

By inductive hypothesis,
$$p_1, p_2, \ldots, p_i \cong q_1, q_2, \ldots, q_i,$$
$$p_{i+1}, p_{i+2}, \ldots, p_{m+1} \cong q_{i+1}, q_{i+2}, \ldots, q_{m+1},$$
and consequently *equal* associated matrices
$$(\epsilon_{st} \cos p_s p_t / r) = (\epsilon_{st} \cos q_s q_t / r)$$
exist for $(s, t = 1, 2, \ldots, i)$ and $(s, t = i+1, i+2, \ldots, m+1)$.

Since $p_s p_t = q_s q_t = \tfrac{1}{2}\pi r$ $(s = 1, 2, \ldots, i;\ t = i+1, i+2, \ldots, m+1)$, we obtain from the above matrices the equal matrices
$$(\epsilon_{st} \cos p_s p_t / r), \qquad (\epsilon_{st} \cos q_s q_t / r) \quad (s, t = 1, 2, \ldots, m+1),$$
where $\epsilon_{st} = \epsilon_{ts} = 1$ $(s = 1, 2, \ldots, i;\ t = i+1, i+2, \ldots, m+1)$, which are readily seen to be associated matrices of $p_1, p_2, \ldots, p_{m+1}$ and $q_1, q_2, \ldots, q_{m+1}$ respectively. It follows that those two $(m+1)$-tuples are f-superposable.

Case 2. *The points of* P^* *may not be labelled as in Case* 1. We assert that in this case an ordering of the points of P^* exists such that (1) no point is a pole of each preceding point, and (2) for each integer i $(1 \leqslant i \leqslant n+1)$ the first i points determine a unique $\mathscr{E}_{i-1,r}$ containing them. For if $p_1, p_2, \ldots, p_j$ are j points of P^* with properties (1) and (2), $j \leqslant n$, then for at least one other point, say p_{j+1}, of P^*, the sequence $p_1, p_2, \ldots, p_j, p_{j+1}$ enjoys these properties also, since if the contrary be assumed it follows from the hypothesis of Case 2 that each of the remaining $m-j+1$ points of P^* is either contained in the $\mathscr{E}_{j-1,r}$ determined uniquely by $p_1, p_2, \ldots, p_j$ or has distance $\tfrac{1}{2}\pi r$ from each of those j points. A point with the latter property is a pole of $\mathscr{E}_{j-1,r}(p_1, p_2, \ldots, p_j)$ and has distance $\tfrac{1}{2}\pi r$ from each point of that subspace. Since $j \leqslant n$, P^* is not contained in $\mathscr{E}_{j-1,r}(p_1, p_2, \ldots, p_j)$, and it is clear that a labelling of P^* contrary to the condition of Case 2 must exist.

Now a sequence $p_1, p_2, \ldots, p_j$ of points of P^* with properties (1), (2) obviously exists for $j = 2$ and so the assertion is established by complete induction.

Let, then, $p_1, p_2, \ldots, p_{m+1}$ be an arrangement of the points of P^* with properties (1), (2) and label the points of Q^* so that
$$q_i = f(p_i) \quad (i = 1, 2, \ldots, m+1).$$
Then $\quad p_1, p_2, \ldots, p_n, p_{n+1}, \ldots, p_{m+1} \approx_f q_1, q_2, \ldots, q_n, q_{n+1}, \ldots, q_{m+1}$
with both $(m+1)$-tuples in left-hand *apolar order*. By the inductive hypothesis a motion Γ exists such that
$$\Gamma(q_i) = p_i \quad (i = 1, 2, \ldots, m-1, m+1).$$

If we denote $\Gamma(q_m)$ by $\bar{p}_m$, then

$$p_1, p_2, \ldots, p_{m-1}, p_m \cong_f q_1, q_2, \ldots, q_{m-1}, q_m \cong_f p_1, p_2, \ldots, p_{m-1}, \bar{p}_m. \quad (\dagger)$$

Since $p_1, p_2, \ldots, p_{n+1}$ do not lie in any $\mathscr{E}_{n-1,r}$ and $m-1 \geqslant n+1$, the points $p_1, p_2, \ldots, p_{m-1}$ are contained *irreducibly* in $\mathscr{E}_{n,r}$. It follows now, from the apolar ordering of $p_1, p_2, \ldots, p_{m+1}$, that the only motion of $\mathscr{E}_{n,r}$ which superposes the first and third of the m-tuples in ($\dagger$) is the identity. Hence $\bar{p}_m = p_m$ and $P^* \cong_f Q^*$.

EXERCISE

Let $p_1, p_2, \ldots, p_{n+1}$ be contained irreducibly in $\mathscr{E}_{n,r}$ and suppose the $(n+1)$-tuple is in left-hand apolar order. Show that the identity is the only motion of $\mathscr{E}_{n,r}$ that leaves each of the points p_i ($i = 1, 2, \ldots, n+1$) fixed. (*Hint.* If $p_1 p_2 \neq \tfrac{1}{2}\pi r$, a motion Γ with $\Gamma(p_i) = p_i$ ($i = 1, 2$) leaves $\mathscr{E}_{1,r}(p_1, p_2)$ pointwise invariant.)

95. Two lemmas

We prove now two lemmas that are used to establish the final reduction theorem (from $n+2$ to $n+1$) obtained in the next section.

LEMMA 95.1. *Let P be a set of $n+2$ points of $\mathscr{E}_{n,r}$, with no $(n+1)$-tuple in any $\mathscr{E}_{n-1,r}$. A right-hand apolar ordering of P exists such that the distance between the third and the last points is not $\tfrac{1}{2}\pi r$.*

Proof. To show that at least one right-hand apolar ordering of P exists, let $p_{i+1}, p_{i+2}, \ldots, p_{n+1}, p_{n+2}$ be $n-i+2$ points of P in right-hand apolar order. Then for at least one of the remaining points of P (label it p_i) the sequence $p_i, p_{i+1}, p_{i+2}, \ldots, p_{n+1}, p_{n+2}$ is in right-hand apolar order, for in the contrary case each of the remaining i points of P has distance $\tfrac{1}{2}\pi r$ from each of the points p_j ($j = i+1, i+2, \ldots, n+2$) and hence lies in the $\mathscr{E}_{i-2,r}$ of $(i-2)$ dimensions which is the absolute polar of the *unique* $\mathscr{E}_{n-i+1,r}$ containing $p_{i+1}, p_{i+2}, \ldots, p_{n+2}$.† But if i points of P lie in an $\mathscr{E}_{i-2,r}$, then any $(n+1)$-tuple of P containing them lies in an $\mathscr{E}_{n-1,r}$, contrary to hypothesis.

It is obvious that the $\mathscr{E}_{n,r}$ does not contain an equilateral $(n+2)$-tuple with edge $\tfrac{1}{2}\pi r$ (no epsilon determinant of such an $(n+2)$-tuple vanishes) and so two points of P may be labelled p_{n+1}, p_{n+2} with the pair in right-hand apolar order. It follows by complete induction that a right-hand apolar ordering of P exists.

Let, now, $p_1, p_2, p_3, p_4, \ldots, p_{n+1}, p_{n+2}$ be a right-hand apolar ordering of P. Then at least one of the points $p_4, p_5, \ldots, p_{n+2}$ has distance from p_3 different from $\tfrac{1}{2}\pi r$. Let p_{i_1} be the point with greatest index

† Since every $(n+1)$-tuple of P is contained irreducibly in $\mathscr{E}_{n,r}$, each $(j+1)$-tuple of P ($j = 1, 2, \ldots, n$) determines a unique subspace of j dimensions containing it.

($3 \leqslant i_1 \leqslant n+2$) having this property. If $i_1 < n+2$, denote by p_{i_2} ($i_1 < i_2 \leqslant n+2$) the point of greatest index with $p_{i_1} p_{i_2} \neq \frac{1}{2}\pi r$.

After m applications of this process, the point p_{n+2} is reached as the point of greatest index such that $p_{i_m} p_{n+2} \neq \frac{1}{2}\pi r$. Then the ordering

$$p_1, p_2, p_3, \ldots, p_{i_1-1}, p_{i_1+1}, \ldots, p_{i_2-1}, p_{i_2+1}, \ldots$$
$$\ldots, p_{i_m-1}, p_{i_m+1}, \ldots, p_{n+2}, p_{i_m}, p_{i_{m-1}}, \ldots, p_{i_2}, p_{i_1}$$

is readily seen to be right-handed apolar, and since $p_3 p_{i_1} \neq \frac{1}{2}\pi r$ the lemma is proved.

LEMMA 95.2. *Let P be a set of $n+2$ points of $\mathscr{E}_{n,r}$, with no $(n+1)$-tuple in any $\mathscr{E}_{n-1,r}$, and let Q be a subset of $\mathscr{E}_{n,r}$ whose points are in a one-to-one correspondence f with those of P. If* (i) *each two f-corresponding $(n+1)$-tuples of P and Q are f-superposable, and* (ii) *P and Q are not f-superposable, then in each right-hand apolar ordering of P either the first point or the second point has distance $\frac{1}{2}\pi r$ from each of the last $n-1$ points.*

Proof. Let $p_1, p_2, \ldots, p_n, p_{n+1}, p_{n+2}$ be any right-hand apolar ordering of P and suppose $\Gamma(q_i) = p_i$ ($i = 2, 3, \ldots, n+2$), while $\Gamma(q_1) = \bar{p}_1 \neq p_1$. Since

$$p_1, p_3, \ldots, p_n, p_{n+1}, p_{n+2} \cong q_1, q_3, \ldots, q_n, q_{n+1}, q_{n+2}$$
$$\cong \bar{p}_1, p_3, \ldots, p_n, p_{n+1}, p_{n+2},$$

it follows from the apolar ordering of P that the hyperplane

$$\mathscr{E}_{n-1,r}(p_3, p_4, \ldots, p_{n+1}, p_{n+2})$$

remains pointwise fixed under a motion which superposes the first and third of the above $(n+1)$-tuples. Hence this hyperplane is part of the locus of points of $\mathscr{E}_{n,r}$ equidistant from p_1 and $\bar{p}_1$.

In similar manner

$$p_1, p_2, p_4, \ldots, p_{n+1}, p_{n+2} \cong \bar{p}_1, p_2, p_4, \ldots, p_{n+1}, p_{n+2},$$

and *if p_2 does not have distance $\frac{1}{2}\pi r$ from each of the points $p_4, p_5, \ldots, p_{n+2}$,* then $\mathscr{E}_{n-1,r}(p_2, p_4, \ldots, p_{n+1}, p_{n+2})$ also forms part of the same equidistant locus. Since these two hyperplanes are distinct (otherwise $p_2, p_3, \ldots, p_{n+1}, p_{n+2}$ lie in an $\mathscr{E}_{n-1,r}$), they intersect in the

$$\mathscr{E}_{n-2,r}(p_4, p_5, \ldots, p_{n+2})$$

which is the absolute polar subspace of the line $\mathscr{E}_{1,r}(p_1, \bar{p}_1)$. Hence

$$p_1 p_4 = p_1 p_5 = \ldots = p_1 p_{n+2} = \tfrac{1}{2}\pi r,$$

and the lemma is established.

Remark. Though the hypothesis of Lemma 95.2 assumes that each two corresponding $(n+1)$-tuples of P and Q are f-superposable, the proof makes use of only three pairs of f-superposable $(n+1)$-tuples. Hence a somewhat stronger property than that asserted is proved.

EXERCISE

Let P and Q be two $(n+1)$-tuples of $\mathscr{E}_{n,r}$ $(n > 1)$ whose points are in a one-to-one correspondence f such that each two corresponding n-tuples are f-superposable. If P and Q are not f-superposable, and no n-tuple of either set lies in an $\mathscr{E}_{n-2,r}$, show that (1) a right-hand apolar ordering of P exists for which the distance between the third and last points is not $\frac{1}{2}\pi r$, and (2) in each right-hand apolar ordering of P either the first point or the second point has distance $\frac{1}{2}\pi r$ from each of the last $n-2$ points.

96. Second reduction theorem

We are now in position to establish the final reduction theorem.

THEOREM 96.1. *Two subsets of $\mathscr{E}_{n,r}$, whose points are in a one-to-one correspondence f, are f-superposable if and only if each two corresponding $(n+1)$-tuples are f-superposable.*

Proof. The necessity is clear. To prove the sufficiency it suffices to show that each two f-corresponding $(n+2)$-tuples are f-superposable (Theorem 94.1).

Let, now, P and Q be any two $(n+2)$-tuples of $\mathscr{E}_{n,r}$ whose points are in a one-to-one correspondence f such that each $n+1$ points of P are f-superposable with the corresponding $n+1$ points of Q. We shall show that $P \cong_f Q$.

Make the inductive hypothesis (anchored for $k = 1$) that for every positive integer k $(k \leqslant n-1)$ any two f-corresponding $(k+2)$-tuples of $\mathscr{E}_{k,r}$ are f-superposable provided each of their f-corresponding $(k+1)$-tuples are f-superposable. We distinguish two cases.

Case 1. An $(n+1)$-tuple of P lies in an $\mathscr{E}_{n-1,r}$.

Let the labelling be selected so that $p_1, p_2, ..., p_{n+1}$ is such a subset of P. If these points are contained in an $\mathscr{E}_{n-2,r}$, then clearly the corresponding $(n+1)$-tuple of Q, *superposable with it*, is also contained in an $\mathscr{E}_{n-2,r}$. But then P and Q lie in $(n-1)$-dimensional subspaces and $P \cong_f Q$ by Theorem 94.1. We may suppose, therefore, that $p_1, p_2, ..., p_{n+1}$ are irreducibly contained in an $\mathscr{E}_{n-1,r}$ (determined by them), and that this $\mathscr{E}_{n-1,r}(p_1, p_2, ..., p_{n+1})$ does not contain the remaining point p_0 of P.†

---

† If $p_0 \in \mathscr{E}_{n-1,r}(p_1, p_2, ..., p_{n+1})$, then, since each $(n+1)$-tuple of P is *superposable* with its corresponding $(n+1)$-tuple of Q, every $(n+1)$-tuple of Q lies in an $\mathscr{E}_{n-1,r}$ and it follows readily that Q does also. Theorem 94.1 then ensures $P \cong_f Q$.

Writing $q_i = f(p_i)$ $(i = 0, 1, ..., n+1)$, let p and q denote feet of p_0 and q_0, respectively, on the hyperplanes

$$\mathcal{E}_{n-1,r}(p_1, p_2, ..., p_{n+1}), \qquad \mathcal{E}_{n-1,r}(q_1, q_2, ..., q_{n+1}).$$

If p or q is not unique, then p_0 *and* q_0 are poles of the (respective) hyperplanes, and a motion of $\mathcal{E}_{n,r}$ that superposes the ordered sets $p_1, p_2, ..., p_{n+1}$ and $q_1, q_2, ..., q_{n+1}$ evidently carries p_0 and q_0 into coincidence. Assuming that p and q are unique, we prove the following assertion.

ASSERTION. $p, p_1, p_2, ..., p_{n+1} \cong q, q_1, q_2, ..., q_{n+1}$.

Since these two $(n+2)$-tuples lie in $(n-1)$-dimensional subspaces, the assertion follows from Theorem 94.1 and the inductive hypothesis provided each two corresponding n-tuples are superposable (in the order indicated). Now $p_1, p_2, ..., p_{n+1} \cong q_1, q_2, ..., q_{n+1}$, and so it suffices to consider only those corresponding n-tuples of the two sets in the assertion that contain, respectively, p and q; say

$$p, p_{i_1}, p_{i_2}, ..., p_{i_{n-1}} \quad \text{and} \quad q, q_{i_1}, q_{i_2}, ..., q_{i_{n-1}}.$$

Denote the two remaining points of the $(n+1)$-tuples $p_1, p_2, ..., p_n, p_{n+1}$ and $q_1, q_2, ..., q_{n+1}$ by $p_{i_n}, p_{i_{n+1}}$ and $q_{i_n}, q_{i_{n+1}}$, respectively, where

$$q_{i_j} = f(p_{i_j}) \quad (j = n, n+1).$$

If $p_{i_1}, p_{i_2}, ..., p_{i_{n-1}}, p_{i_j}$ are *not* in an $\mathcal{E}_{n-2,r}$, where $j = n$ or $j = n+1$, a motion Γ such that $\Gamma(p_0) = q_0$, $\Gamma(p_{i_k}) = q_{i_k}$ $(k = 1, 2, ..., n-1)$, $\Gamma(p_{i_j}) = q_{i_j}$, carries $\mathcal{E}_{n-1,r}(p_1, p_2, ..., p_{n+1})$ into $\mathcal{E}_{n-1,r}(q_1, q_2, ..., q_{n+1})$ and hence carries p into q. Thus whenever $n-1$ of the points $p_1, p_2, ..., p_{n+1}$ form with one of the remaining two points an n-tuple irreducibly contained in $\mathcal{E}_{n-1,r}$, the n-tuple consisting of those $n-1$ points and p is superposable with the corresponding n-tuple of $q, q_1, ..., q_n, q_{n+1}$.

We must consider, finally, an $(n-1)$-tuple $p_{i_1}, p_{i_2}, ..., p_{i_{n-1}}$ such that $p_{i_1}, p_{i_2}, ..., p_{i_{n-1}}, p_{i_j}$ belong to $(n-2)$-dimensional subspaces for $j = n$ and $j = n+1$. Then $p_{i_1}, p_{i_2}, ..., p_{i_{n-1}} \in \mathcal{E}_{n-3,r}$ (*irreducibly*) since otherwise it is readily seen that $p_1, p_2, ..., p_{n+1}$ lie in an $\mathcal{E}_{n-2,r}$, contrary to assumption. To show

$$p, p_{i_1}, p_{i_2}, ..., p_{i_{n-1}} \cong q, q_{i_1}, q_{i_2}, ..., q_{i_{n-1}}$$

in this event, it suffices to show that each ordered $(n-1)$-tuple of the first set (which contains p) is superposable with the corresponding $(n-1)$-tuple of the second set (since both n-tuples are contained in $(n-2)$-dimensional subspaces).

Consider, for example, the $(n-1)$-tuples $p, p_{i_1}, p_{i_2}, ..., p_{i_{n-2}}$ and $q, q_{i_1}, q_{i_2}, ..., q_{i_{n-2}}$. The points $p_{i_1}, p_{i_2}, ..., p_{i_{n-2}}, p_{i_n}, p_{i_{n+1}}$ *do not lie in any* $\mathcal{E}_{n-2,r}$, for if they did then $p_{i_{n-1}}$ would belong to that $\mathcal{E}_{n-2,r}$ (since it belongs to an $(n-3)$-dimensional subspace of it) and hence $p_1, p_2,..., p_{n+1}$ would lie in an $\mathscr{E}_{n-2,r}$. It follows that a motion Γ with

$$\Gamma(p_0, p_{i_1},..., p_{i_{n-2}}, p_{i_n}, p_{i_{n+1}}) = (q_0, q_{i_1},..., q_{i_{n-2}}, q_{i_n}, q_{i_{n+1}})$$

(corresponding points as indicated) carries $\mathscr{E}_{n-1,r}(p_1, p_2,..., p_{n+1})$ into $\mathscr{E}_{n-1,r}(q_1, q_2,..., q_{n+1})$ and hence $\Gamma(p) = q$.

The other pairs of corresponding $(n-1)$-tuples (containing p and q) are treated in a similar manner and so

$$p, p_{i_1}, p_{i_2},..., p_{i_{n-1}} \cong q, q_{i_1}, q_{i_2},..., q_{i_{n-1}}$$

for every set of indices $i_1, i_2,..., i_{n-1}$ selected from 1, 2,..., $n+1$. This establishes the assertion.

Let, now, Γ be any motion such that

$$\Gamma(p_i) = q_i \quad (i = 1, 2,..., n+1) \qquad \Gamma(p) = q.$$

This motion carries $\mathscr{E}_{n-1,r}(p_1, p_2,..., p_{n+1})$ into $\mathscr{E}_{n-1,r}(q_1, q_2,..., q_{n+1})$ and p_0 into a point $\Gamma(p_0)$ whose foot on $\mathscr{E}_{n-1,r}(q_1, q_2,..., q_{n+1})$ is q. Consequently $\Gamma(p_0) = q_0$ or $\Gamma(p_0) = \bar{q}_0$, the reflection of q in $\mathscr{E}_{n-1,r}(q_1, q_2,..., q_{n-1})$. In the latter event, following Γ by a rotation of $\mathscr{E}_{n,r}$ 180° about the pole of $\mathscr{E}_{n-1,r}(q_1, q_2,..., q_{n+1})$ superposes $p_0, p_1,..., p_{n+1}$ and $q_0, q_1,..., q_{n+1}$ (in the order indicated) and the theorem is proved in this case.

Case 2. *No* $(n+1)$-*tuple of P lies in an* $\mathscr{E}_{n-1,r}$.

According to Lemma 95.1 a right-hand apolar ordering of P exists such that the distance between the third and the last points is not $\frac{1}{2}\pi r$. Let

$$p_1, p_2, p_{n+2}, p_3,..., p_n, p_{n+1},$$
$$p_{n+1}p_{n+2} \neq \tfrac{1}{2}\pi r, \qquad (*)$$

be such an ordering of P.

Assuming, now, that P and Q are not f-superposable, Lemma 95.2 permits the assertion that either

$$p_1 p_3 = p_1 p_4 = ... = p_1 p_n = p_1 p_{n+1} = \tfrac{1}{2}\pi r, \qquad (**)$$
or $\qquad p_2 p_3 = p_2 p_4 = ... = p_2 p_n = p_2 p_{n+1} = \tfrac{1}{2}\pi r.$

It is clear that if the first set of relations do not hold, then

$$p_2, p_1, p_{n+2}, p_3,..., p_n, p_{n+1}$$

is another right-hand apolar ordering of P with $p_{n+1}p_{n+2} \neq \tfrac{1}{2}\pi r$, with the second set of relations holding. Interchanging the labellings of p_1 and p_2 restores the apolar ordering (*) and relations (**) hold. We may, therefore, assume the labelling of the points of P so that ordering (*) and relations (**) subsist.

If we label the points of Q so that $q_i = f(p_i)$ $(i = 1, 2, ..., n+2)$, it follows easily from a comparison of two associated matrices of P and Q (and the assumption that these two $(n+2)$-tuples are *not* f-superposable) that
$$p_1 p_2 \neq \tfrac{1}{2}\pi r \neq p_1 p_{n+2}.\dagger \qquad (***)$$
Since $p_{n+1} p_{n+2} \neq \tfrac{1}{2}\pi r$, the arrangement
$$p_1, p_2, p_3, ..., p_{n+1}, p_{n+2}$$
is a right-hand apolar ordering, and Lemma 95.2 (together with $p_1 p_{n+2} \neq \tfrac{1}{2}\pi r$) gives
$$p_2 p_4 = p_2 p_5 = ... = p_2 p_{n+1} = p_2 p_{n+2} = \tfrac{1}{2}\pi r,$$
while, as before, $p_2 p_3 \neq \tfrac{1}{2}\pi r$. Similarly, from the apolar ordering $p_2, p_3, ..., p_{n+1}, p_{n+2}, p_1$ it follows that $p_3 p_5 = p_3 p_6 = ... = p_3 p_{n+2} = \tfrac{1}{2}\pi r$ and $p_3 p_4 \neq \tfrac{1}{2}\pi r$.

If we assume $p_1 p_2, p_2 p_3, ..., p_{i-1} p_i$ each different from $\tfrac{1}{2}\pi r$, and each of the points $p_1, p_2, ..., p_{i-1}$ has distance $\tfrac{1}{2}\pi r$ from all points of P except their immediate predecessor and successor ($p_0 = p_{n+2}$), then
$$p_{i-1}, p_i, p_{i+1}, p_{i+2}, ..., p_{n+2}, p_1, p_2, ..., p_{i-2}$$
are in right-hand apolar order, and since $p_{i-1} p_{i-2} \neq \tfrac{1}{2}\pi r$, Lemma 95.2 yields
$$p_i p_{i+2} = p_i p_{i+3} = ... = p_i p_{n+2} = p_i p_1 = p_i p_2 = ... = p_i p_{i-2} = \tfrac{1}{2}\pi r,$$
while, since P and Q are assumed not f-superposable, $p_i p_{i+1} \neq \tfrac{1}{2}\pi r$.

Thus each point p_i of P $(1 \leq i \leq n+2)$ has distance $\tfrac{1}{2}\pi r$ from exactly $n-1$ points of P, and $p_{i-1} p_i \neq \tfrac{1}{2}\pi r \neq p_i p_{i+1}$ ($p_0 = p_{n+2}, p_1 = p_{n+3}$), with the corresponding statement valid for the set Q, since P and Q are f-congruent.

Let $(\epsilon_{ij} \cos p_i p_j / r)$ and $(\epsilon'_{ij} \cos q_i q_j / r)$ $(i, j = 1, 2, ..., n+2)$ be associated matrices of the ordered $(n+2)$-tuples P and Q, respectively. Using the distance relations obtained above, the f-superposability of corresponding $(n+1)$-tuples, and the assumption that P and Q are *not* f-superposable, it is readily seen that these two matrices may be selected so that
$$\epsilon_{ij} = \epsilon'_{ij} = 1 \quad (i, j = 1, 2, ..., n+2),$$
except that $\epsilon'_{1, n+2} = -1 = \epsilon'_{n+2, 1}$. Now
$$|\epsilon_{ij} \cos p_i p_j / r| = 0 = |\epsilon'_{ij} \cos q_i q_j / r| \quad (i, j = 1, 2, ..., n+2),$$

† It might also be observed that if $p_1 p_2 = \tfrac{1}{2}\pi r$, then p_1 is the pole of the $\mathscr{E}_{n-1, r}$ determined by $p_2, p_3, ..., p_{n+1}$, and a motion bringing the two ordered $(n+1)$-tuples $p_2, p_3, ..., p_{n+1}, p_{n+2}$ and $q_2, q_3, ..., q_{n+1}, q_{n+2}$ into coincidence makes p_1 and q_1 coincide.

since P and Q are $(n+2)$-tuples of $\mathscr{E}_{n,r}$. Expanding these determinants and equating, we obtain after easy computation

$$\cos(p_1 p_2/r) \cdot \cos(p_2 p_3/r) \ldots \cos(p_{n+1} p_{n+2}/r) \cdot \cos(p_1 p_{n+2}/r) = 0,$$

which implies that one of the distances $p_i p_{i+1}$ ($i = 1, 2,\ldots, n+2$) equals $\frac{1}{2}\pi r$. This is contrary to what has been shown above, and the proof of the theorem is complete.

The existence in $\mathscr{E}_{2,r}$ of congruent non-superposable subsets of arbitrary power (not exceeding $\mathfrak{c}$), together with the fact that any two f-congruent *pairs* of points are f-superposable (in any $\mathscr{E}_{n,r}$) shows that $n+1$ is the *smallest* integer for which Theorem 96.1 is valid. We shall see in the next section that $n+1$ is minimum for every n.

97. Superposability order. Pseudo f-superposable sets

As a consequence of our investigations of superposability in elliptic space, we are led to a general notion whose connexion with that concept is similar to the relation of congruence order to congruence. It is natural, therefore, to call this notion *superposability order*.

DEFINITION 97.1. *A space has superposability order σ provided any two subsets of the space are superposable whenever a one-to-one correspondence f exists between the points of the two sets such that corresponding σ-tuples are f-superposable.*

For every positive integer n, the euclidean, spherical, and hyperbolic n-dimensional spaces have minimum superposability order 2.

THEOREM 97.1. *For every positive integer n, the n-dimensional elliptic space $\mathscr{E}_{n,r}$ has the minimum superposability order $n+1$.*

Proof. That $\mathscr{E}_{n,r}$ has superposability order $n+1$, for every positive integer n, follows from Theorem 96.1. To show this a minimum it suffices to prove that for every integer n, $\mathscr{E}_{n,r}$ contains two *non-superposable* $n+1$-tuples $p_1, p_2,\ldots, p_{n+1}$ and $q_1, q_2,\ldots, q_{n+1}$ such that

$$p_1, p_2,\ldots, p_{i-1}, p_{i+1},\ldots, p_{n+1} \cong q_1, q_2,\ldots, q_{i-1}, q_{i+1},\ldots, q_{n+1}$$

$$(i = 1, 2,\ldots, n+1).$$

Consider the two determinants $\Delta = |d_{ij}|$, $d_{ij} = d_{ji} = d$, $|i-j| = 1$ or n, $d_{ii} = 1$, and $d_{ij} = 0$ for all remaining elements ($i,j = 1, 2,\ldots, n+1$), and $\Delta^* = |d'_{ij}|$, where d'_{ij} satisfy all the above conditions *except that* $d'_{1,n+1} = d'_{n+1,1} = -d$. Since each principal minor in each of these determinants equals 1 for $d = 0$, continuity considerations imply the existence of a number d ($0 < d < 1$) such that *every principal minor of each determinant is positive*.

Hence congruent $(n+1)$-tuples $p_1, p_2, \ldots, p_{n+1}$ and $q_1, q_2, \ldots, q_{n+1}$ of $\mathscr{E}_{n,r}$ exist with spher $\mathrm{dist}(p_1, p_{n+1}) = \sup$ spher $\mathrm{dist}(q_1, q_{n+1})$ and

$$p_i p_j = r \arccos d = q_i q_j, \qquad |i-j| = 1 \text{ or } n,$$
$$p_i p_j = \tfrac{1}{2}\pi r = q_i q_j, \qquad |i-j| \neq 1, n \quad (i,j = 1,2,\ldots,n+1; \ i \neq j).$$

The matrices of Δ and Δ^* are associated matrices of these congruent $(n+1)$-tuples, respectively, and since those matrices are clearly not strictly equivalent, the two $(n+1)$-tuples are *not superposable*. But for each two corresponding n-tuples of the sets we have

$$p_1, p_2, \ldots, p_{i-1}, p_{i+1}, \ldots, p_{n+1} \cong q_1, q_2, \ldots, q_{i-1}, q_{i+1}, \ldots, q_{n+1}$$
$$(i = 1, 2, \ldots, n+1),$$

since examination of the corresponding principal minors of Δ and Δ^* shows that equal associated matrices may be found for the two n-tuples. This establishes the theorem.

DEFINITION 97.2. *Two subsets of $\mathscr{E}_{n,r}$ ($n > 1$) are called pseudo f-superposable provided a one-to-one correspondence f exists between their points such that* (1) *each two corresponding n-tuples are f-superposable, and* (2) *the two subsets are not f-superposable.*

THEOREM 97.2. *For each integer n greater than 1 pseudo f-superposable subsets of $\mathscr{E}_{n,r}$ exist.*

Proof. Such sets were constructed in the preceding theorem.

Remark 1. If $p_1, p_2, \ldots, p_{n+1}$ and $q_1, q_2, \ldots, q_{n+1}$ are pseudo f-superposable subsets of $\mathscr{E}_{n,r}$, then no n-tuple of either set lies in an $\mathscr{E}_{n-2,r}$, for if either set contained such an n-tuple, then that n-tuple of the other set superposable with it would also lie in an $\mathscr{E}_{n-2,r}$, and consequently both $(n+1)$-tuples would be subsets of $(n-1)$-dimensional subspaces. Superposing those two subspaces yields two $(n+1)$-tuples of $\mathscr{E}_{n-1,r}$ with corresponding n-tuples f-superposable, and hence $p_1, p_2, \ldots, p_{n+1}$ and $q_1, q_2, \ldots, q_{n+1}$ are f-superposable (Theorem 96.1) contrary to assumption.

Remark 2. If $p_1, p_2, \ldots, p_{n+1}$ and $q_1, q_2, \ldots, q_{n+1}$ are two pseudo f-superposable subsets of $\mathscr{E}_{n,r}$, one of these $(n+1)$-tuples might be contained in an $\mathscr{E}_{n-1,r}$.

Pseudo f-superposable $(n+1)$-tuples of $\mathscr{E}_{n,r}$ are characterized in the following theorem.

THEOREM 97.3. *Two $(n+1)$-tuples $p_1, p_2, \ldots, p_{n+1}$ and $q_1, q_2, \ldots, q_{n+1}$ of $\mathscr{E}_{n,r}$ ($n > 1$) are pseudo f-superposable ($q_i = f(p_i)$ ($i = 1, 2, \ldots, n+1$)) if and only if associated matrices*

$$(\phi_{ij}) = (\epsilon_{ij} \cos p_i p_j / r), \qquad (\phi'_{ij}) = (\epsilon'_{ij} \cos q_i q_j / r) \qquad (i,j = 1,2,\ldots,n+1)$$

of the respective ordered $(n+1)$-tuples exist such that

$$\phi_{ij} = \phi'_{ij} > 0, \quad |i-j| = 1, \quad \phi_{ij} = \phi'_{ij} = 0, \quad |i-j| \neq 1$$
$$(i \neq j; i,j = 1,2,\ldots,n+1; i,j \neq 1, n+1),$$
$$0 < \phi_{1,n+1} = -\phi'_{1,n+1}.$$

Proof. The sufficiency is established in the last part of the proof of Theorem 97.1. To prove the necessity it is observed that no n-tuple of either of the two $(n+1)$-tuples (assumed pseudo f-superposable) lies in an $\mathscr{E}_{n-2,r}$ (Remark 1). It follows (Exercise, § 95) that a right-hand apolar ordering of one of the $(n+1)$-tuples exists for which the distance of the third and the last points is not $\frac{1}{2}\pi r$, and either the first point or the second point has distance $\frac{1}{2}\pi r$ from each of the last $n-2$ points.

The argument used in Case 2 of Theorem 96.1 may now be applied to obtain the desired result.

Thus if P and Q are two pseudo f-superposable $(n+1)$-tuples of $\mathscr{E}_{n,r}$, their points may be ordered so that *none* of the sides $p_i p_{i+1}$ of the ordered $(n+1)$-gon $p_1, p_2, \ldots, p_{n+1}$ equals $\frac{1}{2}\pi r$, while *each* of the diagonals does. The ordered $(n+1)$-gon $q_1, q_2, \ldots, q_{n+1}$ forming the set Q is congruent to $p_1, p_2, \ldots, p_{n+1}$ and each associated *spherical distance* (q_i, q_j) equals the spherical distance (p_i, p_j) *except for* (q_1, q_{n+1}) *and* (p_1, p_{n+1}). These two spherical distances are unequal and supplementary.

It has been emphasized throughout our discussion of elliptic space that the property of free superposability, which we found to be so useful in euclidean and spherical spaces, is not a feature of that space. The question arises whether a vestige of free superposability is present in elliptic space in that any two *superposable* subsets of $\mathscr{E}_{n,r}$ are *freely* superposable; that is, if $P, Q \subset \mathscr{E}_{n,r}$, does $P \cong Q$ and $P \approx_f Q$ imply $P \cong_f Q$?

That this question must be answered in the *negative* is shown by the interesting example of two quintuples in $\mathscr{E}_{3,r}$ which are g-superposable but not f-superposable—indeed, the two quintuples are pseudo f-superposable. The reader may easily show that the $S_{3,r}$ contains two quintuples $p'_1, p'_2, \ldots, p'_5$ and $q'_1, q'_2, \ldots, q'_5$ with

$$\cos(p'_1 p'_2/r) = \cos(p'_1 p'_4/r) = \cos(p'_2 p'_3/r) = a, \quad 0 < a < 1,$$

$$\cos(p'_1 p'_3/r) = \cos(p'_1 p'_5/r) = \cos(p'_2 p'_4/r) = \cos(p'_3 p'_5/r) = 0,$$

$$\cos(p'_2 p'_5/r) = \cos(p'_3 p'_4/r) = b, \quad 0 < b < 1,$$

$$\cos(p'_4 p'_5/r) = -a,$$

and mutual distances of $q'_1, q'_2, ..., q'_5$ such that $q'_i q'_j = p'_i p'_j$, except that $\cos(q'_1 q'_2/r) = -a$, while $3a^2 + b^2 = 1$, $a \neq b$.

Passing to the elliptic space $\mathscr{E}_{3,r}$ yields two quintuples $p_1, p_2, ..., p_5$ and $q_1, q_2, ..., q_5$ such that the congruence $p_1, p_2, ..., p_5 \approx_f q_1, q_2, ..., q_5$ *cannot be extended to the whole space* (that is, the two quintuples are *not f*-superposable) while, on the other hand, the congruence

$$p_1, p_2, p_3, p_4, p_5 \approx_g q_1, q_4, q_5, q_2, q_3$$

can be extended to the whole $\mathscr{E}_{3,r}$.

Thus the two quintuples are superposable, but not freely superposable since the particular congruence f between the sets cannot be extended to a motion. It is observed that each two f-corresponding triples are f-superposable, so the g-superposable quintuples are pseudo f-superposable.

EXERCISES

1. Prove that the two pseudo f-superposable quintuples discussed above are subsets of $\mathscr{E}_{3,r}$, and show that they are g-superposable.
2. If two pseudo f-superposable subsets of $\mathscr{E}_{3,r}$ are such that no quadruple of either set is contained in an $\mathscr{E}_{2,r}$, does each set consist of exactly four points?

REFERENCE

§§ 87–97. Blumenthal [20].

CHAPTER XI

METRIC-THEORETIC PROPERTIES OF THE ELLIPTIC PLANE. CONGRUENCE ORDER

98. Preliminary remarks

A SUBSET of a semimetric space is freely movable in the space provided it is superposable with each subset with which it is congruent. This chapter is concerned principally with establishing that certain configurations which are of basic importance in a metric study of the elliptic plane are freely movable. The results are applied to yield a congruence order for the $\mathscr{E}_{2,r}$ with respect to the class $\{S\}$ of semimetric spaces.

From the way in which elliptic n-space has been studied in preceding chapters we see that each point of $\mathscr{E}_{2,r}$ may be regarded as having two (diametral) images in the $S_{2,r}$, and the distance pq of two points in $\mathscr{E}_{2,r}$ is merely the smallest of the distances in $S_{2,r}$ determined by the two pairs of spherical images of p and q. The collection of all spherical images of the points of a subset of $\mathscr{E}_{2,r}$ is the *complete spherical image* of the subset. Moreover, a subset S of $S_{2,r}$ is a spherical image (associated spherical set) of a subset E of $\mathscr{E}_{2,r}$ provided each point of S is a spherical image of a point of E, while for each point of E one and only one of its spherical images belongs to S. It is clear that two subsets of $\mathscr{E}_{2,r}$ are superposable if and only if they possess congruent spherical images.

EXERCISES

1. Show that the angle of intersection of two lines of $\mathscr{E}_{2,r}$ equals the angle of intersection of their complete spherical images.
2. What is the complete spherical image of (*a*) two mutually perpendicular lines of $\mathscr{E}_{2,r}$, (*b*) a 'circle' of $\mathscr{E}_{2,r}$, (*c*) an elliptic segment ?

99. Orthocentric quadruples

Two points of an elliptic line are end-points of two arcs of the line. The point of each arc that is equidistant from the end-points of the arc is called the *internal* equidistant point of that arc and the *external* equidistant point of the other.

We consider now the equidistant locus of two points in $\mathscr{E}_{n,r}$ (cf. § 79, Exercise 5).

THEOREM 99.1. *The locus of points of the $\mathscr{E}_{n,r}$ equidistant from two given distinct points is the 'cross' consisting of the two $(n-1)$-dimensional subspaces perpendicular to the $\mathscr{E}_{1,r}$ joining the two points, at the respective equidistant points of the arcs determined on this line by the two points.*

Proof. Denote the two given points by p_1, p_2, and let p be any point of $\mathscr{E}_{n,r}$ with $p_1 p = p p_2$. If this distance is $\tfrac{1}{2}\pi r$, then p is a point of that hyperplane $\mathscr{E}_{n-1,r}$ which bisects perpendicularly the *segment* with endpoints p_1, p_2. If $p_1 p = p p_2 \neq \tfrac{1}{2}\pi r$, let q be the foot of the unique perpendicular from p to $\mathscr{E}_{1,r}(p_1, p_2)$. The right triangles (p_1, p, q) and (p_2, p, q) are congruent and $p_1 q = q p_2$. Hence q is one of the two equidistant points of p_1, p_2 on $\mathscr{E}_{1,r}(p_1, p_2)$ and so p is on one of the hyperplanes perpendicular to $\mathscr{E}_{1,r}(p_1, p_2)$ at these equidistant points. Since any point on either of these hyperplanes is equidistant from p_1 and p_2, the theorem is established.

Each of the two $(n-1)$-dimensional subspaces of $\mathscr{E}_n$ forming a cross is called an *arm* of the cross, and the points of the $\mathscr{E}_{n-2,r}$ in which the arms intersect are *vertices* of the cross. A cross of $\mathscr{E}_{n,r}$ is denoted by $X_{n,r}$.

Three points of $\mathscr{E}_{2,r}$ are vertices of a proper apolar triangle provided (1) the points are not on any $\mathscr{E}_{1,r}$, and (2) there exists an apolar ordering of the three points.† The vertices of a proper apolar triangle of $\mathscr{E}_{2,r}$ lie on at most three distinct crosses.‡ If a proper apolar triangle is right-angled, there will be but two crosses containing the vertices.

DEFINITION 99.1. *Four pairwise distinct points p_1, p_2, p_3, p_4 of $\mathscr{E}_{2,r}$ form an orthocentric quadruple provided the elliptic lines $\mathscr{E}_{1,r}(p_i, p_j)$ and $\mathscr{E}_{1,r}(p_k, p_m)$ are mutually perpendicular for every permutation i, j, k, m of the indices* 1, 2, 3, 4.

THEOREM 99.2. *An orthocentric quadruple is freely movable.*

Proof. Let $p_1, p_2, p_3, p_4 \approx p'_1, p'_2, p'_3, p'_4$ be two orthocentric quadruples of $\mathscr{E}_{2,r}$.

Case 1. *Two corresponding triples are superposable.* Assume the labelling so that
$$p_1, p_2, p_3 \cong p'_1, p'_2, p'_3.$$
If $p_1, p_2, p_3 \in \mathscr{E}_{1,r}$, then $p'_1, p'_2, p'_3 \in \mathscr{E}_{1,r}$, and these lines have p_4 and p'_4 as their respective poles. Hence any motion superposing p_1, p_2, p_3 and p'_1, p'_2, p'_3 evidently superposes p_4 and p'_4.

If $p_1 p_2 = p_2 p_3 = p_1 p_3 = \tfrac{1}{2}\pi r$, the superposability of the two quadruples follows at once from Theorem 91.1, while if exactly two of the three distances $p_i p_j$ ($i, j = 1, 2, 3$, $i \neq j$) equal $\tfrac{1}{2}\pi r$, say
$$p_1 p_3 = p_2 p_3 = \tfrac{1}{2}\pi r,$$

---

† See § 87 for definition of 'apolar order'. Clearly an apolar ordering of a triple exists if and only if no point of the triple has distance $\tfrac{1}{2}\pi r$ from each of the other two points.

‡ By a 'cross' we shall mean from now on a θ-cross with $\theta = \tfrac{1}{2}\pi$ (§ 61); that is, the figure formed by two mutually perpendicular elliptic lines. We denote it by $X_{2,r}$ or X.

then $p_4 \in \mathscr{E}_{1,r}(p_1, p_2)$. It follows that $p_1, p_2, p_4 \cong p_1', p_2', p_4'$ and the argument used above applies.

Suppose, finally, that p_1, p_2, p_3 are a proper apolar triple, and let Γ be a motion such that $\Gamma(p_i') = p_i$ ($i = 1, 2, 3$), $\Gamma(p_4') = p_4''$. If $p_4 \neq p_4''$, then p_1, p_2, p_3 lie on the cross $X_{2,r}$ which is the equidistant locus of p_4 and p_4''. But every cross containing the proper apolar triple p_1, p_2, p_3 contains p_4, and since clearly $p_4 \bar{\in} X_{2,r}$, we obtain the desired contradiction which completes the proof of Case 1.

Case 2. None of the corresponding triples are superposable. Then some pair of associated matrices of the two quadruples have determinants

$$\begin{vmatrix} 1 & 12 & 13 & 14 \\ 12 & 1 & 23 & 24 \\ 13 & 23 & 1 & 34 \\ 14 & 24 & 34 & 1 \end{vmatrix} = 0, \quad \begin{vmatrix} 1 & -12 & -13 & -14 \\ -12 & 1 & -23 & -24 \\ -13 & -23 & 1 & -34 \\ -14 & -24 & -34 & 1 \end{vmatrix} = 0,$$

where we have written ij for $\cos(q_i q_j/r)$ with $q_i q_j$ *the distance of spherical images* of p_i and p_j ($i, j = 1, 2, 3, 4$). The vanishing of the two determinants yields, upon expansion,

$$12 \cdot 13 \cdot 23 + 23 \cdot 24 \cdot 34 + 13 \cdot 34 \cdot 14 + 12 \cdot 14 \cdot 24 = 0. \quad (*)$$

Since the angle θ between opposite sides of a spherical quadruple is given by

$$\cos \theta = \frac{12 \cdot 34 - 23 \cdot 14}{\sin(q_1 q_3/r) \sin(q_2 q_4/r)},$$

we have for an orthocentric quadruple

$$12 \cdot 34 = 23 \cdot 14 = 13 \cdot 24 = k. \quad (\dagger)$$

From the assumption distinguishing this case it follows that

$$p_i p_j \neq \tfrac{1}{2} \pi r \quad (i, j = 1, 2, 3) \quad \text{(Lemma 81.1)}.$$

But relations ($\dagger$), ($*$) yield

$$12^2 \cdot 23^2 \cdot 13^2 + k^2 \cdot (12^2 + 23^2 + 13^2) = 0,$$

which contradicts this and completes the proof of the theorem.

EXERCISES

1. Define two elliptic n-tuples to be *completely complementary* provided there exists a congruence between them such that no two corresponding triples are superposable. Show that the $\mathscr{E}_{2,r}$ contains completely complementary quadruples.
2. If two quadruples of $\mathscr{E}_{2,r}$ are completely complementary, may there exist a congruence between them with two corresponding triples superposable?

3. Let $p_1, p_2, p_3, p_4 \approx p_1', p_2', p_3', p_4'$ be two completely complementary quadruples of $\mathscr{E}_{2,r}$. If $\mathscr{E}_{1,r}(p_1, p_2)$ is perpendicular to $\mathscr{E}_{1,r}(p_3, p_4)$, show that the corresponding lines of the 'primed' quadruple are mutually perpendicular.

4. Prove that no quintuple of $\mathscr{E}_{2,r}$ is completely complementary to five points of a cross.

100. Singular loci

We recall that a subset B of a semimetric space S forms a metric basis for the space if $x, y \in S$ and $px = py$ for every element p of B implies $x = y$. If the distances px of a point x of S from each point p of a subset S be referred to as the metric coordinates of x with respect to that subset, then no two points of S have the same metric coordinates with respect to a metric basis of S.

DEFINITION 100.1. *Let E be a subset of a semimetric space S. The set of all points of S not uniquely determined by their metric coordinates with respect to E is the singular locus of E.*

THEOREM 100.1. *The singular locus of a cross in $\mathscr{E}_{2,r}$ consists of those points of the polar of its vertex which are not on the cross.*

Proof. Let p_1, p_2 be distinct points of $\mathscr{E}_{2,r}$ with the same metric coordinates with respect to the cross $X_{2,r}$. Then $X_{2,r}$ is the equidistant locus of p_1, p_2 and it readily follows that the vertex of $X_{2,r}$ has distance $\frac{1}{2}\pi r$ from each of the points p_1, p_2.

THEOREM 100.2. *If a subset of $\mathscr{E}_{2,r}$ lies on one and only one cross, its singular locus is identical with that of the cross.*

Proof. The proof is clear.

Remark. There are exactly two points of the singular locus of a cross with the same metric coordinates with respect to the cross, and they are reflections of one another in either arm of the cross.

THEOREM 100.3. *The singular locus of the set of vertices of a proper apolar triangle, with no angle a right angle, is a subset of the three lines which are the polars of the feet of the altitudes of the given triangle.*

Proof. If p_1, p_2 have the same metric coordinates with respect to the set of vertices of the triangle, then these vertices lie on the equidistant locus of p_1, p_2, which must coincide with one of the three crosses determined by the three vertices. Hence the line $\mathscr{E}_{1,r}(p_1, p_2)$ is the polar of the foot of one of the altitudes, and the theorem follows.

THEOREM 100.4. *A subset of $\mathscr{E}_{2,r}$ forms a metric basis for the plane if and only if it is not contained in a cross.*

Proof. If the subset is contained in a cross, then any point p of the singular locus of the cross has the same metric coordinates with respect to the set as the reflection of p in an arm of the cross, and so the subset does not form a metric basis. Conversely, if the subset does not form a metric basis, it is part of the singular locus of an existing pair of points with the same metric coordinates with respect to the set; that is, the subset is contained in a cross.

THEOREM 100.5. *Four points containing the vertices of a proper apolar triangle determine at most one cross unless they form an orthocentric quadruple.*

Proof. The vertices of a proper apolar triangle determine at most three crosses. The only points common to any two of these crosses are the vertices and orthocentre of the triangle, and hence a fourth point, not the orthocentre, can lie on at most one of these crosses.

EXERCISES

1. What points of the polars of the feet of the altitudes of a proper apolar triangle T are not part of the singular locus of the triangle's vertices?
2. How are the three mutual intersections of the three polars in Exercise 1 related to the vertices and orthocentre of triangle T?
3. Show that the three points of intersection of the three polars may be joined by arcs of elliptic lines whose perpendicular bisectors meet in the orthocentre of T. The triangle formed by these arcs is called the *singular triangle* of T.

101. Freely movable quintuples

We shall need the following lemma.

LEMMA 101.1. *Let the perpendicular bisector of side AB of the proper spherical triangle ABC intersect sides AC and BC ($AC > BC$) in Q and R, respectively, and denote by Q' and R' the reflections of Q and R in the mid-points of AC and BC, respectively. If RQ is either equal or supplementary to $R'Q'$, then $Q = Q'$.*

Proof. Assume the sphere containing triangle ABC to be of unit radius. Denoting the mid-point of AB by P, let $QC = x$, $RC = y$, $QP = q$, $RP = r$. Then from spherical trigonometry

$$\tan q = \sin \tfrac{1}{2}c \tan A, \qquad \tan r = \sin \tfrac{1}{2}c \tan B, \qquad (1)$$

$$\sin(b-x) = \sin q / \sin A, \qquad \sin(a+y) = \sin r / \sin B, \qquad (2)$$

$$\cos(b-x) = \cos \tfrac{1}{2}c \cos q, \qquad \cos(a+y) = \cos \tfrac{1}{2}c \cos r. \qquad (3)$$

Now

$$\cos RQ = \cos(r-q) = \cos r \cos q + \sin r \sin q,$$

$$\cos R'Q' = \cos(b-x)\cos(a+y) + \sin(b-x)\sin(a+y)\cos C,$$

and using (2), (3),

$$\cos R'Q' = \cos q \cos r \cos^2 \tfrac{1}{2}c + \sin q \sin r \cos C / \sin A \sin B.$$

If $RQ = R'Q'$, then one easily obtains

$$\sin^2 \tfrac{1}{2}c = \tan q \tan r \left[\frac{\cos C}{\sin A \sin B} - 1 \right],$$

and using (1),

$$\cos C = \sin A \sin B + \cos A \cos B = \cos(B-A).$$

It follows that $B = A + C$, and since $QB = QA$, then angle $QBA = A$ and so angle $QBC = C$. Hence $QC = QB = QA$ and Q is the mid-point of AC; that is, $Q = Q'$.

If $R'Q'$ is the supplement of RQ, the above procedure is followed to obtain

$$-\cos C = \sin A \sin B + \cos A \cos B [(1+\cos^2 \tfrac{1}{2}c)/\sin^2 \tfrac{1}{2}c]. \qquad (4)$$

Substituting in (4) $\cos C = -\cos A \cos B + \sin A \sin B \cos c$ leads after easy reductions to $\cos C = \cos(B-A)$ and $Q = Q'$ as before.

Can the reader supply a purely geometrical proof for this lemma?

THEOREM 101.1. *Five points of a cross of $\mathscr{E}_{2,r}$, no four of which form an orthocentric quadruple, are freely movable.*

Proof. Suppose $p_1, p_2, \ldots, p_5 \approx q_1, q_2, \ldots, q_5$ ($p_i, q_i \in \mathscr{E}_{2,r}$ ($i = 1, 2, \ldots, 5$)) with the first quintuple a subset of a cross $X_{2,r}$ and none of its quadruples orthocentric. The two quintuples are not completely complementary (Exercise 4, §99) and so some pair of corresponding triples are superposable. We may assume the labelling so that $p_1, p_2, p_3 \cong q_1, q_2, q_3$.

Case 1. The points p_1, p_2, p_3 lie on a line. Now one of the points p_4, p_5 (say p_4) is not the pole of $\mathscr{E}_{1,r}(p_1, p_2, p_3)$, and since $q_1, q_2, q_3 \in \mathscr{E}_{1,r}$, there is a motion Γ of $\mathscr{E}_{2,r}$ sending q_i into p_i ($i = 1, 2, 3, 4$). If $p_4 \in \mathscr{E}_{1,r}(p_1, p_2, p_3)$, then $q_5' = \Gamma(q_5)$ either coincides with p_5 or is its reflection in $\mathscr{E}_{1,r}(p_1, p_2, p_3)$. In the latter case another motion (a rotation of 180° about the pole of $\mathscr{E}_{1,r}(p_1\, p_2, p_3)$) sends q_5' into p_5, leaving the other points fixed, and the superposition of the quintuples is accomplished.

If $p_4 \bar{\in} \mathscr{E}_{1,r}(p_1, p_2, p_3)$, then p_1, p_2, p_3, p_4 lie on just the one cross X and if $q_5' \neq p_5$, X is the equidistant locus of p_5, q_5'. But this is impossible since $p_5 \in X$.

Case 2. The points p_1, p_2, p_3 do not lie on a line. In the event that either (i) one of the points p_1, p_2, p_3 has distance $\tfrac{1}{2}\pi r$ from each of the remaining points, or (ii) two of the lines determined by the three points are mutually perpendicular, the superposition of the two quintuples is left to the reader as an exercise. We suppose in the following that neither (i) nor (ii) subsists (i.e. triangle (p_1, p_2, p_3) is proper apolar without right angles).

Let us observe, now, that the superposability of the two quintuples follows readily from that of any quadruple p_1, p_2, p_3, p_k ($k = 4$ or 5) with its corresponding one. For by Theorem 100.5 and the assumption concerning the triple (p_1, p_2, p_3), the quadruple lies on just the one cross X and the argument concluding Case 1 is applicable.

Subcase α. *The points* p_4, $p_5 \in \mathscr{E}_{1,r}(p_1, p_2)$. Then q_1, q_2, q_4, q_5 are on an elliptic line (Theorem 88.1) and p_1, p_2, p_4, $p_5 \cong q_1$, q_2, q_4, q_5. The argument used in the first part of Case 1 may now be employed.

Subcase β. *The points* p_4, p_5 *are on the arm of X not containing p_1, p_2.* Let Γ be a motion with $\Gamma(q_i) = p_i$ ($i = 1, 2, 3$). If

$$\Gamma(q_4) = p_4 \quad \text{or} \quad \Gamma(q_5) = p_5,$$

the observation contained in the paragraph preceding Subcase α suffices to obtain the desired result. If, on the other hand, $\Gamma(q_4) \neq p_4$ and $\Gamma(q_5) \neq p_5$, then p_4 and $\Gamma(q_4)$ are reflections of each other in the mid-point of a side of the singular triangle of (p_1, p_2, p_3), while p_5 and $\Gamma(q_5)$ are similarly related with respect to the mid-point of another side of that triangle.

The preceding lemma is now used to conclude that $p_4 p_5 \not\cong q_4 q_5$, which contradicts the initial assumption of the congruence of the two quintuples.

Subcase γ. *The points p_4, p_5 are on different arms of X.* Suppose $p_4 \in \mathscr{E}_{1,r}(p_1, p_2)$ and let Γ be a motion with $\Gamma(q_i) = p_i$ ($i = 1, 2, 3$). Then $\Gamma(q_4) = p_4$, for in the contrary case p_4 is on the singular locus of (p_1, p_2, p_3), which is impossible since $p_4 \in \mathscr{E}_{1,r}(p_1, p_2)$. Hence

$$p_1, p_2, p_3, p_4 \cong q_1, q_2, q_3, q_4$$

and the two quintuples are superposable.

An examination of the proof shows that the statement of the theorem can be weakened by admitting orthocentric quadruples with three points in an $\mathscr{E}_{1,r}$. Calling all other orthocentric quadruples *proper* we have:

THEOREM 101.2. *Five points of a cross of $\mathscr{E}_{2,r}$ are freely movable if no four of the points form a proper orthocentric quadruple.*

COROLLARY. *A subset of a cross containing more than five points is freely movable.*

Proof. Let C be a subset of a cross containing more than five points and let C' be a subset of $\mathscr{E}_{2,r}$ congruent to C. If each arm of the cross contains at least two points of C, some five points $p_1, p_2,..., p_5$ of C lie on at most one cross and satisfy the conditions of the previous theorem. Let $p'_1, p'_2,..., p'_5$ be the corresponding points of C', and let Γ be a motion such that $\Gamma(p'_i) = p_i$ ($i = 1, 2,..., 5$). If $\Gamma(p') = p''$, $p' \in C'$, and p is the element of C corresponding to p' in the congruence of the two sets, then $p'' = p$ since otherwise the cross would be the equidistant locus of p and p''. This is impossible since p is a point of the cross.

EXERCISE

Let p_i, q_i ($i = 1, 2,..., 5$) be congruent quintuples of $\mathscr{E}_{2,r}$ with $p_i \sim q_i$ ($i = 1, 2,..., 5$). If $p_1, p_2,..., p_5 \in X$, a cross of $\mathscr{E}_{2,r}$, and $p_1, p_2, p_3 \approx q_1, q_2, q_3$, show that $p_1, p_2,..., p_5 \cong q_1, q_2,..., q_5$ in each of the two following cases:
  (i) p_3 is the pole of the line $\mathscr{E}_{1,r}(p_1, p_2)$,
  (ii) triangle (p_1, p_2, p_3) is a right triangle and no one of the points p_1, p_2, p_3 is the pole of the line containing the other two.

102. A 'crowding' theorem for the elliptic plane

Since the elliptic plane is compact, it is clear that many points of the space cannot be selected without a kind of crowding together of certain of the points. A measure of this crowding and the minimum number of points that ensure it play an interesting and important role in the geometry of the plane.

Our object is to show that *every set of eight points of $\mathscr{E}_{2,r}$ contains a triple with length sum less than πr*. The proof of the theorem is quite elementary in character, but involves a great deal of computational detail. The procedure employed is *fully illustrated* in the computationally simpler, but essentially similar, theorem concerning *nine* points instead of eight.

In what follows it is found convenient to refer to the arc distances on a sphere in terms of degrees as well as linear units. Necessary computations will be carried to the nearest minute or two. This degree of accuracy is more than adequate for our purposes.

THEOREM 102.1. *Every set of nine points of the $\mathscr{E}_{2,r}$ contains a triple with perimeter less than πr.*

Proof. Let $p_1, p_2,..., p_9$ be a set of nine points of $\mathscr{E}_{2,r}$ with d the minimum of the distances determined by the points.

Case I. $d \leqslant 52°$. One of the spherical images of $p_1, p_2,..., p_9$ can be chosen (subject to a rotation and proper labelling) so that (1) $p_1 p_2 = d$, (2) the mid-point of $p_1 p_2$ is the north pole of $S_{2,r}$, and (3) all the points are on the northern hemisphere. If the perimeter of each triangle equals or exceeds πr, then none of the points $p_3, p_4,..., p_9$ lies inside or on the 'metric ellipse' having p_1 and p_2 as foci and constant sum $\pi r - d$. A weaker, but more convenient, observation is that none of these points lies inside the circle on the minor axis of the ellipse as diameter (minor circle). This circle has radius 60° 49′ and so it follows that the points lie on or above the equator and on or below latitude 29° 11′.

Since the spherical projections of the points $p_3, p_4, \ldots, p_9$ onto the equator from the north pole correspond to seven points (not necessarily distinct) on the elliptic line, it follows that some three of them are on the spherical image of an elliptic segment whose length is less than or equal to $\frac{2}{7}\pi r$. This implies that some one of the spherical images of some triple lies in or on the boundary of a spherical 'box' extending north and south of the equator 29° 11′ and having a width of $\frac{2}{7}\pi r$ at the equator. A straightforward computation shows that the perimeter of each triple of vertices is less than πr, from which it readily follows that each three points contained in this box has this property. Thus three of the nine points have a perimeter less than πr.

Case II. $52° < d \leqslant 55°$. The proof in this case proceeds in exactly the same way. The spherical box has now dimensions $\frac{2}{7}\pi r$ by 31° 22′ and while it is true that there is a triangle interior to this box whose perimeter is greater than πr, there is none whose perimeter is greater than πr and all of whose sides exceed 52°.

Case III. $d > 55°$. In this case we consider the spherical image with p_1 at the north pole and all points on the northern hemisphere. Points $p_2, p_3, \ldots, p_9$ must lie on or above the equator and below lat. 35°. The projections of $p_2, p_3, \ldots, p_9$ on the equator from the north pole correspond to eight points of the $\mathscr{E}_{1,r}$, and it follows as before that there must be three points in a spherical box of dimensions $\frac{1}{4}\pi r$ by 35°. Thus there are two points in either the northern or southern half of the box. But the diameter of either half is less than 55°. This completes the proof.

Remark. Examination of the analysis in Case III leads to the conclusion that there exists no 9-tuple of the elliptic plane $\mathscr{E}_{2,r}$ all of whose distances are greater than 55°.

THEOREM 102.2. *Every set of eight points of the $\mathscr{E}_{2,r}$ contains a triple with perimeter less than πr, and eight is the smallest number with this property.*

Proof. The proof of this theorem is similar in plan to that of the nine-point theorem. There is again a division into three cases, namely

(I) $d \leqslant 45°$, (II) $45° < d \leqslant 58°$, (III) $d > 58°$.

Cases (I) and (III) are exactly analogous to the corresponding cases of Theorem 102.1. The boxes in the present case, however, have dimensions $\frac{2}{3}\pi r$ by 24° 29′ and $\frac{2}{7}\pi r$ by 32° respectively. Case (II) provides the principal difficulty and it is found convenient to break it down into one-degree subcases. We suppress the details. An example of a septuple of $\mathscr{E}_{2,r}$ each triple of which has perimeter greater than or equal to πr (or even greater than πr) is easily constructed.

By Lemma 81.1 and the concluding remarks of § 81, each triple of $\mathscr{E}_{2,r}$ with perimeter less than πr is freely movable, and so the preceding theorem has for us the important consequence:

THEOREM 102.3. *Each subset of eight points of $\mathscr{E}_{2,r}$ contains a freely movable triple.*

103. Congruence order of $\mathscr{E}_{2,r}$ with respect to the class of semimetric spaces

The objective of this section is to prove that the $\mathscr{E}_{2,r}$ has congruence order eight with respect to the class of semimetric spaces.

LEMMA 103.1. *If congruent mappings of three distinct subsets of $n-1$ points of a semimetric n-tuple into the semimetric space S have k corresponding points in common which form a metric basis for S, then the semimetric n-tuple may be congruently imbedded in S.*

Proof. Let Γ_1, Γ_2, Γ_3 be congruences that map $p_2, p_3, ..., p_n$; $p_1, p_3, ..., p_n$; $p_1, p_2, p_4, ..., p_n$, respectively, into subsets of S. Γ_i ($i = 1, 2, 3$) will refer both to the mapping and to the image under the mapping.

Since Γ_1 and Γ_2 have a metric basis in common,
$$\Gamma_1(p_i) = \Gamma_2(p_i) = p'_i \quad (i = 3, 4, ..., n).$$
There is thus defined in S an n-tuple of points
$$p'_1 = \Gamma_2(p_1), \quad p'_2 = \Gamma_1(p_2), \quad p'_3, \quad ..., \quad p'_n$$
with $p'_i p'_j = p_i p_j$ ($i, j = 1, 2, 3, 4, ..., n$) with the possible exception of the combination $i = 1, j = 2$. To prove that $p'_1 p'_2 = p_1 p_2$, suppose that Γ_3 takes p_1 and p_2 into p''_1 and p''_2, respectively. Since, however, Γ_1 and Γ_3 have a metric basis in common, $p''_2 = p'_2$. Similarly, $p''_1 = p'_1$, and the theorem is proved.

LEMMA 103.2. *If congruent mappings of four distinct subsets of $n-1$ points of a semimetric n-tuple into the $\mathscr{E}_{2,r}$ have k corresponding points in common which lie on one and only one cross, then the n-tuple may be congruently imbedded in the $\mathscr{E}_{2,r}$.*

Proof. Using the notation of the previous lemma, let Γ_1, Γ_2, Γ_3, Γ_4 be the congruences taking the respective $(n-1)$-tuples into the $\mathscr{E}_{2,r}$. If none of the images of the various points lies on P, the singular locus of the cross, Lemma 103.1 implies the theorem, since the k points of the cross form a metric basis for $\mathscr{E}_{2,r} - P$.

Suppose that the images of some points lie on P. At least two of these images coincide. No loss of generality is entailed if these are assumed to be $\Gamma_1(p_4)$ and $\Gamma_2(p_4)$. Then $\Gamma_1(p_i) = \Gamma_2(p_i) = p'_i$ ($i = 3, 4, ..., n$) and there is thus defined in $\mathscr{E}_{2,r}$ an n-tuple of points

$$p'_1 = \Gamma_2(p_1), \quad p'_2 = \Gamma_1(p_2), \quad p'_3, \quad ..., \quad p'_n$$

having $p'_i p'_j = p_i p_j$ ($i, j = 1, 2, ..., n$) except possibly for $i = 1, j = 2$.

To see that $p'_1 p'_2 = p_1 p_2$, suppose that Γ_3 takes p_1 and p_2 into p''_1 and p''_2 respectively. If $\Gamma_3(p_4) = p'_4$, then, since the k common points together with p'_4 form a metric basis for the $\mathscr{E}_{2,r}$ (Theorem 100.4), $p''_1 = p'_1$ and $p''_2 = p'_2$. If $\Gamma_3(p_4) \neq p'_4$, then these two points are reflections in either arm of the cross. If p''_1 (or p''_2) is not on the cross, then it is the reflection of p'_1 (or p'_2) in either arm of the cross, since otherwise $\Gamma_3(p_4) = p'_4$. Thus in any event $p'_1 p'_2 = p''_1 p''_2 = p_1 p_2$.

LEMMA 103.3. *If congruent mappings of three distinct subsets of $n-1$ points of a semimetric n-tuple into the $\mathscr{E}_{2,r}$ have k corresponding points in common which lie on one and only one cross and, in addition, the three images of some point of the n-tuple do not lie on this cross, then the n-tuple is congruently contained in the $\mathscr{E}_{2,r}$.*

Proof. Let Γ_1, Γ_2, and Γ_3 be the mappings and p a point whose images fail to lie on the cross. If $\Gamma_1(p) = \Gamma_2(p) = \Gamma_3(p)$, then the proposition follows from Lemma 103.1. Otherwise at least two of the images coincide, say $\Gamma_1(p) = \Gamma_2(p)$, and the proof proceeds exactly as in Lemma 103.2.

THEOREM 103.1. *Nine points of a semimetric space S are congruently contained in $\mathscr{E}_{2,r}$ if and only if each eight of the points are imbeddable in $\mathscr{E}_{2,r}$.*

Proof. The necessity is obvious. To prove the sufficiency, let the points of S be $p_1, p_2, ..., p_9$, and denote the mappings taking

$$p_1, p_2, ..., p_{i-1}, p_{i+1}, ..., p_9$$

into $\mathscr{E}_{2,r}$ by Γ_i ($i = 1, 2, ..., 9$). By Theorem 102.3 the image Γ_9 contains a triple, say $\Gamma_9(p_1, p_2, p_3)$, with perimeter less than πr and which is consequently freely movable in $\mathscr{E}_{2,r}$. Hence we may assume that

$$\Gamma_i(p_k) = \Gamma_j(p_k)$$

for $k = 1, 2, 3$ and $(i, j = 4, 5, ..., 9)$.

Case I. $\Gamma_9(p_1, p_2, p_3) \subset \mathscr{E}_{1,r}$. If all the points of all the images Γ_i ($i = 4, 5, ..., 9$) are on this line, or if the only point of these images not on the line is the pole of that line, the theorem is an immediate consequence of Lemma 103.1.

Assuming that $\Gamma_9(p_4)$ is neither a point of the line nor its pole, we see that $\Gamma_i(p_1, p_2, p_3, p_4) \simeq \Gamma_j(p_1, p_2, p_3, p_4)$ $(i,j = 5, 6,..., 9)$. We may thus assume that $\Gamma_i(p_1, p_2, p_3, p_4) = \Gamma_j(p_1, p_2, p_3, p_4)$. These points define uniquely a cross, and the theorem is then a consequence of Lemma 103.2.

Case II. $\Gamma_9(p_1, p_2, p_3) \not\subset \mathscr{E}_{1,r}$. Since the perimeter of $\Gamma_9(p_1, p_2, p_3)$ is less than πr, the points in this case form the vertices of a proper apolar triangle. We again take $\Gamma_i(p_1, p_2, p_3) = \Gamma_j(p_1, p_2, p_3)$ $(i,j = 4, 5,..., 9)$ and select, if possible, an image which is neither on the singular locus nor the orthocentre of $\Gamma_9(p_1, p_2, p_3)$. If the only image not on the singular locus is the orthocentre (other than the images of p_1, p_2, p_3, of course), call this point $\Gamma_9(p_4)$. Then it is asserted that $\Gamma_4 + \Gamma_9(p_4)$ is the desired mapping. The only distances open to question are

$$\Gamma_9(p_4)\Gamma_4(p_i) \quad (i = 1, 2, 3, 5,..., 9).$$

$\Gamma_9(p_4)\Gamma_9(p_i) = \Gamma_9(p_4)\Gamma_4(p_i)$ $(i = 1, 2, 3, 5,..., 8)$, since $\Gamma_9(p_i)$ and $\Gamma_4(p_i)$ either coincide or are reflections of one another in one of the altitudes. Finally, $\Gamma_8(p_4) = \Gamma_9(p_4)$ (Theorem 99.2) and we readily obtain

$$\Gamma_8(p_4)\Gamma_8(p_9) = \Gamma_9(p_4)\Gamma_4(p_9).$$

If there is an image, say $\Gamma_9(p_4)$, which is neither on the singular locus nor the orthocentre, it may or may not lie on one of the crosses defined by the triple $\Gamma_9(p_1, p_2, p_3)$. If it does not, then the theorem follows by Lemma 103.1, while if it does, the theorem follows by Lemma 103.2.

There remains to consider the case in which all the images, other than $\Gamma_9(p_1, p_2, p_3)$, lie on the singular locus of $\Gamma_9(p_1, p_2, p_3)$. Certainly not all the image points are vertices of the singular triangle. Suppose then that $\Gamma_9(p_4)$ is not a vertex. Then at least three of the points $\Gamma_i(p_4)$ $(i = 5, 6,..., 9)$ coincide. Assume $\Gamma_9(p_4) = \Gamma_8(p_4) = \Gamma_7(p_4) = p'_4$. In the event that this point is not a point of any cross containing

$$\Gamma_9(p_1, p_2, p_3) = p'_1, p'_2, p'_3,$$

the theorem is proved (Lemma 103.1).

If any other image of p_4 coincides with these, the theorem is a consequence of Lemma 103.2. As a matter of fact, if any four images of p_i $(i = 4, 5,..., 9)$ coincide, the theorem follows similarly. Furthermore, if p'_1, p'_2, p'_3, p'_4 are on a cross (and hence, since they are not orthocentric, on only one cross) and either $\Gamma_9(p_5)$ or $\Gamma_9(p_6)$ is not on this cross, then Lemma 103.3 implies the theorem.

The only remaining possibility is that $\Gamma_9(p_5)$ and $\Gamma_9(p_6)$ are points of the cross defined by p_1', p_2', p_3', p_4'. It is easy to see that none of the quadruples of the quintuple $\Gamma_9(p_1, p_2, p_3, p_4, p_6)$ is orthocentric. Thus by Theorem 101.1 all the images $\Gamma_i(p_1, p_2, p_3, p_4, p_6)$ ($i = 5, 7, 8, 9$) are superposable and the theorem follows from the opening remarks of the previous paragraph.

Examination of the foregoing reasoning shows that the theorem is valid if 'nine, each eight' be replaced by 'n, each $n-1$, $n \geqslant 9$'. Thus we have the following:

THEOREM 103.2. *If each eight points of a semimetric space are congruently contained in the $\mathscr{E}_{2,r}$, then every finite subset is congruently contained in the $\mathscr{E}_{2,r}$.*

LEMMA 103.4. *Corresponding to each positive integer k and each positive $\epsilon \leqslant \frac{1}{2}\pi r$ there is a positive integer n such that any n points of $\mathscr{E}_{2,r}$ contain a subset of k points with diameter less than ϵ.*

Proof. It follows from the compactness of $\mathscr{E}_{2,r}$ that the space may be covered by a finite number, n_1, of open sets of diameter less than ϵ. If n is selected such that $n \geqslant n_1 \cdot k$, the lemma is established.

THEOREM 103.3. *The $\mathscr{E}_{2,r}$ has congruence order eight with respect to the class of all semimetric spaces.*

Proof. Let S be a semimetric space with each eight points congruently contained in the $\mathscr{E}_{2,r}$. By Theorem 103.2 each finite subset of S is congruently contained in the $\mathscr{E}_{2,r}$. Thus if S is itself a finite set, the theorem is established. Suppose S infinite. The fact that each finite subset of S is congruently contained in the $\mathscr{E}_{2,r}$ together with the lemma above, implies the existence of a quintuple $p_1, p_2, ..., p_5$ of S of diameter less than, say, $\frac{1}{6}\pi r$. All congruent images in $\mathscr{E}_{2,r}$ of such a quintuple are superposable since they are 'essentially spherical'. Suppose $p_1', p_2', ..., p_5'$ a quintuple of the $\mathscr{E}_{2,r}$ congruent to $p_1, p_2, ..., p_5$. We distinguish three cases.

Case I. The quintuple $p_1', p_2', ..., p_5'$ forms a metric basis for $\mathscr{E}_{2,r}$. If $p \in S$, then $p_1, p_2, ..., p_5, p \approx p_1'', p_2'', ..., p_5'', p''$ of $\mathscr{E}_{2,r}$. Since

$$p_1', p_2', ..., p_5' \approx p_1, p_2, ..., p_5 \approx p_1'', p_2'', ..., p_5'',$$

and all congruent images in $\mathscr{E}_{2,r}$ of $p_1', p_2', ..., p_5'$ are superposable, there is a motion Γ of $\mathscr{E}_{2,r}$ transforming p_i'' into p_i' ($i = 1, 2, ..., 5$), and since $p_1', p_2', ..., p_5'$ forms a metric basis, Γ is unique. If p'' is carried into p', there is thus defined a mapping of S into $\mathscr{E}_{2,r}$ which is clearly a congruence, for if $p, q \in S$ and p', q' are, respectively, the corresponding elements of $\mathscr{E}_{2,r}$, then $p_1, p_2, ..., p_5, p, q \approx \bar{p}_1, \bar{p}_2, ..., \bar{p}_5, \bar{p}, \bar{q}$ of $\mathscr{E}_{2,r}$. The motion that carries $\bar{p}_i$ into p'_i ($i = 1, 2,..., 5$) necessarily transforms p, q into p', q', respectively (since $p'_1, p'_2,..., p'_5$ is a metric basis) and hence

$$pq = \bar{p}\bar{q} = p'q'.$$

Case II. *The set $p'_1, p'_2,..., p'_5$ lies on one and only one cross X of $\mathscr{E}_{2,r}$.* If S contains an element p_6 such that

$$p_1, p_2,..., p_5, p_6 \approx p''_1, p''_2,..., p''_5, p''_6 \approx p'_1, p'_2,..., p'_5, p'_6$$

with p'_6 neither a point of the cross X nor its singular locus L, then $p'_1, p'_2,..., p'_5, p'_6$ clearly forms a freely movable metric basis for $\mathscr{E}_{2,r}$ and the method of Case I applies. That method applies equally well in case for each $p \in S$, $p_1, p_2,..., p_5, p \approx \bar{p}_1, \bar{p}_2,..., \bar{p}_5, \bar{p} \approx p'_1, p'_2,..., p'_5, p'$ with $p' \bar{\in} L$, since the quintuple forms a metric basis for all points of $\mathscr{E}_{2,r}$ that do not lie on L.

In the event that neither of the above possibilities subsists, then for each point p of S the image point p' lies on X, if unique, and on L if not unique, with the latter alternative holding for at least one point p_6 of S. Then $p_1, p_2,..., p_5, p_6 \approx p'_1, p'_2,..., p'_5, p'_6$, with $p'_6 \in L$. If now $p \in S$, then $p_1, p_2,..., p_5, p_6, p \approx p'''_1, p'''_2,..., p'''_5, p'''_6, p'''$. A motion sending $p'''_1, p'''_2,..., p'''_5$ into $p'_1, p'_2,..., p'_5$, respectively, carries p'''_6 into p'_6 or p^*_6, its reflection in the cross. In the first case, define p' (the correspondent of p) to be the uniquely determined point p^* of $\mathscr{E}_{2,r}$ into which p''' is carried. If, on the other hand, the motion carries p'''_6 into p^*_6, let p' be the reflection in X of p^*.

To show that this mapping is a congruence, let p, q be elements of S and p', q' their corresponding points in $\mathscr{E}_{2,r}$. Now

$$p_1, p_2,..., p_5, p_6, p, q \approx \bar{p}_1, \bar{p}_2,..., \bar{p}_5, \bar{p}_6, \bar{p}, \bar{q}$$

of $\mathscr{E}_{2,r}$ and hence $p'_1, p'_2,..., p'_5 \approx \bar{p}_1, \bar{p}_2,..., \bar{p}_5$. If a motion sending $\bar{p}_i$ into p'_i ($i = 1, 2,..., 5$) sends $\bar{p}_6$ into p'_6, then $\bar{p}, \bar{q}$ are carried into p', q' respectively, and $pq = \bar{p}\bar{q} = p'q'$, while if $\bar{p}_6$ goes into p^*_6, then

$$pq = \bar{p}\bar{q} = p^*q^* = p'q'.$$

Case III. *The points $p'_1, p'_2,..., p'_5$ lie on more than one cross.* Then either four of the five points, say p'_1, p'_2, p'_3, p'_4, lie in an $\mathscr{E}_{1,r}$ whose pole is the fifth point, or the five points lie on an $\mathscr{E}_{1,r}$. In either case if $p \in S$ implies $p_1, p_2, p_3, p_4, p \approx p'_1, p'_2, p'_3, p'_4, p'$ with $p' \in \mathscr{E}_{1,r}(p'_1, p'_2, p'_3, p'_4)$, then the argument of Case I may be applied since p'_1, p'_2, p'_3, p'_4 form a metric basis for an $\mathscr{E}_{1,r}$. On the other hand, if p' is neither an element of $\mathscr{E}_{1,r}(p'_1, p'_2, p'_3, p'_4)$ nor its pole, points $p'_1, p'_2, p'_3, p'_4, p'$ lie on one and only one cross and the argument proceeds as in Case II.

Thus, any semimetric space whatever is congruently contained in the elliptic plane whenever each octuple of the space has that property. But there is nothing in our proof of Theorem 103.3 which would justify concluding that the number 'eight' is the smallest for which the statement is valid. That number arose in our work as the smallest integer k such that each k points of $\mathscr{E}_{2,r}$ contains a triple with perimeter less than πr (and hence is freely movable). The presence of a freely movable triple was exploited to yield the brief proof of the property of congruence order eight for the $\mathscr{E}_{2,r}$.

A procedure that has much in common with the one employed in this chapter, but which is not based upon a 'crowding' theorem, has been used by Haantjes and Seidel to show that Theorem 103.3 remains valid when 'eight' is replaced by 'seven'; that is, the elliptic plane has congruence indices $(7, 0)$ with respect to the class $\{S\}$ of semimetric spaces. Since $\mathscr{E}_{2,r}$ contains an equilateral sextuple (Theorem 80.2), it follows that for no number k does $\mathscr{E}_{2,r}$ have congruence indices $(6, k)$ with respect to $\{S\}$, and so the result of Haantjes and Seidel is the best possible.†

Neither the foregoing procedure nor the method of Haantjes and Seidel is practicable for ascertaining a finite congruence order of $\mathscr{E}_{n,r}$ for any $n > 2$. Since $\mathscr{E}_{n,r}$ is compact metric, it has finite or hyperfinite congruence order with respect to the class of all separable semimetric spaces (Theorem 60.2), and that is the best that can be said at this time. More powerful methods must be developed before the general problem can be approached.

EXERCISES

1. A semimetric set is pseudo-$\mathscr{E}_{1,r}$ provided each of its triples is imbeddable in $\mathscr{E}_{1,r}$ while the whole set is not. Prove that pseudo-$\mathscr{E}_{1,r}$ quadruples exist that are imbeddable in $\mathscr{E}_{2,r}$ and obtain a characterization of all such quadruples.
2. Denote by $\{S^*\}$ a class of semimetric spaces S^* with the property that if $p_1, p_2, p_3 \in S^*$, then $p_1, p_2, p_3 \subseteqq \mathscr{E}_{1,r}$ or at least two of the distances $p_i p_j$ $(i, j = 1, 2, 3)$ equal $\frac{1}{2}\pi r$. Prove that $\mathscr{E}_{2,r}$ has the smallest congruence order 5 with respect to the class $\{S^*\}$.
3. Does $\mathscr{E}_{2,r}$ contain a pseudo-$\mathscr{E}_{1,r}$ quintuple?
4. Define and investigate pseudo-$\mathscr{E}_{2,r}$ sets.
5. Let S be a semimetric space which it is desired to imbed in $\mathscr{E}_{2,r}$. Investigate the possibility of relaxing the condition that *each* septuple of S be imbeddable in $\mathscr{E}_{2,r}$; i.e., determine whether free septuples may be permitted in S (see § 47). Would it suffice, for example, to require merely that each septuple of S *containing a given quadruple* be imbeddable in $\mathscr{E}_{2,r}$?
6. Show that if a quadruple Q of $\mathscr{E}_{2,r}$ has an associated epsilon matrix (ϵ_{ij}), $\epsilon_{ij} = -1$ $(i, j = 1, 2, 3, 4; i \neq j)$, $\epsilon_{ii} = 1$, then Q is superposable with each quadruple of $\mathscr{E}_{2,r}$ that is congruent with it.

† Of course, the indices $(7, 0)$ are not best with respect to certain restricted classes of semimetric spaces; see, for example, Exercise 1, § 86, and Exercise 2 of this section.

104. Congruence invariance of metric bases. Congruence indices of the cross

Two matters of obvious interest and importance in a metric study of elliptic space are raised by the following queries.

1. *Is the property of being a metric basis a congruence invariant?* If B is a metric basis for an elliptic space, is any subset of the space congruent with B necessarily a metric basis?

An easy example shows that this query must be given a negative answer. We shall, however, see that for $\mathscr{E}_{2,r}$ any metric basis which is not a congruence invariant consists of at most five points.

2. *What congruence indices does the cross have with respect to the class $\{S\}$ of semimetric spaces?* The cross, as the equidistant locus of a point-pair, plays an important role in elliptic geometry. Congruence order of the space might be approached by finding first a congruence order for the cross. This section obtains two independent congruence symbols for the cross in $\mathscr{E}_{2,r}$.

THEOREM 104.1. *The property of being a metric basis for the $\mathscr{E}_{2,r}$ is not a congruence invariant.*

Proof. Let p_1, p_2, p_3 be a non-isosceles, proper triple of $\mathscr{E}_{2,r}$ with no angle a right angle, and let $A_1 A_2 A_3$ be its singular triangle. If the altitude through p_1 intersects the side $A_1 A_3$ in p_4, and p_4' is the reflection of this point in the mid-point of side $A_1 A_3$, the metric basis p_1, p_2, p_3, p_4' is congruent to the four points p_1, p_2, p_3, p_4 which do not form a metric basis.

Remark. If the orthocentre p_5 of p_1, p_2, p_3 is adjoined, two congruent quintuples result, one of which is a metric basis while the other is not.

THEOREM 104.2. *The property of being a metric basis for the $\mathscr{E}_{2,r}$ is a congruence invariant if the basis contains more than five points.*

Proof. If B is a metric basis of the $\mathscr{E}_{2,r}$ containing more than five points and B' a subset of the $\mathscr{E}_{2,r}$ congruent to it, then B' must itself be a metric basis. For in the contrary case B' is a subset of a cross, and (corollary to Theorem 101.2) B is then also a subset of a cross and hence not a metric basis.

This theorem cannot be extended to higher dimensions, as the following example indicates. Consider the configuration p_1, p_2, p_3, p_4' and p_1, p_2, p_3, p_4 of the previous example. Erect perpendicular lines to the $\mathscr{E}_{2,r}$ at the points p_1, p_2, p_3, respectively. The set G consisting of these lines and the point p_4' forms a metric basis for the $\mathscr{E}_{3,r}$ and is congruent to the set G' consisting of the same lines together with the point p_4. But G' is not a metric basis of the $\mathscr{E}_{3,r}$. Thus, since G has the power of the continuum, it is not possible by restricting the power of the metric basis to ensure that sets congruent to it will likewise be metric bases.

We turn now to the second of the questions.

THEOREM 104.3. *A cross X in $\mathscr{E}_{2,r}$ has congruence indices $(6, 2)$ with respect to the class of semimetric spaces.*

Proof. Let S be a semimetric space having more than eight points, each six of which are congruently contained in X.

Case I. *Three points p_1, p_2, p_3 of S are linear.* Then each triple of $\mathscr{E}_{2,r}$ congruent with p_1, p_2, p_3 is linear. If every quadruple of S containing p_1, p_2, p_3 is congruently contained in an $\mathscr{E}_{1,r}$, it follows that S is congruent to a subset of the $\mathscr{E}_{1,r}$. If the only point p of S which together with p_1, p_2, p_3 is not congruently contained in an $\mathscr{E}_{1,r}$ is such that

$$pp_1 = pp_2 = pp_3 = \tfrac{1}{2}\pi r,$$

the same result follows. Assume, then, a point p_4 such that

$$p_1, p_2, p_3, p_4 \approx p_1', p_2', p_3', p_4' \subset X$$

lies on only one cross. The quadruple p_1', p_2', p_3', p_4' forms a metric basis for X and is, in addition, freely movable in X.

We define now a mapping of S into X. If p is any element of S, there is a subset $p_1', p_2', p_3', p_4', p'$ of X congruent to p_1, p_2, p_3, p_4, p. Thus to each point p of S there corresponds one and (since p_1', p_2', p_3', p_4' is a metric basis of the cross) only one point p' of X. To see that this mapping is a congruence, let p_α and p_β be any two points of S and consider the congruence Γ such that $\Gamma(p_1, p_2, p_3, p_4, p_\alpha, p_\beta) \approx p_1', p_2', p_3', p_4', p_\alpha'', p_\beta''$. Since p_1', p_2', p_3', p_4' is a metric basis for X, $p_\alpha'' = p_\alpha'$, $p_\beta'' = p_\beta'$, and

$$p_\alpha' p_\beta' = p_\alpha'' p_\beta'' = p_\alpha p_\beta.$$

Case II. *No triple of S is linear.* Let $p_1, p_2, \ldots, p_6$ be six points of S and $p_1', p_2', \ldots, p_6'$ a subset of X congruent to them. No four of these points of X can be on one arm else a linear triple results. It is readily seen that some five points of the sextuple of X contain no orthocentric quadruple and thus, by Theorem 101.1, are freely movable in the plane. Since these five points are on only one cross, they are freely movable in the cross as well. Let p_1', p_2', p_3' be on one arm, and p_4' and p_5' on the other.

We define now a mapping of S into X as follows. Let p be any point of S and consider the points $p_1', p_2', \ldots, p_5', p'$ of X congruent to $p_1, p_2, \ldots, p_5, p$. Since $p_1', p_2', \ldots, p_5'$ form a metric basis for X, the point p' is uniquely defined. Furthermore, no two distinct points of S can map into the orthocentre of one of the triangles $p'_4 p'_i p'_j$ ($i,j = 1, 2, 3$). For suppose p, q two points of S mapping into p' the orthocentre of $p'_4 p'_i p'_j$, and let p''_1, p''_2, p''_3, p'', q'' be five points of X congruent to the points p_1, p_2, p_3, p, q of S. Since p''_1, p''_2, p''_3, p'' and p''_1, p''_2, p''_3, q'' are both superposable with p'_1, p'_2, p'_3, p', this would imply $pq = 0$.

Now since S contains more than eight points, some two points p_α, p_β of S map into points of X one of which (p'_α) is not the orthocentre of any of the triangles $p'_4 p'_i p'_j$ ($i,j = 1, 2, 3$). It is asserted that

$$p'_1, p'_2, ..., p'_5, p'_\alpha, p'_\beta \approx p_1, p_2, ..., p_5, p_\alpha, p_\beta.$$

The only distance open to question is $p'_\alpha p'_\beta$. Consider the points $p''_1, p''_2, ..., p''_4, p''_\alpha, p''_\beta$ of X congruent to $p_1, p_2, ..., p_4, p_\alpha, p_\beta$ of S. Since $p''_1, p''_2, p''_3, p''_4, p''_\alpha$ contain no properly orthocentric quadruple,

$$p''_1, p''_2, p''_3, p''_4, p''_\alpha \cong p'_1, p'_2, p'_3, p'_4, p'_\alpha \quad \text{(Theorem 101.2)}.$$

Let the superposition of the first quintuple with the second take p''_β into p'''_β. Again, since $p'_1, p'_2, p'_3, p'_4, p'_\beta \approx p'_1, p'_2, p'_3, p'_4, p'''_\beta$ and p'_1, p'_2, p'_3, p'_4 is a metric basis for X, $p'_\beta = p'''_\beta$ and so $p'_\alpha p'_\beta = p_\alpha p_\beta$. Thus

$$p_1, p_2, p_3, p_4, p_5, p_\alpha, p_\beta$$

are congruently contained in the cross, with $p_4, p_5, p_\alpha, p_\beta$ congruently contained in an $\mathscr{E}_{1,r}$. But this implies a linear triple and so Case II is impossible.†

THEOREM 104.4. *The cross has congruence order seven with respect to the class of semimetric spaces.*

Proof. Since each seven points of S are now congruently imbeddable in X, then certainly a linear triple occurs. The proof, then, exactly parallels that of Case I of Theorem 104.3.

REFERENCES

§§ 98–104. Blumenthal [19] and Kelly. See Haantjes [3] and Seidel, and Seidel [2] for the investigations referred to in § 103.

SUPPLEMENTARY PAPERS

Seidel [2]—an algebraic-geometric proof of the congruence order 7 property of the elliptic plane.

† See Exercise 4, § 79.

CHAPTER XII

GENERALIZED HYPERBOLIC SPACE

105. Introductory remarks

THE n-dimensional hyperbolic space $\mathscr{H}_{n,r}$ of curvature $-1/r^2$, $r > 0$, is defined in §9 (Example 4) and shown there to be metric. Though there are marked metric differences between $\mathscr{H}_{n,r}$ and the euclidean n-space E_n for $n > 1$, hyperbolic space possesses so many of those euclidean properties upon which our metric study of E_n was based that no essentially different methods are needed here. In contrast to elliptic space whose metric behaviour we have found to be quite abnormal (that is, very unlike the behaviour of E_n) the $\mathscr{H}_{n,r}$ offers little of novelty. We shall, therefore, greatly curtail our discussion of the imbedding and characterization problems on the one hand, and, on the other, generalize the situation somewhat by considering a class of spaces to which the hyperbolic spaces belong.

EXERCISES

1. Show that the hyperbolic and euclidean lines are congruent, that is, $\mathscr{H}_{1,r} \approx E_1$.
2. If $p_0, p_1, ..., p_m \in \mathscr{H}_{n,r}$, the '$m$-bein' formed by the rays joining p_i to p_0 ($i = 1, 2, ..., m$) is isogonally imbeddable in E_n. (*Hint*. Consider a conformal model of $\mathscr{H}_{n,r}$ in E_n.)
3. List some of the metric properties that $\mathscr{H}_{n,r}$ and E_n have in common (e.g. any congruence between any two sets is extendible to a motion, points of the space congruently contained in a k-dimensional subspace lie in a k-dimensional subspace). What can you say concerning a metric basis for the hyperbolic plane?

106. The $(n+2)$-point relation in $\mathscr{H}_{n,r}$

In 1870 Schering gave in determinant form a relation satisfied by the ten mutual distances of five points in hyperbolic three-space, and later stated the analogous relation for $n+2$ points in $\mathscr{H}_{n,r}$. In neither case, however, did he indicate how his relation was obtained. A derivation of Schering's five-point relation was published by Mansion some twenty years later.

Schering's $(n+2)$-point relation is obtained in this section in a manner completely different from that used by Mansion, and we shall derive an inequality satisfied by the mutual distances of the points of each $(n+1)$-tuple of $\mathscr{H}_{n,r}$ that does not lie in an $\mathscr{H}_{n-1,r}$. This 'simplicial inequality' (the generalization of the triangle inequality), of basic importance in our work, was apparently not considered by Schering or Mansion.

Calling an $(m+1)$-tuple of $\mathcal{H}_{n,r}$ dependent or independent according as it is or is not contained in a subspace $\mathcal{H}_{k,r}$ with $k < m$, the desired $(n+2)$-point relation, as well as the simplicial inequality, is given by the following theorem.

Theorem 106.1. *If $p_1, p_2, ..., p_{m+1}$ are $m+1$ points of $\mathcal{H}_{n,r}$, the determinant $|\cosh p_i p_j / r|$ $(i,j = 1, 2, ..., m+1)$ has the sign of $(-1)^m$ or vanishes according as the $(m+1)$-tuple is independent or dependent.*

Proof. Multiplying the first column of the determinant by $\cosh p_1 p_k/r$ and subtracting the result from the kth column, $k = 2, 3, ..., m+1$, we find, after applying the law of cosines for hyperbolic geometry,

$$|\cosh p_i p_j/r| = (-1)^m \left(\prod_{k=2}^{m+1} \sinh^2 p_1 p_k/r\right)|\cos p_1 : p_i, p_j|$$

$$(i,j = 2, 3, ..., m+1),$$

where $p_1 : p_i, p_j$ denotes the angle formed at p_1 by the rays $\overrightarrow{p_1 p_i}, \overrightarrow{p_1 p_j}$. It remains to show that the mth-order determinant $|\cos p_1 : p_i, p_j|$ $(i,j = 2, 3, ..., m+1)$ is positive if $p_1, p_2, ..., p_{m+1}$ are independent points and zero otherwise.

Now the m-bein formed by the m rays $\overrightarrow{p_1 p_k}$ $(k = 2, 3, ..., m+1)$ of $\mathcal{H}_{n,r}$ is *isogonally* imbeddable in an n-dimensional euclidean space (Exercise 2, § 105). If μ denotes the m-bein of E_n isogonal with the m-bein of $\mathcal{H}_{n,r}$, with the ray of μ corresponding to $\overrightarrow{p_1 p_k}$ denoted by ρ_k $(k = 2, 3, ..., m+1)$, let ρ_k intersect the unit sphere $S_{n-1,1}$ with centre at the vertex of μ in p'_k $(k = 2, 3, ..., m+1)$. Then

$$p'_i p'_j = \text{spher dist}(p'_i, p'_j) = \angle p_1 : p_i, p_j \quad (i,j = 2, 3, ..., m+1),$$

and so $\quad |\cos p_1 : p_i, p_j| = |\cos p'_i p'_j| \quad (i,j = 2, 3, ..., m+1).$

Now by Theorem 63.1 $|\cos p'_i p'_j|$ $(i,j = 2, 3, ..., m+1)$ is positive unless $p'_2, p'_3, ..., p'_{m+1}$ are spherically dependent, in which event the determinant vanishes. But the spherical dependence of $p'_2, p'_3, ..., p'_{m+1}$ is equivalent to the m rays $\rho_2, \rho_3, ..., \rho_{m+1}$ being contained in an E_{m-1}, which occurs if and only if the points $p_1, p_2, ..., p_{m+1}$ lie in an $\mathcal{H}_{m-1,r}$ (i.e. if and only if $p_1, p_2, ..., p_{m+1}$ form a dependent set).

Corollary. *If $p_1, p_2, ..., p_{n+2} \in \mathcal{H}_{n,r}$, then*

$$|\cosh p_i p_j/r| \quad (i,j = 1, 2, ..., n+2)$$

vanishes and each non-vanishing principal minor of order $k+1$ has the sign of $(-1)^k$.

The determinant
$$\Lambda_m(p_1, p_2, ..., p_m) = |\cosh p_i p_j/r| \quad (i,j = 1, 2, ..., m)$$
has the same importance in the metric theory of hyperbolic space as the determinants Δ_m and D have in spherical and euclidean spaces, respectively.

EXERCISES

1. Let p_1, p_2, p_3, p be four non-collinear points of $\mathcal{H}_{2,r}$ with $p_1, p_2, p_3 \in \mathcal{H}_{1,r}$ and p_2 between p_1, p_3. Apply the hyperbolic law of cosines to the two triangles (p_1, p_2, p), (p_2, p_3, p), and obtain

$$\sinh(p_2 p_3/r)\cosh(p_1 p/r) - \sinh(p_1 p_3/r)\cosh(p_2 p/r) +$$
$$+ \sinh(p_1 p_2/r)\cosh(p_3 p/r) = 0 \qquad (*)$$

Letting p approach p_1, p_2, p_3 in turn, use (*) to show that $|\cosh p_i p_j/r|$ $(i,j = 1, 2, 3)$ vanishes.

2. If $p_1, p_2, p_3, p_4 \in \mathcal{H}_{2,r}$, the labelling may be assumed so that the lines $\mathcal{H}_{1,r}(p_1, p_2)$, $\mathcal{H}_{1,r}(p_3, p_4)$ intersect in a point q. Apply the procedure of Exercise 1 to show that $|\cosh p_i p_j/r|$ $(i,j = 1, 2, 3, 4)$ vanishes.

3. Extend the method of Exercises 1, 2 to show that
$$|\cosh p_i p_j/r| \quad (i,j = 1, 2, ..., n+2)$$
vanishes for $p_1, p_2, ..., p_{n+2}$ points of $\mathcal{H}_{n,r}$.

107. Generalized hyperbolic space. Three postulates

In § 67 we remarked upon our need of properties of the convex n-sphere that, if previously known, are not readily found in the literature. The desired properties were listed and it was stated that they are all consequences of five basic properties.

In a study of hyperbolic n-space we are faced with a similar difficulty in satisfying our needs from the existing literature. There has been no systematic study of hyperbolic n-space and so it is necessary to develop for ourselves those elementary properties of the space that are useful in our work. It was found in doing this that (1) all of those properties are consequences of a very few basic ones, and (2) the hyperbolic cosine function that is characteristic of hyperbolic space might be replaced by any one of a large class of functions. These considerations give rise to the definition of a class of semimetric spaces containing $\mathcal{H}_{n,r}$ and suggest the desirability of developing the metric properties of that class.

To this end, let $\phi(x; r)$ be a real single-valued function of the real variable x and the parameter r, with $\phi(0; r) \geqslant 0$. If $\phi(x; r)$ is defined for every $x = p_i p_j$ (p_i, p_j points of a semimetric space S and r fixed) denote by $\Lambda_m(p_1, p_2, ..., p_m)$ the symmetric determinant $|\phi(p_i p_j; r)|$ $(i,j = 1, 2, ..., m)$, and let n be a pre-assigned positive integer.

POSTULATE I. *The determinant Λ_{n+2} vanishes for every set of $n+2$ points of S.*

DEFINITION 107.1. *A set of k points of S ($k > 1$) is dependent or independent according as the determinant Λ_k of the k points vanishes or not. A set consisting of exactly one point is independent.*†

Since Λ_k is obviously a congruence invariant, so are 'dependence' and 'independence' of k-tuples.

POSTULATE II. *There exists at least one $(n+1)$-tuple of S with non-vanishing determinant Λ_{n+1}.*

POSTULATE III. *The determinant Λ_{k+1} of $k+1$ independent points of S, $k > 0$, has the sign $(-1)^k$.*

Let $\mathscr{H}_{n,r}^{\phi}$ denote any semimetric space S for which Postulates I, II, III are valid. Then (Theorem 106.1, corollary) hyperbolic n-space $\mathscr{H}_{n,r}$ is a member of the class $\{\mathscr{H}_{n,r}^{\phi}\}$ with $\phi(x;r) = \cosh x/r$.‡

EXERCISES

1. If $p_1, p_2, ..., p_{k+1}, p \in \mathscr{H}_{n,r}^{\phi}$ and the $(k+1)$-tuple $p_1, p_2, ..., p_{k+1}$ is a dependent set, then the $(k+2)$-tuple is dependent. Prove this and note the following corollaries:
    (1) each subset of an independent set is an independent set,
    (2) if an m-tuple is independent, then $m \leqslant n+1$,
    (3) for every positive integer $k \leqslant n$, $\mathscr{H}_{n,r}^{\phi}$ contains at least one independent $(k+1)$-tuple.
2. What use is made of the assumption that $\phi(0;r) \geqslant 0$? (*Hint.* Consider the proof of the theorem in Exercise 1 for $k = 1$.)

108. Subspaces of the $\mathscr{H}_{n,r}^{\phi}$

Let $p_1, p_2, ..., p_{k+1}$ be an independent $(k+1)$-tuple of $\mathscr{H}_{n,r}^{\phi}$ ($k \geqslant 0$) and denote by $\Omega(p_1, p_2, ..., p_{k+1})$ the set of all those points ω of $\mathscr{H}_{n,r}^{\phi}$ for which $\Lambda_{k+2}(p_1, p_2, ..., p_{k+1}, \omega) = 0$. Clearly

$$p_i \in \Omega(p_1, p_2, ..., p_{k+1}) \quad (i = 1, 2, ..., k+1).$$

If now $\omega_1, \omega_2, ..., \omega_s$ are any s points of $\Omega(p_1, p_2, ..., p_{k+1})$, $s \geqslant 1$, then $\Lambda_{k+s+1}(p_1, p_2, ..., p_{k+1}, \omega_1, \omega_2, ..., \omega_s) = 0$, for $p_1, p_2, ..., p_{k+1}, \omega_1$ is a dependent set and hence so is $p_1, p_2, ..., p_{k+1}, \omega_1, \omega_2, ..., \omega_s$ (Exercise 1, § 107).

THEOREM 108.1. *Each set of $k+2$ points of $\Omega(p_1, p_2, ..., p_{k+1})$ is a dependent set.*

Proof. If $\omega_1, \omega_2, ..., \omega_{k+2}$ are $k+2$ points of $\Omega(p_1, p_2, ..., p_{k+1})$, the determinant $\Lambda_{2k+3}(p_1, p_2, ..., p_{k+1}, \omega_1, \omega_2, ..., \omega_{k+2})$ has the $(k+1)$th-order

---
† For a justification of this definition, see Exercise 1.
‡ Compare with §§ 67, 68.

principal minor $\Lambda_{k+1}(p_1, p_2,..., p_{k+1})$ different from zero, while by the remark preceding this theorem, every $(k+2)$th- and every $(k+3)$th-order principal minor containing $\Lambda_{k+1}(p_1, p_2,..., p_{k+1})$ vanishes. It follows that the rank of the $(2k+3)$th-order determinant is $k+1$ and hence $\Lambda_{k+2}(\omega_1, \omega_2,..., \omega_{k+2}) = 0$.

Since for each positive integer i not exceeding k, $\Omega(p_1, p_2,..., p_{k+1})$ contains a set of $i+1$ independent points (e.g. an $(i+1)$-tuple of $p_1, p_2,..., p_{k+1}$) while each $(k+2)$-tuple of $\Omega(p_1, p_2,..., p_{k+1})$ is a dependent set, we are justified in referring to $\Omega(p_1, p_2,..., p_{k+1})$ as *a k-dimensional subspace* of $\mathcal{H}^\phi_{n,r}$ and shall denote it by $\mathcal{H}^\phi_{k,r}$. The points $p_1, p_2,..., p_{k+1}$ are called *base points* of $\mathcal{H}^\phi_{k,r}$. Clearly $\mathcal{H}^\phi_{n,r}$ is itself an n-dimensional subspace.

If $p_1, p_2,..., p_{k+1}$ is an independent $(k+1)$-tuple of $\mathcal{H}^\phi_{n,r}$ ($k < n$), there is at least one point p^* of $\mathcal{H}^\phi_{n,r}$ such that $\Lambda_{k+2}(p_1, p_2,..., p_{k+1}, p^*) \neq 0$, for in the contrary case each point of $\mathcal{H}^\phi_{n,r}$ is contained in the k-dimensional subspace with base points $p_1, p_2,..., p_{k+1}$ and every $k+2$ points of $\mathcal{H}^\phi_{n,r}$ are dependent. Since $k < n$, this contradicts Postulate II. Other obvious properties of subspaces are (i) $\mathcal{H}^\phi_{k,r}$ is uniquely determined by its base points, and (ii) for every positive integer k not exceeding n, k-dimensional subspaces exist.

THEOREM 108.2. *Any independent $(k+1)$-tuple of a k-dimensional subspace may be taken as its base points.*

Proof. Let $\mathcal{H}^\phi_{k,r}(q_1, q_2,..., q_{k+1})$ denote the k-dimensional subspace whose base points $q_1, q_2,..., q_{k+1}$ form an independent $(k+1)$-tuple of the subspace $\mathcal{H}^\phi_{k,r}(p_1, p_2,..., p_{k+1})$. Since each $k+2$ points of the latter subspace have a vanishing Λ-determinant, then $\Lambda_{k+2}(q_1, q_2,..., q_{k+1}, p_i) = 0$ ($i = 1, 2,..., k+1$) and so $p_1, p_2,..., p_{k+1}$ are points of $\mathcal{H}^\phi_{k,r}(q_1, q_2,..., q_{k+1})$. If, now, $q^* \in \mathcal{H}^\phi_{k,r}(q_1, q_2,..., q_{k+1})$, $q^* \neq p_i$ ($i = 1, 2,..., k+1$), then

$$\Lambda_{k+2}(p_1, p_2,..., p_{k+1}, q^*) = 0$$

and so $q^* \in \mathcal{H}^\phi_{k,r}(p_1, p_2,..., p_{k+1})$. Similarly, it is seen that each point of $\mathcal{H}^\phi_{k,r}(p_1, p_2,..., p_{k+1})$ belongs to $\mathcal{H}^\phi_{k,r}(q_1, q_2,..., q_{k+1})$, and the theorem is proved.

COROLLARY. *Two distinct k-dimensional subspaces can have at most k independent points in common.*

We have so far subjected $\phi(x; r)$ only to the very mild requirements of being defined over the distance set of S and non-negative for $x = 0$. In all that follows we shall assume that $\phi(x; r)$ *is a monotonic (strictly)* *increasing function of x and $\phi(0; r) > 0$.† No additional assumptions on $\phi(x; r)$ are needed. Since the ϕ-function for $\mathcal{H}_{n,r}$ (namely $\cosh x/r$) satisfies these conditions, we have not 'lost' the hyperbolic n-space as a member of the class $\{\mathcal{H}_{n,r}^{\phi}\}$. We make use of our assumptions concerning ϕ in the following theorem.

THEOREM 108.3. *If p_i, $\bar{p}_i$ ($i = 1, 2, ..., k+1$) are two congruent sets of $k+1$ independent points of two (coincident or distinct) k-dimensional subspaces $\mathcal{H}_{k,r}^{\phi}$, $\bar{\mathcal{H}}_{k,r}^{\phi}$, respectively, and if $p, q \in \mathcal{H}_{k,r}^{\phi}$, $\bar{p}, \bar{q} \in \bar{\mathcal{H}}_{k,r}^{\phi}$ such that*

$$p_1, p_2, ..., p_{k+1}, p \approx \bar{p}_1, \bar{p}_2, ..., \bar{p}_{k+1}, \bar{p},$$
$$p_1, p_2, ..., p_{k+1}, q \approx \bar{p}_1, \bar{p}_2, ..., \bar{p}_{k+1}, \bar{q},$$

then $pq = \bar{p}\bar{q}$.

Proof. If $\Lambda_{k+3}(p_1, p_2, ..., p_{k+1}, p, q; x)$ denotes the determinant obtained from $\Lambda_{k+3}(p_1, p_2, ..., p_{k+1}, p, q)$ upon replacing pq by x, then the two congruences of the hypothesis give

$$\Lambda_{k+3}(p_1, p_2, ..., p_{k+1}, p, q; x) \equiv \Lambda_{k+3}(\bar{p}_1, \bar{p}_2, ..., \bar{p}_{k+1}, \bar{p}, \bar{q}; x).$$

From this, Theorem 108.1, and the first corollary of the theorem in Exercise 1, §107, we conclude $\Lambda_{k+3}(p_1, p_2, ..., p_{k+1}, p, q; x)$ vanishes for $x = pq$ and $x = \bar{p}\bar{q}$.

Since $\Lambda_{k+2}(p_1, p_2, ..., p_{k+1}, p) = 0$, it follows that

$$\Lambda_{k+3}(p_1, p_2, ..., p_{k+1}, p, q; x) = [A \cdot \phi(x; r) + B]^2, \quad A \neq 0,$$

and so, from the monotonicity of $\phi(x; r)$, the determinant on the left has a unique zero; that is, $pq = \bar{p}\bar{q}$.

The theorem has two useful corollaries.

COROLLARY 1. *If p_i, $\bar{p}_i$ ($i = 1, 2, ..., k+1$) are independent $(k+1)$-tuples of $\mathcal{H}_{k,r}^{\phi}$, $\bar{\mathcal{H}}_{k,r}^{\phi}$, respectively, then corresponding to each point p of $\mathcal{H}_{k,r}^{\phi}$ there is at most one point $\bar{p}$ of $\bar{\mathcal{H}}_{k,r}^{\phi}$ with*

$$p_1, p_2, ..., p_{k+1}, p \approx \bar{p}_1, \bar{p}_2, ..., \bar{p}_{k+1}, \bar{p}.$$

COROLLARY 2. *Any $k+1$ independent points of $\mathcal{H}_{k,r}^{\phi}$ form a metric basis for the subspace.*

EXERCISES

1. Determine A in the expression for $\Lambda_{k+3}(p_1, p_2, ..., p_{k+1}, p, q; x)$.
2. Show that any two points of $\mathcal{H}_{n,r}^{\phi}$ are independent if and only if they are distinct.

109. Some metric properties of $\mathcal{H}_{n,r}^{\phi}$

To solve the subset problem for generalized hyperbolic space and determine its congruence indices we show that $\mathcal{H}_{n,r}^{\phi}$ possesses properties

† The need for assuming $\phi(0; r) > 0$ will appear later.

analogous to those used in solving these problems for euclidean space. When this has been done, it will not be necessary to repeat the arguments already applied in the earlier case in order to draw analogous conclusions. We shall, therefore, seek to establish in this and the next section those desired properties.

THEOREM 109.1. *If $p_1, p_2, ..., p_{k+1}$ is an independent set of $\mathcal{H}^\phi_{n,r}$ and $p \in \mathcal{H}^\phi_{n,r}$ such that for each $i = 1, 2, ..., k$ the points*

$$p_1, p_2, ..., p_{i-1}, p, p_{i+1}, ..., p_k, p_{k+1}$$

are dependent, then $p = p_{k+1}$.

Proof. The argument is inductive, anchored for $k = 1$. Since $p_1, p_2, ..., p_{k+1}$ are independent points, (i) each k-tuple determines a unique $(k-1)$-dimensional subspace, and (ii) the k subspaces

$$\mathcal{H}^\phi_{k-1,r}(p_2, p_3, ..., p_{k+1}), \; \mathcal{H}^\phi_{k-1,r}(p_1, p_3, ..., p_{k+1}), ..., \; \mathcal{H}^\phi_{k-1,r}(p_1, ..., p_{k-1}, p_{k+1})$$

are pairwise distinct. Hence each pair can have at most $k-1$ independent points in common (corollary, Theorem 108.2), and so the k points $p_1, p_2, ..., p_{i-1}, p, p_{i+1}, ..., p_{j-1}, p_{j+1}, ..., p_k, p_{k+1}$ which the ith and jth of these subspaces have in common are *dependent* and

$$\Lambda_k(p_1, ..., p_{i-1}, p, p_{i+1}, ..., p_{j-1}, p_{j+1}, ..., p_k, p_{k+1}) = 0$$

$$(i, j = 1, 2, ..., k, i < j).$$

Suppose, now, i is fixed, say $i = 1$. Then

$$\Lambda_k(p, p_2, ..., p_{j-1}, p_{j+1}, ..., p_k, p_{k+1}) = 0 \quad (j = 2, 3, ..., k)$$

and so $p, p_2, ..., p_{j-1}, p_{j+1}, ..., p_k, p_{k+1}$ are dependent sets for every integer $j = 2, 3, ..., k$. Applying the inductive hypothesis to p and the independent k-tuple $p_2, p_3, ..., p_{k+1}$ gives $p = p_{k+1}$.

COROLLARY. *If $p_1, p_2, ..., p_{k+1}$ are independent points of $\mathcal{H}^\phi_{k,r}$ the k-subspaces $\mathcal{H}^\phi_{k-1,r}(p_1, ..., p_{i-1}, p_{i+1}, ..., p_{k+1})$ $(i = 1, 2, ..., k)$ have exactly the one point p_{k+1} in common.*

THEOREM 109.2. *For each integer $k < n$ and each k-dimensional subspace $\mathcal{H}^\phi_{k,r}$ there exists a $(k+1)$-dimensional subspace of which $\mathcal{H}^\phi_{k,r}$ is a proper subset.*

THEOREM 109.3. *For each integer $k \leq n$, each $\mathcal{H}^\phi_{k,r}$ contains every $\mathcal{H}^\phi_{m,r}$ determined by every independent $(m+1)$-tuple of its points.*

The proofs are left to the reader.

THEOREM 109.4. *A k-dimensional subspace $\mathcal{H}^\phi_{k,r}$ does not contain an equilateral set of $k+2$ distinct points.*

Proof. If $p_1, p_2, \ldots, p_{k+2}$ be assumed a subset of $\mathscr{H}^{\phi}_{k,r}$ with $p_i p_j = a > 0$ ($i, j = 1, 2, \ldots, k+2$; $i \neq j$), then

$$\Lambda_{k+2}(p_1, \ldots, p_{k+2}) = [\phi(0; r) - \phi(a; r)]^{k+1} \cdot [\phi(0; r) + (k+1)\phi(a; r)],$$

which, according to the assumptions made on ϕ, fails to vanish, in contradiction to Theorem 108.1.

THEOREM 109.5. *If p, q are distinct points of $\mathscr{H}^{\phi}_{k,r}$ containing the independent set $p_1, p_2, \ldots, p_k$ with $p_i p = p_i q$ ($i = 1, 2, \ldots, k$), the $(k-1)$-dimensional subspace $\mathscr{H}^{\phi}_{k-1,r}(p_1, p_2, \ldots, p_k)$ is the locus of points of $\mathscr{H}^{\phi}_{k,r}$ equidistant from p and q.*

Proof. If $p^* \in \mathscr{H}^{\phi}_{k,r}$ equidistant from p and q, then

$$\Lambda_{k+1}(p_1, p_2, \ldots, p_k, p^*) = 0,$$

for in the contrary case the independence of $p_1, p_2, \ldots, p_k, p^*$ together with

$$p_1, p_2, \ldots, p_k, p^*, p \approx p_1, p_2, \ldots, p_k, p^*, q$$

implies $p = q$ (Corollary 2, Theorem 108.3). Hence each point of $\mathscr{H}^{\phi}_{k,r}$ equidistant from p and q belongs to the subspace $\mathscr{H}^{\phi}_{k-1,r}(p_1, p_2, \ldots, p_k)$.

Further, if $p^* \in \mathscr{H}^{\phi}_{k-1,r}(p_1, p_2, \ldots, p_k)$, then $pp^* = qp^*$, for the equation $\Lambda_{k+2}(p_1, \ldots, p_k, p^*, p; x) = 0$ is satisfied for $x = pp^*$ and $x = qp^*$. But since

$$\Lambda_{k+2}(p_1, p_2, \ldots, p_k, p^*, p; x) = -[k+1, k+2]^2/\Lambda_k(p_1, p_2, \ldots, p_k),$$

it has only one zero, and the proof is complete.

THEOREM 109.6. *If q, s, t are any three distinct points of $\mathscr{H}^{\phi}_{k,r}$ with $qs = st = qt$, and $p_1, p_2, \ldots, p_k \in \mathscr{H}^{\phi}_{k,r}$ such that*

$$p_i q = p_i s = p_i t \quad (i = 1, 2, \ldots, k),$$

then $p_1, p_2, \ldots, p_k$ are dependent.

Proof. For $k = 2$ suppose the points p_1, p_2 are independent (that is, distinct). Then p_1, p_2, q are *independent*, since in the contrary case

$$\Lambda_3(p_1, p_2, q) = \Lambda_3(p_1, p_2, s) = \Lambda_3(p_1, p_2, t) = 0$$

and $q, s, t \in \mathscr{H}^{\phi}_{1,r}(p_1, p_2)$, which is impossible since that subspace does not contain an equilateral triple (Theorem 109.4). But the independence of p_1, p_2, q and the congruence $p_1, p_2, q, s \approx p_1, p_2, q, t$ imply $s = t$ (Theorem 108.3, Corollary 2). From this contradiction we conclude p_1, p_2 are dependent and the theorem is proved for $k = 2$.

Suppose, now, the theorem true for all subspaces of dimensions less than k ($k > 2$). If the k points $p_1, p_2, \ldots, p_k$ of $\mathscr{H}^{\phi}_{k,r}$ contain a dependent $(k-1)$-tuple, they are dependent, so we may assume each $(k-1)$-tuple independent. Now $p_1, p_2,..., p_{k-1}, q$ are *independent*, for in the contrary case $p_1, p_2,..., p_{k-1}, q$ together with $p_1, p_2,.., p_{k-1}, s$ and $p_1, p_2, .., p_{k-1}, t$ are dependent and q, s, t lie in a $(k-2)$-dimensional subspace containing $p_1, p_2,..., p_{k-1}$. But then $p_1, p_2,..., p_{k-1}$ are dependent.

Hence $p_1, p_2,..., p_{k-1}, q;\ p_1, p_2,..., p_{k-1}, s;\ p_1, p_2,..., p_{k-1}, t$ are independent k-tuples determining three $(k-1)$-dimensional subspaces

$$\mathcal{H}^\phi_{k-1,r}(p_1,p_2,...,p_{k-1},q), \qquad \mathcal{H}^\phi_{k-1,r}(p_1,p_2,...,p_{k-1},s)$$
$$\mathcal{H}^\phi_{k-1,r}(p_1,p_2,...,p_{k-1},t).$$

If any two of these subspaces coincide, the dependence of $p_1, p_2,..., p_{k-1}$ follows from the inductive hypothesis. If the subspaces are pairwise distinct, each contains p_k (Theorem 109.5) and the dependence of $p_1, p_2,..., p_k$ follows from the corollary of Theorem 108.2.

THEOREM 109.7. *Let $p_1, p_2,..., p_k$ be k independent points of $\mathcal{H}^\phi_{k,r}$ and let $p, q; p^*, q^*$ be two pairs of points of $\mathcal{H}^\phi_{k,r}$ such that*

$$\text{either} \quad p^* \in \mathcal{H}^\phi_{k-1,r}(p_1,...,p_k) \quad \text{or} \quad p^* \neq p$$
and
$$\text{either} \quad q^* \in \mathcal{H}^\phi_{k-1,r}(p_1,...,p_k) \quad \text{or} \quad q^* \neq q.$$
If
$$p_1, p_2,..., p_k, p^* \approx p_1, p_2,..., p_k, p,$$
$$p_1, p_2,..., p_k, q^* \approx p_1, p_2,..., p_k, q,$$
then $pq = p^*q^*$.

Proof. If both of the first alternatives subsist, then $p = p^*$ and $q = q^*$. Examining the other possibilities, two cases are distinguished.

Case 1. Exactly one of the points p^*, q^* is an element of
$$\mathcal{H}^\phi_{k-1,r}(p_1,p_2,...,p_k).$$
Assuming $\Lambda_{k+1}(p_1,p_2,...,p_k,p^*) = 0$, the determinant
$$\Lambda_{k+2}(p_1,p_2,...,p_k,p^*,q^*; x)$$
obtained from $\Lambda_{k+2}(p_1,p_2,...,p_k,p^*,q^*)$ by replacing p^*q^* (and q^*p^*) by x vanishes for $x = pq$ and $x = p^*q^*$. Since $\Lambda_k(p_1,p_2,...,p_k) \neq 0$, it is seen (as in the proof of Theorem 108.3) that $\Lambda_{k+2}(p_1,p_2,...,p_k,p^*,q^*)$ has at most one real zero and so $pq = p^*q^*$.

Case 2. Both of the second alternatives hold. Considering again the function $\Lambda_{k+2}(p_1,p_2,...,p_k,p^*,q^*; x)$ (which does not vanish identically) it is seen that each of the numbers pq, p^*q^*, pq^*, p^*q is a zero of the function. But since

$$\Lambda_k(p_1,p_2,...,p_k)\Lambda_{k+2}(p_1,p_2,...,p_k,p^*,q^*; x)$$
$$= \Lambda_{k+1}(p_1,p_2,...,p_k,p^*)\Lambda_{k+1}(p_1,p_2,...,p_k,q^*) - [A\phi(x;r)+B]^2,$$

with $A = -\Lambda_k(p_1, p_2,..., p_k)$ and B free from x, it follows from the monotonic character of $\phi(x; r)$ that $\Lambda_{k+2}(p_1, p_2,..., p_k, p^*, q^*; x)$ vanishes for at most two real values of x. Hence at most two of the numbers pq, p^*q^*, pq^*, p^*q are distinct.

Now if $pq = p^*q$, then the congruence

$$p_1, p_2,..., p_k, q, p \approx p_1, p_2,..., p_k, q, p^*,$$

together with the independence of $p_1, p_2,..., p_k, q$ implies $p = p^*$, contrary to assumption. Similarly, it is seen that $pq \neq pq^*$. It follows, then, that $pq = p^*q^*$ and the proof is complete.

110. Additional metric properties of $\mathscr{H}_{k,r}^{\phi}$

It is necessary to establish a few more properties of generalized hyperbolic space that are needed to obtain its congruence order.

THEOREM 110.1. *If $p_1, p_2,..., p_{k+1}$ are an independent subset of $\mathscr{H}_{k,r}^{\phi}$, the $k+1$ $(k-1)$-dimensional subspaces*

$$\mathscr{H}_{k-1,r}^{\phi}(p_1, p_2,..., p_{i-1}, p_{i+1},..., p_{k+1}) \quad (i = 1, 2,..., k+1)$$

have no point in common.

Proof. If the contrary be assumed and p^* is a point of each subspace, then $\Lambda_{k+1}(p_1, p_2,..., p_{i-1}, p^*, p_{i+1},..., p_{k+1}) = 0$ $(i = 1, 2,..., k+1)$. Since $p^* \in \mathscr{H}_{k,r}^{\phi}(p_1, p_2,..., p_{k+1})$, then $\Lambda_{k+2}(p_1, p_2,..., p_{k+1}, p^*) = 0$. Now each element in the last column of this determinant (with the exception of the last element) has its co-factor equal to zero, and consequently

$$\phi(0; r) \cdot \Lambda_{k+1}(p_1, p_2,..., p_{k+1}) = \Lambda_{k+2}(p_1, p_2,..., p_{k+1}, p^*) = 0.$$

But $\phi(0; r) \neq 0$ and so $\Lambda_{k+1}(p_1, p_2,..., p_{k+1}) = 0$, in contradiction to the independence of $p_1, p_2,..., p_{k+1}$.

LEMMA 110.1. *If $p_1, p_2,..., p_k, p_{k+1}$ form an independent subset of $\mathscr{H}_{k,r}^{\phi}$, that subspace contains at most one point equidistant from them.*

Proof. Let p^* be a point of $\mathscr{H}_{k,r}^{\phi}$ with $p^*p_i = x$ $(i = 1, 2,..., k+1)$. From $\Lambda_{k+2}(p_1, p_2,..., p_{k+1}, p^*) = 0$ follows easily

$$-\Lambda_{k+1}^*(p_1, p_2,..., p_{k+1}) \cdot \phi^2(x; r) = \Lambda_{k+1}(p_1, p_2,..., p_{k+1}) \cdot \phi(0; r),$$

where $\Lambda_{k+1}^*(p_1, p_2,..., p_{k+1})$ denotes the determinant obtained by annexing to $\Lambda_{k+1}(p_1, p_2,..., p_{k+1})$ a row and a column whose intersection element is 0 and all remaining elements are 1.

Since $\phi > 0$ throughout its domain of definition, there is at most one x satisfying the above equation, and it follows from Theorem 108.3 (Corollary 2) that *at most one point p^* of $\mathscr{H}_{k,r}^{\phi}$ is equidistant from $p_1, p_2,..., p_{k+1}$.*

A characterization of those independent $(k+1)$-tuples of $\mathscr{H}_{k,r}^\phi$ for which an equidistant point p^* of $\mathscr{H}_{k,r}^\phi$ exists is of interest (see Exercise 1).

THEOREM 110.2. *Let $p_1, p_2, \ldots, p_{k+1}$ be an independent subset of $\mathscr{H}_{k,r}^\phi$ and p an element of $\mathscr{H}_{k,r}^\phi$ not common to any two of the $k+1$ subspaces*

$$\mathscr{H}_{k-1,r}^\phi(p_1, \ldots, p_{i-1}, p_{i+1}, \ldots, p_{k+1}) \quad (i = 1, 2, \ldots, k+1).$$

Denote by p^i a point of $\mathscr{H}_{k,r}^\phi$ such that

$$p_1, p_2, \ldots, p_{i-1}, p^i, p_{i+1}, \ldots, p_{k+1} \approx p_1, p_2, \ldots, p_{i-1}, p, p_{i+1}, \ldots, p_{k+1}$$

$$(i = 1, 2, \ldots, k+1),$$

where $p^i \neq p$ if $p \bar{\in} \mathscr{H}_{k-1,r}^\phi(p_1, \ldots, p_{i-1}, p_{i+1}, \ldots, p_{k+1})$. Then there exists at most one point p^* of $\mathscr{H}_{k,r}^\phi$ with

$$p^*p^1 = p^*p^2 = \ldots = p^*p^{k+1}.$$

Proof. The proof is organized by a consideration of two cases.

Case 1. The point p belongs to one of the $(k-1)$-dimensional subspaces

$$\mathscr{H}_{k-1,r}^\phi(p_1, p_2, \ldots, p_{i-1}, p_{i+1}, \ldots, p_{k+1}) \quad (i = 1, 2, \ldots, k+1).$$

Assuming the labelling so that $p \in \mathscr{H}_{k-1,r}^\phi(p_2, \ldots, p_{k+1})$ then clearly $p = p^1$ and $p \neq p^i$ ($i = 2, 3, \ldots, k+1$). Since each of the points

$$p_1, \ldots, p_{i-1}, p_{i+1}, \ldots, p_{k+1}$$

is equidistant from p and p^i ($i = 2, 3, \ldots, k+1$), then

$$\mathscr{H}_{k-1,r}^\phi(p_1, \ldots, p_{i-1}, p_{i+1}, \ldots, p_{k+1})$$

is the locus of points of $\mathscr{H}_{k,r}^\phi$ equidistant from p and p^i for each $i = 2, 3, \ldots, k+1$ (Theorem 109.5). Hence p^* belongs to each of these k subspaces which, by the corollary to Theorem 109.1, have exactly the point p_1 in common. Consequently $p^* = p_1$ and the theorem is proved in this case.

Case 2. The point p does not belong to any of the subspaces

$$\mathscr{H}_{k-1,r}^\phi(p_1, \ldots, p_{i-1}, p_{i+1}, \ldots, p_{k+1}) \quad (i = 1, 2, \ldots, k+1).$$

If the $k+1$ points p^i ($i = 1, 2, \ldots, k+1$) form an independent set, the theorem follows from the preceding lemma. The proof is completed by showing that if the set $p^1, p^2, \ldots, p^{k+1}$ is dependent, then no point of $\mathscr{H}_{k,r}^\phi$ is equidistant from p^i ($i = 1, 2, \ldots, k+1$). Suppose $p^* \in \mathscr{H}_{k,r}^\phi$ with $p^1 p^* = p^2 p^* = \ldots = p^{k+1} p^*$.

Let $\bar{r}$ be the cardinal number of the maximal independent subset of $p^1, p^2, \ldots, p^{k+1}$ and assume the labelling so that $p^1, p^2, \ldots, p^{\bar{r}}$ are independent $(2 \leqslant \bar{r} \leqslant k)$. Since $p^i \neq p$ ($i = 1, 2, \ldots, k+1$), each of the subspaces $\mathscr{H}^{\phi}_{k-1,r}(p_1,...,p_{i-1},p_{i+1},...,p_{k+1})$ is the locus of points of $\mathscr{H}^{\phi}_{k,r}$ equidistant from p and p^i ($i = 1,2,...,k$) and consequently p_{k+1} is equidistant from $p^1, p^2,..., p^k$. Writing

$$p^1 p_{k+1} = p^2 p_{k+1} = ... = p^k p_{k+1} = d,$$

we shall show that $p^{k+1}p_{k+1} = d$.

From the vanishing determinant $\Lambda_{\bar{r}+2}(p^1,p^2,...,p^{\bar{r}},p^{k+1},p_{k+1})$ we obtain

$$\begin{vmatrix} \phi_{11} & \phi_{12} & \cdots & \phi_{1\bar{r}} & \phi_{1,k+1} \\ \phi_{21} & \phi_{22} & \cdots & \phi_{2\bar{r}} & \phi_{2,k+1} \\ \cdot & \cdot & \cdots & \cdot & \cdot \\ \phi_{\bar{r}1} & \phi_{\bar{r}2} & \cdots & \phi_{\bar{r}\bar{r}} & \phi_{\bar{r},k+1} \\ \delta & \delta & \cdots & \delta & \phi(p^{k+1}p_{k+1};r) \end{vmatrix} = 0, \quad (*)$$

where $\phi_{ij} = \phi(p^i p^j;r)$, $\phi_{ii} = \phi(0;r)$ ($i = 1,2,...,\bar{r}$; $j = 1,2,...,\bar{r},k+1$), and $\delta = \phi(d;r)$. In a similar manner the vanishing of the determinant $\Lambda_{\bar{r}+2}(p^1,p^2,...,p^{\bar{r}},p^{k+1},p^*)$ shows that the determinant obtained from the one above by replacing each element in the last row by 1 is equal to zero.

Since $\Lambda_{\bar{r}}(p^1,p^2,...,p^{\bar{r}}) = |\phi_{ij}|$ ($i,j = 1,2,...,\bar{r}$) is not zero, the function obtained from the determinant in (*) upon replacing $p^{k+1}p_{k+1}$ by x is linear in $\phi(x;r)$. But this function has $x = p^{k+1}p_{k+1}$ as well as $x = d$ as zeros, and it follows that $p^{k+1}p_{k+1} = d$. But this is impossible, for it implies that p_{k+1} is a point of $\mathscr{H}^{\phi}_{k-1,r}(p_1,p_2,...,p_k)$, the locus of points equidistant from p and p^{k+1}, and hence $p_1, p_2,..., p_k, p_{k+1}$ are dependent contrary to hypothesis.

EXERCISES

1. Let $p_1,p_2,...,p_{k+1}$ be an independent subset of $\mathscr{H}^{\phi}_{k,r}$ and suppose $\phi(0;r) = 1$. Show that a necessary condition that for each $i = 2,3,...,k+1$ the $\mathscr{H}^{\phi}_{i-1,r}(p_1,p_2,...,p_i)$ contain a point p_i^* equidistant from $p_1,p_2,...,p_i$ is that the semimetric $(k+1)$-tuple $q_1,q_2,...,q_{k+1}$ with

$$q_s q_t = [\phi(p_s p_t;r)-1]^{\frac{1}{2}} \quad (s,t = 1,2,...,k+1)$$

be congruent with an independent *euclidean* $(k+1)$-tuple.

2. Let the semimetric space S, the 'ground-space' of $\mathscr{H}^{\phi}_{n,r}$ be the euclidean plane E_2 and let $\phi(x) = 1+x^2$. Show that Postulates I, II, III are satisfied with $n = 3$ and consequently the space so obtained is a generalized hyperbolic space of three dimensions. What are the one-dimensional and two-dimensional subspaces?

111. Two existence postulates. Imbedding theorems for $\mathscr{H}^{\phi}_{n,r}$

The three-dimensional generalized hyperbolic space defined in Exercise 2 of the preceding section shows that Postulates I, II, III do not ensure the congruence of any two subspaces of the same dimensionality.

It is desirable, therefore, to adjoin a fourth postulate from which this important property may be derived.

POSTULATE IV. *If p_i, $\bar{p}_i$ $(i = 1, 2, ..., k+1)$ are two congruent sets of $k+1$ points (not necessarily pairwise distinct) of two (coincident or distinct) k-dimensional subspaces $\mathscr{H}^{\phi}_{k,r}, \overline{\mathscr{H}}^{\phi}_{k,r}$, respectively, $k \leqslant n$, then to each point p of $\mathscr{H}^{\phi}_{k,r}$ there corresponds at least one point $\bar{p}$ of $\overline{\mathscr{H}}^{\phi}_{k,r}$ such that*

$$p_1, p_2, ..., p_{k+1}, p \approx \bar{p}_1, \bar{p}_2, ..., \bar{p}_{k+1}, \bar{p}.$$

Remark. It follows from Theorem 108.3 that the point $\bar{p}$ of Postulate IV is unique when the points $p_1, p_2, ..., p_{k+1}$ form an *independent* $(k+1)$-tuple.

THEOREM 111.1. *Each two k-dimensional subspaces of $\mathscr{H}^{\phi}_{n,r}$ are congruent.*†

Proof. By Postulate IV it is clear that $\mathscr{H}^{\phi}_{k,r}$ contains $k+1$ independent points congruent with $k+1$ points of $\overline{\mathscr{H}}^{\phi}_{k,r}$. A mapping of $\mathscr{H}^{\phi}_{k,r}$ into $\overline{\mathscr{H}}^{\phi}_{k,r}$ exists by Postulate IV and the remark following it, which Theorem 108.3 shows to be a congruence. If $\mathscr{H}^{*\phi}_{k,r}$ denotes the subspace of $\overline{\mathscr{H}}^{\phi}_{k,r}$ that is congruent to $\mathscr{H}^{\phi}_{k,r}$, it follows from Theorem 108.2 that

$$\mathscr{H}^{*\phi}_{k,r} = \overline{\mathscr{H}}^{\phi}_{k,r}$$

and the theorem is proved.

THEOREM 111.2. *A semimetric space $S \subseteq \mathscr{H}^{\phi}_{n,r}$ if and only if there exists an integer k $(0 \leqslant k \leqslant n)$ such that (a) S contains $k+1$ points $s_0, s_1, ..., s_k$ congruent with an independent $(k+1)$-tuple of an $\mathscr{H}^{\phi}_{k,r}$, and (b) each $(k+3)$-tuple of S containing these $k+1$ points is congruently imbeddable in an $\mathscr{H}^{\phi}_{k,r}$. Then S is irreducibly congruently contained in an $\mathscr{H}^{\phi}_{k,r}$.*

Proof. Use of the properties of $\mathscr{H}^{\phi}_{n,r}$ developed in the preceding sections permits an argument so similar to that employed in establishing Theorem 38.1 that we leave the details to the reader.

COROLLARY. *Generalized hyperbolic space $\mathscr{H}^{\phi}_{n,r}$ has congruence order $n+3$ with respect to the class $\{S\}$ of semimetric spaces.*

A final postulate endows $\mathscr{H}^{\phi}_{n,r}$ with a reflection property.

POSTULATE V. *If $p_1, p_2, ..., p_k$ are k independent points of*

$$\mathscr{H}^{\phi}_{k,r} \quad (1 \leqslant k \leqslant n),$$

then there exists for each point p of $\mathscr{H}^{\phi}_{k,r}$ independent of them at least one point p^* of $\mathscr{H}^{\phi}_{k,r}$ $(p^* \neq p)$ such that $p_1, p_2, ..., p_k, p \approx p_1, p_2, ..., p_k, p^*$.

† From now on $\mathscr{H}^{\phi}_{n,r}$ denotes a generalized hyperbolic space in which Postulate IV is valid.

Remark. It follows from Theorem 109.7 that the point p^* of Postulate V is unique.

A careful examination of §46 reveals that all those properties of E_n that enter into the proof that E_n has quasi congruence order $n+2$ have been either assumed in the five postulates of $\mathscr{H}_{n,r}^{\phi}$ or have been derived as consequences of those postulates. We may, therefore, regard the following theorem as established:

THEOREM 111.3. *Generalized hyperbolic space $\mathscr{H}_{n,r}^{\phi}$ has quasi congruence order $n+2$ with respect to the class $\{S\}$ of semimetric spaces.*

Since $\mathscr{H}_{n,r}^{\phi}$ might possess an equilateral $(n+1)$-tuple, the congruence symbol $(n+2,1)$ is, in general, the best possible.

The reader will have no difficulty in proving the following theorems using the familiar procedures of earlier sections (e.g. §§42, 45).

THEOREM 111.4. *A semimetric space S is (irreducibly) congruently imbeddable in $\mathscr{H}_{n,r}^{\phi}$ if and only if S contains $n+1$ points $s_0, s_1,..., s_n$ such that* (i) $s_0, s_1,..., s_n$ *are congruent with an independent $(n+1)$-tuple of $\mathscr{H}_{n,r}^{\phi}$,* (ii) *for each point s of S, $s_0, s_1,..., s_n, s \mathrel{\underset{\approx}{\subset}} \mathscr{H}_{n,r}^{\phi}$, and* (iii) *if $s, s' \in S$, $\Lambda_{n+3}(s_0, s_1,..., s_n, s, s') = 0$.*

THEOREM 111.5. *If $s_0, s_1,..., s_{n+2}$ is a pseudo-$\mathscr{H}_{n,r}^{\phi}$ set, then*

$$\operatorname{sgn} \Lambda_{n+3}(s_0, s_1,..., s_{n+2}) = (-1)^{n+1}.$$

THEOREM 111.6. *A semimetric $(n+2)$-tuple $s_0, s_1,..., s_{n+1}$ is congruently imbeddable in $\mathscr{H}_{n,r}^{\phi}$ if and only if each of its $(n+1)$-tuples has that property and $\Lambda_{n+2}(s_0, s_1,..., s_{n+1}) = 0$.*

On the other hand, an imbedding theorem analogous to Theorem 42.2 cannot be proved without assuming additional properties for $\mathscr{H}_{n,r}^{\phi}$. It is clear, for example, that the simplest question that might be asked—namely, what distance relation must be satisfied by a semimetric *pair* of points in order that the pair may be congruently imbeddable in a one-dimensional space $\mathscr{H}_{1,r}^{\phi}$—cannot be answered on the present basis.

EXERCISES

1. Investigate pseudo-$\mathscr{H}_{n,r}^{\phi}$ $(n+3)$-tuples (that is, sets of $n+3$ points that are not imbeddable in $\mathscr{H}_{n,r}^{\phi}$ although each of their $(n+2)$-tuples are).
2. Prove Theorems 111.4–111.6.
3. Formulate an additional postulate for $\mathscr{H}_{n,r}^{\phi}$ that will ensure the imbeddability in $\mathscr{H}_{n,r}^{\phi}$ of a semimetric $(n+1)$-tuple $s_0, s_1,..., s_n$ if and only if

$$\operatorname{sgn} \Lambda_{j+1}(s_0, s_1,..., s_j) = (-1)^j \quad (j = 1, 2,..., n).$$

Ch. XII, § 111   GENERALIZED HYPERBOLIC SPACE   287

4. Prove that for the ordinary hyperbolic space Postulates IV, V are valid (Postulates I, II, III were seen to be valid in § 107).
5. Assuming $\mathscr{H}_{n,r}^{\phi}$ complete, convex, and externally convex, solve the space problem for $\mathscr{H}_{n,r}^{\phi}$.
6. If $\phi(0; r) = 1$ and $p_1, p_2, p_3 \in \mathscr{H}_{1,r}^{\phi}$ implies $p_1, p_2, p_3 \subsetneq E_1$, what can be said about the function $\phi(x; r)$?
7. If $\phi(x; r) = 1 + x^2/2r^2 +$ terms of higher order in x/r, show that for a semimetric m-tuple $p_1, p_2, ..., p_m$
$$\lim_{r \to \infty} (2r^2)^{m-1} \cdot \Lambda_m(p_1, p_2, ..., p_m) = -D(p_1, p_2, ..., p_m).$$

REFERENCES

§§ 106. Schering [1], Mansion [1], Blumenthal [21], [29].
§§ 107–11. Blumenthal [22].

PART IV
APPLICATIONS OF DISTANCE GEOMETRY

FOREWORD

THE foregoing parts of this book have been devoted to a systematic development of distance geometry and the applications of metric methods to geometry. We have seen, for example, how the local metric properties of curves and surfaces might be studied by purely metric methods, and differential geometry freed from its age-old servitude to analysis (with its dependence on coordinates and equations) that leads frequently to the imposition of hypotheses which, though necessary for the proper functioning of the analytical machinery, are extraneous to the geometric nature of the problems considered.

By metrizing the fundamental notions of classical differential geometry (such as curvature and torsion for curves of E_3 and the Gauss curvature of surfaces) a unification of their treatments in euclidean and non-euclidean spaces is obtained as well as the ready means for extending these notions to spaces in which their classical analytical formulations are meaningless. One is assured, moreover, that while the metric proofs of theorems may seem more complicated than are those of their analogues in euclidean space that involve the machinery of the calculus, the *hypotheses* are all geometrically essential and put in evidence precisely those metric properties of the space upon which the theorems depend. An excellent example of this is Theorem 33.3 which characterizes a metric segment among all arcs of a metric ptolemaic space by the vanishing of the Menger curvature.

But metric methods are extremely useful in parts of mathematics other than the one that traditionally we call 'geometry'. By 'metric methods' we mean the association of 'distances' with pairs of elements wherever such association is possible and useful—the 'distances' being themselves elements of a distance set that usually, but by no means invariably, is a set of real numbers. Perhaps in most parts of modern mathematics occasional use is made of this device. Modern calculus of variations has been extensively studied in this manner and the generality of the problems treated in that subject has been enormously increased. Thus, for example, metric methods permit consideration of integrals along each continuous curve, rectifiable or not, and existence theorems are obtained 'whose generality surpasses in three directions that of Tonelli's most general theorems'. Such theorems are, moreover, proved in a short and simple manner by interpreting the integrals occurring in variation problems as arc lengths in a suitably defined distance space. The existence theorems thus become theorems concerning the existence of geodesics in that distance space.

A detailed account of the application of distance geometry to calculus of variations would, however, require a larger amount of space in this book than the somewhat specialized nature of the results obtained would seem to justify. We shall, therefore, turn our attention to more accessible illustrations of the use of metric methods and shall find, rather surprisingly, that some parts of classical algebra as well as parts of modern (abstract) algebra furnish excellent material for the exploitation of such a *modus operandi*.

CHAPTER XIII

METRIC METHODS IN DETERMINANT THEORY

112. Cayley–Menger determinants

DUE to the convenient determinant form in which we have chosen to express the distance relations that enter into the many metric characterizations of spaces and sets given in the preceding pages, these characterizations are easily made to yield new determinant theorems which in some cases differ markedly from those encountered in the classical theory. Just as the characterizations themselves might have been stated in terms of distance relations expressed as properties of matrices or quadratic forms, the new determinant theorems they inspire are readily translated into quadratic form or matrix theorems.

It should be observed that in some instances (to be pointed out later) determined efforts to give purely algebraic proofs of determinant theorems that follow so easily when metric methods are applied have been quite unsuccessful. Nor is it likely that those theorems would ever have been conjectured without the metric motivation.

We are concerned in this section with the class of symmetric determinants that can be put in the form

$$\begin{vmatrix} 0 & 1 & 1 & \cdots & 1 \\ 1 & 0 & r_{12} & \cdots & r_{1m} \\ 1 & r_{21} & 0 & \cdots & r_{2m} \\ \cdot & \cdot & \cdot & \cdots & \cdot \\ 1 & r_{m1} & r_{m2} & \cdots & 0 \end{vmatrix},$$

where $r_{ij} = r_{ji} > 0$ $(i,j = 1,2,\ldots,m;\ i \neq j)$. We shall denote such determinants by D, refer to them as Cayley–Menger determinants, and at times write them symbolically as

$$\begin{vmatrix} 0 & 1 \\ 1 & r_{ij} \end{vmatrix} \quad (i,j = 1,2,\ldots,m).$$

THEOREM 112.1. *Let D be of order $m+1 > n+4$ (n a positive integer), and suppose* (i) *for every positive integer $k \leqslant n+1$, each bordered principal minor of order $k+1$ has the sign of $(-1)^k$ or vanishes,* (ii) *each bordered principal minor of order $n+3$ vanishes; then the rank of D is at most $n+2$.*

Proof. Consider a semimetric m-tuple $p_1, p_2, ..., p_m$ ($m > n+3$) with $p_i p_j^2 = r_{ij}$ ($i,j = 1, 2, ..., m$). From (i) and the corollary to Theorem 42.3 it follows that each $n+1$ of these points are imbeddable in E_n. Now (ii) ensures that *each $n+2$ of the m points are imbeddable in E_n*, for this result follows from Theorem 45.1 for those $(n+2)$-tuples all of whose $(n+1)$-tuples are congruently contained in E_{n-1}, and is a consequence of Lemma 42.1 for an $(n+2)$-tuple containing an $(n+1)$-tuple irreducibly imbeddable in E_n. Since $m > n+3$, it follows from Theorem 46.1 that $p_1, p_2, ..., p_m$ are imbeddable in E_n and $D = D(p_1, p_2, ..., p_m)$ has rank $r+2$, where r is the dimension of the smallest dimensional euclidean space congruently containing the m-tuple (Theorem 42.3). Since $r \leqslant n$, the theorem is established.

The theorems of § 47 permit a weakening of the hypotheses of the foregoing theorem as well as a generalization of it. Before stating these theorems we observe that the existence of pseudo-E_n $(n+3)$-tuples and the non-vanishing of the determinant D for such $(n+3)$-tuples (Theorem 45.1) imply that the number $n+4$ of Theorem 112.1 *cannot be reduced*. It is also remarked that condition (ii) cannot be weakened to suppose merely that each bordered $(n+3)$th-order principal minor *containing a given non-vanishing $(n+2)$th-order bordered principal minor* vanishes (assuming one such to exist). This is seen by the example given in the third paragraph of § 47.

THEOREM 112.2. *Let D be of order $m+1 > n+4$ (n a positive integer). If there exists a bordered principal minor M of order $n+3$ such that* (i) *at least one bordered $(n+2)$th-order principal minor of M is not zero, while* (ii) *every $(n+3)$th-order bordered principal minor of D with an n-th-order unbordered principal minor in common with M vanishes and has all of its non-vanishing bordered $(k+1)$th-order principal minors with the sign of $(-1)^k$ ($k = 2, 3, ..., n+1$), then the rank of D is $n+2$.*

Proof. Introducing as before the semimetric m-tuple

$$S = (p_1, p_2, ..., p_m)$$

with $p_i p_j^2 = r_{ij}$ ($i,j = 1, 2, ..., m$) it is seen that (1) S contains at least $n+4$ points, (2) S contains a set Q of $n+2$ points of which at least one $(n+1)$-tuple is not imbeddable in E_{n-1} (namely the $(n+2)$-tuple whose Cayley–Menger determinant is M), and (3) by hypothesis (ii) each $n+2$ points of S with at least n points in common with Q are imbeddable in E_n. Application of Theorem 47.1 shows that S is irreducibly congruently contained in E_n and the conclusion of our theorem now follows from Theorem 42.3.

THEOREM 112.3. *Let D be of order $m+1 > n+j+4$ (n a positive integer, j a non-negative integer). If apart from at most j bordered $(n+3)$th-order principal minors of D every such minor vanishes and each of its non-vanishing bordered principal minors of order $k+1$ has the sign of $(-1)^k$ ($k = 2, 3, ..., n+1$), then the rank of D is at most $n+2$.*

Proof. The reader will have no difficulty in recognizing this theorem as an algebraic translation of Theorem 47.2.

Efforts to provide purely algebraic proofs for the preceding three determinant theorems have been quite unsuccessful. An algebraic proof involving much 'case analysis' has been obtained for the first link of the determinant-chain Theorem 112.1 (namely for the case $n = 1$) by methods which, however, cannot be applied for $n > 1$. The theorems are not related in any obvious manner to those already in the extensive literature of determinant theory and any algebraic proof of them is likely to be quite difficult.

Interesting theorems concerning Cayley–Menger determinants are obtained by utilizing results concerning metric transforms of metric and euclidean subsets. We have seen (Theorem 52.1) that the metric transform $\phi(M)$ of any metric space M by $\phi(x) = x^\alpha$, $0 \leqslant \alpha \leqslant \frac{1}{2}$, has the euclidean four-point property, and that the bounds for α are sharp. This yields the following determinant theorem.

THEOREM 112.4. *If the determinant*

$$D = \begin{vmatrix} 0 & 1 \\ 1 & r_{ij} \end{vmatrix} \quad (i,j = 1, 2, 3, 4)$$

has each bordered fourth-order principal minor negative or zero, then for $0 \leqslant \alpha \leqslant \frac{1}{2}$ the determinant

$$D^{(\alpha)} = \begin{vmatrix} 0 & 1 \\ 1 & r_{ij}^\alpha \end{vmatrix} \quad (i,j = 1, 2, 3, 4)$$

has the following properties:

  (i) *each bordered fourth-order principal minor is negative*,
  (ii) $D^{(\alpha)} \geqslant 0$,
  (iii) $D^{(\alpha)} > 0$ *if* $0 \leqslant \alpha < \frac{1}{2}$,
  (iv) $D^{(\alpha)} = 0$ *if and only if* $\alpha = \frac{1}{2}$ *and D does not vanish, while all of its bordered fourth-order principal minors do*,
  (v) $\alpha = \frac{1}{2}$ *is the greatest exponent for which* (i) *and* $D^{(\alpha)} \geqslant 0$ *are valid*.

Proof. If p_1, p_2, p_3, p_4 is a semimetric quadruple with

$$p_i p_j^2 = r_{ij} \quad (i,j = 1, 2, 3, 4),$$

then the hypothesis implies the quadruple is metric and by Theorem 52.1 points p_1', p_2', p_3', p_4' of E_3 exist with $p_i' p_j' = (p_i p_j)^\alpha$ $(i,j = 1, 2, 3, 4)$. Then $(p_i' p_j')^2 = r_{ij}^\alpha$ $(i,j = 1, 2, 3, 4)$ and no one of the twelve angles determined by the euclidean quadruple is obtuse.† This establishes (i) and (ii), while (v) is a consequence of the remark immediately following the proof of Theorem 52.1.

It was shown in the proof of Theorem 52.1 that $D^{(\alpha)}$ does not vanish for $0 \leqslant \alpha < \tfrac{1}{2}$, and so (iii) is valid.

To establish (iv) we show that the metric transform of a metric quadruple by $\phi(x) = x^{\frac{1}{2}}$ is imbeddable in E_2 if and only if the quadruple is pseudo-linear. If p_1, p_2, p_3, p_4 is pseudo-linear, the labelling may be assumed so that (Exercise 1, §45)

$$p_1 p_2 = p_3 p_4 = a > 0, \qquad p_2 p_3 = p_1 p_4 = b > 0,$$

$$p_1 p_3 = p_2 p_4 = a + b,$$

and the metric transform of this quadruple by $\phi(x) = x^{\frac{1}{2}}$ is clearly congruent with the quadruple of E_2 forming the vertices of a rectangle with sides $\sqrt{a}$ and $\sqrt{b}$.

Conversely, suppose that the planar quadruple p_1', p_2', p_3', p_4' is the metric transform of the metric quadruple p_1, p_2, p_3, p_4 by $\phi(x) = x^{\frac{1}{2}}$; that is, $p_i' p_j' = (p_i p_j)^{\frac{1}{2}}$ $(i,j = 1, 2, 3, 4)$. Then no one of the twelve angles determined by the planar quadruple is obtuse and since not all of these angles are acute (Exercise 1, §52), at least one is a right angle. It follows readily that p_1', p_2', p_3', p_4' are the vertices of a rectangle‡ and consequently p_1, p_2, p_3, p_4 has all of its triples linear. But p_1, p_2, p_3, p_4 is not linear, for the metric transform of any linear quadruple by $\phi(x) = x^{\frac{1}{2}}$ evidently has two points each of which is the vertex of two right angles, which is not the case for the planar quadruple p_1', p_2', p_3', p_4'. Hence p_1, p_2, p_3, p_4 is pseudo-linear, and the proof is complete.

The reader will observe that the hypothesis of Theorem 112.4 makes no assumption concerning the sign of D; that is, the quadruple p_1, p_2, p_3, p_4 is assumed merely metric. If it be assumed *euclidean* (that is, if it be

---

† See the remark immediately preceding the proof of Theorem 52.1.
‡ We have, incidentally, the little geometric theorem that four points of a plane are the vertices of a rectangle if and only if none of the twelve angles they determine is obtuse.

supposed that $D \geqslant 0$), then conclusions (i) and (iii) are valid for
$$0 \leqslant \alpha < 1.$$
Without additional assumptions the preceding theorem cannot be extended to determinants D of orders exceeding 5. We do, however, have the following analogue for determinants of arbitrary order.

THEOREM 112.5. *Let D be of order $n+2$ and suppose that there exists an integer r ($r \leqslant n$) such that* (i) *the bordered principal minors*
$$\begin{vmatrix} 0 & 1 \\ 1 & r_{ij} \end{vmatrix} \quad (i,j = 1, 2, \ldots, k+1)$$
of D have the sign of $(-1)^{k+1}$ for every k ($1 \leqslant k \leqslant r$), and (ii) *the rank of D is $r+2$. Then for each α ($0 \leqslant \alpha \leqslant 1$) every bordered principal minor of $D^{(\alpha)}$ of order $i+1$ vanishes or has the sign of $(-1)^i$ for $i = 1, 2, \ldots, n+1$.*

Proof. The hypotheses imply that the auxiliary semimetric $(n+1)$-tuple $p_0, p_1, \ldots, p_n$ is imbeddable (irreducibly) in E_r (Theorem 42.3) and hence the metric transform of the $(n+1)$-tuple by $\phi(x) = x^\alpha$ ($0 \leqslant \alpha \leqslant 1$) is imbeddable in Hilbert space (Theorem 54.1), and the conclusion follows (Theorem 40.1).

EXERCISES

1. If p_1, p_2, p_3, p_4 is pseudo-linear with $p_1 p_2 = p_3 p_4 = a$, $p_2 p_3 = p_1 p_4 = b$, $p_1 p_3 = p_2 p_4 = a+b$, show that
$$D(p_1, p_2, p_3, p_4) = -32 a^2 b^2 (a+b)^2 = -32 (\text{product of the 'sides'}).$$
The value of the determinant D of the most general pseudo-planar *quintuple* is not known. Find D for several special pseudo-planar quintuples (for example, suppose the triangle p_1, p_2, p_3 is equilateral and take p_4 in several special positions).
2. Prove that if the order of D is 5, then D vanishes whenever each of its fourth-order principal minors does. (*Hint.* Show that the unbordered fourth-order principal minor in the Cayley–Menger determinant of a pseudo-linear quadruple does not vanish.)
3. Investigate non-vanishing determinants D of order 5 with four fourth-order principal minors vanishing. Show that $D = -32 a^2 b^2 c^2$, where
$$a^2 + b^2 + c^2 \pm 2ab - 2bc - 2ac = 0.$$
4. Is Theorem 112.1 valid for determinants of form D but with the weaker restriction r_{ij} real instead of $r_{ij} > 0$ for every pair of distinct indices i, j? Note that hypothesis (i) of the theorem implies for $k = 2$ that $r_{ij} \geqslant 0$.

113. Determinants of type Δ

A determinant of type Δ is a symmetric determinant $|r_{ij}|$, $r_{ij} = r_{ji}$, with $-1 \leqslant r_{ij} \leqslant 1$ ($i \neq j$) and $r_{ii} = 1$ ($i,j = 1, 2, \ldots, m$). Such determinants are of frequent occurrence in statistics. Our characterization of pseudo-$S_{n,r}$ sets of more than $n+3$ points leads to a remarkable property of these determinants.

THEOREM 113.1. *Let Δ be of order $m > n+3$ (n a positive integer) with $-1 < r_{ij} < 1$ ($i,j = 1, 2,..., m; i \neq j$). If*

(i) *every principal minor of order less than $n+2$ is non-negative,*
(ii) *every principal minor of order $n+2$ vanishes,*
(iii) *at least one principal minor of order $n+3$ does not vanish, then*

(1) *the absolute value of every element of Δ (not on the principal diagonal) is $1/(n+1)$,*
(2) *after multiplying appropriate rows and the same numbered columns of Δ by -1, each element (not on the principal diagonal) is $-1/(n+1)$,*
(3) *for each positive integer k ($1 \leq k \leq m$) and each k-th-order principal minor Δ_k of Δ,*

$$\Delta_k = \frac{1}{n+1}\left[\frac{n+2}{n+1}\right]^{k-1}(n-k+2).$$

Proof. Put $r_{ij} = \cos\alpha_{ij}$, $0 < \alpha_{ij} < \pi$ ($i,j = 1, 2,..., m; i \neq j$) and introduce a semimetric m-tuple $p_1, p_2,..., p_m$ with

$$p_i p_j = \alpha_{ij} \quad (i,j = 1, 2,..., m; i \neq j).$$

Then $\Delta = \Delta_m(p_1, p_2,..., p_m) = |\cos p_i p_j|$ ($i,j = 1, 2,..., m$).

Since $p_i p_j \leq \pi$ for each pair of indices i, j, the hypotheses of the theorem imply (Theorem 63.1, corollary) that each $n+2$ of the m points are congruently imbeddable in $S_{n,1}$ but the m points are *not* (since by (iii) and the theorem just cited, the m points contain at least one $(n+3)$-tuple which is not imbeddable in $S_{n,1}$). Hence $p_1, p_2,..., p_m$ form a pseudo-$S_{n,1}$ set of more than $n+3$ points, with no two diametral, and conclusions (1), (2) follow at once from Theorems 71.2, 71.3, respectively, while (3) is obtained from (*) of Lemma 69.1.

The author knows of no other theorem in determinant theory in which the hypotheses determine (in a non-trivial manner) the absolute value of every element of the determinant! The theorem is sharp with respect to (1) the order m of Δ, and (2) the bounds of r_{ij}. No purely algebraic proof of the theorem exists despite efforts to obtain one.

The characterization theorem for general pseudo-$S_{n,r}$ sets of more than $n+3$ points (Theorem 72.1) yields the following generalization and extension of the preceding theorem to a wider class of Δ determinants.

THEOREM 113.2. *Let Δ be of order $m > n+k+2$ (n, k positive integers) with $-1 \leqslant r_{ij} < 1$. If Δ satisfies hypotheses* (i), (ii), (iii) *of Theorem* 113.1, *and if* (iv) -1 *occurs in at most $2(k-1)$ rows of Δ, then after an appropriate symmetric shifting of rows and columns of Δ and after multiplying certain rows and the same numbered columns of Δ by -1, each element of the determinant is either 1 or $-1/(n+1)$, and every element not on the principal or secondary diagonal is $-1/(n+1)$. If -1 occurs in exactly $2t$ rows of Δ ($t \leqslant k-1$), the rank of Δ is $m-t$.*

Proof. We see, as before, that a semimetric m-tuple $p_1, p_2, \ldots, p_m$ with $p_i p_j = \alpha_{ij}$ (where $r_{ij} = \cos \alpha_{ij}$ and $0 < \alpha_{ij} \leqslant \pi$) ($i, j = 1, 2, \ldots, m; i \neq j$) is a pseudo-$S_{n,1}$ set which, by hypothesis (iv), contains at most $k-1$ diametral point-pairs. Since $p_i p_j > 0$ for $i \neq j$, and $m > n+k+2$, the m-tuple has at least $n+k+3$ pairwise distinct points and so by Theorem 72.1 the distance of any two distinct points of the m-tuple is either π or $\cos^{-1}[\pm 1/(n+1)]$. Consequently each element of Δ (not on the principal diagonal) is $1/(n+1)$ or $-1/(n+1)$ or -1.

Suppose, now, that $p_1, p_2, \ldots, p_m$ contains exactly t diametral point-pairs ($t \leqslant k-1$) and assume the labelling so that

$$p_1 p_m = p_2 p_{m-1} = \ldots = p_t p_{m-(t-1)} = \pi.†$$

By Lemma 72.1 each element of the ith row (column) of $\Delta = \Delta(p_1, \ldots, p_m)$ is now the negative of the corresponding element in the $(m-i+1)$th row (column) for each $i = 1, 2, \ldots, t$.

Multiplication of the last row and column by -1 makes the first and last rows (columns) of the resulting determinant identical. Multiplying the $(m-1)$th row and column of this determinant by -1 makes them identical with the second row and column, respectively, and *again the elements in the last row (column) are equal to their corresponding elements in the first row (column)*. Applying this procedure to each of the *last t rows (columns) in turn (from the mth to the $(m-t+1)$th) yields a determinant in which the rows (columns) numbered $1, 2, \ldots, t$ are identical with the rows (columns) numbered $m, m-1, \ldots, m-t+1$, respectively.

The principal minor of order $m-t$ in the upper left-hand corner of Δ is clearly not affected by the operations described above. This minor is the determinant $\Delta(p_1, p_2, \ldots, p_{m-t})$ of a pseudo-$S_{n,1}$ set (Lemma 72.2) *without diametral point-pairs* and since $m-t \geqslant m-(k-1) > n+3$, the preceding theorem ensures that after appropriate rows and the same numbered columns of the minor have been multiplied by -1, all of its elements outside the principal diagonal are $-1/(n+1)$.

† This amounts to an appropriate (symmetric) shifting of rows and columns of Δ.

Now the multiplication by -1 of any of the rows and same numbered columns of this minor which are numbered from $t+1$ to $m-t$ inclusive will not change the pairwise identity already established between the first and last t rows (columns) since any altered row or column will have its last t elements as well as its first t elements changed in sign. If, finally, it is necessary to multiply one of the first t rows and columns of the minor by -1, say the qth row and column ($1 \leqslant q \leqslant t$), multiplication by -1 of the row and column numbered $m-q+1$ will restore the identity of the jth and $(m-j+1)$th rows (columns) for every $j = 1, 2, ..., t$.

The conclusions of the theorem are now apparent.

It will be observed that hypothesis (iv) of the preceding theorem is needed only to ensure that $m-t > n+3$ so that Theorem 113.1 might be applied to the upper left hand principal minor of Δ of order $m-t$. The reader may easily verify that if $m > 2(n+3)$ and -1 occurs in exactly $2t$ rows of Δ, then $m-t > n+3$ so that hypothesis (iv) might be dropped. We have then the following companion theorem.

THEOREM 113.3. *Let Δ be of order $m > 2(n+3)$ (n a positive integer) with $-1 \leqslant r_{ij} < 1$. If Δ satisfies hypotheses* (i), (ii), (iii) *of Theorem* 113.1, *then, after multiplication by -1 of appropriate rows and the same numbered columns, each element is either 1 or $-1/(n+1)$. If -1 occurs in exactly $2t$ rows, the rank of Δ is $m-t$.*

It is remarked that Theorem 113.2 rather than Theorem 113.3 is the generalization and extension of Theorem 113.1, for it reduces to that theorem when $k = 1$. Its proof, however, leans heavily upon Theorem 113.1.

EXERCISES

1. State and prove an analogue of Theorem 113.2 for Δ determinants with $-1 < r_{ij} \leqslant 1$ ($i \neq j$).
2. What is an analogue of Theorem 113.3 for Δ determinants with $-1 \leqslant r_{ij} \leqslant 1$ ($i \neq j$)?

114. Determinants of types Δ_{NP}, Δ_{NN}, and Λ

The subclass of Δ determinants for which $-1 \leqslant r_{ij} \leqslant 0$ ($i \neq j$) is denoted by Δ_{NP}. For determinants of this subclass a theorem somewhat stronger than Theorem 113.1 can be proved under weaker hypotheses.

THEOREM 114.1. *Let Δ_{NP} be of order $m > n+3$. If* (i) *every principal minor of order less than $n+1$ is non-negative, while every principal minor of order $n+1$ is positive, and* (ii) *every principal minor of order $n+2$ vanishes, then every element of Δ_{NP} outside the principal diagonal has the value $-1/(n+1)$.*

Proof. The conclusion follows from Theorem 113.1 when it is shown that (a) $-1 < r_{ij}$ $(i,j = 1, 2,..., m)$, and (b) at least one $(n+3)$th-order principal minor of Δ_{NP} does not vanish.

If $S = (p_1, p_2,..., p_m)$ is a semimetric m-tuple with $r_{ij} = \cos(p_i p_j)$ and $\frac{1}{2}\pi \leqslant p_i p_j \leqslant \pi$ $(i,j = 1, 2,..., m; i \neq j)$, then by (i) and (ii) each $(n+2)$-tuple of S is congruently imbeddable in $S_{n,1}$ *and every $(n+1)$-tuple of S is $S_{n,1}$-independent* (§ 67). It follows that no two points of S are diametral and so $r_{ij} > -1$ for every pair of indices.

Suppose, now, that every $(n+3)$th-order principal minor of Δ_{NP} vanishes and consequently the m-tuple S is imbeddable in $S_{n,1}$. Let $s_1, s_2,..., s_m$ be m points of $S_{n,1}$ with $p_1, p_2,..., p_m \approx s_1, s_2,..., s_m$. Then $\frac{1}{2}\pi \leqslant s_i s_j < \pi$ $(i,j = 1, 2,..., m; i \neq j)$ and the determinant Δ_{n+1} formed for each $n+1$ of the m points $s_1, s_2,..., s_m$ is positive. We shall obtain the desired contradiction by showing that the $S_{n,1}$ does *not contain such an m-tuple even for $m = n+3$*.

Denote by $v^{(i)}$ $(i = 1, 2,..., n+3)$ the vectors $\overrightarrow{os_i}$ drawn from the centre o of $S_{n,1}$ to the points s_i of $S_{n,1}$. Since each $n+1$ of the points $s_1, s_2,..., s_{n+3}$ is an independent set, the same is true of each $n+1$ of the unit vectors $v^{(1)}, v^{(2)},..., v^{(n+3)}$, and a cartesian coordinate system may be introduced in the E_{n+1} containing $S_{n,1}$ so that the direction cosines (components) of the vectors $v^{(i)}$ $(i = 1, 2,..., n+3)$ are given by the following table.

Vector	Direction cosines					
$v^{(1)}$	1,	0,	0,	. .	0,	
$v^{(2)}$	$v_1^{(2)},$	$v_2^{(2)},$	0,	. .	0,	$v_2^{(2)} > 0$
$v^{(3)}$	$v_1^{(3)},$	$v_2^{(3)},$	$v_3^{(3)},$	. .	0,	$v_3^{(3)} > 0$
.	.	.	.	. .	.	.
$v^{(n+1)}$	$v_1^{(n+1)},$	$v_2^{(n+1)},$	$v_3^{(n+1)},$	. .	$v_{n+1}^{(n+1)}$	$v_{n+1}^{(n+1)} > 0$
$v^{(n+2)}$	$v_1^{(n+2)},$	$v_2^{(n+2)},$	$v_3^{(n+2)},$	. .	$v_{n+1}^{(n+2)},$	
$v^{(n+3)}$	$v_1^{(n+3)},$	$v_2^{(n+3)},$	$v_3^{(n+3)},$	. .	$v_{n+1}^{(n+3)},$	

with the inequalities on the right subsisting.

Denoting the inner product of vectors $v^{(i)}$, $v^{(j)}$ by $(v^{(i)}, v^{(j)})$ and remembering that $\frac{1}{2}\pi \leqslant s_i s_j < \pi$, we have

$$(v^{(i)}, v^{(j)}) = \cos s_i s_j \leqslant 0 \quad (i,j = 1, 2,..., n+3; i \neq j).$$

This inequality applied successively for $i = 1, 2,..., n+1$, together with the inequalities exhibited on the right of the table, show that all elements in the table that lie *below* the diagonal of the first $n+1$ rows and columns are negative or zero. But then

$$0 \geq (v^{(n+2)}, v^{(n+3)}) = v_1^{(n+2)}v_1^{(n+3)} + \ldots + v_{n+1}^{(n+2)}v_{n+1}^{(n+3)}$$

implies that each summand is zero. Consequently either $v_{n+1}^{(n+2)} = 0$ or $v_{n+1}^{(n+3)} = 0$ and so $v^{(n+2)}$ or $v^{(n+3)}$ forms with the first n vectors $v^{(1)}, v^{(2)}, \ldots, v^{(n)}$ a dependent set. This contradiction establishes the theorem.

Remark 1. In proving the above theorem we have incidentally established a geometrical fact of much interest in itself; namely that $n+2$ *is the highest power of a subset S of an n-sphere such that no distance is less than a quadrant and each $n+1$ points of S are independent.* For the equilateral $(n+2)$-tuple of $S_{n,r}$ is such a set and we have seen above that no such $(n+3)$-tuple exists.

Remark 2. Hypothesis (i) may be weakened to require that all principal minors of order $n+1$ or less are non-negative, and at least one non-vanishing principal minor of order $n+1$ exists such that no principal minor of that order having an nth-order principal minor in common with it vanishes. This is clearly sufficient to obtain the contradiction based on the table, and it is easily shown to imply that $r_{ij} \neq -1$ for every pair of indices.

It is observed that for $n < 4$, Theorem 114.1 is valid if hypothesis (i) is further weakened to require merely that each principal minor of order less than $n+2$ be non-negative, *provided no element of Δ_{NP} is -1.* This follows upon showing that for $n = 1, 2, 3$ the $S_{n,1}$ does not contain $n+4$ points with each distance at least $\tfrac{1}{2}\pi$ and less than π. This might also be the case for $n = 4$, but for $n = 5$ it is not true. The reader may easily verify that the $S_{5,1}$ contains the nine points whose coordinates in the E_6 with origin at the centre of $S_{5,1}$ are

$(1, .\,, .\,, .\,, .\,, .\,)$, $(.\,, 1, .\,, .\,, .\,, .\,)$, $(.\,, .\,, 1, .\,, .\,, .\,)$
$(.\,, .\,, .\,, .\,, .\,, 1)$, $(-1/\sqrt{2}, .\,, .\,, .\,, -1/\sqrt{2}, .\,, .\,)$,
$(-1/\sqrt{2}, .\,, .\,, 1/\sqrt{2}, .\,, .\,)$, $(.\,, -1/\sqrt{2}, .\,, .\,, .\,, .\,, -1/\sqrt{2})$,
$(.\,, .\,, -1/\sqrt{2}, .\,, -1/\sqrt{2}, .\,)$, $(.\,, .\,, .\,, -1/\sqrt{5}, .\,, 2/\sqrt{5}, .\,)$,

where zero coordinates are indicated by dots. Each two of these points has distance at least $\tfrac{1}{2}\pi$ and *less than* π.

If the order m of a determinant Δ_{NP} is sufficiently larger than n, does the conclusion of Theorem 114.1 follow upon assuming merely that each principal minor of order $n+1$ or less is non-negative and each $(n+2)$th-order principal minor vanishes?

An argument similar to that based on the table in the proof of Theorem 114.1 shows that (1) if $n+2$ points of $S_{n,1}$ have every distance greater than $\frac{1}{2}\pi$, then each $(n+1)$-tuple is an independent set, and (2) the n-sphere does not contain a subset of more than $n+2$ points with every distance exceeding $\frac{1}{2}\pi$. These results lead to the following theorem.

THEOREM 114.2. *Let Δ_{NP} be of order $m > n+3$ with $-1 \leqslant r_{ij} < 0$. If every principal minor of order less than $n+2$ is non-negative and every principal minor of order $n+2$ vanishes, then every element of Δ_{NP} not on the principal diagonal has the value $-1/(n+1)$.*

A determinant of type Δ with $0 \leqslant r_{ij} < 1$ $(i,j = 1, 2, ..., m; i \neq j)$ is denoted by Δ_{NN}.

THEOREM 114.3. *Let Δ_{NN} be of order $m > n+3$. If* (i) *every principal minor of order less than $n+2$ is non-negative, and* (ii) *every principal minor of order $n+2$ vanishes, then the rank of Δ_{NN} does not exceed $n+1$.*

Proof. It evidently suffices to prove that each $(n+3)$th-order principal minor is zero. Writing $\Delta_{NN} = \Delta_m(p_1, p_2, ..., p_m)$, suppose an $(n+3)$th-order principal minor does not vanish. Then $p_1, p_2, ..., p_m$ form a pseudo-$S_{n,1}$ set of more than $n+3$ points with no pair diametral, and it follows from Theorem 113.1 and the hypothesis that each element of Δ_{NN} outside the principal diagonal has the value $1/(n+1)$. This is impossible, however, since then each $(n+2)$th-order principal minor equals $2[n/(n+1)]^{n+1}$ and consequently fails to vanish.

Let Λ denote a determinant $|r_{ij}|$, $r_{ij} = r_{ji} > 1$ $(i,j = 1, 2, ..., m; i \neq j)$, $r_{ii} = 1$ $(i = 1, 2, ..., m)$.

THEOREM 114.4. *Let Λ be of order $m > n+3$. If* (i) *for each integer $1 \leqslant k \leqslant n$, each $(k+1)$th-order principal minor vanishes or has the sign of $(-1)^k$, and* (ii) *each principal minor of order $n+2$ vanishes, then the rank of Λ does not exceed $n+1$.*

Proof. Introduce a semimetric set $S = (p_1, p_2, ..., p_m)$ with

$$r_{ij} = \cosh p_i p_j \quad (i,j = 1, 2, ..., m).$$

Using theorems of § 111 and the procedures of §§ 42, 45 it is easily shown that hypotheses (i), (ii) imply that each $n+2$ points of S are congruently imbeddable in the n-dimensional hyperbolic space $\mathscr{H}_{n,1}$. It follows from Theorem 111.3 that $S \subseteqq \mathscr{H}_{n,1}$ and from Theorem 106.1 that the rank of Λ is at most $n+1$.

115. Quasi rank of Δ_{NN} determinants

Imbedding and congruence order theorems for elliptic spaces yield determinant theorems of a novel kind.

DEFINITION 115.1. *A symmetric determinant* $A = |a_{ij}|$, $a_{ij} = a_{ji}$ $(i,j = 1, 2,..., m)$ *has quasi rank* r^* *provided an epsilon matrix*

$$\epsilon = (\epsilon_{ij}), \qquad \epsilon_{ij} = \epsilon_{ji} = \pm 1, \quad \epsilon_{ii} = 1 \quad (i,j = 1, 2,..., m)$$

exists such that the determinant $|\epsilon_{ij} a_{ij}|$ $(i,j = 1, 2,..., m)$ *has rank* r^*. *If, moreover, all non-vanishing principal minors of* $|\epsilon_{ij} a_{ij}|$ $(i,j = 1, 2,..., m)$ *are positive, the determinant A is called quasi positive definite of rank* r^*.†

In terms of the notion just defined Theorem 78.2 states that a semimetric m-tuple $p_1, p_2,..., p_m$ is congruently imbeddable in the elliptic n-space $\mathscr{E}_{n,r}$ if and only if (i) no distance exceeds $\frac{1}{2}\pi r$, and (ii) the determinant $|\cos p_i p_j / r|$ $(i,j = 1, 2,..., m)$ is quasi positive definite of rank at most $n+1$.

THEOREM 115.1. *Let Δ_{NN} be of order $m > 3$. If each principal minor of order 4 has quasi rank 2, then so has Δ_{NN}, and 4 is the smallest number with this property.*

Proof. Write $r_{ij} = \cos p_i p_j$ $(i,j = 1, 2,..., m)$, where $p_1, p_2,..., p_m$ is a semimetric set and $0 < p_i p_j \leqslant \frac{1}{2}\pi$ for $i \neq j$. It follows from the imbedding theorem stated above that each quadruple of the m points is imbeddable in $\mathscr{E}_{1,r}$, and since $\mathscr{E}_{1,r} \approx S_{1,\frac{r}{2}}$ (Exercise 2, §78) which has congruence order 4, then $p_1, p_2,..., p_m \Subset \mathscr{E}_{1,r}$ and another application of the imbedding theorem completes the proof.

It is easily shown (Exercise 4) that the theorem is not valid if the number 4 is replaced by 3.

THEOREM 115.2. *Let Δ_{NN} be of order $m > 4$ with*

$$0 < r_{ij} \quad (i,j = 1, 2,..., m).$$

If each third-order principal minor has quasi rank 2, then Δ_{NN} has also, unless each element has either the value $\frac{1}{2}$ or $\frac{1}{2}\sqrt{3}$.

Proof. Introducing the semimetric m-tuple $p_1, p_2,..., p_m$ as before (in this case $p_i p_j < \frac{1}{2}\pi$ $(i,j = 1, 2,..., m)$) then each three of the m points are imbeddable in $\mathscr{E}_{1,r}$ and hence also in $S_{1,\frac{1}{2}}$. If $p_1, p_2,..., p_m \Subset \mathscr{E}_{1,r}$, then $\Delta_{NN} = \Delta(p_1, p_2,..., p_m)$ has quasi rank 2, while in the contrary event $p_1, p_2,..., p_m$ form a pseudo-$S_{1,\frac{1}{2}}$ set of more than four points with no pair

† The concept of quasi rank may, of course, be given a much more general form, but we shall utilize it here only as defined above.

diametral. Hence $\cos 2(p_i p_j) = \pm \frac{1}{2}$ for each pair of distinct indices, and since $r_{ij}^2 = \cos^2 p_i p_j = \frac{1}{2}[1+\cos 2(p_i p_j)]$ $(r_{ij} > 0)$, we obtain $r_{ij} = \frac{1}{2}$ or $\frac{1}{2}\sqrt{3}$ $(i \neq j)$ and the theorem is established.

Remark. The requirement $r_{ij} > 0$ in the hypothesis of Theorem 115.2 served merely to ensure that if $p_1, p_2, \ldots, p_m$ is pseudo-$S_{1,\frac{1}{2}}$, no two of the points are diametral. By virtue of Theorem 113.3 that demand may be suppressed if $m > 8$ or (Theorem 113.2) if $m > k+3$ (k a positive integer) and 0 occurs in at most $2(k-1)$ rows of Δ_{NN}.

A comparison of Theorem 115.2 with the case $n = 1$ of Theorems 113.1 and 114.3 illustrates some effects of substituting for the classical notion of 'rank' the new concept of 'quasi rank'.

The two preceding theorems are first links in two chains of determinant theorems concerning quasi rank. The second link in the chain containing Theorem 115.1 has been obtained (geometrically) with very great difficulty, while not even a conjecture concerning the fundamental constant of the third link seems justified at this time. The second link of the more striking but geometrically less important chain containing Theorem 115.2 is not yet known.

The succeeding links of the chains referred to contain an explicit hypothesis that does not appear in the first links but which is satisfied in those cases. This hypothesis is that the principal minors concerned be quasi positive definite—a condition that, for example, each fourth-order principal minor of quasi rank 2 obviously satisfies since

$$1-r_{ij}^2 > 0 \quad (i,j = 1,2,\ldots,m; \; i \neq j).$$

We are led, therefore, to the following problems: *What is the smallest positive integer k such that if Δ_{NN} has order at least k and every k-th-order principal minor of Δ_{NN} is quasi positive definite of rank $n+1$ or less (n a given positive integer), then (1) Δ_{NN} is quasi positive definite of rank $n+1$ or less, (2) Δ_{NN} is quasi positive definite?*†

Problem (1) is solved in Theorem 115.1 for $n = 1$, which shows that $k = 4$. For $n = 2$ we have the following solution of that problem.

THEOREM 115.3. *Let Δ_{NN} be of order $m > 6$. If each principal minor of order 7 is quasi positive definite of rank 3 or less, then so is Δ_{NN}, and 7 is the smallest number with that property.*

Proof. It was shown in Theorem 103.3 that the $\mathscr{E}_{2,r}$ has congruence order 8 with respect to all semimetric spaces, and this result would lead

---

† It should be noted that a kth-order principal minor of Δ_{NN} that is quasi positive definite of rank $n+1$ or less might also be quasi positive definite with rank exceeding $n+1$. This possibility is not excluded.

at once by way of the formulation of the elliptic imbedding theorem given in the beginning of this section to the validity of the first part of the conclusion were the number 7 replaced by 8. But it was stated in § 103 that Haantjes and Seidel have shown that the best congruence order of the elliptic plane with respect to the class of semimetric spaces is 7, and the theorem is an immediate consequence of that result.

The characterization of pseudo-$\mathscr{E}_{2,r}$ sets would furnish the first extension of Theorem 115.2. This study has been started but is not yet completed.

It would be interesting, indeed, to obtain a purely algebraic proof for Theorem 115.3, the geometric proof of which is so very complicated. Were such a proof forthcoming it might point the way towards the solutions of the general problems (1) and (2) for which no 'reasonable' geometric approach has so far been devised.

EXERCISES

1. If $\Delta_{NN} = |r_{ij}|$ $(i,j = 1, 2,..., m; m > 3)$ has each fourth-order principal minor of quasi rank 2, show that the determinant obtained by bordering the determinant $|r_{ij}^2|$ $(i,j = 1, 2,..., m)$ with a row and column of 1's, with intersection element 2, has rank 3.
2. Investigate the distribution of the elements $\frac{1}{2}$ and $\frac{1}{2}\sqrt{3}$ in a determinant Δ_{NN} satisfying the hypotheses of Theorem 115.2. What can you say concerning the value of Δ_{NN}?
3. A determinant $\Delta_{NN} = |r_{ij}|$ $(i,j = 1, 2,..., m)$ is *intrinsically* quasi positive definite of rank r provided that for each epsilon matrix $\epsilon = (\epsilon_{ij})$ $(i,j = 1, 2,..., m)$ such that $|\epsilon_{ij} r_{ij}|$ $(i,j = 1, 2,..., m)$ has all principal minors non-negative, the quasi rank of Δ_{NN} is r. Give examples of such determinants.
4. Show that the validity of Theorem 115.1 is destroyed if the number 4 is replaced by 3.

REFERENCES

§§ 112–15. Some of the theorems are in Blumenthal [21, 23, 26]; others are formulated here for the first time.

CHAPTER XIV

METRIC METHODS IN LINEAR INEQUALITIES

116. Coefficient sets C and solution sets $\Sigma(C)$

THE basis for the application of metric methods to the theory of linear inequalities is provided by the results obtained in Chapter VIII concerning congruence indices of hemispheres and small caps, as well as the intersection theorems from which those results ensued. The geometrization of the theory in the manner presented here has several advantages. It serves to make intuitive some of the classical results and to suggest new ones, while the distinction between finite and infinite systems of inequalities commonly made in earlier theories is entirely lacking in this development. Other advantages will, it is hoped, become evident later.

Let $f_i(t)$, $i = 1, 2, ..., n+1$, be $n+1$ real functions defined on an arbitrary (non-null) subset T of the euclidean line E_1, $n \geqslant 1$, which do not all vanish for any one element of T. If $x_1, x_2, ..., x_{n+1}$ are $n+1$ indeterminates let (I) denote the system of inequalities

$$\sum_{i=1}^{n+1} f_i(t) x_i \geqslant 0 \quad (t \in T \subset E_1).$$

The system (I) is finite, denumerable, or non-denumerable according to the power of the linear subset T. For each element t of T the point $c(t)$ with cartesian coordinates $(c_1(t), c_2(t), ..., c_{n+1}(t))$, where

$$c_i(t) = f_i(t) \Big/ \Big[\sum_{i=1}^{n+1} f_i^2(t) \Big]^{\frac{1}{2}} \quad (i = 1, 2, ..., n+1),$$

has unit distance from the origin and hence is a point of $S_{n,1}$, the *convexly metrized* unit n-sphere with centre at the origin.

DEFINITION 116.1. *The subset* $\{c(t)\}$, $t \in T$, *of* $S_{n,1}$ *defined by a system* (I) *of inequalities is called the normalized or associated coefficient set of the system, and is denoted by* C.

It is important to observe that (1) each solution of system (I) is a solution of the normalized system of inequalities (NI)

$$\sum_{i=1}^{n+1} c_i(t) x_i \geqslant 0 \quad (t \in T),$$

and conversely, and (2) the correspondence of the points of C with the members of system (*I*) is one-to-one if and only if no member of the system is a positive constant times another.

Now apart from the trivial solution $x_1 = x_2 = \ldots = x_{n+1} = 0$, there corresponds to each solution of (*I*) a normalized solution
$$x_i = s_i, \quad i = 1, 2, \ldots, n+1,$$
which is representative of the ray of solutions $x_i = k \cdot s_i, i = 1, 2, \ldots, n+1$, $k > 0$. Thus to each non-trivial solution of (*I*) there corresponds a point $s = (s_1, s_2, \ldots, s_{n+1})$ of $S_{n,1}$.

DEFINITION 116.2. *The subset $\{s\}$ of points of $S_{n,1}$ (possibly vacuous) consisting of the normalized solutions of a system (I) of inequalities is called the normalized or associated solution set of the system, and is denoted by $\Sigma(C)$.*

It is proposed to study system (*I*), as well as other kinds of systems of inequalities, by applying the devices and results of distance geometry to investigate the mutual relations of the two subsets C and $\Sigma(C)$ of $S_{n,1}$. The system of inequalities itself, its algebraic nature, etc., have fulfilled their functions as soon as these two subsets have been defined and the metric considerations become operative.

If the system (*I*) can be transformed by a non-singular linear transformation on its indeterminates into one in which fewer than $n+1$ indeterminates appear, it is said to be *reducible*. In such an event it would clearly be more convenient to work with the reduced system. *We shall agree, therefore, that system (I) is irreducible.* It follows from this agreement that the coefficient set C is *irreducibly contained* in $S_{n,1}$ and hence *the associated solution set $\Sigma(C)$ of system (I) never contains a diametral point-pair*. Also, since $n \geqslant 1$, *the associated coefficient set C never consists of a diametral point-pair*.

EXERCISES

1. The system (*I*) is homogeneous. Show that a non-homogeneous system of inequalities in m indeterminates is *equivalent* to a homogeneous system in $m+1$ indeterminates.
2. Prove the two properties of C and $\Sigma(C)$ that are said above to follow from the agreement concerning the irreducibility of system (*I*).

117. Some elementary properties of C and $\Sigma(C)$

Since all of the hemispheres $H_{n,r}$ with which we are concerned in this chapter belong to the unit n-sphere $S_{n,1}$, we shall denote them by H_n, and write $H_n(p)$ to indicate the hemisphere with centre p.

THEOREM 117.1. *The solution set $\Sigma(C)$ consists of all points x of $S_{n,1}$ such that $H_n(x) \supset C$.*

Proof. If $x \in S_{n,1}$ and $c \in C$, then
$$\sum_{i=1}^{n+1} c_i(t)x_i = \cos xc(t),$$
where $xc(t)$ denotes the distance in $S_{n,1}$ of points x, $c(t)$. Hence $x \in \Sigma(c)$ if and only if $xc(t) \leqslant \tfrac{1}{2}\pi$; that is, $C \subset H_n(x)$.

The set functional Σ has many of the properties of set complement as shown in the next theorem.† In order that $\Sigma(C)$ might be defined for every subset C of $S_{n,1}$ it is convenient to put $\Sigma(0) = S_{n,1}$.

THEOREM 117.2. *The functional $\Sigma(C)$ is defined for every subset C of $S_{n,1}$ and has the following properties*: (i) $\Sigma(0) = S_{n,1}$, $\Sigma(S_{n,1}) = 0$, (ii) *if* $C_1 \subset C_2$, *then* $\Sigma(C_1) \supset \Sigma(C_2)$, (iii) $\Sigma^2(C) = \Sigma(\Sigma(C)) \supset C$, (iv) $\Sigma^3(C) = \Sigma(C)$, (v) $\Sigma(C_1 + C_2) = \Sigma(C_1).\Sigma(C_2)$, (vi) $\Sigma(C_1.C_2) \supset \Sigma(C_1) + \Sigma(C_2)$.

Proof. Part (i) is obvious. If $p \in \Sigma(C_2)$, then $H_n(p) \supset C_2 \supset C_1$; that is, $p \in \Sigma(C_1)$ and (ii) is proved. Now $p \in C$ implies $H_n(p) \supset \Sigma(C)$ and consequently $p \in \Sigma^2(C)$, which establishes (iii).

If $p \in \Sigma(C)$, then $H_n(p) \supset \Sigma^2(C)$ and so $p \in \Sigma^3(C)$; that is, $\Sigma(C) \subset \Sigma^3(C)$. But (ii) applied to (iii) gives $\Sigma^3(C) \subset \Sigma(C)$, and so (iv) is proved. The proofs of (v) and (vi) are immediate.

THEOREM 117.3. *The set $\Sigma(C)$ is a closed and (metrically) convex subset of $S_{n,1}$.*

Proof. If $C = 0$, then $\Sigma(0) = S_{n,1}$ is metrically convex since $n > 0$. For $C \neq 0$ the metric convexity of $\Sigma(C)$ follows from § 73 (Exercise 2) upon showing that $\Sigma(C)$ is a product of hemispheres and is different from $S_{0,1}$.

If $c \in C$, then $H_n(c) \supset \Sigma(C)$ and so $\Sigma(C) \subset \prod_{c \in C} H_n(c)$. But clearly each point of $\prod_{c \in C} H_n(c)$ has distance at most $\tfrac{1}{2}\pi$ from each point c of C and consequently the product set is part of $\Sigma(C)$. Thus $\Sigma(C) = \prod_{c \in C} H_n(c)$ and since $\Sigma(C)$ does not contain a pair of diametral points for $C \neq 0$ (according to the agreement of the preceding section) then $\Sigma(C) \neq S_{0,1}$ and $\Sigma(C)$ is metrically convex. As a product of closed sets, $\Sigma(C)$ is, moreover, closed.

Considered as a space, $\Sigma(C)$ is, then, a compact and convex metric space.‡ Denoting the passing points of $\Sigma(C)$ by $P(\Sigma(C))$ and the terminal points by $T(\Sigma(C))$ (see § 20),
$$\Sigma(C) = P(\Sigma(C)) + T(\Sigma(C)), \qquad P(\Sigma(C)).T(\Sigma(C)) = 0.$$

† The properties of Σ bear an even stronger resemblance to those of 'product complement' in a distributive lattice.

‡ Unless stated otherwise, 'convex' as used in this chapter means 'metrically convex'. Compare with § 73.

The first summand in this decomposition of $\Sigma(C)$ into two mutually exclusive sets is called the *ordinary part* of $\Sigma(C)$, and its elements are *ordinary* solutions of (I), while the second summand forms the *fundamental* part of $\Sigma(C)$.

The properties of the set of ordinary solutions listed below are immediate consequences of Theorems 20.1, 20.2.

THEOREM 117.4. *The ordinary part $P(\Sigma(C))$ of each solution set $\Sigma(C)$ is* (i) *an F_σ,* (ii) *segmentally connected,* (iii) *dense-in-itself, and* (iv) *if not null, everywhere dense in $\Sigma(C)$.*

The set $P(\Sigma(C))$ is closed if and only if $T(\Sigma(C)) = 0$ or $\Sigma(C)$ consists of a single point. If $P(\Sigma(C)) = 0$ while $\Sigma(C) \neq 0$, then since $\Sigma(C)$ is convex, $T(\Sigma(C))$ must consist of a single point.

Each subset S of the compact convex metric space $S_{n,1}$ $(n \geqslant 1)$ has a convex extension (Theorem 19.1) which is unique if and only if $S \neq S_{0,1}$. For if $S = S_{0,1}$, any half great circle joining the two points of S is a convex extension of S, while if $S \neq S_{0,1}$ the product of all closed and convex subsets of $S_{n,1}$ containing S is easily seen to constitute the unique convex extension of S. We shall denote a convex extension of S by S^*. Since the associated coefficient set C of a system (I) is different from $S_{0,1}$, C^* is unique and is the product of all closed and convex subsets of $S_{n,1}$ containing C. Hence C^* is part of every hemisphere that contains C.

THEOREM 117.5. *For each associated set C with convex extension C^*, $\Sigma(C) = \Sigma(C^*)$.*

Proof. If $s \in \Sigma(C)$, then $H_n(s) \supset C$. But since $H_n(s)$ is closed and convex, it follows that $H_n(s) \supset C^*$ and consequently $s \in \Sigma(C^*)$.

On the other hand, $s \in \Sigma(C^*)$ implies $C \subset C^* \subset H_n(s)$ and so $s \in \Sigma(C)$.

Remark 1. The system (NI) of inequalities may be called *closed and convex* provided its coefficient set is a closed and convex subset of $S_{n,1}$. By Theorem 117.5 there corresponds to each system (NI) a closed and convex system (NI) with the same set of solutions. Hence each system (I) of inequalities has the same set of solutions as the closed and (spherically) convex system (NI) whose coefficient set is the convex extension on $S_{n,1}$ of the normalized coefficient set of (I).

If an associated set C is contained in a hemisphere, its convex extension C^* is the product P of all the *hemispheres* containing C, for since $C \neq S_{0,1}$, this product is closed, convex, contains C, and consequently contains C^*. But if $p \in P$ and $p \bar{\in} C^*$ then (since C^* is closed and convex) an $S_{n-1,1}$ exists which is the rim of a hemisphere that contains C^* and does not contain p. Since this hemisphere contains C, it is a member of the family of hemispheres whose product is P, and it follows that $p \,\bar{\in}\, P$.

Remark 2. The convex extension C^* is the maximal subset of $S_{n,1}$ that contains C and has its (normalized) solution set identical with $\Sigma(C)$. For if $p \in S_{n,1}$ and $p \,\bar{\in}\, C^*$, then from the preceding observation there exists a hemisphere that contains C (and consequently C^*) but does not contain p. Denoting the centre of such a hemisphere by q then $pq > \frac{1}{2}\pi$ and so $q \in \Sigma(C^*)$ but $q \,\bar{\in}\, \Sigma(C^*+(p))$.

Remark 3. If S is any subset of $S_{n,1}$ with $C \subset S \subset C^*$, then
$$\Sigma(S) = \Sigma(C^*);$$
in particular $\Sigma(\bar{C}) = \Sigma(C^*)$. The proof follows immediately from (ii) of Theorem 117.2 and Theorem 117.5.

Though the associated solution set $\Sigma(C)$ of a system (I) may, of course, be null, for each associated coefficient set C, $\Sigma^2(C) \neq 0$. Whenever C contains a diametral point-pair, $\Sigma(C)$ is part of an $S_{n-1,1}$ which, according to our agreement, is never the case for C.

EXERCISE

Use well-known properties of convex bodies in E_{n+1} to show that if p is a boundary point of a closed, convex, proper subset S of $S_{n,1}$ there is at least one great hypersphere $S_{n-1,1}$ through p which is the rim of a hemisphere H_n containing S. Such an $S_{n-1,1}$ is called a *supporting hypersphere* of S at p.

118. The generalized Minkowski theorem

In the classical investigations of Minkowski on finite systems of (homogeneous) linear inequalities, a non-trivial solution is called *fundamental* provided it cannot be expressed as the sum of two essentially different non-trivial solutions (two solutions not being considered essentially different if one is a positive constant times the other), and a set of fundamental solutions is complete whenever each fundamental solution is essentially the same as a member of the set. The result of central importance in those investigations is Minkowski's theorem that the general solution of any irreducible finite system (I) of inequalities is a linear homogeneous combination of the solutions forming a complete set of fundamental solutions, with coefficients arbitrary non-negative numbers.

We shall obtain the extension of the Minkowski theorem to *arbitrary* systems (I) (that is, finite, denumerable, or non-denumerable) as a corollary of the next theorem.

THEOREM 118.1. *If S is any closed, convex subset of $S_{n,1}$, and S contains no diametral point-pairs, then S is the convex extension of its set of terminal points; that is, $S = T^*(S)$.*

Proof. The theorem being clearly valid for $n = 1$, we make the inductive assumption of its validity for subsets of $S_{n-1,1}$, $n > 1$, and show first that the boundary $B(S)$ of S is part of $T^*(S)$.

Let $S_{n-1,1}(p)$ denote a supporting great hypersphere of S at p, a point of $B(S)$, and put $R = S \cdot S_{n-1,1}(p)$. Then R is a closed, convex subset of $S_{n-1,1}$, with no diametral point-pairs, and hence $R = T^*(R)$. But since R consists of boundary points of S, it is clear that $T(R) \subset T(S)$. Consequently $T^*(R) \subset T^*(S)$ and so $p \in T^*(S)$. It follows that

$$B(S) \subset T^*(S).$$

If, now, q is a point of S, not belonging to $B(S)$, then q is a passing point of S and clearly lies between two points x, y of $B(S)$. But then $q \in T^*(S)$ since it is between two points of $T^*(S)$. We have shown that $S \subset T^*(S)$.

The proof is completed upon observing that $T(S) \subset S$, closed and convex, implies $T^*(S) \subset S$.

COROLLARY 1. (*Minkowski.*) *If C is the associated coefficient set of a system (I) of inequalities, then $\Sigma(C) = T^*(\Sigma(C))$.*

COROLLARY 2. *If the associated coefficient set C of a system (I) of inequalities is closed, convex, and free from diametral point-pairs, then $\Sigma(C) = \Sigma(T(C))$.*

For $C = T^*(C)$ and hence $\Sigma(C) = \Sigma(T^*(C)) = \Sigma(T(C))$.

COROLLARY 3. *If the convex extension C^* of a coefficient set C contains no diametral point-pairs, then $\Sigma(C) = \Sigma(T(C^*))$.*

For then $C^* = T^*(C^*)$ and $\Sigma(C) = \Sigma(C^*) = \Sigma(T^*(C^*)) = \Sigma(T(C^*))$.

It was proved in Theorem 117.2 that the functional Σ satisfies the first De Morgan formula (see (1), (2) of §2) for two (and hence for any finite number of) coefficient sets—this is part (v) of the theorem—but only 'half' of the second De Morgan formula was established in part (vi), which states that $\Sigma(C_1 C_2) \supset \Sigma(C_1) + \Sigma(C_2)$. There is no difficulty in finding examples to show that the inclusion sign cannot be reversed, and so Σ does not have the property expressed by the second De Morgan formula (even when the coefficient sets are closed and convex).

But for closed and convex coefficient sets Σ does satisfy an analogue of that important formula which is, indeed, the analogue that is satisfied by the operation of product complement mentioned above. This, as well as an additional property of Σ (which further accentuates the similarity between Σ and the operation of product complement), is a corollary of the following important and easily established theorem.

THEOREM 118.2. *An associated coefficient set C satisfies the relation $\Sigma^2(C) = C$ if and only if C is closed and convex.*

Proof. The necessity follows from Theorem 117.3. To prove the sufficiency it is observed that $\Sigma^3(C) = \Sigma(C)$ (part (iv), Theorem 117.2) implies that $\Sigma^2(C)$ has the same solution set as C, a closed and convex subset of $\Sigma^2(C)$ (part (iii), Theorem 117.2). Whence, according to Remark 2, § 117, $C \subset \Sigma^2(C) \subset C^* = C$, and the proof is complete.

COROLLARY 1. *If C_1, C_2 are closed and convex associated coefficient sets, then* (i) $\Sigma(C_1 C_2) = \Sigma^2[\Sigma(C_1) + \Sigma(C_2)]$, *and* (ii) $\Sigma^2(C_1 . C_2) = \Sigma^2(C_1) . \Sigma^2(C_2)$.

For
$$\Sigma^2[\Sigma(C_1) + \Sigma(C_2)] = \Sigma(\Sigma[\Sigma(C_1) + \Sigma(C_2)])$$
$$= \Sigma(\Sigma^2(C_1) . \Sigma^2(C_2))$$
$$= \Sigma(C_1 . C_2).$$

Then

$$\Sigma^2(C_1 C_2) = \Sigma^3[\Sigma(C_1) + \Sigma(C_2)] = \Sigma[\Sigma(C_1) + \Sigma(C_2)] = \Sigma^2(C_1) . \Sigma^2(C_2).$$

COROLLARY 2. *If C is an associated coefficient set, then $\Sigma^2(C) = C^*$.*

For $\Sigma^2(C) = \Sigma(\Sigma(C)) = \Sigma(\Sigma(C^*)) = \Sigma^2(C^*) = C^*$.

Deviating a little from the classical terminology, we call a system of homogeneous linear inequalities with coefficient set $\Sigma(C)$ the *canonical polar system* of the system with coefficient set C, and using $\Sigma^2(C)$ as a coefficient set gives the *canonical form* of the system with associated set C. Part (iv) of Theorem 117.2 generalizes a result that Dines and McCoy established for finite systems, while Corollary 2 above is the extension to arbitrary systems of the Minkowski consequence theorem.

In view of the analogy that has been pointed out between Σ and the operation of set complementation or product complement—an analogy that is particularly striking when Σ operates on closed and convex sets—it is apropos to inquire concerning the product of C and $\Sigma(C)$, which is null in the case of set complementation as well as product complement. For C arbitrary, easy examples exist with $C . \Sigma(C) = 0$ and $C . \Sigma(C) \neq 0$, but for C closed, convex, and different from $S_{n,1}$, *the product of C and $\Sigma(C)$ is never null.*†

EXERCISE

Let C be a closed, convex subset of $S_{2,1}$ different from a great circle. Prove that $C . \Sigma(C) \neq 0$.

† It is recalled that by our agreement $C \not\subset S_{n-1,1}$. If $C = S_{n-1,1}$ for example, then $\Sigma(C)$ consists of the two poles of C and so $C . \Sigma(C) = 0$.

119. Coincidence of C and $\Sigma(C)$

We have seen that for each coefficient set C, $\Sigma^3(C) = \Sigma(C)$ while if C is closed and convex, then $\Sigma^2(C) = C$. It is natural to inquire what conditions are necessary and sufficient in order that $\Sigma(C) = C$.

If $p \in S$, a subset of $S_{n,1}$, $\operatorname{diam}_p S = \text{l.u.b.}_{x \in S} px$ is called the diameter of S relative to p, and S is said to have *constant diameter d* provided for each point p of S, $\operatorname{diam}_p S = d$.

THEOREM 119.1. *A necessary and sufficient condition that $C = \Sigma(C)$ is that C be closed and convex and the boundary $B(C)$ of C have constant diameter $\frac{1}{2}\pi$.*

Proof. If $C = \Sigma(C)$, then C is closed and convex, $B(C) \subset C$, and each point of $B(C)$ is the centre of a hemisphere containing $B(C)$; that is $\operatorname{diam}_p B(C) \leqslant \frac{1}{2}\pi$, $p \in B(C)$.

The centre q of a hemisphere $H_n(q)$ that contains C and whose rim is a supporting $S_{n-1,1}$ of C at p is evidently a point of $\Sigma(C)$ and hence of C. If q were an interior point of C, then the segment joining p and q could be prolonged through q to a point r of C so that pqr holds. But this implies $pr > \frac{1}{2}\pi$ and contradicts $p \in \Sigma(C)$; consequently $q \in B(C)$ and $\operatorname{diam}_p B(C) \geqslant pq = \frac{1}{2}\pi$. Thus the necessity is proved.

Assume, now, that C is closed, convex, and $B(C)$ has constant diameter $\frac{1}{2}\pi$. Then $B(C)$ is neither null nor an $S_{k,1}$, $k \leqslant n$, so $C \neq S_{n,1}$. It follows that C is contained in a hemisphere. It is clear that $\operatorname{diam}_p B(C) = \frac{1}{2}\pi$, $p \in B(C)$, implies $\operatorname{diam}_p C = \frac{1}{2}\pi$ and so $C \subset H_n(p)$. Thus each element of $B(C)$, and consequently each element of C, belongs to $\Sigma(C)$.

We show now that $\Sigma(C) \subset C$. Each point p^* of $B(C)$ is the centre of a hemisphere $H_n(p^*)$ that contains C and has a point p of $B(C)$ on the $S_{n-1,1}$ forming its rim. The set C is the product of all these hemispheres $H_n(p^*)$, $p^* \in B(C)$, and consequently if $q \bar{\in} C$, there exists a point p^* of $B(C)$ such that $H_n(p^*)$ does not contain q. Hence $p^*q > \frac{1}{2}\pi$ and $q \bar{\in} \Sigma(C)$. It follows that $\Sigma(C) \subset C$, and the proof is complete.

EXERCISES

1. Let S be a closed, convex, proper subset of $S_{2,1}$, not part of a great circle, and let p be a point of $B(S)$. Denote by q a point on a supporting great circle Γ of S at p with $pq = \frac{1}{2}\pi$ and let Γ' be another supporting great circle of S that passes through q. The angle made by Γ, Γ' is called the *breadth of S relative to p*. If the breadth is constant for every point p of $B(S)$, the curve $B(S)$ is said to be of constant (spherical) breadth.

Prove that if C is the associated coefficient set of a system (I) of inequalities in three indeterminates and $\Sigma(C) = C$, then $B(C)$ is of constant spherical breadth $\frac{1}{2}\pi$.

2. Prove that conversely, if $B(C)$ has constant spherical breadth $\frac{1}{2}\pi$ and C is closed, convex, and has diameter at most $\frac{1}{2}\pi$, then $\Sigma(C) = C$.
3. Give examples of subsets C of $S_{2,1}$ with $\Sigma(C) = C$.

120. Existence theorems for solutions of a system (I)

Perhaps the most fundamental problem in the relationship of C and $\Sigma(C)$—the one most studied in the classical theory—is the determination of properties of C which ensure that $\Sigma(C)$ is not null (that is, that the system (I) have a non-trivial solution). It is clear from our approach that $\Sigma(C) \neq 0$ if and only if C is contained in a hemisphere H_n, and so the problem becomes one of determining criteria that a subset of $S_{n,1}$ may be covered by a hemisphere—one of the problems studied in § 76.

THEOREM 120.1. *Each system (I) consisting of at most $n+1$ members has a non-trivial solution.*

Proof. The associated coefficient set C contains at most $n+1$ points and consequently lies on an $S_{n-1,r}$, $r < 1$. The centre of the small cap K_n whose rim is this $S_{n-1,r}$ furnishes a non-trivial solution of (I).

THEOREM 120.2. *If the associated set C of a system (I) has more than $2n+2$ pairwise distinct points, then the system has a non-trivial solution if and only if each sub-system of $2n+1$ of the inequalities has a non-trivial solution.*

Proof. According to Theorem 76.1, the subset C of $S_{n,1}$ is congruently contained in (that is, C is coverable by) a hemisphere.

COROLLARY. *If each sub-system of $2n+2$ members of a system (I) has a non-trivial solution, then so has the entire system.*

Remark. If the agreement concerning the irreducibility of system (I) is not made, Theorem 120.2 is still valid since Theorem 76.1, on which it is based, places no restriction on the class $\{S\}$ of semimetric spaces. If the maximum number of linearly independent members of (I) is $k+1$, $k \leqslant n$, then Theorem 120.2 as well as the theorems that follow in this section, are valid with k in place of n.

We call a subset C of $S_{n,1}$ *global* provided $\Sigma(C) = 0$, and *irreducibly global* if it is global and no proper subset has this property. A system (I) is *pseudo-solvable* provided its coefficient set C is global but no subsystem of $2n+1$ members is global. The following theorem characterizes pseudo-solvable systems (I).

THEOREM 120.3. *A system (I) is pseudo-solvable if and only if the normalized system may be transformed by a non-singular linear transformation into the system $x_i \geqslant 0$, $-x_i \geqslant 0$, $i = 1, 2, ..., n+1$.*

Proof. The proof follows at once from Theorem 120.2 and Remark 2, § 76.

Irreducibly global sets contain at most $2n+2$ points (corollary, Theorem 120.2)—for example, the coefficient set of a pseudo-solvable system—and at least $n+2$ points (Theorem 120.1).

Lemma 74.3 is readily applied to prove the following existence theorem that contains the corollary to Theorem 120.2 as a special case.

THEOREM 120.4. *A system (I) has a non-trivial solution if and only if each sub-system of $n+k+1$ members has a non-trivial solution in common with a fixed, arbitrarily selected sub-system of $n-k+1$ linearly independent members of (I), where k may be any one of the numbers $0, 1, 2,..., n+1$.*

EXERCISE
Show that the number $2n+1$ of Theorem 120.2 cannot be reduced no matter how great the cardinal number of C.

121. Systems of strict inequalities

Denote by (SI) a system of strict inequalities
$$\sum_{i=1}^{n+1} f_i(t)x_i > 0 \quad (t \in T)$$
whose left-hand members are subjected to the same conditions as before. The solutions of system (SI) are identical with those of the normalized system (NSI). Denote again by C the subset of $S_{n,1}$ obtained from a system (NSI). If C is closed, it is clear that a point s of $S_{n,1}$ is a solution of (SI) if and only if s is the centre of a small cap $K_{n,\rho}(s)$ of spherical radius ρ less than $\tfrac{1}{2}\pi$ that contains C.

THEOREM 121.1. *Let the associated coefficient set C of a system (SI) be closed. The system has a solution if and only if the convex extension C^* of C does not contain a diametral point-pair.*

Proof. If C^* does not contain a diametral point-pair, it is contained in a small cap (Lemma 74.1) which contains C and whose centre is a solution of (SI). If, on the other hand, system (SI) has a solution, then so has system (NSI) and consequently C is contained in a small cap K_n which clearly contains C^*. Then C^* has no diametral point-pairs.

THEOREM 121.2. *Any irreducible system (SI) of at most $n+1$ inequalities has a solution.*

The proof is clear.

THEOREM 121.3. *If a system (SI) has its associated set C closed, then the system has a solution if and only if each sub-system of $n+2$ inequalities has a solution, and the number $n+2$ cannot be reduced.*

Proof. The necessity is obvious. To prove the sufficiency it is observed that each subset of $n+2$ points of C is coverable by a small cap of radius ρ less than $\frac{1}{2}\pi$ and a standard compactness argument shows that the least upper bound ρ^* of these numbers ρ is also less than $\frac{1}{2}\pi$.

Consider now the family of small caps K_{n,ρ^*} with the centres at the points of C. Each $n+2$ of these caps intersect and since no one of these caps contains a diametral point-pair, the case $k=0$ of Theorem 75.2 permits the conclusion that there is a point common to all the caps of the family. Hence C is contained in a small cap and system (SI) has a solution.

Since $S_{n,1}$ contains an equilateral $(n+2)$-tuple which is irreducibly global, the number $n+2$ of the theorem cannot be reduced.

EXERCISES

1. Supply the details of the 'compactness argument' referred to in Theorem 121.3.
2. Show that if K is a constant greater than $\frac{1}{3}$, the normalized system of inequalities

$$\sum_{i=1}^{3} c_i(t)x_i \geqslant K \quad (t \in T)$$

has a normalized solution if and only if each sub-system of three inequalities has a normalized solution.

REFERENCES

§§ 116–21. Blumenthal [24]. See Dines [1] and McCoy for the result referred to in § 118.

SUPPLEMENTARY PAPER

Motzkin, T. *Beiträge zur Theorie der Linearen Ungleichungen*, Dissertation, Basel, 1933 (Jerusalem, 1936).

CHAPTER XV

METRIC METHODS IN LATTICE THEORY

122. Introduction

THE reader is referred to §§ 2, 3, and 9 (Example 8) for the definitions of lattice, distributive lattice, normed lattice, and the metric space $D(L)$ associated with a normed lattice L. In those sections the lattice product of two elements a, b of a lattice is denoted by ab, and the distance of these elements in $D(L)$ by (a, b). *We shall adhere to that notation throughout this chapter.*

The metric space $D(L)$ arises by imposing a metric structure on the algebraic structure formed by the lattice L. Metric concepts and relations in $D(L)$ lead, therefore, to lattice concepts and relations, some of which might not have been formulated or observed without the motivation furnished by a study of $D(L)$. Thus, for example, lattice equivalences in *normed* lattices L of metric betweenness in $D(L)$ provide means of defining several kinds of 'betweenness' in *arbitrary* lattices and suggest a study of lattice relations that, in the general case, are quite divorced from their metric genesis. Of equal interest is the determination of lattice properties of L necessary and sufficient to ensure that $D(L)$ have a desired metric property. What lattice conditions, for example, are necessary and sufficient in order that $D(L)$ might be congruently imbeddable in Hilbert space?

It seems, therefore, that application of the metric programme to lattice theory, via the associated metric space $D(L)$, introduces new and interesting concepts and problems into that theory, and adds further riches to an already opulent subject.

The metric programme may be applied to complemented distributive lattices (Boolean algebras) in quite another manner. Instead of attaching to elements a, b a real number (a, b) (the distance of a, b in $D(L)$) one associates with the pair *an element $d(a, b)$ of the lattice*; that is, the distance set is a subset of the Boolean algebra itself. A study of these so-called *autometrized* Boolean algebras has just begun. Since the element attached to a, b as distance is the ring sum of a and b, the results obtained are immediately translatable into theorems concerning the ring sum in a Boolean ring with a unit.

123. Lemmas from lattice theory

We shall need some elementary concepts and results of lattice theory which, for the sake of completeness, are presented in this section. A lattice L is called a *modular* or *Dedekind* lattice provided the following weak distributive property is valid: *if $a, b, c \in L$ and $a \prec c$, then*

$$(a+b)c = a+bc.$$

It is clear that Definition 2.1 of a lattice is unaltered by an interchange of 'sum' and 'product', and it readily follows that a *duality principle* is valid in a general lattice L, according to which one may derive from any identity in the elements of L or any property of the general lattice L a valid (dual) identity or property by the interchange of $+, \cdot, \prec$ with $\cdot, +, \succ$, respectively, where $a \succ b$ if and only if $b \prec a$. It is important to observe the modularity condition defined above is *self-dual* and consequently the duality principle is valid also in modular lattices.

A very weak form of cancellation law holds in modular lattices and is indeed equivalent to the weak distributive property used to define such lattices. Let (C) denote the following weak cancellation law: *if $a, b, c \in L$, $a \prec b$ and $ac = bc$, $a+c = b+c$, then $a = b$.*

LEMMA 123.1. *A lattice L is modular if and only if (C) is valid in L.*

Proof. If L is modular and $a, b, c \in L$ with $a \prec b$, $ac = bc$, $a+c = b+c$, then $a+bc = a+ac = a$; also $(a+c)b = (b+c)b = b$ (see Exercise 1, § 2). But from modularity, $a \prec b$ implies $(a+c)b = a+bc$, and so $a = b$; that is, (C) is valid in L.

Suppose now that (C) holds in L and put $p = a+bc$, $q = (a+b)c$, where $a, b, c \in L$ with $a \prec c$. Then $a+bc \prec a+b$ and $a+bc \prec c$; consequently $a+bc \prec (a+b)c$ and so $p \prec q$. It follows that $bp \prec bq$ and $b+p \prec b+q$. But $bq = b(a+b)c = bc \prec bp$ and $b+q \prec a+b \prec b+p$, and consequently $bp = bq$, $b+p = b+q$. Applying (C) to p, q, b yields $p = q$; that is, $(a+b)c = a+bc$ and the lemma is proved.

LEMMA 123.2. *A normed lattice is modular.*

Proof. Let a, b, c be elements of a normed lattice and suppose $a \prec b$, $ac = bc$, $a+c = b+c$. Then from

$$|a+c|+|ac| = |a|+|c| \quad \text{and} \quad |b+c|+|bc| = |b|+|c|$$

follows $|a| = |b|$ and hence (since $a \prec b$) $a = b$. Thus property (C) is valid and consequently the lattice is modular.

EXERCISES

1. Show that if the distributive property $(a+b)c = ac+bc$ is valid for each three elements a, b, c of a lattice L, then so is the dual property $ab+c = (a+c)(b+c)$, and conversely. Hence the duality principle is valid in distributive lattices.

2. Prove that the linear subspaces of a projective space form a lattice L_P if lattice product ab is defined as the set-theoretic product (intersection) of the subspaces a, b and lattice sum $a+b$ is the subspace of smallest dimension containing a and b. Note that $a \prec b$ if and only if $a \subset b$.
3. Show that the lattice defined in Exercise 2 is modular but not distributive. Verify that the lattice may be normed by putting $|a| = 1 + \dim a$.
4. Show that the set of all ordered pairs (x_1, x_2) of non-negative real numbers form a lattice L_C with respect to the (partial) order relation $(x_1, x_2) \prec (y_1, y_2)$ if and only if $x_1 \leqslant y_1$ and $x_2 \leqslant y_2$. Is L_C modular; distributive? Show that L_C is normed by defining $|x| = x_1 + x_2$, where $x = (x_1, x_2) \in L_C$.

124. Lattice characterizations of metric betweenness

The reader will recall that a point c is metrically between two points a, b (symbolized by writing $B(a, c, b)$) provided $a \neq c \neq b$ and

$$(a, c) + (c, b) = (a, b).$$

It is convenient in a purely metric theory to require that a point c which is between two points a, b does not coincide with either a or b. In the present discussion, however, it is preferable to use the term in a wider sense and *to define $B(a, c, b)$ to mean merely $(a, c) + (c, b) = (a, b)$*.†

How may metric betweenness in $D(L)$ be recognized in the normed lattice L? It is clear that if an element c of L is order-between elements a, b (that is, $a \prec c \prec b$ or $b \prec c \prec a$), then $B(a, c, b)$ holds, for

$$(a, c) = |a+c| - |ac| = \pm[|c|-|a|], \qquad (c, b) = \pm[|b|-|c|],$$
$$(a, b) = \pm[|b|-|a|]$$

with either all plus signs or all minus signs chosen, and consequently $(a, c) + (c, b) = (a, b)$. But this sufficient condition is not necessary. On the other hand, we shall soon see that if $B(a, c, b)$ subsists, then

$$ab \prec c \prec a+b,$$

but this necessary condition is not sufficient. The following necessary *and* sufficient condition is due to Glivenko.

Theorem 124.1. *If a, b, $c \in L$, a normed lattice, then $B(a, c, b)$ holds in $D(L)$ if and only if*

$$ac + cb = c = (a+c)(c+b). \qquad \text{(G)}$$

Proof. Assuming $B(a, c, b)$ and using part (2) of the definition of norm (§ 9, Example 8) one readily obtains $-|ab| = |c| - |ac| - |bc|$. Since

---

† Whenever '+' operates on real numbers it signifies, as above, ordinary addition. There seems little danger of any confusion in this double use of the symbol +. Note that since the notation for metric betweenness used in the earlier part of the book (namely acb) might be confused with the lattice product of the elements a, c, b, the notation $B(a, c, b)$ is employed.

$(ac)(bc) = (ab)c$, $-|(ab)c| = |ac+cb|-|ac|-|bc|$. Now $(ab)c \prec ab$ implies $|(ab)c| \leqslant |ab|$ and consequently $|c| \leqslant |ac+cb|$.

But $|ac+cb| \leqslant |c|$ since $ac+cb \prec c$, and so $|c| = |ac+cb|$. It follows now from part (1) of the norm definition that $c = ac+cb$, and by the duality principle, $c = (a+c)(c+b)$. Hence $B(a,c,b)$ implies
$$ac+cb = c = (a+c)(c+b).$$

Suppose now that the above lattice relation holds. From
$$ab \prec (a+c)(c+b) = c$$
follows $ab = (ac)(bc)$. Dually, $a+b \succ ac+cb = c$ implies
$$a+b = (a+c)+(c+b).$$
Then
$$|a+b|-|ab| = |(a+c)+(c+b)|-|(ac)(bc)|$$
$$= |(a+c)+(c+b)|+|(a+c)(c+b)|-|ac+cb|-|(ac)(bc)|$$
$$= |a+c|+|c+b|-|ac|-|cb|;$$
and so $(a,b) = (a,c)+(c,b)$.

Remark. If $B(a,c,b)$ holds, then by (G) $c = ac+cb \prec a+b$ and $ab \prec (a+c)(c+b) = c$. Hence the relation $ab \prec c \prec a+b$ is a necessary condition for betweenness $B(a,c,b)$ as remarked above.

It is observed that Glivenko's lattice characterization (G) of metric betweenness (in the wide sense) is self-dual. Two new lattice characterizations have recently been obtained, neither of which is self-dual. They follow from (G) upon establishing the next lemma.

LEMMA 124.1. *In any modular lattice the following relations are equivalent*:
$$ac+cb = c = (a+c)(c+b), \tag{G}$$
$$ac+cb = c = ab+c, \tag{G*}$$
$$(a+c)(c+b) = c = (a+b)c. \tag{G**}$$

Proof. (G) implies (G*), for clearly $ab \prec (a+c)(c+b) = c$ and so $c = ab+c$. Assuming (G*), $ac+cb = c$ yields
$$b+c = b+ac+cb = b+ac.$$
Hence, by the modularity condition
$$a(b+c) = a(b+ac) = ac+ab.$$
This, together with $ab \prec c$ by (G*), gives $a(b+c) \prec c$ and so $c = c+a(b+c)$. Hence
$$ac+cb = c = c+a(b+c) = (a+c)(c+b),$$
the last equality resulting from modularity and $c \prec c+b$.

Since (G**) is the dual of (G*), its equivalence with (G) may be established by dualizing the above argument (or invoking the duality principle) and the lemma is proved.

According to Lemma 123.2, each normed lattice is modular and we have the following theorem.

THEOREM 124.2. *Each of the lattice relations* (G), (G*), (G**) *is a necessary and sufficient condition for metric betweenness* $B(a, c, b)$.

COROLLARY 1. *If* $a, b, c \in L$, *a normed lattice, then* $B(a, c, b)$ *holds in* $D(L)$ *if and only if there exists an element* x *of* L *such that*
$$c = ab + xa + xb.$$

Proof. Assuming an element x of L such that $c = ab + xa + xb$, then by modularity
$$a(ab + xa + xb) = ab + a(xa + xb) = ab + xa + xab.$$
Similarly
$$(ab + xa + xb)b = ab + (xa + xb)b = ab + xb + xab.$$
Hence $ac + cb = ab + xa + xb = c$. In like manner it is seen that $c = (a+c)(c+b)$ and from (G) we conclude $B(a, c, b)$ holds.

On the other hand, if $B(a, c, b)$ subsists, then from (G*)
$$ac + cb = c = ab + c.$$
Then $ab + ca + cb = ab + c = c$ and c is itself the desired element x.

Dually we can show that $B(a, c, b)$ holds if and only if
$$c = (a+b)(a+x)(b+x)$$
for some element x of the lattice.

COROLLARY 2. *If* $a, b, c \in L$, *a normed lattice, then* $B(a, c, b)$ *holds in* $D(L)$ *if and only if* $a(b+c) \prec c \prec a+bc$.

Proof. If $B(a, c, b)$ subsists, then $a(b+c) \prec (a+c)(c+b) = c$ by (G), while $c = ac + cb \prec a + bc$ by (G*).

Assume now that $a(b+c) \prec c \prec a + bc$. Then
$$a + c = a + bc + c = a + bc \quad \text{and} \quad ac = a(b+c)c = a(b+c).$$
It follows that
$$(a+c)(c+b) = (a+bc)(c+b) = c + (a+bc)b = c + bc + ab = c + ab = c,$$
since $ab \prec a(b+c) \prec c$. Dualizing yields $ac + cb = c$ and $B(a, c, b)$ holds by (G).

Remark 1. If $B(a, c, b)$ holds, then c distributes with a and b; that is, $(a+b)c = ac + bc$. This follows from (G) and (G**).

Remark 2. Putting $x = ab$ in $ab+xa+xb$ gives ab and putting $x = a+b$ in the same expression gives $a+b$. Hence by Corollary 1, ab and $a+b$ are each between a, b; that is, $B(a, ab, b)$ and $B(a, a+b, b)$ hold. Since $ab \prec a \prec a+b$, $ab \prec b \prec a+b$, then $B(ab, a, a+b)$ and $B(ab, b, a+b)$ subsist. It follows readily that the four points a, b, ab, $a+b$ of $D(L)$ form a *pseudo-linear* quadruple if pairwise distinct (Exercise 1, § 45).

THEOREM 124.3. *A quadruple of $D(L)$ is pseudo-linear if it forms a sub-lattice of L with two elements non-comparable.*

Thus pseudo-linear quadruples generally abound in $D(L)$. We shall see later that the absence of these quadruples imposes a severe restriction on L and its associated metric space.

EXERCISES

1. Determine whether or not the relation $ac+cb = c = (a+b)c$ is necessary and sufficient for $B(a, c, b)$, where a, b, c are elements of a normed lattice L.
2. Examine the betweenness relation in the lattice L_C (Exercise 4, § 123). What is the locus of points of $D(L_C)$ metrically between two non-comparable elements of L_C?

125. Betweenness in normed distributive lattices. Betweenness in arbitrary lattices

We have observed that in normed lattices the lattice relation $ab \prec c \prec a+b$ is a necessary condition for metric betweenness $B(a, c, b)$. It is, however, not sufficient; for if a, b, c denote three pairwise distinct concurrent lines in the lattice of linear subspaces of the projective plane (see Exercises 2, 3, § 123), then $ab = bc = ac$ (the point of concurrency) and $a+b = b+c = a+c$ (the plane). Then $ab \prec c \prec a+b$ but

$$(a, b) = (b, c) = (a, c) = 2$$

and so $B(a, c, b)$ does not hold. We shall prove in this section that if L is distributive, then the condition $ab \prec c \prec a+b$ *is* necessary and sufficient for $B(a, c, b)$ to subsist, and that, furthermore, this is a characteristic property of distributive lattices.

LEMMA 125.1. *If L is a normed distributive lattice and $a, b, c \in L$, then $B(a, c, b)$ holds if and only if $ab \prec c \prec a+b$.*

Proof. Assuming $ab \prec c \prec a+b$, then from $ab \prec c$ follows $c = ab+c$, and from $c \prec a+b$ follows $c = c(a+b) = ac+cb$. Hence

$$ac+cb = c = ab+c$$

and $B(a, c, b)$ holds by (G*). Since the necessity of the condition has already been proved, the lemma is established.

LEMMA 125.2. *A normed lattice L is distributive if and only if for each three elements a, b, c of L*

$$2|a+b+c| - 2|abc| = |a+b| + |b+c| + |a+c| - |ab| - |bc| - |ac|.$$

Proof. Since $ac+bc \prec (a+b)c$ and $ab+c \prec (a+c)(b+c)$, then the distributive laws will hold in the normed lattice L if and only if at least one (and hence both) of the following conditions is satisfied:

$$|ac+bc| = |(a+b)c|, \tag{1}$$
$$|ab+c| = |(a+c)(b+c)|. \tag{2}$$

Using part (2) of the definition of norm, (1), (2) take the forms

$$|ac| + |bc| - |abc| = |a+b| + |c| - |a+b+c|$$
$$= |a| + |b| + |c| - |ab| - |a+b+c| \tag{3}$$

and

$$|a| + |b| - |a+b| + |c| - |abc| = |a+c| + |b+c| - |a+b+c|. \tag{4}$$

Hence the normed lattice L is distributive if and only if at least one (and hence both) of the following two norm relations holds:

$$|a+b+c| - |abc| = |a| + |b| + |c| - |ab| - |bc| - |ac|, \tag{*}$$
$$|a+b+c| - |abc| = |a+b| + |b+c| + |a+c| - |a| - |b| - |c|. \tag{**}$$

Adding, we obtain the norm relation stated in the lemma as a necessary condition for distributivity of L. The condition is also sufficient, for assuming it and writing the norms of the sums appearing in the right-hand member as the sums of the norms minus the norms of the products, we obtain

$$|a+b+c| - |abc| = |a| + |b| + |c| - |ab| - |bc| - |ac|,$$

which is the relation (*).

THEOREM 125.1. *A normed lattice L is distributive if and only if the lattice relation $ab \prec c \prec a+b$ is a necessary and sufficient condition for the metric relation $B(a,c,b)$.*

Proof. By Lemma 125.1, distributivity of L implies the equivalence of $ab \prec c \prec a+b$ and $B(a,c,b)$. Assume, now, the equivalence of these two relations. Then from $ab \prec ab+c(a+b) \prec a+b$ we may conclude the betweenness relation $B(a, ab+c(a+b), b)$; that is,

$$|a+b| - |ab| = |a+ab+c(a+b)| - |a(ab+c(a+b))| +$$
$$+ |ab+c(a+b)+b| - |b(ab+c(a+b))|.$$

It follows that

$$|a+b| - |ab| = |a+c(a+b)| - |a(ab+c(a+b))| +$$
$$+ |b+c(a+b)| - |b(ab+c(a+b))|,$$

and by modularity,

$$|a+b|-|ab| = |(a+b)(a+c)|-|ab+a(a+b)c|+$$
$$+|(a+b)(b+c)|-|ab+b(a+b)c|$$
$$= |(a+b)(a+c)|-|ab+ac|+|(a+b)(b+c)|-|ab+bc|.$$

Hence

$$|a+b|-|ab| = |a+b|+|a+c|-|a+b+c|-|ab|-|ac|+|abc|+$$
$$+|a+b|+|b+c|-|a+b+c|-|ab|-|bc|+|abc|,$$

and consequently

$$2|a+b+c|-2|abc| = |a+b|+|b+c|+|a+c|-|ab|-|bc|-|ac|.$$

But this norm relation implies distributivity of L (Lemma 125.2) and the theorem is proved.

Glivenko's lattice characterization (G) of metric betweenness in normed lattices was adopted by Pitcher and Smiley as a definition of betweenness in an *arbitrary* lattice. Thus if a, b, c are elements of any lattice L, then c is defined to be lattice-between a and b, symbolized by writing $LB(a,c,b)$, provided $ac+cb = c = (a+c)(c+b)$. It is easily seen that lattice-betweenness has many of the properties of metric betweenness (§§ 12, 13). Thus, for example, the relation $LB(a,c,b)$ is symmetric with respect to the outer points, has special inner point ($LB(a,c,b)$ and $LB(a,b,c)$ imply $b=c$), and enjoys the transitive property $LB(a,b,c)$ and $LB(a,d,b)$ imply $LB(d,b,c)$. The five-point transitivity $LB(a,b,c)$, $LB(a,d,b)$, and $LB(a,c,e)$ imply $LB(d,c,e)$ is also valid.

Modular and distributive lattices may be characterized by the possession by the relation $LB(a,b,c)$ of certain transitive properties. A lattice is modular, for example, if and only if the betweenness relation has the property that $LB(a,b,c)$ and $LB(a,d,b)$ imply $LB(a,d,c)$; it is distributive if and only if $LB(a,b,c)$, $LB(a,d,c)$, and $LB(b,e,d)$ imply $LB(a,e,c)$.† The latter result is an extension of a metric characterization of distributive lattices due to Glivenko, in which lattice-betweenness is replaced by metric betweenness. Similarly, Pitcher and Smiley give a purely lattice proof of Theorem 125.1 and hence prove that *any* lattice is distributive provided for each three of its elements a, b, c the relation $ab \prec c \prec a+b$ is equivalent to $LB(a,c,b)$.

The lattice betweenness $LB(a,b,c)$ is *characterized* among all triadic relations $T(a,b,c)$ defined for all triples of every lattice by the following properties: (1) *symmetry in outer points*, (2) *special inner point*, (3) *the transitivity $T(a,b,c)$ and $T(a,d,b)$ imply $T(d,b,c)$*, (4) *if c is order-*

---

† See § 12; in particular, Exercises 2, 3.

between a, b then $T(a, c, b)$ holds, (5) for $a, b \in L$, $T(a, a+b, b)$ and $T(a, ab, b)$ hold, (6) if $a, b, c \in L$ such that $LB(a, b, c)$ holds, and if x, y, z, w are any four elements of the sub-lattice generated by a, b, c, then $T(x, y, z)$ and $T(x, w, y)$ imply $T(x, w, z)$.

Studies similar to that of Pitcher and Smiley might be made for each of the lattice relations (G*), (G**) which are equivalent to (G) in normed lattices but not in *every* lattice. The various transitivities enjoyed by the notions of lattice betweenness based on them, and their characterizations among all triadic relations should be investigated.

126. Lattice characterization of pseudo-linear quadruples

We have seen (Theorem 124.3) that if a, b are non-comparable elements of a normed lattice L (that is, a, b, ab, and $a+b$ are pairwise distinct), then $a, b, ab, a+b$ form a pseudo-linear quadruple in $D(L)$. But the lattice L_P of linear subspaces of a projective three-space furnishes an example to show that this sufficient condition for a pseudo-linear quadruple is not necessary. For if a, b are two skew lines, each of which intersects each of two skew lines c, d, then

$$(a, b) = |a+b| - |ab| = 4 = (c, d), \quad (b, c) = (b, d) = (a, d) = (a, c) = 2$$

and the four elements are pseudo-linear without being a sub-lattice of L_P. Since pseudo-linear quadruples play an important role in metric aspects of lattice theory, it is desirable to characterize them in terms of lattice relations alone. We give first a simple necessary condition, which for *distributive* lattices is also sufficient.

THEOREM 126.1. *A necessary condition that four distinct points form a pseudo-linear quadruple is that the labelling may be so chosen that*

$$a+c = b+d, \quad ac = bd. \tag{N}$$

Proof. If a, b, c, d form a pseudo-linear quadruple, the labelling may be assumed so that $B(a, b, c)$, $B(b, c, d)$, $B(c, d, a)$, $B(d, a, b)$ hold. Then by (G*) and (G**)

$$\begin{aligned} b(a+c) &= b = b+ac, \\ c(b+d) &= c = c+bd, \\ d(a+c) &= d = d+ac, \\ a(b+d) &= a = a+bd. \end{aligned} \tag{\dagger}$$

It follows that

$$ac \prec b \prec a+c, \quad ac \prec d \prec a+c,$$
$$bd \prec a \prec b+d, \quad bd \prec c \prec b+d,$$

from which one readily obtains

$$ac \prec bd, \quad b+d \prec a+c, \quad bd \prec ac, \quad a+c \prec b+d.$$

Hence $b+d = a+c$, $bd = ac$.

THEOREM 126.2. *If L is a distributive lattice, then a necessary and sufficient condition that four distinct points form a pseudo-linear quadruple is that the labelling may be so chosen that $a+c = b+d$ and $ac = bd$.*

Proof. By the preceding theorem, the condition is necessary. If a, b, c, d are four distinct points of L such that $a+c = b+d$ and $ac = bd$, then the equalities (†) are valid. Since L is distributive, these equalities imply (G*), which in turn implies $B(a, b, c)$, $B(b, c, d)$, $B(c, d, a)$, $B(d, a, b)$. Hence the four distinct points form a pseudo-linear quadruple.

Remarks. If L is not distributive, the necessary condition (N) is not sufficient. In the lattice L_P let a be a plane, c a point not on a, and b, d two skew lines, neither of which contains c. Then $a+c = b+d$ and $ac = bd$, but a, b, c, d are not pseudo-linear; indeed, no three of the elements are linear.

If a, b, c, d are four pairwise distinct elements of L for which condition (N) holds for two pairings of the four elements, then a, b, c, d are surely *not* pseudo-linear. For suppose $a+b = c+d$, $ab = cd$, and $a+c = b+d$, $ac = bd$. Then clearly $c \prec a+b$ and so $a+c \prec a+b$. But also $b \prec a+c$ and hence $a+b \prec a+c$. Consequently $a+c = a+b$, and dually, $ac = ab$. It follows that $(a, b) = (c, d) = (a, c) = (b, d)$, and linearity of a, b, c together with $(a, b) = (a, c)$ imply $(c, a)+(a, b) = (b, c)$; that is, $ca+ab = a = ab+c$. Examining b, c, d we see that $B(b, d, c)$ must subsist, and so $bd+dc = d$. But then $a = d$, contrary to assumption.

If (N) is satisfied for all three pairings of four distinct elements, then the quadruple is *equilateral*. Each pair of the elements has the same product (sum).

Though condition (N) is necessary and sufficient for pseudo-linearity of four distinct elements whenever the normed lattice L is distributive, there are non-distributive normed lattices with the same property. The class of lattices characterized by the equivalence of (N) with pseudo-linearity contains the class of distributive lattices and is a proper subclass of the class of modular lattices. It would be interesting to obtain a lattice characterization of that class.

THEOREM 126.3. *A necessary and sufficient condition that four distinct points of a normed lattice L form a pseudo-linear quadruple is that the labelling may be chosen so that*

$$a+c = b+d, \qquad a = (a+b)(a+d), \qquad b = (b+c)(b+a),$$
$$c = (c+d)(c+b), \qquad d = (d+a)(d+c). \tag{NS}$$

Proof. Four distinct points form a pseudo-linear quadruple provided the labelling may be so chosen that $B(a, b, c)$, $B(b, c, d)$, $B(c, d, a)$, $B(d, a, b)$ subsist; that is (applying (G)), provided a choice of labelling exists such that

$$ab+bc = b = (a+b)(b+c), \qquad bc+cd = c = (b+c)(c+d),$$
$$ad+dc = d = (a+d)(d+c), \qquad ba+ad = a = (b+a)(a+d).$$

These relations clearly imply (NS). Conversely, they are implied by (NS). We show, for example, that (NS) implies $b = ab+bc$.

From (NS) $bc = (b+c)(b+a)c = (a+b)c$, and so

$$a+bc = a+(a+b)c = (a+c)(a+b)$$

by modularity. But since $b+d = a+c$, then $b \prec (a+c)(a+b)$ and so $b \prec a+bc$. Consequently $b = b(a+bc) = ab+bc$ by modularity.

The other relations are established in a similar manner.

Dually, the relations $ac = bd$, $a = ba+ad$, $b = ab+bc$, $c = bc+cd$, $d = ad+dc$ are also necessary and sufficient for pseudo-linearity of (pairwise distinct) elements a, b, c, d.

EXERCISE

Let L be a normed lattice in which condition (N) is not sufficient for pseudo-linearity of four distinct elements a, b, c, d. Show that L contains a sub-lattice of five elements $p_1, p_2, \ldots, p_5$ with p_1, p_2, p_3 pairwise non-comparable and

$$p_4 = p_i p_j, \qquad p_5 = p_i + p_j \quad (i, j = 1, 2, 3;\ i \neq j).$$

127. Congruent imbedding of normed lattices in convex normed lattices

A one-to-one mapping f of a lattice L into a lattice L^* is a *lattice imbedding* provided $f(ab) = f(a)f(b)$ and $f(a+b) = f(a)+f(b)$ for every pair of elements a, b of L. If both L and L^* are normed, and $|f(a)| = |a|$ for every element a of L, then clearly $D(L)$ is congruently mapped by f into $D(L^*)$. Lattice L is then said to be *congruently lattice imbeddable* in L^*.

A normed lattice L is *convex* provided its associated metric space $D(L)$ is (metrically) convex; that is, for each two distinct elements a, b of L, an element c exists, *distinct from a and b*, such that $B(a, c, b)$ holds. The question of congruently imbedding an arbitrary normed lattice in a convex normed lattice was considered in 1928 by Aronszajn, to whom the following theorem is due.

THEOREM 127.1. *Every normed lattice L may be congruently lattice imbedded in a convex normed lattice.*

Proof. Let L be a normed lattice and consider the set $L \times L$ (cartesian product) of all ordered pairs $[x, y]$ of elements of L.† Setting $[a, b] = [c, d]$ if and only if $a = c$ and $b = d$, it is easily seen that $L \times L$ is a normed lattice L_1 under the following definitions of lattice product, lattice sum, and norm:

$$[a, b][c, d] = [ac, bd],$$
$$[a, b]+[c, d] = [a+c, b+d],$$
$$|[a, b]| = \tfrac{1}{2}[|a|+|b|].$$

It follows that $[a, b] \prec [c, d]$ is equivalent to $a \prec c$ and $b \prec d$.

Now the mapping $f(a) = [a, a]$ of L into L_1 is clearly a lattice imbedding and since $|f(a)| = |[a, a]| = |a|$, then L is congruently lattice imbeddable in L_1.

The procedure by which the normed lattice L_1 was obtained from the normed lattice L may be repeated to obtain an infinite sequence $L, L_1, L_2, \ldots, L_n, \ldots$ of normed lattices, each of which is congruently lattice imbeddable in its successor in the sequence.

Now let L^* denote the set sum (union) of $L = L_0, L_1, L_2, \ldots, L_n, \ldots$ and identify each element α of L_i with the element $[\alpha, \alpha]$ of L_{i+1} ($i = 0, 1, 2, \ldots$). It follows that $L_0 \subset L_1 \subset L_2 \ldots \subset L_n \subset \ldots$ and so if $x, y \in L^*$ a smallest index k exists such that $x, y \in L_k$. Defining as lattice sum and product of x, y in L^* their lattice sum and product, respectively, in L_k makes L^* a lattice; while attaching to an element x of L^* as norm the same number attached to it as norm in the lattice L_k of smallest index containing x makes L^* a normed lattice.

The normed lattice L^* is convex, for let L_k denote the lattice of smallest index containing a given pair of distinct elements x, y of L^*. Then $[x, y] \in L_{k+1}$ distinct from x and y. We have

$$(x, [x, y]) = |x+[x, y]|-|x[x, y]|$$
$$= |[x, x]+[x, y]|-|[x, x][x, y]|$$
$$= |[x, x+y]|-|[x, xy]|$$
$$= \tfrac{1}{2}[|x|+|x+y|]-\tfrac{1}{2}[|x|+|xy|]$$
$$= \tfrac{1}{2}[|x+y|-|xy|] = \tfrac{1}{2}(x, y).$$

Similarly, $([x, y], y) = \tfrac{1}{2}(x, y)$ and so the point $[x, y]$ of $D(L^*)$ is metrically between the points x and y. Hence $D(L^*)$ is metrically convex and L^* is a convex lattice.

Finally, it is clear that the mapping $f(a) = [a, a]$ is a norm-preserving lattice imbedding of L into L^*, and the proof is complete.

† The term 'ordered' does not, of course, refer to any precedence relation in L or $L \times L$. It is used here in its metamathematical sense.

Remarks. If L is distributive, so is L^*.

In a manner quite similar to the Hausdorff 'completion' process of extending a metric space to a complete metric space, it may be shown that *every normed lattice is congruently lattice imbeddable in a metrically complete normed lattice* (a normed lattice L being called metrically complete provided $D(L)$ is complete).

Imbedding congruently a normed lattice L in the convex normed lattice L^*, and applying the completion process to L^* a lattice L^{**} is obtained that is easily seen to be convex. *Hence any normed lattice L is congruently lattice imbeddable in a complete and convex normed lattice.*

If a normed lattice L is called *separable* provided $D(L)$ is separable, the reader's attention is directed to the following query: *Does there exist a separable, complete, convex normed lattice into which every separable distributive normed lattice is congruently lattice imbeddable?*

EXERCISES

1. Let f be a one-to-one mapping of a lattice L onto a lattice L'. Show that if f preserves sums (that is, if $f(a+b) = f(a)+f(b)$), then f preserves products and hence is an isomorphism. (*Hint.* Show first that f and f^{-1} are order-preserving.) Dually, if f preserves products, it preserves sums. Is the 'onto' property of f essential?

2. A one-to-one mapping f of a *normed* lattice L onto a normed lattice L' is said to have property (S) if it preserves sums, property (D) if it preserves distances, and property (N) if it preserves norms (*modulo* a constant).
Prove that if f has any two of these properties it has also the third.

128. Congruence of $D(L)$ with a euclidean subset

The imbedding of any normed lattice in a complete and convex normed lattice established in the preceding section suggests the problem of determining what normed lattices have their associated metric spaces congruently imbeddable in a euclidean or Hilbert space. In obtaining the simple but highly restrictive necessary and sufficient condition for this to occur, it is found that the pseudo-linear quadruples play an unexpectedly decisive role.

THEOREM 128.1. *The associated metric space $D(L)$ of a normed lattice L is congruent with a subset of Hilbert space if and only if it is congruently contained in the euclidean line. The space $D(L) \subseteqq E_1$ if and only if it contains no pseudo-linear quadruple; that is, if and only if L is simply ordered.*

Proof. If $D(L)$ is congruent with a subset of Hilbert space, then $D(L)$ does not contain a pseudo-linear quadruple. (Why?) Hence if $a, b \in L$, $a \neq b$, then the four elements $a, b, ab, a+b$ are not pairwise distinct

(Theorem 124.3). But $ab = a+b$ contradicts $a \neq b$ and so one of the four equalities $a = ab$, $a = a+b$, $b = ab$, $b = a+b$ must hold; that is, $a \prec b$ or $b \prec a$. Consequently each two elements of L are comparable and L is simply ordered.

It follows easily that if $a, b, c \in L$, then one of these elements is order-between the other two and hence (§ 124) $B(a,b,c)$ or $B(a,c,b)$ or $B(b,c,a)$ subsists. Thus $D(L)$ has each of its triples linear, and since $D(L)$ contains no pseudo-linear quadruple, we may conclude that $D(L) \subseteqq E_1$.

To complete the proof it is noted that if L is simply ordered, it does not contain a pseudo-linear quadruple. For in the contrary case, a labelling of such a quadruple may be chosen so that $a+c = b+d$, $ac = bd$ (Theorem 126.1). But since L is simply ordered, $a+c$ is either a or c and $b+d$ is either b or d; that is, a or c coincides with b or d, which is impossible.

Remark. Though the associated metric space $D(L)$ of a normed lattice L is congruently imbeddable in euclidean (Hilbert) space if and only if it is imbeddable in the straight line, subsets of $D(L)$ may be congruently contained in euclidean spaces other than the line.

EXERCISE

How may planar quadruples of $D(L)$ be characterized in L?

129. Properties of the associated metric space of a normed lattice

We have seen that only in the relatively trivial case of a simply ordered lattice (a chain) is the associated metric space congruently imbeddable in a euclidean or Hilbert space, and then it is imbeddable in the E_1. This suggests that the metric space $D(L)$ of any normed lattice possesses certain metric characteristics inherent in its genesis, and it is desirable to ascertain what those characteristics are. Considered in another way, the problem is that of characterizing metrically those metric spaces M in which a partial order may be defined, with respect to which the points are the elements of a normed lattice L whose associated metric space $D(L)$ is the space M.

If L has a first element, the problem was solved by Glivenko, and later by Smiley and Transue. The characterization of M obtained by the latter two workers is in terms of metric singularities and certain five-point betweenness transitivities. Essentially, it differs but little from the earlier solution. The same five-point transitivity is used in both treatments (implicitly by Glivenko), and there is a close relationship between the second property of Smiley and Transue (involving the metric singularity—a pseudo-linear quadruple) and Property II of Glivenko, to be stated later. We shall adopt Glivenko's procedure.

Let, now, L denote a normed lattice with a first element o; that is, $o \prec x$ for every element x of L. There is no loss of generality in assuming that $|o| = 0$, for if $|o| = N \neq 0$, the real function $\mu(x) = |x| - N$, $x \in L$, is easily seen to be a *norm* with $\mu(o) = 0$. The associated metric space $D(L)$ is, moreover, invariant under this change of norm, since

$$\mu(a+b) - \mu(ab) = |a+b| - |ab|, \qquad a, b \in L.$$

If a, b are points of $D(L)$ with $B(o, a, b)$ holding (that is,

$$(o, a) + (a, b) = (o, b)),$$

it is convenient in what follows to say that 'a is nearer o than b' or simply 'a is nearer than b'.† If a is nearer than b we say that b is farther than a.

Remark. If $a, b \in D(L)$, then a is nearer than b if and only if $a \prec b$. For $(o, a) + (a, b) = (o, b)$ is equivalent by (G) to

$$oa + ab = a = (o+a)(a+b);$$

that is, to $ab = a$ or $a \prec b$.

LEMMA 129.1. *The associated metric space $D(L)$ of a normed lattice L with first element has the following Property* I: *if $a, b, x \in D(L)$ and each of a, b is nearer (farther) than x, then each element y of $D(L)$ metrically between a, b (in the wide sense) is nearer (farther) than x.*‡

Proof. If a is nearer than x and b is nearer than x, then, by the preceding remark, $a \prec x$ and $b \prec x$. Then $ay \prec x$, $by \prec x$, and so $ay + by \prec x$. Since $B(a, y, b)$ holds, $ay + yb = y$ and consequently $y \prec x$. But this implies $B(o, y, x)$; that is, y is nearer than x.

Similarly, if x is nearer than a and nearer than b, then $x \prec a$, $x \prec b$. Hence $x \prec a + y$, $x \prec b + y$ and so $x \prec (a+y)(b+y) = y$ (since $B(a, y, b)$ subsists); that is, $x \prec y$ and y is farther than x.

LEMMA 129.2. *The associated metric space $D(L)$ of a normed lattice with a first element has Property* II: *if $a, b \in L$, the subset $B(a, b)$ of points of $D(L)$ metrically between a, b (in the wide sense) contains a point nearest o and a point farthest from o.*§

Proof. We shall show that ab is the point of $B(a, b)$ nearest o and $a+b$ is the point of $B(a, b)$ farthest from o. It has already been observed that

---

† Since $B(o, a, b)$ does not preclude $a = b$, it would be more accurate to say 'a is no farther from o than b'.

‡ Compare Property I with the five-point betweenness transitivity: oax, obx, ayb imply oyx (reading 'nearer'). What is the analogous transitivity, reading 'farther'?

§ It is clear that the 'nearest' and 'farthest' points postulated in Property II are unique.

ab and $a+b$ are between a, b (remark, § 124). If, now, $c \in L$ with $B(a,c,b)$ holding, then by a remark of § 124, $ab \prec c \prec a+b$ and consequently ab is nearer than c while $a+b$ is farther than c.

130. Characterization of $D(L)$

We have seen that the associated metric space of a normed lattice with a first element has the metric properties described in Property I and Property II. It remains to show that these two properties are characteristic of $D(L)$.

THEOREM 130.1. *Every metric space D with Properties I and II is the associated metric space of a normed lattice with a first element.*

Proof. Let o denote a point of the metric space D with respect to which 'nearer than' is defined. We show first that a partial ordering is established in D by writing $a \prec b$ if and only if $B(o,a,b)$ holds $(a, b \in D)$. Clearly $a \prec a$, while addition of the two distance relations represented by $B(o,a,b)$ and $B(o,b,a)$ gives $(a,b) = 0$; that is, $a \prec b$ and $b \prec a$ are equivalent to $a = b$.

If $a, b, c \in D$ with $a \prec b$ and $b \prec c$, the relation $a \prec c$ obviously holds in case the points o, a, b are not pairwise distinct, while if o, a, b are pairwise distinct, then $B(o,a,b)$, $B(o,b,c)$ represent metric betweenness in the strict sense and they imply $B(o,a,c)$ or $a \prec c$ (see (3) of Theorem 12.1). Hence D is partially ordered.

Properties I, II are used to show that D contains for each two of its elements a greatest lower and least upper bound. If $a, b \in D$, denote by ab the unique point of the set $B(a,b)$ of between-points of a, b (in the wide sense) that is nearest o, and by $a+b$ the unique point of $B(a,b)$ farthest from o. These points exist by Property II.

The element ab is the greatest lower bound of a, b, for since $a, b \in B(a,b)$, then $B(o,ab,a)$, $B(o,ab,b)$ hold and so $ab \prec a$, $ab \prec b$. If, now, $x \in D$ and $x \prec a$, $x \prec b$, then, by Property I, each point of $B(a,b)$ is farther than x and so $x \prec ab$.

In a similar manner it is seen that $a+b$ is the least upper bound of a, b. Consequently D forms a lattice L with respect to the precedence relation $\prec$ defined above, and the elements ab, $a+b$ are the lattice product and sum, respectively, of the elements a, b (§ 3, Exercise 6). Evidently the element o of L is the first element in the lattice.

The lattice L is normed by defining $|a| = (o,a), a \in L$. If $a, b \in L$ ($a \neq b$) and $a \prec b$, then $B(o,a,b)$ holds and $(o,a) + (a,b) = (o,b)$. The metricity of D, together with $a \neq b$, yields $(o,a) < (o,b)$ and so $|a| < |b|$.

If $a, b \in L$, from $ab \prec a$ and $ab \prec b$ follows
$$2(o, ab)+(ab, a)+(ab, b) = (o, a)+(o, b).$$
Since ab is between a, b this equality becomes
$$2(o, ab)+(a, b) = (o, a)+(o, b);$$
that is, $\qquad |ab| = \tfrac{1}{2}[|a|+|b|-(a, b)].$ $\qquad$ (*)

Proceeding from $a \prec a+b$, $b \prec a+b$, and $B(a, a+b, b)$ we obtain
$$|a+b| = \tfrac{1}{2}[|a|+|b|+(a, b)], \qquad (**)$$
and consequently $|a+b|+|ab| = |a|+|b|$.

Finally, subtracting (*) from (**) gives
$$(a, b) = |a+b|-|ab|,$$
and so D is the associated metric space of the normed lattice L with first element o. From the definition of norm, $|o| = 0$.

Combining this theorem with the two lemmas of the preceding section yields a characterization of those metric spaces that are associated metric spaces of normed lattices with first elements.

131. Autometrized Boolean algebras. Elementary properties of distance

A lattice L with both a first element o and a last element 1 ($x \prec 1$ for every element x of L) is *complemented* provided it contains for each element a an element a' such that $a+a' = 1$ and $aa' = o$. If a distributive lattice is complemented, it is called a Boolean algebra (Definition 3.2). In every Boolean algebra complements are unique and the formulae of De Morgan are valid (§ 3, Exercise 3).

Among all the distance spaces studied in the foregoing sections of this book the euclidean line E_1 is unique in that the distance set of the space is a subset of the space itself; that is, there is associated with each pair of elements of E_1 *an element of E_1 as distance*. We shall introduce 'distance' in a Boolean algebra in, broadly speaking, a similar manner; namely, by associating with each pair a, b of its elements an element $d(a, b)$ *of the algebra* as distance.

DEFINITION 131.1. *If a, b are elements of a Boolean algebra B, their distance $d(a, b)$ is defined by* $d(a, b) = ab'+a'b.$

It is observed that the element set of B forms a commutative ring, where ring-product $a \otimes b$ is identified with lattice-product ab and the ring-sum $\oplus$ of a, b is defined to be $d(a, b)$. This ring, called the *associated ring* of B and denoted by $R(B)$, has the first element o of B as the unit element of its additive group and the last element 1 as the identity element for multiplication. Since $a \otimes a = aa = a$, $R(B)$ is a Boolean ring. Hence every theorem concerning distance in B gives rise to a corresponding theorem concerning a Boolean ring with unity. In this manner metric methods may serve as a useful device in ring theory.

Remark 1. From $d(a,b) = ab'+a'b = (a+b)(a'+b') = (a+b)(ab)'$ follows $d' = (ab')'(a'b)' = (a'+b)(a+b') = ab+a'b'$.

Clearly $d(a,b) = d(a',b')$, and since $d(o,a) = a$, the first element o of B acts as an origin.

The following theorem establishes the interesting fact that the distance function $d(a,b)$ has *formally* all of the properties characterizing the general metric (§ 9) except, of course, the property of being a non-negative real number.

THEOREM 131.1. *In any Boolean algebra B the distance function $d(a, b)$ has the following properties*:

(i) *if $a = b$, then $d(a,b) = o$,*
(ii) *if $a \neq b$, then $d(a,b) \neq o$,*
(iii) $d(a,b) = d(b,a)$,
(iv) $d(a,b)+d(b,c) \succ d(a,c)$.

Proof. Properties (i), (iii) are obvious. To prove (ii) assume $d(a,b) = o$. Then from $ab' \prec ab'+a'b = o$ follows $ab' = o$ and, taking complements, $a'+b = 1$.† But this, together with $a'b = o$ (which is obtained in the same manner as $ab' = o$) and the uniqueness of complements, gives $a = b$.

To establish the 'triangle inequality', we have
$$[d(a,b)+d(b,c)]d(a,c) = (ab'+a'b+bc'+b'c)(ac'+a'c)$$
$$= ab'c'+abc'+a'bc+a'b'c$$
$$= ac'(b+b')+a'c(b+b')$$
$$= ac'+a'c = d(a,c),$$
and consequently $d(a,b)+d(b,c) \succ d(a,c)$.

THEOREM 131.2. *If $a, b \in B$ with $d(a,b) = c$, then $d(a,c) = b$ and $d(b,c) = a$.*

Proof. From $a'b+ab' = c$ and $ab+a'b' = c'$ follow $a'b = a'c$ and $ab = ac'$. Hence $b = a'b+ab = a'c+ac' = d(a,c)$. In a similar manner it is seen that $d(b,c) = a$.

COROLLARY. *If $a, c \in B$, there is exactly one element b of B with*
$$d(a,b) = c.$$

† It follows readily from the uniqueness of complements that $(b')' = b$.

For since

$$d(a, d(a,c)) = ad'(a,c) + a'd(a,c)$$
$$= a(ac+a'c') + a'(ac'+a'c) = ac+a'c = c,$$

then $d(a,c)$ is one such element b; while if $x \in B$ and $d(a,x) = c$, then by the preceding theorem $x = d(a,c) = b$.

Remark. The above corollary shows that *any given element of B forms a metric basis for B.* Thus B does not contain an isosceles triple of pairwise distinct points. This has important consequences in the study of the group of motions of B with which we shall soon be concerned. It is noted also that, according to the corollary, B contains an element with arbitrarily assigned distance from any chosen element of B.

EXERCISES

1. Establish the facts stated above concerning $R(B)$.
2. If $x, y, z \in B$, show that $d(x, d(y,z)) = d(z, d(x,y))$.

132. Betweenness in autometrized Boolean algebras

In an autometrized Boolean algebra betweenness (in the wide sense) with respect to the distance function is defined in the customary way; that is, $B(a,c,b)$ holds provided $d(a,c) + d(c,b) = d(a,b)$. We recall that an element c of a lattice L is lattice-between two of its elements a, b (that is, $LB(a,c,b)$ holds) provided the condition

$$ac + cb = c = (a+c)(c+b) \tag{G}$$

is satisfied (§ 125). In normed lattices the two kinds of betweennesses are, as we have seen, equivalent. We show in this section that the equivalence is valid in Boolean algebras also.†

THEOREM 132.1. *In an autometrized Boolean algebra lattice-betweenness $LB(a,c,b)$ is equivalent to distance betweenness $B(a,c,b)$.*

Proof. Assume lattice-betweenness $LB(a,c,b)$; that is,

$$ac + cb = c = (a+c)(c+b).$$

Then $ab \prec (a+c)(c+b) = c$ implies $ab = (ac)(cb)$ and

$$a+b \succ ac+cb = c$$

---

† It might be observed that if the norm of an element x of a Boolean algebra be defined by $|x| = d(o,x) = x$, the analogues of the two conditions that enter into the definition of a norm are satisfied, as well as the definition of distance in terms of norm, for the relation $d(a,b) = |a+b| - |ab|$ becomes in Boolean algebra

$$d(a,b) = |a+b| \cdot |ab|' = (a+b)(a'+b').$$

gives $a+b = (a+c)+(c+b)$. Consequently
$$(a+b)(ab)' = [(a+c)+(c+b)][(ac)(bc)]'$$
$$= (a+c)(ac)'+(c+b)(bc)'+ab'+a'b+bc'+b'c+ac'+a'c;$$
that is, $\qquad d(a,b) = d(a,c)+d(c,b)+d(a,b)$
and $\qquad d(a,b) \succ d(a,c)+d(c,b).$

Use of the triangle inequality (Theorem 131.1) yields
$$d(a,b) = d(a,c)+d(c,b).$$

Supposing, now, that distance betweenness $B(a,c,b)$ holds, then
$$(a+c)(ac)'+(c+b)(cb)' = (a+b)(ab)'. \qquad (*)$$
Adding $(a+c)ac$, $(c+b)cb$ to the left side and their respective equivalents ac, cb to the right side of the above equality, we obtain (applying distributivity)
$$(a+c)(ac+(ac)')+(c+b)(cb+(cb)') = (a+b)(ab)'+ac+cb.$$
Hence $c \prec a+c+b = (a+b)(ab)'+ac+cb \prec a+b$ and so
$$c = (a+b)c = ac+cb.$$

Complementation of $(*)$ yields $(ac+a'c')(cb+c'b') = ab+a'b'$, and dualizing the above procedure gives
$$[ac+(a+c)][ac+a'c'][cb+(c+b)][cb+c'b'] = (ab+a'b')(a+c)(c+b),$$
$$[ac+(a+c)a'c'][cb+(c+b)c'b'] = (ab+a'b')(a+c)(c+b),$$
$$c \succ (ac)(cb) = (ab+a'b')(a+c)(c+b) \succ ab.$$

Hence $c = c+ab = (a+c)(c+b)$, and the proof is complete.

The relations (G), (G*), (G**) are, of course, equivalent in any distributive lattice and hence in any Boolean algebra. From the preceding theorem, each is equivalent to distance betweenness in an autometrized Boolean algebra.

133. The group of motions of B

A congruent (that is, distance-preserving) mapping of B onto itself is a *motion*. Since $d(a,b) = o$ if and only if $a = b$, *a motion of B is a biuniform mapping*. The next few lemmas establish some elementary properties of a motion.

LEMMA 133.1. *Any congruent mapping f of B into itself is involutory.*

Proof. If $x \in B$, denote $f(f(x))$ by y. Since f is a congruence,
$$d(x, f(x)) = d(f(x), y)$$
and so each of the elements x, y has the same distance from the element

$f(x)$. But $f(x)$ forms a metric basis for B (remark following the corollary to Theorem 131.2) and consequently $y = x$. Thus for each element x of B, $f(f(x)) = x$ and the lemma is proved.

COROLLARY. *An autometrized Boolean algebra B is monomorphic; that is, any congruent mapping f of B into itself is a motion.*

Proof. The corollary is established when it is shown that f is an 'onto' mapping; that is, that each element x of B is the image by f of an element of B. But it follows at once by the above lemma that if $x \in B$, then $x = f(f(x))$.

LEMMA 133.2. *There is at most one motion f of B taking an assigned element a into an assigned element b.*

Proof. Suppose that g is a motion of B with $g(a) = f(a) = b$. If g is different from f, an element x of B exists ($x \neq a$) such that $f(x) \neq g(x)$. Since f and g are congruences, $d(a,x) = d(b,f(x)) = d(b,g(x))$ and $b, f(x), g(x)$ form an isosceles triple of pairwise distinct elements, which is impossible.

COROLLARY. *If a motion of B leaves one element fixed, it is the identity.*

LEMMA 133.3. *If f is any motion of B and $x \in B$, then $f(x') = f'(x)$; that is, any motion of B is complement-preserving.*

Proof. Since f is a congruence, $d(f(x), f(x')) = d(x, x') = 1$ and so $f(x)f'(x') + f'(x)f(x') = 1$. Taking the lattice product of both sides with $f(x)$ and with $f'(x)$ gives

$$f(x)f'(x') = f(x), \qquad f'(x)f(x') = f'(x).$$

Complementation of the second equality gives $f(x) + f'(x') = f(x)$, which, together with the first equality, yields the lemma.

LEMMA 133.4. *If f is any motion of B and $x \in B$, then $f(x) = d(f(o), x)$ and $d(x, f(x)) = d(o, f(o))$.*

Proof. Since $x = d(o, x) = d(f(o), f(x))$, then $f(x) = d(f(o), x)$ and $d(x, f(x)) = f(o) = d(o, f(o))$ by Theorem 131.2.

Remark 1. By the above lemma, a motion is completely determined when the element into which it takes the first element o is known. Further, each motion is translatory since the distance $d(x, f(x))$ each element is moved is constant.

Remark 2. Although the group of motions of B and the group of automorphisms of the Boolean algebra B are both subgroups of the group of biuniform, complement-preserving mappings of the element set of B onto itself, they have only the identity mapping in common since every automorphism of B leaves o fixed and only the identity motion has that property.

LEMMA 133.5. *If $a \in B$, there is exactly one motion f of B such that $f(o) = a$.*

Proof. By Lemma 133.2 there is at most one such motion. Now the mapping $f(x) = d(x,a) = a'x+ax'$ of B into itself has $f(o) = a$. It is easily seen to be a congruence, for

$$d(f(x), f(y)) = d(a'x+ax', a'y+ay')$$
$$= (a+x')(a'+x)(a'y+ay')+(a'x+ax')(a+y')(a'+y)$$
$$= a'x'y+axy'+a'xy'+ax'y$$
$$= (xy'+x'y)(a+a') = d(x,y).$$

Hence f is a motion (corollary, Lemma 133.1) and the lemma is proved.

Remark. In the proof of the above lemma there is established the distance formula $d[d(x,a), d(y,a)] = d(x,y)$. Let the reader show that this formula may be obtained by a double application of the relation $d(x, d(y,z)) = d(z, d(x,y))$ of Exercise 2, § 131.

We are now in position to establish the main result of this section.

THEOREM 133.1. *The group of motions $M(B)$ of an autometrized Boolean algebra B is isomorphic to the additive group of the associated Boolean ring $R(B)$ of B.*†

Proof. Associate with each element a of $R(B)$ the unique motion f that takes o into a; namely, $f(x) = d(x,a)$ (Lemma 133.5).‡ Since for each element f of $M(B)$ an element a of B exists with $f(o) = a$, this association maps $R(B)$ in a one-to-one manner *onto* $M(B)$.

If, now, $f, g \in M(B)$ that correspond, respectively, to elements a, b of $R(B)$, we show that the element $a \oplus b = d(a,b)$ of $R(B)$ is associated with the 'product' $fg = f(g)$ in $M(B)$ of f and g. To do this it suffices to show that $f(g(o)) = a \oplus b$. But this is immediate since

$$f(g(o)) = f(b) = d(b,a) = a \oplus b,$$

and the theorem is proved.

134. Free mobility in an autometrized Boolean algebra

We have seen that the mapping $f(x) = d(x,a)$ is a motion of B that interchanges the elements o and a. If this motion is followed by the motion $g(x) = d(x,b)$ that carries o to b, the result is the motion

$$h(x) = g(f(x)) = d(d(x,a), b) = d(x, d(a,b))$$

† See § 131 for definition of $R(B)$.
‡ $R(B)$ and B have the same element-set.

that maps element a onto element b. Hence if a, b are any two elements of B, there exists a motion of B that takes a into b.

THEOREM 134.1. *The group of motions of an autometrized Boolean algebra is transitive.*

The notion of free mobility in an autometrized Boolean algebra B is defined in the same way as in semimetric spaces (§ 98); that is, a subset E of B is *freely movable* provided any congruence between E and any subset F of B may be extended to a motion. A distance space has the property of *free mobility* provided each of its subsets is freely movable.

THEOREM 134.2. *An autometrized Boolean algebra B has the property of free mobility.*

Proof. Let E be any subset of B and let g denote a congruence of E with a subset F of B. We show that g may be extended to a motion.

If $a \in E$ and $b = g(a)$ is the element of F corresponding to a by the congruence g, consider the unique motion $f(x) = d(x, d(a, b))$ of B that maps a into b. This motion is identical with the congruence g for all elements of E, for if $x \in E$, then $d(a, x) = d[g(a), g(x)]$ by the congruence g, and $d(a, x) = d[f(a), f(x)]$ by the motion f. But $f(a) = b = g(a)$ and consequently each of the elements $f(x), g(x)$ has the same distance from b. Since b is a metric basis for B, it follows that $f(x) = g(x)$, and the theorem is proved.

EXERCISE

If $a, b, m, n \in B$ with $a, b \approx m, n$, exhibit a motion f with $f(a) = m, f(b) = n$.

135. Congruence order of B with respect to the class of B-metrized spaces

An abstract set Σ forms a *B-metrized space* provided there is attached to each pair α, β of its elements an element $d(\alpha, \beta)$ of a Boolean algebra B, with $d(\alpha, \beta) = d(\beta, \alpha)$ and $d(\alpha, \beta) = o$ if and only if $\alpha = \beta$. The manner in which an element of B is attached to a pair of its elements in order to make B autometrized (namely, $d(a, b) = ab' + a'b$) conforms to those two requirements (Theorem 131.1) and hence an autometrized Boolean algebra is a B-metrized space. We seek its best congruence indices with respect to the class of all B-metrized spaces.

THEOREM 135.1. *The autometrized Boolean algebra has the best congruence indices (3, 0) with respect to the class of B-metrized spaces.*

Proof. Since an autometrized Boolean algebra B does not contain an equilateral triple of distinct points (remark following Theorem 131.2), it is clear that for no number k are the indices $(2, k)$ valid. It remains to show that B does have congruence indices $(3, 0)$.

Let Σ be any B-metrized space with each three of its elements congruently imbeddable in B. Let α be any fixed element of Σ and a any fixed element of B. If $\zeta \in \Sigma$, elements a_1, x_1 of B exist with $\alpha, \zeta \approx a_1, x_1$, and the unique motion f of B which takes a_1 into a carries x_1 into an element x. Clearly $\alpha, \zeta \approx a, x$ and a mapping $x = x(\zeta)$ of Σ into B is established. *This mapping is a congruence*; for let ζ, η be any two elements of Σ and x, y their respective corresponding elements of B by means of the above mapping. By hypothesis, elements a^*, x^*, y^* of B exist with $\alpha, \zeta, \eta \approx a^*, x^*, y^*$, and the unique motion that takes a^* into a evidently carries x^* into x and y^* into y. Hence $\alpha, \zeta, \eta \approx a^*, x^*, y^* \approx a, x, y$ and so $d(\zeta, \eta) = d(x^*, y^*) = d(x, y)$, completing the proof of the theorem.

EXERCISES

1. Let $g(a, b)$ be any mapping of the set of element-pairs a, b of a Boolean algebra B into B. If g has properties (i)–(iv) of Theorem 131.1 and is a group composition in B, show that $g(a, b)$ is identical with $d(a, b)$.
2. Characterize geometrically those subsets of B that are sub-groups with respect to the group composition $d(a, b)$.
3. Show that B is ptolemaic in the metric d.

REFERENCES

§ 124. Glivenko [1], Blumenthal [25] and Ellis.
§ 125. Glivenko [1], Pitcher [1] and Smiley.
§ 126. Blumenthal [25] and Ellis.
§ 127. Glivenko [1].
§ 128. Blumenthal [25] and Ellis.
§§ 129, 130. Glivenko [2, 3], Smiley [2] and Transue.
§ 131. Ellis [2].

SUPPLEMENTARY PAPER

L. M. Blumenthal, 'Boolean geometry, I', *National Bureau of Standards Report*, No. 1482, February, 1952.

BIBLIOGRAPHY

ALEXITS, G.
1. 'La Torsion des espaces distanciés', *Compositio Mathematicae*, **6** (1938–9), 471–7.

ALT, F.
1. 'Über eine metrische Definition der Krümmung einer Kurve', Vienna Dissertation, 1931 (unpublished). See also 'Zur Theorie der Krümmung', *Ergebnisse eines mathematischen Kolloquiums* (Wien), **4** (1932), 4.

ARONSZAJN, N.
1. 'Neuer Beweis der Streckenverbundenheit vollständiger konvexer Räume', ibid. **6** (1935), 45–56.

BING, R. H.
1. 'Partitioning a set', *Bulletin of the American Mathematical Society*, **55** (1949), 1101–9.

BIRKHOFF, GARRETT
1. *Lattice Theory*, American Mathematical Society Colloquium Publications, **25**, 1948.
2. 'Metric foundations of geometry, I', *Transactions of the American Mathematical Society*, **55** (1944), 465–92.

BLANC, E.
1. 'Les Espaces métriques quasi convexes', *Annales scientifiques de l'École Normale Supérieure*, Troisième Série, **55** (1938), 1–82.

BLUMENTHAL, L. M.
1. 'Distance geometries', *University of Missouri Studies*, **13** (1938), no. 2.
2. (with Robinson, C. V.) 'A new characterization of the straight line', *Reports of a Mathematical Colloquium*, Second Series, Issue 2 (1940), 25–27.
3. 'New characterizations of segments and arcs', *Proceedings of the National Academy of Sciences*, **29** (1943), 107–9.
4. 'Remarks on a weak four-point property', *Revista de Ciencias*, **45** (1943), 183–93.
5. 'A new concept in distance geometry, with applications to spherical subsets', *Bulletin of the American Mathematical Society*, **47** (1941), 435–43.
6. 'Methods and problems of distance geometry', *Memorias de Matemática del Instituto 'Jorge Juan'*, **5**, Madrid (1948), pp. 1–40.
7. 'On the four-point property', *Bulletin of the American Mathematical Society*, **39** (1933), 423–6.
8. 'Kurzer Beweis eines Satzes von Menger', *Ergebnisse eines mathematischen Kolloquiums* (Wien), **7** (1936), 6–7.
9. 'Distance geometry notes', *Bulletin of the American Mathematical Society*, **50** (1944), 235–41.
10. 'Some imbedding theorems and characterization problems of distance geometry', ibid. **49** (1943), 321–38.
11. 'Remarks concerning the euclidean four-point property', *Ergebnisse eines mathematischen Kolloquiums* (Wien), **7** (1936), 7–10.
12. 'Generalized euclidean space in terms of a quasi inner product', *American Journal of Mathematics*, **72** (1950), 686–98.

13. (with Garrett, G. A.) 'Characterization of spherical and pseudo-spherical sets of points', *American Journal of Mathematics*, **55** (1933), 619–40.
14. 'Concerning spherical spaces', ibid. **57** (1935), 51–61.
15. 'La caracterización métrica de espacios ϕ-esféricos', *Revista de la Universidad Nacional del Tucumán*, Serie A, **5** (1946), 69–93.
16. (with Thurman, G. R.) 'The characterization of pseudo-spherical sets', *American Journal of Mathematics*, **62** (1940), 835–54.
17. (with Stamey, W. L.) 'Characterization of pseudo-$S_{n,r}$ sets', *Bulletin of the American Mathematical Society*, **56** (1950), 361 (abstract 370).
18. 'Metric characterization of elliptic space', *Transactions of the American Mathematical Society*, **59** (1946), 381–400.
19. (with Kelly, L. M.) 'New metric-theoretic properties of elliptic space', *Revista de la Universidad Nacional del Tucumán*, Serie A, **7** (1949), 81–107.
20. 'Congruence and superposability in elliptic space', *Transactions of the American Mathematical Society*, **62** (1947), 431–51.
21. 'New theorems and methods in determinant theory', *Duke Mathematical Journal*, **2** (1936), 396–404.
22. 'The geometry of a class of semimetric spaces', *Tohoku Mathematical Journal*, **43** (1937), 205–24.
23. 'Metric methods in determinant theory', *American Journal of Mathematics*, **61** (1939), 912–22.
24. 'Metric methods in linear inequalities', *Duke Mathematical Journal*, **15** (1948), 955–66.
25. (with Ellis, D. O.) 'Notes on lattices', ibid. **16** (1949), 585–90.
26. 'Notes on an extension of matrix rank', *Revista de la Universidad Nacional del Tucumán*, Serie A, **4** (1944), 235–41.
27. 'Note on an arc without tangents', *Reports of a Mathematical Colloquium*, Second Series, Issue 4 (1942), 1–3.
28. (with Gillam, B. E.) 'Distribution of points in n-space', *American Mathematical Monthly*, **50** (1943), 181–5.
29. 'Metric foundation of hyperbolic geometry', *Revista de Ciencias*, **40** (1938), 3–20.

BUSEMANN, H.
1. 'On Leibnitz's definition of planes', *American Journal of Mathematics*, **63** (1941), 101–11.
2. 'Metrically homogeneous spaces', ibid. **68** (1946), 340–4.

COXETER, H. S. M.
1. *Non-euclidean Geometry*, Mathematical Expositions, No. 2. University of Toronto Press, Toronto, 1942.

VAN DANTZIG, D.
1. (with van der Waerden, B. L.) 'Über metrisch homogene Räume', *Hamburger Abhandlungen*, **6** (1928), 374–6.

DINES, L. L.
1. (with McCoy, N. H.) 'On linear inequalities', *Proceedings and Transactions of the Royal Society of Canada*, Section III: Mathematical, Physical and Chemical Sciences (3), **27** (1933), 37–70.

ELLIS, D. O.
1. See Blumenthal [25].
2. 'Autometrized Boolean algebras, I', *Canadian Journal of Mathematics*, **3** (1951), 87–93.

FRÉCHET, M.
1. 'Sur quelques points du calcul fonctionnel', *Rendiconti del Circolo Matematico di Palermo*, **22** (1906), 1–74.
2. *Les Espaces abstraits*, Gauthier-Villars, Paris, 1928.

GARRETT, G. A. 1. See Blumenthal [13].

GILLAM, B.
1. 'A new set of postulates for euclidean geometry' (University of Missouri Doctoral Dissertation), *Revista de Ciencias*, **42** (1940), 869–99.

GLIVENKO, V.
1. *Théorie des structures*, Paris, 1938.
2. 'Géométrie des systèmes de choses normées', *American Journal of Mathematics*, **58** (1936), 799–828.
3. 'Contributions à l'étude des systèmes de choses normées', ibid. **59** (1937), 941–56.

HAANTJES, J.
1. 'Distance geometry. Curvature in abstract metric spaces', *Proceedings, Akademie van Wetenschappen*, Amsterdam, **50** (1947), 496–508.
2. 'A characteristic local property of geodesics in certain spaces', ibid. **54** (1951), 66–73.
3. (with Seidel, J.) 'The congruence order of the elliptic plane', ibid. **50** (1947), 403–5.
4. 'Equilateral point-sets in elliptic two- and three-dimensional spaces', *Nieuw Archief voor Wiskunde* (2), **22** (1948), 355–62.

HAUSDORFF, F.
1. *Grundzüge der Mengenlehre*, Leipzig, 1914.

HUNTINGTON, E. V.
1. (with Kline, J. R.) 'Sets of independent postulates for betweenness', *Transactions of the American Mathematical Society*, **18** (1917), 301–25.
2. 'A new set of postulates for betweenness with proof of complete independence', ibid. **26** (1924), 257–82.

KELLY, L. M. 1. See Blumenthal [19].

KLINE, J. R. 1. See Huntington [1].

LINDENBAUM, A.
1. 'Contributions à l'étude de l'espace métrique, I', *Fundamenta Mathematicae*, **8** (1926), 209–22.

MANSION, P.
1. 'Relation entre des distances de cinq points en géométrie non-euclidienne' *Annales de la Société Scientifique de Bruxelles*, **15** (1890–1), 8–11; **19** (1894–5) 189–93.

McCOY, N. H. See Dines [1].

MENGER, K.
1. 'Untersuchungen über allgemeine Metrik', *Mathematische Annalen*, **100** (1928), 75–163.

2. (with Milgram, A. N.) 'On linear sets in metric spaces', *Reports of a Mathematical Colloquium*, Second Series, Issue 1 (1939), 16–17.
3. 'Zur Metrik der Kurven', *Mathematische Annalen*, **103** (1930), 466–501.
4. 'New foundation of euclidean geometry', *American Journal of Mathematics*, **53** (1931), 721–45.
5. 'Die Metrik des Hilbertschen Raumes', *Anzeiger der Akademie der Wissenschaften in Wien, mathematisch-naturwissenschaftliche Klasse*, **65** (1928), 159–60.

MILGRAM, A. N.
1. See Menger [2].
2. 'Some metric topological invariants', *Reports of a Mathematical Colloquium*, Second Series, Issues 5–6 (1944), 25–35.

VON NEUMANN, J.
1. 'Fourier integrals and metric geometry', *Transactions of the American Mathematical Society*, **50** (1941), 226–51.

PAUC, C.
1. 'Courbure dans les espaces métriques', *Rendiconti della Accademia Nazionale dei Lincei*, **24** (1936), 109–15.

PEPPER, P. M.
1. (with Topel, B. J.) 'Imbedding theorems under weakened hypotheses', Part I', *Reports of a Mathematical Colloquium*, Second Series, Issue 4 (1943), 31–55.
2. 'A new method in imbedding theorems', ibid., Issue 8 (1948), 39–48.
3. 'Nearly euclidean imbedding spaces for pseudo-E_n sets', ibid., Issue 3 (1941), 34–46.

PITCHER, E.
1. (with Smiley, M. F.) 'Transitivities of betweenness', *Transactions of the American Mathematical Society*, **52** (1942), 95–114.

ROBINSON, C. V.
1. See Blumenthal [2].
2. 'A simple way of computing the Gauss curvature of a surface', *Reports of a Mathematical Colloquium*, Second Series, Issue 5–6 (1944), 16–24.
3. *Contributions to Distance Geometry: (1) Congruence Order of Euclidean and Spherical Subsets; (2) Helly Theorems on the Sphere*, University of Missouri Doctoral Dissertation, 1940.
4. 'A characterization of the disc', *Bulletin of the American Mathematical Society*, **47** (1941), 818–19.
5. 'Spherical theorems of Helly type and congruence indices of spherical caps', *American Journal of Mathematics*, **64** (1942), 260–72.

SCHERING, E.
1. 'Die Schwerkraft im Gaussischen Raume', *Göttinger Nachrichten*, 1870, pp. 311–21; ibid. 1873, pp. 13–21.

SCHOENBERG, I. J.
1. 'On metric arcs of vanishing Menger curvature', *Annals of Mathematics*, **41** (1940), 715–26.
2. 'Remarks to Maurice Fréchet's article "Sur la définition axiomatique d'une classe d'espaces vectoriels distanciés applicables vectoriellement sur l'espace de Hilbert"', *Annals of Mathematics*, **36** (1935), 724–32.

3. 'Metric spaces and positive definite functions', *Transactions of the American Mathematical Society*, **44** (1938), 522–36.
4. 'On certain metric spaces arising from euclidean spaces by a change of metric and their imbedding in Hilbert space', *Annals of Mathematics*, **38** (1937), 787–93.
5. 'Metric spaces and completely monotone functions', ibid. **39** (1938), 811–41.
6. See von Neumann [1].

SEIDEL, J.
1. See Haantjes [3].
2. *De congruentie-orde van het elliptische vlak*, Thesis, University of Leiden, 1948. iv + 71 pp.

SIERPIŃSKI, W.
1. *Introduction to General Topology*, University of Toronto Press, Toronto, 1934.

SMILEY, M. F.
1. See Pitcher [1].
2. (with Transue, W. R.) 'Applications of transitivities of betweenness in lattice theory', *Bulletin of the American Mathematical Society*, **49** (1943), 280–7.

STAMEY, W. L. 1. See Blumenthal [17].

THURMAN, G. R. 1. See Blumenthal [16].

DE TILLY, J.
1. 'Essai de géométrie analytique générale', Mémoires couronnés et autres mémoires publiés par l'Académie Royale de Belgique, **47** (1892–3), mémoire 5 See also Mathesis, 1893, Supplement.

TRANSUE, W. R.
1. See Smiley [2].

VAIDYANATHASWAMY, R.
1. *Treatise on Set Topology*, Indian Mathematical Society, Madras, 1947.

VILLE, J.
1. 'Sur une proposition de M. L. M. Blumenthal', *Ergebnisse eines mathematischen Kolloquiums* (Wien), **7** (1936), 10–11.

VAN DER WAERDEN, B. L.
1. See van Dantzig [1].

WALD, A.
1. 'Axiomatik des Zwischenbegriffes in metrischen Räumen', *Mathematische Annalen*, **104** (1931), 476–84.
2. 'Begründung einer koordinatenlosen Differentialgeometrie der Flächen', *Ergebnisse eines mathematischen Kolloquiums* (Wien), **7** (1936), 24–46.

WHYBURN, G. T.
1. *Analytic Topology*, American Mathematical Society Colloquium Publications, **28**, 1942.

WILSON, W. A.
1. 'On semimetric spaces', *American Journal of Mathematics*, **53** (1931), 361–73.
2. 'On the imbedding of metric sets in euclidean space', ibid. **57** (1935), 322–6.
3. 'A relation between metric and euclidean spaces', ibid. **54** (1932), 505–17.
4. 'On certain types of continuous transformations of metric spaces', ibid. **57** (1935), 62–68.

INDEX

abstract cosine, 174.
accumulation point, 10.
Alexits, G., 88, 339.
algebraic middle-element, 142.
algebraically linear, 141.
almost congruent, 111.
Alt, F., 73, 75, 76, 78, 339.
apolar order, 231.
arc, 13, 59; length of, 59, 62, 64, 67, 68; lower semi-continuity of arc length, 65; rectifiable, 59.
Aronszajn, N., 41, 58, 325, 339.
associated coefficient set C, 304.
— matrices of elliptic m-tuple, 237.
— n-sphere, 231.
— solution set $\Sigma(C)$, 305.
— spherical $2m$-tuple, 236.

Beer, G., 73.
betweenness, 33.
— characterization of, in arbitrary lattice, 322–3; in metric space, 36.
— in arbitrary lattices, 322.
— in autometrized Boolean algebra, 333.
— lattice equivalents of, 317, 319.
— properties of, in metric space, 33; in normed distributive lattices, 320–2.
Biedermann, M., 90.
Bing, R. H., 58, 339.
Birkhoff, G., 31, 147, 339.
Blanc, E., 58, 339.
Blichfeldt, H. F., 33, 58.
Blumenthal, L. M., 31, 58, 89, 99, 120, 121, 134, 147, 161, 191, 230, 254, 272, 287, 303, 314, 338, 339, 340, 341, 342, 343.
Boolean algebra, 4, 315.
— — autometrized, 315, 331; distance in, 331, 332; free mobility in 336–7; group of motions in, 334–6; indices in, 337.
breadth: relative to p, 311; (constant) spherical, 311.
Busemann, H., 147, 340.

Canonical form, 310; polar system, 310.
Cantor product theorem, 51.
Carathéodory, C., 51.
Cauchy, A. L.: inequality, 16; generalization of, 17.
— sequence, 8.
chain, 328.
compact, 29; finitely, 29; locally, 72.
complete, 28; almost, 72.
— spherical image, 255.

completely associative, 3; commutative, 3; complementary, 257.
congruence indices, 91.
— — of autometrized Boolean algebra, 337; of hemisphere, 202–3; of spherical caps, 192, 203–5.
— — complete set of, 92.
— — $C_{1,r}$, 154–5; D_r, 159; $E_{n,r}$, 95; 118; $\mathscr{E}_{2,r}$, 267, 269; $\mathscr{E}_{n,r}$, 230; $\mathscr{H}_{n,r}$, 97; $\mathscr{H}^{\phi}_{n,r}$, 285–6; P_n, 153; $S_{n,r}$, 97; $X_{2,r}$, 271–2.
— order, 92; hyperfinite, 147; quasi, 92; transfinite, 151.
— symbol, 92; preferential ordering of, 92.
congruences, 32.
congruent imbedding in convex n-sphere, 162, 163; in elliptic space, 208; in euclidean space, 90, 106, 107; in Hilbert space, 132, 133, 134, 136; in generalized hyperbolic space, 285.
— sets, 35.
conic, 154.
continuity of distance function, 9.
continuous curve, 68.
— — path-length of, 69.
convex circle, 10.
— closed subset of E_n, 21, 40.
— extension, 5.
— n-sphere, 16.
— polyhedron, 196.
— strictly, 22.
— subpolyhedron, 196.
— subset of $S_{n,r}$, 193.
— surface, 21.
convexification, 72.
coverable, 152.
Coxeter, H. S. M., 232, 340.
cross, 255.
'crowding', 262.
curvature (metric definitions of) by Alt, 76; by Haantjes, 76; by Menger, 75; by Wald, 88, 89.

van Dantzig, D., 241, 242, 254, 340, 343.
de Donder, T., 33, 58.
δ-dense, 61.
δ-supplement, 207.
De Morgan formulae, 3, 309, 331.
determinant: Cayley–Menger, 97, 99, 290–4; $\Delta_k(p_1, p_2,..., p_k)$, 162; $\Delta_m(a)$, 175; $\Delta^*(p_1, p_2, p_3)$, 209, 215; Δ, 294–7; Δ_{NN}, 300, 301–3; Δ_{NP}, 297–9; Gram, 139; Λ, 300; $\Lambda_m(p_1, p_2,..., p_m)$, 275.
diametral, 164.

INDEX 345

Dines, L. L., 51, 310, 314, 340.
disk, 159.
— characterization of, 160; congruence order of, 159.
distributivity, 4.
domain, 158; simply connected, 160.

Ellis, D. O., vi, 338, 340.
ϵ-chain, 61; homogeneous, 60.
epsilon matrix, 208.
equilateral subsets of $\mathscr{E}_{n,r}$, 211–14.

Finsler, P., 75, 76.
foot, 46, 124, 129.
Fréchet, M., iv, 5, 8, 31, 341.
free $(n+2)$-tuples, 118, 119, 120; $(n+3)$-tuples, 93.
freely superposable, 253.
fundamental part of $\Sigma(C)$, 307; solution, 308.
f-superposable, 231.

Gaddum, J. W., vi, 68.
Garrett, G. A., 191, 340, 341.
generalized hyperbolic space, 275.
geodesic: arc, 70; distance, 67.
geometry, 32; distance, v, 32.
Gillam, B. E., 99, 147, 340, 341.
Glivenko, V., 317, 322, 328, 329, 338, 341.
Graustein, W. C., 78.
ground-space, 284.

Haantjes, J., 75, 76, 77, 78, 84, 89, 230, 269, 272, 303, 341, 343.
Hausdorff, F., 6, 16, 31, 65, 341; completion process, 327.
Helly, E., 160, 192; number, 202.
hemisphere, 192.
— dependent (independent) set of, 193; half-space of, 195.
Hilbert, D., 72; see space.
homeomorphic, 31.
homeomorphism, 31.
Huntington, E. V., 58, 341.

idempotent, 2.
imbeddable, irreducibly, 99.
independent, 93; $(m+1)$-tuple of $\mathscr{H}_{n,r}$, 274; of $\mathscr{H}^\phi_{n,r}$, 276; of $S_{n,r}$, 172; $S_{n,r}$, 172.
inner product, 138.
— — quasi, 139.
isogonal conjugate, 113, 115, 154.
— spherical, 179.
isometric extension, 41, 43; function, 41; imbedding problem, 91, 101, 104; mapping, 35.

Jordan, P., 138.

Kelly, L. M., vi, 230, 272, 340, 341.
Kline, J. R., 58, 341.

lattice, 2; cancellation law in, 316; complemented, 4; convex, 325; distributive, 4, duality principle in, 316; first element of, 329; imbedding, 325–7; last element of, 331; L_C, 317; L_P, 317; metric space associated with, 25; modular (Dedekind), 316; normed, 25.
Lindenbaum, A., 31, 161, 341.
line: directed straight, 7; existence of, 56; metric characterization of, 56; straight, 9; uniqueness of, 57.
— metric, 56.
linear content, 63, 64.
— graph, 63.

Mansion, P., 33, 58, 273, 287, 341.
mapping, 30.
— bicontinuous, 31; biuniform, 30; congruent (isometric), 35; continuous, 30, 31; inverse, 30.
'm-bein', 273.
McCoy, N. H., 310, 314, 340, 341.
Menger, K., v, 31, 33, 41, 51, 58, 59, 72, 74, 75, 76, 79, 89, 90, 93, 99, 118, 120, 122, 147, 288, 341, 342.
metric about an element, 141.
— basis, 95.
— characterization, 90; of E_n, 122, 128, 129; of Hilbert space, 136; of line, 56.
— convexity, 41.
— Hausdorff, 23.
— lattice, 25.
— segment, 41.
— space, 14; bounded, 23.
— transforms, 130, 134.
— uniform continuity of, 15.
Milgram, A. N., 58, 89, 341, 342.
Mimura, Y., 67.
Minkowski, H., 40, 51, 308, 309, 310.
— consequence theorem, 310; theorem 308, 309.
motion, 93.
movable (freely), 255.
Motzkin, T., 314.
Myers, S. B., 72.

nearer than, 329.
von Neumann, J., 138, 147, 342, 343.
n-lattice, 73.
norm, 137, 145.
normalized coefficient set C, 304.
— solution set $\Sigma(C)$, 305.
n-point relations, 32.
n-sphere chord $C_{n,r}$, 41, 154–6.
— convex $S_{n,r}$, 16.

346   INDEX

One-to-one correspondence, 30.
order-between, 317.
ordered associated spherical m-tuple, 236.
ordinary part of $\Sigma(C)$, 307; solutions of (I), 307.

Pauc, C., 76, 77, 83, 89, 342.
Pepper, P. H., 120, 121, 342, 343.
Pitcher, E., 322, 323, 338, 342.
points, 7; passing, 53, 306; terminal, 53, 306.
pole of $\mathscr{E}_r^1$, 224.
product complement, 306, 309.
property, euclidean $(k+1)$-point, 127; pythagorean, 129, 137; spherical $(k+1)$-point, 170; weak euclidean four-point, 123; weak spherical four-point, 164.
pseudo-E_n $(n+3)$-tuples, 110; characterization of, 114.
pseudo-f-superposable, 252.
pseudo-$S_{n,r}$ sets, 171; characterization of, 180, 188, 189, 191.
pseudo-$\mathfrak{S}_{n,\rho}$ sets, 174.
ptolemaic inequality, 78; space, 78.

Quadruple, imbedding curvature of, 88.
— orthocentric, 154, 255; proper, 261.
— pseudo-linear, 114, 131, 293, 320; lattice characterization of, 324.
— quasi independent, 85.
quasi positive definite, 301; intrinsically, 303.
quasi-rank, 301.

Rademacher, H., 160.
Riesz, F., 192.
ring, 331; associated, 331.
Robinson, C. V., 58, 89, 161, 205, 339, 342.
Russell paradox, 2.

Sawyer, J. W., 87.
scalar multiple, 140.
Schering, E., 273, 287, 342.
Schoenberg, I. J., 73, 79, 105, 106, 121, 133, 135, 138, 147, 160, 342, 343.
Schwarz inequality, 16, 17, 23, 139; postulates, 139.
segment (metric), 41; characteristic properties of, 44, 45, 48, 60, 84; existence of, 41; unique, 49, 50.
Seidel, J., 269, 272, 303, 341, 343.
separable, 30.
sequence: Cauchy (convergent), 8; limit, 8.
set, abstract, 1.
— closed, 10.
— closure of, 10.
— complement, 3.

set, connected, 160.
— countable (denumerable), 29.
— derived, 10.
— diameter of, 46; relative to p, 311.
— distance, 7.
— F_σ, 53.
— geodesic ordering of, 59.
— G_δ, 53.
— global, 312.
— monomorphic, 157.
— necessary criteria for, 1.
— normally ordered, 59.
— open, 6, 10.
— partially ordered, 3.
— product (intersection, meet), 2.
— simply ordered, 328.
— subset of, 1.
— sum, 2.
Sierpiński, W., 31, 343.
singular locus, 258; triangle, 259.
Smiley, M. F., 322, 323, 328, 338, 342, 343.
space: abstract, 5; Banach, 138; C, 26; distance, 7; $D(L)$, 25; $\mathscr{E}_r$, 215, 216; elliptic $\mathscr{E}_{n,r}$, 18; euclidean E_n, 16; E_ω, 26; generalized euclidean, 138, 144; Hausdorff topological, 6; Hilbert, 23; $\mathscr{H}_{n,r}^\phi$, 276; hyperbolic $\mathscr{H}_{n,r}$, 19 (see $\mathscr{H}_{n,r}^\phi$); I_n, 26; K, 6; L, 5; metric, 14; Minkowski, 21; $M_n^{(p)}$, 26; normed linear, 137; ptolemaic, 78; pseudo-E_n, 109, 110; semimetric, 7; Σ_r, 164; $\mathfrak{S}_{n,\rho}$, 174; $S_{n,r}$, 16; subset of, 9.
spherical cap, 192; base plane of, 193; rim of, 192.
— neighbourhood, 10.
spread, 68.
Stamey, W. L., 191, 340, 343.
strictly elementary transformation, 236.
— equivalent matrices, 236.
subspace, 9, 125, 172.
superposability order, 251; of $\mathscr{E}_{n,r}$, 251.
superposable, 93; subsets of $\mathscr{E}_{n,r}$, 231; triples of $\mathscr{E}_{2,r}$, 215.
system of inequalities: (I), 304; (NI), 304; (SI), 313; (NSI), 313; irreducible (reducible), 305; pseudo-solvable, 312.

θ-cross, 152; congruence order of, 153.
Thurman, G. R., 191, 340, 343.
de Tilly, J., 32, 33, 58, 343.
Tonelli, L., 289.
Topel, B. J., 120, 121, 342, 343.
topology, 5, 31; closure, 5; distance, 8; limit, 5; metric, v, 32; open set, 5.
torsion of a metric space, 87; of four points, 86; $t^*(p)$, 87.
transcendental number, 1.

INDEX

transform, 31; metric, 130.
Transue, W. R., 328, 338, 343.
triangle inequality postulate, 14.
triple: equilateral, 46, 47; linear, 56; two — property, 56.
truncated icosahedron, 213.

Vaidyanathaswamy, R., 31, 343.
vectorial application, 147.

vertices of proper apolar triangle, 256.
Ville, J., 132, 147, 343.

van der Waerden, B. L., 241, 242, 254, 340, 343.
Wald, A., 36, 58, 88, 89, 343.
Whyburn, G. T., 31, 343.
Wilson, W. A., 31, 109, 121, 122, 128, 147, 343.

CHELSEA SCIENTIFIC BOOKS

CHELSEA SCIENTIFIC BOOKS

THEORIE DES OPERATIONS LINEAIRES
By S. BANACH
—1933-63. xii + 250 pp. 5⅜x8. 8284-0110-1.   **$4.95**

DIFFERENTIAL EQUATIONS
By H. BATEMAN
CHAPTER HEADINGS: I. Differential Equations and their Solutions. II. Integrating Factors. III. Transformations. IV. Geometrical Applications. V. Diff. Eqs. with Particular Solutions of a Specified Type. VI. Partial Diff. Eqs. VII. Total Diff. Eqs. VIII. Partial Diff. Eqs. of the Second Order. IX. Integration in Series. X. The Solution of Linear Diff. Eqs. by Means of Definite Integrals. XI. The Mechanical Integration of Diff. Eqs.

—1917-67. xi + 306 pp. 5⅜x8. 8284-0190-X.   **$4.95**

MEASURE AND INTEGRATION
By S. K. BERBERIAN

A highly flexible graduate level text. Part I is designed for a one-semester introductory course; the book as a whole is suitable for a full-year course. Numerous exercises.

Partial Contents: PART ONE: I. Measures. II. Measurable Functions. III. Sequences of Measurable Functions. IV. Integrable Functions. V. Convergence Theorems. VI. Product Measures. VII. Finite Signed Measures. PART TWO: VIII. Integration over Locally compact Spaces (... The Riesz-Markoff Representation Theorem, ...). IX. Integration over Locally Compact Groups (Topological Groups, ..., Haar Integral, Convolution, The Group Algebra, ...). BIBLIOGRAPHY. INDEX.

—1965-70. xx + 312 pp. 6x9. 8284-0241-8.   **$7.95**

L'APPROXIMATION
By S. BERNSTEIN and CH. de LA VALLÉE POUSSIN

TWO VOLUMES IN ONE:

Leçons sur les Propriétés Extrémales et la Meilleure Approximation des Fonctions Analytiques d'une Variable Réelle, *by Bernstein*.

Leçons sur l'approximation des Fonctions d'une Variable Réelle, *by Vallée Poussin*.

—1925/19-69. 363 pp. 6x9. 8284-0198-5. 2 v. in 1.   **$7.95**

CALCUL DES PROBABILITES
By J. BERTRAND

A well-known work.

—2nd ed. 1907-71. lvii + 322 pp. 5⅜x8.   **In prep.**

ABHANDLUNGEN
By F. W. BESSEL

—1875-1971. 1,354 pp. 6x9. Three vols in one.   **In prep.**

OPERE MATEMATICHE
By E. BETTI

—1903/13-71. Approx. 1,100 pp. 6x9.   **In prep.**

CHELSEA SCIENTIFIC BOOKS

CONFORMAL MAPPING
By L. BIEBERBACH

Translated from the fourth German edition by F. STEINHARDT.

Partial Contents: I. Foundations; Linear Functions (Analytic functions and conformal mapping, Integral linear functions, $w = 1/z$, Linear functions, Groups of linear functions). II. Rational Functions. III. General Considerations (Relation between c. m. of boundary and of interior of region, Schwarz's principle). IV. Further Study of Mappings ($w = z + 1/z$, Exponential function, Trigonometric functions, Elliptic integral of first kind). V. Mappings of Given Regions (Illustrations, Vitali's theorem, Proof of Riemann's theorem, Actual constructions, Potential-theoretic considerations, Distortion theorems, Uniformization, Mapping of multiply-connected plane regions onto canonical regions).

"Presented in very attractive and readable form."—*Math. Gazette.*

—1952. vi + 234 pp. 4½x6½. 8284-0090-3. Cloth **$3.50**
8284-0176-4. Paper **$1.50**

BASIC GEOMETRY
By G. D. BIRKHOFF and R. BEATLEY

"is in accord with the present approach to plane geometry. It offers a sound mathematical development ... and at the same time enables the student to move rapidly into the heart of geometry."—*The Mathematics Teacher.*

"should be required reading for every teacher of Geometry."—*Mathematical Gazette.*

—3rd ed. 1959. 294 pp. 5⅜x8. 8284-0120-9. **$6.00**
TEACHER'S MANUAL. 160 pp. 5⅜x8. **$2.00**
ANSWER BOOK. 38 pp. 5⅜x8. **$0.95**

VORLESUNGEN UEBER DIFFERENTIALGEOMETRIE, Vols. I, II
By W. BLASCHKE

TWO VOLUMES IN ONE.

Partial Contents: VOL. I. 1. Theory of Curves. 2. Extremal Curves. 3. Strips. 4. Theory of Surfaces. 5. Invariant Derivatives on a Surface. 6. Geometry on a Surface. 7. On the Theory of Surfaces in the Large. 8. Extremal Surfaces. 9. Line Geometry.

VOL. II. *Affine Differential Geometry.* 1. Plane Curves in the Small. 2. Plane Curves in the Large. 3. Space Curves. 5. Theory of Surfaces. 6. Extremal Surfaces. 7. Special Surfaces.

—3rd ed. (Vol. I); 2nd ed. (Vol. II). 1930/23-67. 589 pp.
—6x9. 8284-0202-7. Two vols. in one. **$12.00**

KREIS UND KUGEL
By W. BLASCHKE

Isoperimetric properties of the circle and sphere, the (Brunn-Minkowski) theory of convex bodies, and differential-geometric properties (in the large) of convex bodies. A standard work.

—x + 169 pp. 5½x8½. 8284-0059-8. Cloth **$3.50**
8284-0115-2. Paper **$1.50**

CHELSEA SCIENTIFIC BOOKS

MATHEMATICAL PAPERS
By W. K. CLIFFORD

One of the world's major mathematicians, Clifford's papers cover only a 15-year span, for he died at age 34. [Included in this volume is Clifford's English translation of an important paper of Riemann.]

—1882-67. 70 + 658 pp. 5⅜x8.   8284-0210-8.   **$15.00**

ESSAI SUR L'APPLICATION DE L'ANALYSE AUX PROBABILITES
By M. J. CONDORCET

A photographic reproduction of a very rare and historically important work in the Theory of Probability. An original copy brings many hundreds of dollars in the rare book market.

—1785. Repr. 1971. 191 + 304 pp. 6x9.   **In prep**

MODERN PURE SOLID GEOMETRY
By N. A. COURT

In this second edition of this well-known book on synthetic solid geometry, the author has supplemented several of the chapters with an account of recent results.

—2nd ed. 1964. xiv + 353 pp. 5½x8¼. 8284-0147-0.   **$7.50**

SPINNING TOPS AND GYROSCOPIC MOTION
By H. CRABTREE

Partial Contents: Introductory Chapter. CHAP. I. Rotation about a Fixed Axis. II. Representation of Angular Velocity. Precession. III. Discussion of the Phenomena Described in the Introductory Chapter. IV. Oscillations. V. Practical Applications. VI-VII. Motion of Top. VIII. Moving Axes. IX. Stability of Rotation. Periods of Oscillation. APPENDICES: I. Precession. II. Swerving of "sliced" golf ball. III. Drifting of Projectiles. IV. The Rising of a Top. V. The Gyro-compass. ANSWERS TO EXAMPLES.

—2nd ed. 1914-67. 203 pp. 6x9.   8284-0204-3.   **$4.95**

THEORIE GENERALE DES SURFACES
By G. DARBOUX

One of the great works of the mathematical literature.

An unabridged reprint of the latest edition of *Leçons sur la Théorie générale des surfaces et les applications géométriques du Calcul infinitésimal.*

—Vol. I (2nd ed.) xii+630 pp. Vol. II (2nd ed.) xvii+584 pp. Vol. III (1st ed.) xvi+518 pp. Vol. IV. (1st ed.) xvi+537 pp.
     8284-0216-7.   **In prep.**

GESAMMELTE MATHEMATISCHE WERKE
By R. DEDEKIND

"The re-issue of these volumes . . . is a mark of the enormous importance to modern mathematical thought of Dedekind's great work."—*Mathematical Gazette.*

Three vols. in two.   **$25.00**

CHELSEA SCIENTIFIC BOOKS

THE DOCTRINE OF CHANCES
By A. DE MOIVRE

In the year 1716 Abraham de Moivre published his *Doctrine of Chances*, in which the subject of Mathematical Probability took several long strides forward. A few years later came his *Treatise of Annuities*. When the third (and final) edition of the *Doctrine* was published in 1756 it appeared in one volume together with a revised edition of the work on Annuities. It is this latter two-volumes-in-one that is here presented in an exact photographic reprint.

—3rd ed. 1756-1967. xi + 348 pp. 6x9. 8284-0200-0.   **$7.95**

DE MORGAN. See D. E. SMITH

COLLECTED MATHEMATICAL PAPERS
By L. E. DICKSON

—1969. 4 vols. Approx. 3,400 pp. 6½x9¼.   **In prep.**

HISTORY OF THE THEORY OF NUMBERS
By L. E. DICKSON

"A monumental work . . . Dickson always has in mind the needs of the investigator . . . The author has [often] expressed in a nut-shell the main results of a long and involved paper *in a much clearer way than the writer of the article did himself*. The ability to reduce complicated mathematical arguments to simple and elementary terms is highly developed in Dickson."—*Bulletin of A. M. S.*

—Vol. I (Divisibility and Primality) xii + 486 pp. Vol. II (Diophantine Analysis) xxv + 803 pp. Vol. III (Quadratic and Higher Forms) v + 313 pp. 5⅜x8. 8284-0086-5.
Three vol. set. **$22.50**

STUDIES IN THE THEORY OF NUMBERS
By L. E. DICKSON

—1930-62. viii + 230 pp. 5⅜x8. 8284-0151-9.   **$4.95**

ALGEBRAIC NUMBERS
By L. E. DICKSON, et al.

TWO VOLUMES IN ONE.

Both volumes of the *Report of the Committee on Algebraic Numbers* are here included, the authors being L. E. Dickson, R. Fueter, H. H. Mitchell, H. S. Vandiver, and G. E. Wahlen.

Partial Contents: CHAP. I. Algebraic Numbers. II. Cyclotomy. III. Hensel's p-adic Numbers. IV. Fields of Functions. I'. The Class Number in the Algebraic Number Field. II'. Irregular Cyclotomic Fields and Fermat's Last Theorem.

—1923/28-67. ii + 211 pp. 5⅜x9. 8284-0211-6.
Two vols. in one. **$4.95**

PLANE TRIGONOMETRY
By L. E. DICKSON

In all his books, advanced and elementary, Professor Dickson is noted for the extraordinary clarity of his writing. This very elementary book is no exception.

CHELSEA SCIENTIFIC BOOKS

"This book introduces at an early stage concrete applications . . . We thereby obtain an abundance of simple problems whose importance is so convincing that they cannot fail to arouse real interest. Actual experience with classes has firmly convinced the author that these practical applications offer the best means to drive home the principles of trigonometry and to make the subject truly vital."—*From Prof. Dickson's Preface.*

—1922-70. xii + 211 pp. 5⅜x8. 8284-0230-2. **$3.95**

INTRODUCTION TO THE THEORY OF ALGEBRAIC EQUATIONS, by L. E. DICKSON. See SIERPINSKI

VORLESUNGEN UEBER ZAHLENTHEORIE
By P. G. L. DIRICHLET and R. DEDEKIND

The fourth (last) edition of this great work contains, in its final form, the epoch-making "Eleventh Supplement," in which Dedekind outlines his theory of algebraic numbers.

"Gauss' *Disquisitiones Arithmeticae* has been called a 'book of seven seals.' It is hard reading, even for experts, but the treasures it contains (and partly conceals) in its concise, synthetic demonstrations are now available to all who wish to share them, largely the result of the labors of Gauss' friend and disciple, Peter Gustav Lejeune Dirichlet (1805-1859), who first broke the seven seals . . . [He] summarized his personal studies and his recasting of the *Disquisitiones* in his *Zahlentheorie*. The successive editions (1863, 1871, 1879, 1893) of this text . . . made the classical arithmetic of Gauss accessible to all without undue labor."—*E. T. Bell*, in *Men of Mathematics* and *Development of Mathematics.*

—4th ed. 1893-1968. xv + 657 pp. 5⅜x8. 8284-0213-2. **$13.50**

WERKE
By P. G. L. DIRICHLET

The mathematical works of P. G. Lejeune Dirichlet, edited by L. Kronecker.

—1889/97-1969. 1,086 pp. 6½x9¼. 8284-0225-6.

Two vols. in one. **$23.50**

THE INTEGRAL CALCULUS
By J. EDWARDS

A leisurely, immensely detailed, textbook of over 1,900 pages, rich in illustrative examples and manipulative techniques and containing much interesting material that must of necessity be omitted from less comprehensive works.

There are forty large chapters in all. The earlier cover a leisurely and a more-than-usually-detailed treatment of all the elementary standard topics. Later chapters include: Jacobian Elliptic Functions, Weierstrassian Elliptic Functions, Evaluation of Definite Integrals, Harmonic Analysis, Calculus of Variations, etc. Every chapter contains many exercises (with solutions).

—2 vols. 1921/22-55. 1,922 pp. 5x8.
8284-0102-0; 8284-0105-5. Each volume **$9.50**

CHELSEA SCIENTIFIC BOOKS

CURVE TRACING
By P. FROST

This much-quoted and charming treatise gives a very readable treatment of a topic that can only be touched upon briefly in courses on Analytic Geometry. Teachers will find it invaluable as supplementary reading for their more interested students and for reference. The Calculus is not used.

Partial Contents: Introductory Theorems. II. Forms of Certain Curves Near the Origin. Cusps. Tangents to Curves. Curvature. III. Curves at Great Distance from the Origin. IV. Simple Tangents. Direction and amount of Curvature. Multiple Points. Curvature at Multiple Points. VI. Asymptotes. VIII. Curvilinear Asymptotes. IX. The Analytical Triangle. X. Singular Points. XI. Systematic Tracing of Curves. XII. The Inverse Process. CLASSIFIED LIST OF THE CURVES DISCUSSED. FOLD-OUT PLATES.

Hundreds of examples are discussed in the text and illustrated in the fold-out plates.

—5th (unaltered) ed. 1960. 210 pp. + 17 fold-out plates.
—5⅜×8. 8284-0140-3. **$4.95**

GESAMMELTE MATHEMATISCHE WERKE
By L. FUCHS

—1904/09-71. L. C. 72-113126. 6x9. 3 vol. set. **In prep.**

LECTURES ON ANALYTICAL MECHANICS
By F. R. GANTMACHER

Translated from the Russian by PROF. B. D. SECKLER, with additions and revisions by Prof. Gantmacher.

Partial Contents: CHAP. I. Differential Equations of Motion of a System of Particles. II. Equations of Motion in a Potential Field. III. Variational Principles and Integral-Invariants. IV. Canonical Transformations and the Hamilton-Jacobi Equation. V. Stable Equilibrium and Stability of Motion of a System (Lagrange's Theorem on stable equilibrium, Tests for unstable E., Theorems of Lyapunov and Chetayev, Asymptotically stable E., Stability of linear systems, Stability on basis of linear approximation, . . .). VI. Small Oscillations. VII. Systems with Cyclic Coordinates. BIBLIOGRAPHY.

—Approx. 300 pp. 6x9.   8284-0175-6. **In prep.**

THE THEORY OF MATRICES
By F. R. GANTMACHER

This treatise, by one of Russia's leading mathematicians gives, in easily accessible form, a coherent account of matrix theory with a view to applications in mathematics, theoretical physics, statistics, electrical engineering, etc. The individual chapters have been kept as far as possible independent of each other, so that the reader acquainted with the contents of Chapter I can proceed immediately to the chapters that especially interest him. Much of the material has been available until now only in the periodical literature.

Partial Contents. VOL. ONE. I. Matrices and Matrix Operations. II. The Algorithm of Gauss and Applications. III. Linear Operators in an n-Dimensional Vector Space. IV. Characteristic Polynomial and Minimal Polynomial of a Matrix (Generalized Bézout Theorem, Method of Faddeev for Simultaneous Computation of Coefficients of Characteristic Polynomial and Adjoint Matrix, . . .). V. Functions of Matrices (Various Forms of the Definition, Components, Application to Integration of System of Linear Differential Eqns, Stability of Motion, . . .). VI. Equivalent Transformations of Polynomial Matrices; Analytic Theory of Elementary Divisors. VII. The Structure of a Linear Operator in an n-Dimensional Space (Minimal Polynomial, Congruence, Factor Space, Jordan Form, Krylov's Method of Transforming Secular Eqn, . . .). VIII. Matrix Equations (Matrix Polynomial Eqns, Roots and Logarithm of Matrices, . . .). IX. Linear Operators in a Unitary Space. X. Quadratic and Hermitian Forms.

VOL. TWO. XI. Complex Symmetric, Skew-symmetric, and Orthogonal Matrices. XII. Singular Pencils of Matrices. XIII. Matrices with Non-Negative Elements (Gen'l and Spectral Properties, Reducible M's, Primitive and Imprimitive M's, Stochastic M's, Totally Non-Negative M's, . . .). XIV. Applications of the Theory of Matrices to the Investigation of Systems of Linear Differential Equations. XV. The Problem of Routh-Hurwitz and Related Questions (Routh's Algorithm, Lyapunov's Theorem, Infinite Hankel M's, Supplements to Routh-Hurwitz Theorem, Stability Criterion of Liénard and Chipart, Hurwitz Polynomials, Stieltjes' Theorem, Domain of Stability, Markov Parameters, Problem of Moments, Markov and Chebyshev Theorems, Generalized Routh-Hurwitz Problem, . . .). BIBLIOGRAPHY.

—Vol. I. 1960. x + 374 pp. 6x9.   8284-0131-4.   **$7.50**
—Vol. II. 1960. x + 277 pp. 6x9.   8284-0133-0.   **$6.50**

UNTERSUCHUNGEN UEBER HOEHERE ARITHMETIK

By C. F. GAUSS

In this volume are included all of Gauss's number-theoretic works: his masterpiece, *Disquisitiones Arithmeticae*, published when Gauss was only 25 years old; several papers published during the ensuing 31 years; and papers taken from material found in Gauss's handwriting after his death.

These papers (pages 457-695 of the present book) include a fourth, fifth, and sixth proof of the Quadratic Reciprocity Law, researches on biquadratic residues, quadratic forms, and other topics.

—1889-65. xv + 695 pp. 6x9. 8284-0191-8.   **$8.75**

COMMUTATIVE NORMED RINGS

By I. M. GELFAND, D. A. RAIKOV, and G. E. SHILOV

Partial Contents: CHAPS. I AND II. General Theory of Commutative Normed Rings. III. Ring of Absolutely Integrable Functions and their Discrete Analogues. IV. Harmonic Analysis on Commutative Locally Compact Groups. V. Ring of Functions of Bounded Variation on a Line. VI. Regular Rings. VII. Rings with Uniform Convergence. VIII. Normed Rings with an Involution and their Representations. IX. Decomposition of

Normed Ring into Direct Sum of Ideals. HISTORICO-BIBLIOGRAPHICAL NOTES. BIBLIOGRAPHY.
—1964. 306 pp. 6x9. 8284-0170-5.　　　　$7.50

THEORY OF PROBABILITY
By B. V. GNEDENKO

This textbook, by a leading Russian probabilist, is suitable for senior undergraduate and first-year graduate courses. It covers, in highly readable form, a wide range of topics and, by carefully selected exercises and examples, keeps the reader throughout in close touch with problems in science and engineering.

"extremely well written . . . suitable for individual study . . . Gnedenko's book is a milestone in the writing on probability theory."—*Science*.

Partial Contents: I. The Concept of Probability (Various approaches to the definition. Space of Elementary Events. Classical Definition. Geometrical Probability. Relative Frequency. Axiomatic construction . . .). II. Sequences of Independent Trials. III Markov Chains IV. Random Variables and Distribution Functions (Continuous and discrete distributions. Multidimensional d. functions. Functions of random variables. Stieltjes integral). V. Numerical Characteristics of Random Variables (Mathematical expectation. Variance...Moments). VI. Law of Large Numbers (Mass phenomena. Tchebychev's form of law. Strong law of large numbers...). VII. Characteristic Functions (Properties. Inversion formula and uniqueness theorem. Helly's theorems. Limit theorems. Char. functs. for multidimensional random variables...). VIII. Classical Limit Theorem (Liapunov's theorem. Local limit theorem). IX. Theory of Infinitely Divisible Distribution Laws. X. Theory of Stochastic Processes (Generalized Markov equation. Continuous S. processes. Purely discontinuous S. processes. Kolmogorov-Feller equations. Homogeneous S. processes with independent increments. Stationary S. process. Stochastic integral. Spectral theorem of S. processes. Birkhoff-Khinchine ergodic theorem). XI. Elements of Queueing Theory (General characterization of the problems. Birth-and-death processes. Single-server queueing systems. Flows. Elements of the theory of stand-by systems). XII. Elements of Statistics (Problems. Variational series. Glivenko's Theorem and Kolmogorov's criterion. Two-sample problem. Critical region . . . Confidence limits). TABLES. BIBLIOGRAPHY. ANSWERS TO THE EXERCISES.

—4th ed. 1968. 527 pp. 6x9.　　　8284-0132-2.　$9.50

TRAITE DES COURBES SPECIALES REMARQUABLES
By F. GOMES TEIXEIRA

A comprehensive treatise, in three volumes, on curves in the plane and in space, and their special properties.

—2nd (corr.) ed. 1908/09/15-71. 1,337 pp. 3 vol. set. **In prep.**

INVARIANTENTHEORIE
By P. A. GORDAN

TWO VOLUMES IN ONE. A classical work.

—1885/7-1971. 583 pp. 5⅜x8. 2 vols. in 1.　　**In prep.**

A SHORT HISTORY OF GREEK MATHEMATICS
By J. GOW

A standard work on the history of Greek mathematics, with special emphasis on the Alexandrian school of mathematics.

—1884-68. xii + 325 pp. 5⅜x8.   8284-0218-3.   **$6.50**

THE ALGEBRA OF INVARIANTS
By J. H. GRACE and A. YOUNG

An introductory account.

Partial Contents: I. Introduction. II. The Fundamental Theorem. III. Transvectants. V. Elementary Complete Systems. VI. Gordan's Theorem. VII. The Quintic. VIII. Simultaneous Systems. IX. Hilbert's Theorem. XI. Apolarity. XII. Ternary Forms. XV. Types of Covariants. XVI. General Theorems on Quantics. APPENDICES.

—1903-65. vii + 384 pp. 5⅜x8.   8284-0180-2.   **$4.95**

DIE AUSDEHNUNGSLEHRE
By H. G. GRASSMANN

The *Ausdehnungslehre* appeared in two different versions, that of 1844 and that of 1862. This work is the first version [with the interpolatory material added by Grassmann in 1878], in a fourth edition.

—4th ed. 1969. xii + 435 pp. 6x9. 8284-0222-1.   **$12.50**

DIE AUSDEHNUNGSLEHRE VON 1862
By H. G. GRASSMANN

This work is the third edition of the 1862 *Ausdehnungslehre*. [See above for the 1844/78 version.]

—1971. vii + 511 pp. 6x9. 8284-0236-1.   **In prep.**

GESAMMELTE MATHEMATISCHE UND PHYSIKALISCHE WERKE
By H. G. GRASSMANN

Volumes Two and Three of Grassmann's collected works. Volume One, part 1 and Volume One, part 2 are the two versions of the *Ausdehnungslehre* [see above].

—1971. 1,495 pp. 6x9. 8284-0236-1. Vols. 2 and 3.   **In prep.**

A TREATISE ON DYNAMICS
By A. GRAY

—1911-71. xv + 626 pp. 5⅜x8.   **In prep.**

LORD KELVIN: An Account of His Scientific Life and Works
By A. GRAY

—1908-71. ix + 309 pp. 5x7⅛.   **Prob. $3.95**

MATHEMATICAL PAPERS
By G. GREEN

The collected papers of a celebrated mathematical physicist.

—1871-1970. xii + 336 pp. 5⅜x8. 8284-0229-9.   **$8.50**

CHELSEA SCIENTIFIC BOOKS

GYROSCOPIC THEORY
By G. GREENHILL

This work is intended to serve as a collection in one place of the various methods of the theoretical explanation of the motion of a spinning body, and as a reference for mathematical formulas required in practical work.

—1914-67. vi + 277 pp. + Fold-out Plates. 6½x10¾.
8284-0205-1.   **$9.50**

LES INTEGRALES DE STIELTJES et leurs Applications aux Problèmes de la Physique Mathématique
By N. GUNTHER

The present work is a reprint of Vol. I of the publications of the V. A. Steklov Institute of Mathematics, in Moscow. The text is in French.

—1932-49. 498 pp. 5⅜x8.   8284-0063-6.   **$6.95**

ONDES: Leçons sur la Propagation des Ondes et les Equations de l'Hydrodynamique
By J. HADAMARD

"[Hadamard's] unusual analytic proficiency enables him to connect in a wonderful manner the physical problem of propagation of waves and the mathematical problem of Cauchy concerning the characteristics of partial differential equations of the second order."—*Bulletin of the A. M. S.*

—Repr. 1949. viii + 375 pp. 5½x8½.   8284-0058-X.   **$6.00**

REELLE FUNKTIONEN. Punktfunktionen
By H. HAHN

—1932-48. xi + 415 pp. 5½x8½.   8284-0052-0.   **$6.00**

ALGEBRAIC LOGIC
By P. R. HALMOS

"Algebraic Logic is a modern approach to some of the problems of mathematical logic, and the theory of polyadic Boolean algebras, with which this volume is mostly concerned, is intended to be an efficient way of treating algebraic logic in a unified manner.

"[The material] is accessible to a general mathematical audience; no vast knowledge of algebra or logic is required . . . Except for a slight Boolean foundation, the volume is essentially self-contained."—*From the Preface.*

—1962. 271 pp. 6x9. 8284-0154-3.   **$4.95**

INTRODUCTION TO HILBERT SPACE AND THE THEORY OF SPECTRAL MULTIPLICITY
By P. R. HALMOS

A clear, readable introductory treatment of Hilbert Space.

—2nd ed. 1957. 120 pp. 6x9. 8284-0082-2.   **$3.50**

CHELSEA SCIENTIFIC BOOKS

LECTURES ON ERGODIC THEORY
By P. R. HALMOS

CONTENTS: Introduction. Recurrence. Mean Convergence. Pointwise Convergence. Ergodicity. Mixing. Measure Algebras. Discrete Spectrum. Automorphisms of Compact Groups. Generalized Proper Values. Weak Topology. Weak Approximation. Uniform Topology. Uniform Approximation. Category. Invariant Measures. Generalized Ergodic Theorems. Unsolved Problems.

"Written in the pleasant, relaxed, and clear style usually associated with the author. The material is organized very well and painlessly presented."
—*Bulletin of the A.M.S.*

—1956-60. viii + 101 pp. 5⅜x8.   8284-0142-X.   **$3.25**

ELEMENTS OF QUATERNIONS
By W. R. HAMILTON

Sir William Rowan Hamilton's last major work, and the second of his two treatises on quaternions.

—3rd ed. 1899/1901-68. 1,185 pp. 6x9. 8284-0219-1.
Two vol. set.   **$29.50**

RAMANUJAN:
Twelve Lectures on His Life and Works
By G. H. HARDY

The book is somewhat more than an account of the mathematical work and personality of Ramanujan; it is one of the very few full-length books of "shop talk" by an important mathematician.

—1940-59. viii + 236 pp. 6x9.   8284-0136-5.   **$4.95**

GRUNDZUEGE DER MENGENLEHRE
By F. HAUSDORFF

The original, 1914 edition of this famous work contains many topics that had to be omitted from later editions, notably, the theories of content, measure, and discussion of the Lebesgue integral. Also, general topological spaces, Euclidean spaces, special methods applicable in the Euclidean plane, the algebra of sets, partially ordered sets, etc.

—1914-49. 484 pp. 5⅜x8.   8284-0061-X.   **$7.50**

SET THEORY
By F. HAUSDORFF

Hausdorff's classic text-book is an inspiration and a delight. The translation is from the Third (latest) German edition.

"We wish to state without qualification that this is an indispensable book for all those interested in the theory of sets and the allied branches of real variable theory."—*Bulletin of A. M. S.*

—2nd ed. 1962. 352 pp. 6x9.   8284-0119-5.   **$7.50**

CHELSEA SCIENTIFIC BOOKS

ELECTRICAL PAPERS
By O. HEAVISIDE

Heaviside's collected works are in five volumes: The two volumes of his *Electrical Papers* and the three volumes of his *Electromagnetic Theory*.

"The [forthcoming publication of my *Electromagnetic Theory*] brought the question of a reprint of the earlier papers to a crisis. For, as the later work grows out of the earlier, it seemed an absurdity to leave the earlier work behind. [It possesses] sufficient continuity of subject-matter and treatment, and even regularity of notation, to justify its presentation in the original form . . . It might be regarded . . . as an educational work for students of theoretical electricity."—*From the Preface*.

—2nd (c.) ed. 1892-1970. 1,183 pp. 5⅜x8. 8284-0235-3.
Two vol. set. **$29.50**

ELECTROMAGNETIC THEORY
By O. HEAVISIDE

Third edition, with an Introduction by B. A. Behrend and with added notes on Heaviside's unpublished writings. A classic since its original publication in 1894-1912.

—3rd ed. 1970. 1,610 pp. 5⅜x8. 8284-0237-X.
Three vol. set **$29.50**

VORLESUNGEN UEBER DIE THEORIE DER ALGEBRAISCHEN ZAHLEN
By E. HECKE

"An elegant and comprehensive account of the modern theory of algebraic numbers."
—*Bulletin of the A. M. S.*

—2nd ed. 1970. viii + 274 pp. 5⅜x8. 8284-0046-6.   **$6.50**

INTEGRALGLEICHUNGEN UND GLEICHUNGEN MIT UNENDLICHVIELEN UNBEKANNTEN
By E. HELLINGER and O. TOEPLITZ

"Indispensable to anybody who desires to penetrate deeply into this subject."—*Bulletin of A.M.S.*

—1928-53. 286 pp. 5⅜x8.   8284-0089-X. **$4.95**

THEORIE DER ALGEBRAISCHE FUNKTIONEN EINER VARIABELN
By K. HENSEL and G. LANDSBERG

Partial Contents: PART ONE (Chaps. 1-8): Algebraic Functions on a Riemann Surface. PART TWO (Chaps. 9-13): The Field of Algebraic Functions. PART THREE (Chaps. 14-22): Algebraic Divisors and the Riemann-Roch Theorem. PART FOUR (Chaps. 23-27): Algebraic Curves. PART FIVE (Chaps. 28-31): The Classes of Algebraic Curves. PART SIX (Chaps. 32-37): Algebraic Relations among Abelian Integrals. APPENDIX: Historical Development. Geometrical Methods. Arithmetical Methods.

—1902-65. xvi + 707 pp. 6x9.   8284-0179-9. **$9.50**

CHELSEA SCIENTIFIC BOOKS

Grundzüge Einer Allgemeinen Theorie der LINEAREN INTEGRALGLEICHUNGEN
By D. HILBERT

—1912-53. 306 pp. 5½x8¼.   8284-0091-1. **$4.95**

GEOMETRY AND THE IMAGINATION
By D. HILBERT and S. COHN-VOSSEN

Translated from the German by P. NEMENYI.

"A fascinating tour of the 20th century mathematical zoo. . . . Anyone who would like to see proof of the fact that a sphere with a hole can always be bent (no matter how small the hole), learn the theorems about Klein's bottle—a bottle with no edges, no inside, and no outside—and meet other strange creatures of modern geometry will be delighted with Hilbert and Cohn-Vossen's book."
—*Scientific American.*

"Should provided stimulus and inspiration to every student and teacher of geometry."—*Nature.*

"A mathematical classic. . . . The purpose is to make the reader *see* and *feel* the proofs. . . . readers can penetrate into higher mathematics with . . . pleasure instead of the usual laborious study."
—*American Scientist.*

"Students, particularly, would benefit very much by reading this book . . . they will experience the sensation of being taken into the friendly confidence of a great mathematician and being shown the real significance of things."—*Science Progress.*

"A person with a minimum of formal training can follow the reasoning. . . . an important [book]."
—*The Mathematics Teacher.*

—1952. 358 pp. 6x9.   8284-0087-3. **$7.50**

GESAMMELTE ABHANDLUNGEN
(Collected Papers)
By D. HILBERT

Volume I (Number Theory) contains Hilbert's papers on Number Theory, including his long paper on Algebraic Numbers. Volume II (Algebra, Invariant Theory, Geometry) covers not only the topics indicated in the sub-title but also papers on Diophantine Equations. Volume III carries the sub-title: Analysis, Foundation of Mathematics, Physics, and Miscellaneous Papers.

—1932/35-66. 1,457 pp. 6x9. 8284-0195-0. Each vol. **$8.95**
                                          Three vol. set **$23.50**

PRINCIPLES OF MATHEMATICAL LOGIC
By D. HILBERT and W. ACKERMANN

"As a text the book has become a classic . . . the best introduction for the student who seriously wants to master the technique. Some of the features which give it this status are as follows:

"The first feature is its extraordinary lucidity. A second is the intuitive approach, with the introduction of formalization only after a full discussion of motivation. Again, the argument is rigorous and exact . . . A fourth feature is the emphasis on general extra-formal principles . . . Finally, the work is relatively free from bias . . . All together, the book still bears the stamp of the genius of one of the great mathematicians of modern times."—*Bulletin of the A.M.S.*

—1959. xii + 172 pp. 6x9.   8284-0069-5. **$3.95**

LEHRBUCH DER THETAFUNKTIONEN
By A. KRAZER

"Dr. Krazer has succeeded in the difficult task of giving a clear deductive account of the complicated formal theory of multiple theta-functions within the compass of a moderate sized text-book . . . Distinguished by clearness of style and general elegance of form."—*Mathematical Gazette.*

—1903-71. xxiv + 509 pp. 5⅜x8. 8284-0244-2.   **$12.50**

WERKE
By L. KRONECKER

—6 vols. in 5. 1895/97/99/1929/30/31-68. 2,530 pp. 6½x8½.
Five vol. set. **$59.50**

GROUP THEORY
By A. KUROSH

Translated from the second Russian edition and with added notes by PROFESSOR K. A. HIRSCH.

Partial Contents: PART ONE: The Elements of Group Theory. Chap. I. Definition. II. Subgroups (Systems, Cyclic Groups, Ascending Sequences of Groups). III. Normal Subgroups. IV. Endomorphisms and Automorphisms. Groups with Operators. V. Series of Subgroups. Direct Products. Defining Relations, etc. PART TWO: Abelian Groups. VI. Foundations of the Theory of Abelian Groups (Finite Abelian Groups, Rings of Endomorphisms, Abelian Groups with Operators). VII. Primary and Mixed Abelian Groups. VIII. Torsion-Free Abelian Groups. Editor's Notes. Bibliography.

Vol. II. PART THREE: Group-Theoretical Constructions. IX. Free Products and Free Groups (Free Products with Amalgamated Subgroup, Fully Invariant Subgroups). X. Finitely Generated Groups. XI. Direct Products. Lattices (Modular, Complete Modular, etc.). XII. Extensions of Groups (of Abelian Groups, of Non-commutative Groups, Cohomology Groups). PART FOUR: Solvable and Nilpotent Groups. XIII. Finiteness Conditions, Sylow Subgroups, etc. XIV. Solvable Groups (Solvable and Generalized Solvable Groups, Local Theorems). XV. Nilpotent Groups (Generalized, Complete, Locally Nilpotent Torsion-Free, etc.). Editor's Notes. Bibliography.

—Vol. I. 2nd ed. 1959. 271 pp. 6x9.   8284-0107-1.   **$6.00**
—Vol. II. 2nd ed. 1960. 308 pp. 6x9.   8284-0109-8.   **$6.00**
—Vol. III. Approx. 200 pp. 6x9.   **In prep.**

LECTURES ON GENERAL ALGEBRA
By A. G. KUROSH

Translated from the Russian by PROFESSOR K. A. HIRSCH, with a special preface for this edition by PROFESSOR KUROSH.

Partial Contents: CHAP. I. Relations. II. Groups and Rings (Groupoids, Semigroups, Groups, Rings, Fields, . . . , Gaussian rings, Dedekind rings). III. Universal Algebras. Groups with Multioperators (. . . Free universal algebras, Free products of groups). IV. Lattices (Complete lat-

tices, Modular lattice, Schmidt-Ore Theorem, ..., Distributive lattices). V. Operator Groups and Rings. Modules. Linear Algebras (... Free modules, Vector spaces over fields, Rings of linear transformations, ..., Derivations, Differential rings). VI. Ordered and Topological Groups and Rings. Rings with a Valuation. BIBLIOGRAPHY.
—1970. 335 pp. 6x9. 8284-0168-3. $6.95

OEUVRES (Collected Works)
By E. LAGUERRE
With a preface by H. POINCARE.
—1898/1905-71. 1,202 pp. 5⅜x8. Two vol. set.   In prep.

DIFFERENTIAL AND INTEGRAL CALCULUS
By E. LANDAU
A masterpiece of rigor and clarity.

"And what a book it is! The marks of Landau's thoroughness and elegance, and of his undoubted authority, impress themselves on the reader at every turn, from the opening of the preface ... to the closing of the final chapter.

"It is a book that all analysts ... should possess ... to see how a master of his craft like Landau presented the calculus when he was at the height of his power and reputation."
—*Mathematical Gazette.*
—3rd ed. 1965. 372 pp. 6x9.   8284-0078-4. $6.95

HANDBUCH DER LEHRE VON DER VERTEILUNG DER PRIMZAHLEN
By E. LANDAU
TWO VOLUMES IN ONE.

To Landau's monumental work on prime-number theory there has been added, in this edition, two of Landau's papers and an up-to-date guide to the work: an Appendix by Prof. Paul T. Bateman.
—2nd ed. 1953. 1,028 pp. 5⅜x8. 8284-0096-2.
Two vols. in one. $16.50

VORLESUNGEN UEBER ZAHLENTHEORIE
By E. LANDAU
—Vol. I, Pt. 2. *(Additive Number Theory) xii + 180 pp. Vol. II. (Analytical Number Theory and Geometrical Number Theory) viii + 308 pp. Vol. III. (Algebraic Number Theory and Fermat's Last Theorem) viii + 341 pp. 5¼x8¼. *(Vol. I, Pt. 1 is issued as **Elementare Zahlentheorie** (in German) or as **Elementary Number Theory** (in English).)   8284-0032-6.
Three vols. in one. $16.50

GRUNDLAGEN DER ANALYSIS
By E. LANDAU
The student who wishes to study mathematical German will find Landau's famous *Grundlagen der Analysis* ideally suited to his needs.

Only a few score of German words will enable him to read the entire book with only an occasional glance at the Vocabulary! [A COMPLETE German-English vocabulary, prepared with the novice especially in mind, has been appended to the book.]
—4th ed. 1965. 173 pp. 5½x8½. 8284-0024-5. Cloth **$3.95**
8284-0141-1. Paper **$1.95**

CHELSEA SCIENTIFIC BOOKS

FOUNDATIONS OF ANALYSIS
By E. LANDAU

"Certainly no clearer treatment of the foundations of the number system can be offered. . . . One can only be thankful to the author for this fundamental piece of exposition, which is alive with his vitality and genius."—*J. F. Ritt, Amer. Math. Monthly.*

—2nd ed. 1960. xiv + 136 pp. 6x9. 8284-0079-2. **$3.95**

ELEMENTARE ZAHLENTHEORIE
By E. LANDAU

"Interest is enlisted at once and sustained by the accuracy, skill, and enthusiasm with which Landau marshals . . . facts and simplifies . . . details."
—*G. D. Birkhoff, Bulletin of the A. M. S.*

—1927-50. vii + 180 + iv pp. 5½x8½. 8284-0026-1. **$4.50**

ELEMENTARY NUMBER THEORY
By E. LANDAU

The present work is a translation of Prof. Landau's famous *Elementare Zahlentheorie*, with added exercises by Prof. Paul T. Bateman.

—2nd ed. 1966. 256 pp. 6x9.   8284-0125-X. **$4.95**

Einführung in die Elementare und Analytische Theorie der ALGEBRAISCHE ZAHLEN
By E. LANDAU

—2nd ed. 1927-49. vii + 147 pp. 5⅜x8. 8284-0062-8 **$2.95**

NEUERE FUNKTIONENTHEORIE, by E. LANDAU.
See WEYL

Mémoires sur la Théorie des SYSTEMES DES EQUATIONS DIFFERENTIELLES LINEAIRES, Vols. I, II, III
By J. A. LAPPO-DANILEVSKII

THREE VOLUMES IN ONE.

A reprint, in one volume, of Volumes 6, 7, and 8 of the monographs of the Steklov Institute of Mathematics in Moscow.

"The theory of [systems of linear differential equations] is treated with elegance and generality by the author, and his contributions constitute an important addition to the field of differential equations."—*Applied Mechanics Reviews.*

—1934/5/6-53. 689 pp. 5⅜x8. 8284-0094-6.
Three vols. in one. **$12.50**

CELESTIAL MECHANICS
By P. S. LAPLACE

One of the landmarks in the history of human thought. Four volumes translated into English by NATHANIEL BOWDITCH, with an extensive running commentary by the Translator, plus a fifth volume in the original French containing historical material and commentary on the earlier volumes. For the most part, this latter is incorporated into Bowditch's systematic commentary.

"Undoubtedly the greatest systematic treatise ever published."—*Bulletin of the Amer. Math. Society.*

"The four superb volumes [are] much more than a translation; indeed, the extent of Bowditch's own contributions equals, or perhaps exceeds, that of the translation proper . . . Bowditch's commentary restores all the intermediate steps omitted by Laplace . . . The notes also contain full accounts of progress subsequent to the publication of the original volumes. The fifth, supplementary, volume of the French edition . . . is not translated, but most of its content is embodied in the translator's notes."
—*Nature.*

—Vol. I: xxiv+136+746 pp. Vol. II: xviii+990 pp. Vol. III: xxx+910 pp.+117 pp. of tables. Vol. IV: xxvi+1,018 pp. 1829; 1832; 1834; 1839. Reprint, 1967. 6½x9¼. 8284-0194-2. ..
Vols. I-IV. **$85.00**
—Vol. V (in French) : ix + 508 pp. 1825-1882. Reprint, 1969. —6½x9¼. 8284-0214-0. **$17.50**
8284-0194-2; 8284-0214-0. Five vol. set. **$99.50**

TOPOLOGY
By S. LEFSCHETZ

CONTENTS: I. Elementary Combinatorial Theory of Complexes. II. Topological Invariance of Homology Characters. III. Manifolds and their Duality Theorems. IV. Intersections of Chains on a Manifold. V. Product Complexes. VI. Transformations of Manifolds, their Coincidences, Fixed Points. VII. Infinite Complexes. VIII. Applications to Analytical and Algebraic Varieties.
—2nd ed. 1930-66. 410 pp. 5⅜x8.   8284-0116-0. **$6.00**

SELECTED PAPERS
By S. LEFSCHETZ

A selection of the major papers of Professor Lefschetz, together with his [French-language] book *L'Analysis Situs et la Géométrie Algébrique.*
—1971. Approx. 512 pp. 6x9. 8284-0234-5.   **$12.00**

ELEMENTS OF ALGEBRA
By HOWARD LEVI

"This book is addressed to beginning students of mathematics. . . . The level of the book, however, is so unusually high, mathematically as well as pedagogically, that it merits the attention of professional mathematicians (as well as of professional pedagogues) interested in the wider dissemination of their subject among cultured people . . . **a closer approximation to the right way to teach mathematics to beginners than anything else now in existence.**"—*Bulletin of the A. M. S.*
—4th ed. 1962. 189 pp. 5⅜x8.   8284-0103-9. **$3.95**

CHELSEA SCIENTIFIC BOOKS

GRAPHICAL METHODS, by RUNGE. See SIERPINSKI

EUCLIDES VINDICATUS
By G. SACCHERI

—2nd ed. 1971. Approx. 300 pp. 5⅜x8. 8284-0212-4. **In prep.**

ANALYTIC GEOMETRY OF THREE DIMENSIONS
By G. SALMON

A rich and detailed treatment by the author of *Conic Sections, Higher Plane Curves*, etc.

Partial Contents: Chap. I. Coordinates. III. Plane and Line. IV-VI. Quadrics. VIII. Foci and Focal Surfaces. IX. Invariants and Covariants of Systems of Quadrics. XI. General Theory of Surfaces. XII. Curves and Developables (Projective properties, non-projective properties, . . .).

Vol. II. Chap. XIII. Partial Differential Equations of Families of Surfaces. XIII (a). Complexes, Congruences, Ruled Surfaces. XIII (b). Triply Orthogonal Systems of Surfaces, Normal Congruences of Curves. XIV. The Wave Surface, The Centro-surface, etc. XV. Surfaces of Third Degree. XVI. Surfaces of Fourth Degree. XVII. General Theory of Surfaces. XVIII. Reciprocal Surfaces.

—Vol. I. 7th ed. 1927-58. xxiv + 470 pp. 5x8. 8284-0122-5. **$4.95**

—Vol. II. 5th ed. 1928-65. xvi + 334 pp. 5x8. 8284-0196-9. **$4.95**

CONIC SECTIONS
By G. SALMON

"The classic book on the subject, covering the whole ground and full of touches of genius."
—*Mathematical Association.*

—6th ed. xv + 400 pp. 5⅜x8.   8284-0099-7. Cloth **$4.95**
                               8284-0098-9. Paper **$1.95**

HIGHER PLANE CURVES
By G. SALMON

CHAPTER HEADINGS: I. Coordinates. II. General Properties of Algebraic Curves. III. Envelopes. IV. Metrical Properties. V. Cubics. VI. Quartics. VII. Transcendental Curves. VIII. Transformation of Curves. IX. General Theory of Curves.

—3rd ed. 1879-1960. xix + 395 pp. 5⅜x8. 8284-0138-1. **$4.95**

LESSONS INTRODUCTORY TO THE MODERN HIGHER ALGEBRA
By G. SALMON

A classical account of the theory of Determinants and Invariants.

—5th ed. 1887-1964. xv + 376 pp. 5¼x8. 8284-0150-0. **$4.95**

CHELSEA SCIENTIFIC BOOKS

MEHRDIMENSIONALE GEOMETRIE
By P. H. SCHOUTE

Vol. I: Die linearen Räume. Vol. II: Die Polytope.
—1902/05-71. 638 pp. 4½x7. Two vol. set.   **In prep.**

PFAFF'S PROBLEM AND ITS GENERALIZATIONS
By J. A. SCHOUTEN and W. v. d. KULK

Partial Contents: Chap. III. The Outer Problem. IV. Classification of Covariant Vector Fields and Pfaffians . . . VI. Contact Transformations. VII. Theory of Vector Manifolds and Element Manifolds . . . X. Solution of Systems of Differential Equations.

—1949-69. xvi + 542 pp. 5⅜x8. 8284-0221-3.   **$12.00**

INTRODUCTION TO MODERN ALGEBRA AND MATRIX THEORY
By O. SCHREIER and E. SPERNER

An English translation of the revolutionary work, *Einführung in die Analytische Geometrie und Algebra*. Chapter Headings: I. Affine Space. Linear Equations. (Vector Spaces). II. Euclidean Space. Theory of Determinants. III. The Theory of Fields. Fundamental Theorem of Algebra. IV. Elements of Group Theory. V. Matrices and Linear Transformations. **The treatment of matrices is especially extensive.**

"Outstanding . . . good introduction . . . well suited for use as a text . . . Self-contained and each topic is painstakingly developed."
—*Mathematics Teacher.*

—2nd ed. 1959. viii + 378 pp. 6x9.   8284-0080-6.   **$6.95**

PROJECTIVE GEOMETRY OF n DIMENSIONS
By O. SCHREIER and E. SPERNER

Translated from the German by Calvin A. Rogers.

Suitable for a one-semester course on the senior undergraduate or first-year graduate level. The background required is minimal: The definition and simplest properties of vector spaces and the elements of matrix theory.

There are exercises at the end of each chapter to enable the student to test his mastery of the material.

Chapter Headings: I. n-Dimensional Projective Space. II. General Projective Coordinates. III. Hyperplane Coordinates. The Duality Principle. IV. The Cross Ratio. V. Projectivities. VI. Linear Projectivities of P_n onto Itself. VII. Correlations. VIII. Hypersurfaces of the Second Order. IX. Projective Classification of Hypersurfaces of the Second Order. X. Projective Properties of Hypersurfaces of the Second Order. XI. The Affine Classification of Hypersurfaces of the Second Order. XII. The Metric Classification of Hypersurfaces of the Second Order.

—1961. 208 pp. 6x9.   8284-0126-8.   **$4.95**

CHELSEA SCIENTIFIC BOOKS

TEXTBOOK OF TOPOLOGY
By H. SEIFERT and W. THRELFALL

A translation of the above.

—Approx. 380 pp. 6x9. **In prep.**

TABLES OF ARC LENGTH
By E. G. SEWOSTER

Full title: EIGHT-PLACE TABLES OF LENGTHS OF CIRCULAR ARCS. The tabulation is for every second of arc, 0° to 45°.

—3rd ed. 1971. 270 pp. 7x10. 8284-0240-X. **Prob. $10.00**

FROM DETERMINANT TO TENSOR,
by W. F. SHEPPARD. See KLEIN

HYPOTHESE DU CONTINU
By W. SIERPINSKI

"One sees how deeply this postulate cuts through all phases of the foundations of mathematics, how intimately many fundamental questions of analysis and geometry are connected with it ... a most excellent addition to our mathematical literature."
—*Bulletin of A. M. S.*

—2nd ed. 1957. xvii + 274 pp. 5⅜x8. 8284-0117-9. **$4.95**

CONGRUENCE OF SETS,
and other monographs
By SIERPINSKI, KLEIN, RUNGE, and DICKSON

FOUR VOLUMES IN ONE.

ON THE CONGRUENCE OF SETS AND THEIR EQUIVALENCE BY FINITE DECOMPOSITION, by *W. Sierpinski.* 1. Congr. of Sets. 2. Translation of Sets. 3. Equiv. of Sets by Finite Decomposition. 4. D. into Two Parts. ... 7. Paradoxical D's. ... 10. The Hausdorff Paradox. 11. Paradox of Banach and Tarski. 12. Banach Measure. The General Problem of Measure. 13. Absolute Measure. 14. Paradox of J. von Neumann.

THE MATHEMATICAL THEORY OF THE TOP, by *F. Klein.* Well-known lectures on the analytical formulas relating to the motion of the top.

GRAPHICAL METHODS, by *C. Runge.*

INTRODUCTION TO THE THEORY OF ALGEBRAIC EQUATIONS, by *L. E. Dickson.* Dickson's earliest (1903) account of the subject, substantially less abstract than his later exposition. *From Dickson's Preface:* "The subject is here presented in the historical order of its development. The First Part (Chaps. I-IV) is devoted to the Lagrange-Cauchy-Abel theory of general algebraic equations. The Second Part (Chaps. V-XI) is devoted to Galois' theory of algebraic equations ... The aim has been to make the presentation strictly elementary, with practically no dependence upon any branch of mathematics beyond elementary algebra. There occur numerous illustrative examples, as well as sets of elementary exercises."

—1954/1897/1912/1903-1967. 461 pp. 5¼x8. 8284-0209-4.
Four vols. in one. **$6.50**